The greatest ambition of any moderately successful nineteenth-century French scientist was to become a member of the Academy of Sciences. *Science under Control* is the first major study, in any language, of this elite institution, in a period which began with such influential figures as Laplace and Cuvier and extends to the time of Louis Pasteur and Henri Poincaré. The book attempts to remove the veil of mystery and misunderstanding which has shrouded this key institution and its procedures. The French government exercised political, financial and bureaucratic control over the Academy and the Academy in turn sat in judgement over all serious scientific production. Only with its approval could the work of French scientists win acceptance and their careers advance. This book examines the politics of science in a historical context drawing on a wealth of original archival and published sources.

The author argues that the Academy was of importance not only nationally but also internationally because of its influence and because of the establishment of certain procedures now considered basic to the organisation of modern science. The book therefore provides a case study of carefully regulated scientific production encouraged yet constrained within a system of reports, prizes and elections. The book will prove to be an invaluable source of information and of discussion on the social, political, religious and cultural dimensions of science in this crucial period.

— Science under control —

The French Academy of Sciences 1795–1914

The meeting place of the Academy

Books by the same author

The Society of Arcueil: A view of French science at the time of Napoleon I

Science in France in the revolutionary era described by Thomas Bugge (ed.)

Historical studies in the language of chemistry

The emergence of science in western Europe (ed.)

Gay-Lussac: Scientist and bourgeois

— SCIENCE UNDER CONTROL —

The French Academy of Sciences 1795–1914

Maurice Crosland

Professor of History of Science,
University of Kent at Canterbury

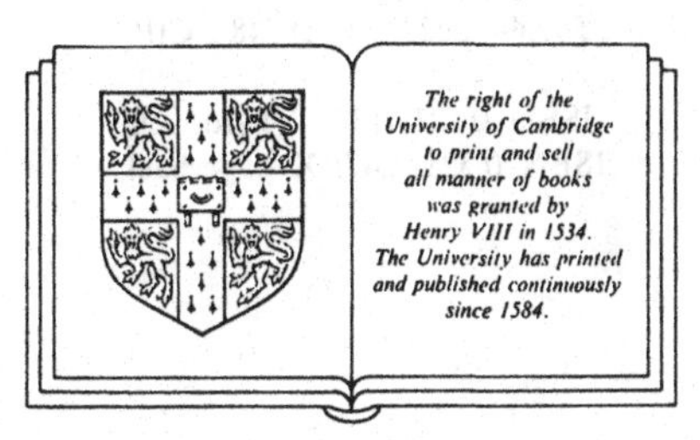

CAMBRIDGE UNIVERSITY PRESS
Cambridge
New York Port Chester Melbourne Sydney

PUBLISHED BY THE PRESS SYNDICATE OF THE UNIVERSITY OF CAMBRIDGE
The Pitt Building, Trumpington Street, Cambridge, United Kingdom

CAMBRIDGE UNIVERSITY PRESS
The Edinburgh Building, Cambridge CB2 2RU, UK
40 West 20th Street, New York NY 10011–4211, USA
477 Williamstown Road, Port Melbourne, VIC 3207, Australia
Ruiz de Alarcón 13, 28014 Madrid, Spain
Dock House, The Waterfront, Cape Town 8001, South Africa

http://www.cambridge.org

First published 1992
First paperback edition 2002

A catalogue record for this book is available from the British Library

Library of Congress Cataloguing in Publication data
Crosland, Maurice P.
Science under control: the French Academy of Sciences, 1795–1914
/ Maurice Crosland.
p. cm.
ISBN 0 521 41373 7 (hardcover)
1. Académie des sciences (France) / Institut de France – History. 2. Science –
France – History. I. Title.
Q46.C77 1992
509.44–dc20 91-28748 CIP

ISBN 0 521 41373 7 hardback
ISBN 0 521 52475 X paperback

CONTENTS

TABLES AND FIGURES

PREFACE

If institutionalised research is to be productive, a fairly wide margin for the autonomy of its practitioners has to be built into the institutions. One of the central problems for scientific establishments financed and controlled by extraneous agencies thus becomes that of the balance between dependence and independence.

(Norbert Elias, 'Scientific establishments' in *Sociology of the Sciences*, **6** (1982), 4.)

From science, all statesmen and politicians want are instrumentalities, powers but not power: weapons, techniques, information, communications, and so on. As for scientists, what have they wanted of government? They expressly have not wished to be politicised. They have wanted support in the obvious form of funds, but also in the shape of institutionalisation and in the provision of authority for the legitimation of their community in its existence and in the activities, or in other words for its professional status.

(Charles C. Gillispie, *Science and polity in France at the end of the old regime*, Princeton, N.J., 1980, p. 549.)

Major institutions deserve their histories no less than leading scientists, although the task may take longer and the interpretation may be more complex. Sometimes an individual catches the imagination of the public as if he were science itself, but scientific organisations in the long run may have greater power and influence than any private individual. Scientific organisations have the power to encourage, to constrain and conceivably even to subvert the scientific endeavour. Controlled and financed by governments in modern times, they are sometimes called upon to unleash and direct powerful natural forces. At the very least they study such forces. If we go back two hundred years the scale was obviously much more modest, but many of the principles were similar. The French Academy of Sciences was one of a very small number of organisations which had a permanent influence on the development of modern science.

If we consider what has been written about such scientific institutions, it is often understandably the foundation of a society which has attracted special attention. Most books and articles on the history of the Royal Society, for example, focus on the early years. Similarly it is the French Academy of Sciences of the *ancien régime*, which has attracted most attention.[1] There is a good modern study of the Academy in the eighteenth century[2] but virtually nothing on the later period.[3] Yet a former secretary of the Academy justly commented that if anyone should have the time and energy to write a book on the Academy, particularly in the early nineteenth century, he would have the privilege of dealing with a very important period in the history of French science.[4] Interestingly, it was assumed that such a historian would be not only French by nationality but also a member of the Academy.

It may, therefore, be necessary for the author to state that this is not an in-house production. Although an independent scholar may occasionally envy someone commissioned to write an 'official history' of an institution under the benevolent patronage of that institution, which might include special facilities and personal encouragement, in the end he may conclude that it is too high a price to pay. But, given that this study is by an outsider and a foreigner to boot, one should not suppose that it constitutes an attack on the Academy. Muck-raking might occasionally be a temptation indulged in by a writer wishing to enliven a boring tale. But when one is privileged to be studying the activities of some of the greatest scientists of all time and their associates, one seeks primarily to understand, which requires some knowledge of the historical context. One starts from a belief in the value of the scientific enterprise but one does not ignore practical problems nor ordinary human failings. Whatever some official eulogies may have said to the contrary,[5] even the greatest French scientists were human beings and not demi-gods.

The author must ask for the indulgence of any reader who has an exclusive

1 See e.g. L. F. Alfred Maury, *L'ancienne Académie des Sciences*, 2nd edn, 1864; Joseph Bertrand, *L'Académie des Sciences et les Académiciens de 1666 à 1793*, 1869; René Taton, *Les origines de l'Académie Royale des Sciences*, 1965; Claire Salomon-Bayet, *L'institution de la science et l'expérience du vivant: Méthode et expérience à l'Académie des Sciences, 1666–1793*, 1978; James McClellan, 'The Académie Royale des Sciences, 1699–1793. A statistical portrait', *Isis*, **72** (1981), 541–67; Dorinda Outram, 'The ordeal of vocation: The Paris Academy of Sciences and the Terror, 1793–5', *History of Sciences*, **21** (1983), 251–73. Most recently we have: Alice Stroup, *A company of scientists: Botany, patronage and community at the seventeenth-century Parisian Royal Academy of Sciences*, Berkeley, Cal., 1990. N.B. All books in English may be understood to be published in London and all books in French in Paris, unless another location is given.

2 Roger Hahn, *The anatomy of a scientific institution. The Paris Academy of Sciences, 1666–1803*, Berkeley, Cal., 1971.

3 A partial exception is a collection written by members of the Academy and spanning three centuries: Institut de France, *Académie des Sciences, Troisième Centenaire, 1666–1966* (2 vols.), 1967. On the foundation of the Institute and the Napoleonic period see the second part of E. Maindron, *L'Académie des Sciences*, 1888, pp. 141ff. These works and several articles by the present writer are naturally referred to in the course of this book. For studies of the Academy prize system see the references in Chapter 7 (pp. 270n., 276n.).

4 'S'il se trouve un jour parmi nous quelqu'un qui veuille consacrer sa vie à écrire l'histoire complète de notre Compagnie, c'est avec une joie patriotique qu'il s'arrêtera sur la période qui comprend la première moitié du XIXe siècle.' G. Darboux, *L'Institut de France*, 1907 p. 37.

5 Berthelot, *M.A.I.*, **47** (1904), lxiii.

interest in one particular Academician and possibly finds that he receives hardly a mention in the text. This is because the focus is on the institution rather than its many members. Of course, a few leading Academicians, who held the key position of permanent secretary or contributed in some major way to the work of the Academy, deserve special treatment. The names of Arago[6], Berthelot[7], Cuvier[8] and Dumas[9] provide a veritable alphabet of the Academy and of French science in the nineteenth century, but this still leaves many names in the background. There may well be a place for further study of members of the Academy and, if this book should provide a springboard for such a study, so much the better.

It is a pleasure to acknowledge the help and support that I have received from the University of Kent at Canterbury and from colleagues there. Thanks to a grant from the Nuffield Foundation going back to 1974, we were able to build up in the university library a collection of nineteenth-century French scientific periodicals, which have been invaluable in the present work, none more so than the comprehensive *Comptes rendus* of the Academy. I must thank the library staff for their general assistance. In the History of Science Unit Alec Dolby has made numerous helpful comments and constructive criticisms over a period of several years. He and Crosbie Smith have recently engaged in discussions on historiography from which I am sure I have benefited. W. M. L. Bell, who teaches a course on the republican tradition in nineteenth-century France, has given me useful advice, which should have prevented me from making too many blunders in this area. Over the years I have also discussed the political and social history of nineteenth-century France with William Fortescue. I have also

[6] D. F. J. Arago (1786–1853), astronomer and physicist, was elected to the Academy at the unusually early age of 23. Although originally a protégé of Laplace, he came to lead a rival group in physical science during the Restoration period. When elected as secretary of the Academy in 1830, he used his position to introduce the *Comptes rendus* and to secure the election of a number of republican scientists to the Academy.

[7] Marcellin Berthelot (1827–1907), elected secretary of the Academy in 1889, became one of the most powerful figures in the history of nineteenth-century French science. Best known for his research on organic chemistry, he produced a record number of publications, many being simply alternative versions of the same work. He became Minister of Education for a short time. He was a prominent member of the anti-clerical group in the Academy.

[8] J. L. N. F. (but often called 'Georges') Cuvier (1769–1832), a Württemberg Lutheran by origin, came to Paris in 1795 and launched himself on a scientific career with spectacular success. He did important work in zoology and comparative anatomy but it was in academic politics that he was almost unrivalled. His politics changed with that of succeeding governments and he eventually was ennobled. He was secretary of the Academy for nearly thirty years, his other centre of power being the Muséum d'Histoire Naturelle.

[9] Jean-Baptiste Dumas (1800–84) was an important organic chemist, who also excelled in administration. He held a number of municipal and government posts in the later part of his life. He founded a major research school of chemistry and became a powerful academic patron from the mid-century. He was a moderate Catholic, who became a senator under the Second Empire. He was a major figure in the politics of science long before his election in 1868 as secretary of the Academy.

profited from discussions with several of my (former) graduate students who have worked on different aspects of the history of science in France: Graham Smith, John Cawood, Margaret Bradley, Susan Court, Keith Boughey, David Bickerton, Leo Klosterman, John L. Davis, Antonio Galvez and Ana Carneiro. Ben Marsden kindly commented on some of the mathematics.

Throughout the book I hope that I have adequately indicated my indebtedness to the published work of other scholars. For the religious perspective, which I had not previously examined, I am indebted to the published work of John McManners, Harry Paul and Mary Jo Nye and I should also like to thank Maurice Larkin for his comments on an earlier draft of one section. I am grateful to an old friend, Ted Caldin, for taking the time to look through much of the draft typescript and making useful comments for clarification and improvement. He also pointed to the desirability of the inclusion of a chronological table outlining the major political changes in nineteenth-century France. Any remaining faults in the book, however, are the responsibility of the author and not that of the people named.

I must thank the permanent secretaries of the Academy of Sciences for permission to use their archives. I have received general assistance over a long period from the archivists: Mme. Pierre Gauja, M. Pierre Berthon and Mme. Demeulenaere-Douyère. Users of the archives are also indebted to Mme. Claudine Pouret and Mlle. Geneviève Darrieus.

Finally I owe a tremendous debt to Irene Crow who, with great efficiency and unfailing good humour, has typed successive drafts of this book.

In the production of the book I must thank Fiona Thomson at the Cambridge University Press and my copy editor Jeremy Smith.

ABBREVIATIONS

A.c.[p.]	*Annales de chimie [et de physique]*, Paris, 1789, 2nd series 1816–
A.I.	Archives of the Institute, Paris
A.N.	Archives Nationales, Paris
A.S.	Archives of the Academy of Sciences, Paris
B.J.H.S.	*British Journal of the History of Science*
B.M.	British Library (British Museum), London
C.R.	*Comptes rendus hebdomadaires des séances de l'Académie des Sciences*, Paris, 1835–
D.S.B.	*Dictionary of Scientific Biography*, ed. Charles C. Gillispie, 16 vols., New York, 1970–80
M.A.I.	*Mémoires de l'Académie des Sciences de l'Institut*, Paris, 1816–
P.V.I.	*Procès-verbaux des séances de l'Académie des Sciences, tenues depuis la fondation de l'Institut jusqu'au mois d'août 1835, publiés conformément à une décision de l'Académie par MM. les secretaires perpetuels*, 10 vols., Hendaye, 1910–22

Table 1. *Chronological table*

		French history			The Academy and French science
1789	July	Fall of the Bastille, the traditional 'beginning' of the French Revolution			
1792	April	Beginning of war (with Austria)			
	September	Declaration of **First Republic**			
1793	(June)–	The Terror	1793	August	All Academies suppressed
1794	(July)				
1795	November	Rule of **Directory** begins	1795	October – December	**Establishment of National Institute with three Classes**
1799	November	Bonaparte seizes power. Beginning of **Consulate**	1796	April	Further regulations for Institute
			1798–9		International Congress on metric standards
1804		**First Empire** under Emperor Napoleon I	1803	January	New constitution of Institute, abolishing Second Class
1812		Napoleon's retreat from Moscow	1808		'University of France', which includes Faculties of Science
1814	March	First Bourbon Restoration, followed by brief return of Napoleon			
1815	July	Second Bourbon **Restoration (Louis XVIII)**	1816		New constitution of Academies. **'First Class' becomes (Royal) Academy of Sciences**
1824		Elections produce an ultra-royalist Parliament	1822		Paris Faculty of Medicine closed temporarily by government
1825		Coronation of **Charles X** at Rheims	1829		Academy begins to make independent use of Montyon funds
			1830	March	Academy debate between Geoffroy Saint-Hilaire and Cuvier
1830	July	Capture of Algiers. Revolution. Constitutional monarchy of **Louis-Philippe**			

1831–2		Cholera epidemic	1835	July	First issue of *Comptes rendus*
			1839	August	Joint meeting of Academy with Académie des Beaux Arts to announce invention of photography by Daguerre
1842		Completion of Auguste Comte, *Cours de philosophie positive*	1840		First allocation of government money to pay for *Comptes rendus*
1848		Revolution in Paris. Provisional Government. **Second Republic**	1847		Government grant for *Comptes rendus* increased from 15 000f. to 30 000f.
1852	December	Proclamation of **Second Empire** under Napoleon III	1855–6		Academy resists extra duties proposed by Minister Fortoul
			1859		Beginning of Pasteur–Pouchet debate
1861		Expedition to Mexico	1863–6		Membership of Geography section increased from 3 to 6
1870		Franco–Prussian war, Government of National Defense, **Third Republic**	1870–1		Academy meetings continue during siege of Paris
1871	March–May	Commune	1871		Phylloxera commission
			1874		Transit of Venus expeditions
1879		Republican majority in Senate	1878		Charles Darwin elected as corresponding member
1887–9		Building of Eiffel Tower			
1889		Centenary celebrations of French Revolution	1889		Berthelot elected permanent secretary of Academy
1898		Zola's *J'accuse* (Dreyfus affair)	1898		Marie Curie announces discovery of radium to Academy
			1901		*Caisse des recherches scientifiques* (independent of Academy)
1904		Law for complete separation of Church and State	1908		Bonaparte Foundation (grants)
			1913		New section in Academy of full non-resident members
1914	August	First World War	1915		First Loutreuil Foundation awards

INTRODUCTION

> The National Institute will be in a way the epitome of the world of learning, the representative body of the republic of letters, the honorable goal of all the ambitions of science and of talent, the most magnificent recompense of great effort and of outstanding success.
>
> (Daunou, Report to the Convention, 19 October 1795.)

> I have no hesitation in saying, after having recently seen the Academy of Sciences at its weekly labours, that it is the noblest and most effective institution that ever was organised for the promotion of science.
>
> (Sir David Brewster, *Report of 20th meeting of the British Association*, Edinburgh, 1850, p. xli.)

> I was not mistaken with respect to the Academy of Sciences and the other academies founded on the same basis; they are bodies depending on the government and functioning by its orders...For thirty years not a single cog in the wheels of the Academy of Sciences has been worn out; it is the same system in operation; it turns around the same axis; the handle of the machine has often been changed but never the spring; it has outlived the downfall of all its masters. It has always...made the sounds required of it, stifling those which did not please it, raising up to the highest place an individual favoured by the government, even if he is an idiot.
>
> (F. V. Raspail, *Nouveau système de chimie organique*, 2nd edn, 1838, vol. 1, pp. xxvi–xxvii.)

The growth of science over the last few centuries has been compared by some critics to the opening of Pandora's box. But if some of the *applications* of science have been harmful, science itself as knowledge has been more generally recognised as good. Science has produced many wonders, but its very success has introduced problems. Science has been particularly successful in exploring new knowledge rather than, like the humanities, re-examining human experience from different standpoints. But, given this increased understanding of the natural world, should it be left, as formerly in Britain, to the private individual to pursue as a hobby or was it important enough to be a matter of government concern? In France from early times the government wanted to be involved and for several reasons. One was obviously that this new knowledge might be of practical use to the Crown. Another might have been that uncontrolled knowledge of nature could constitute a threat to the established order. A third and more positive reason was that Louis XIV wanted to be seen as the patron of learning. As time went on, the benefits to the modern state of the patronage of science became more evident. For example, by the time of the French Revolution of 1789, men of science were able to apply their expertise to national defence. Perhaps science had almost come of age but not without imposing the burden of new responsibilities on its practitioners.

The title of this book refers both to governmental control of the official French body of science, known as the Academy of Sciences for most of our period, and the control, by the Academy, of scientific production by registration and judgement. If the author had been concerned exclusively with relations between the government and the Academy, the book might well have been entitled 'The Academy under control'. However, although this is a part of our concern, we have a particular interest in the control exercised by the Academy over science. We have, therefore, accepted a dual task. The question of the *application* of science to control the natural world has only been dealt with fairly superficially. Although the Academy included some technology among its concerns, this was not its principal purpose.

'Control' is a term capable of a range of different interpretations from the most authoritarian to the more liberal, and certainly government control in any one country showed considerable change and development in the early modern period. Generally speaking, early reliance on command and restraint gave way to more indirect methods of control.[1] So in society generally there was a development of chains of interdependence, binding individuals to each other, which was to affect the emergent scientific community. For men capable of restraining their short-term impulses in the pursuit of long-term goals, changes in society and in the development of the organisation of the scientific enterprise offered social rewards in terms of status or power and, at a comparatively early date in France, in terms of income and careers.

In the eighteenth century France was ruled by an absolute monarchy. With the French Revolution, despite its cry of 'liberty', came the Terror (1793–4) with its draconian measures including price control (maximum price legislation), conscription and mass executions. Under Bonaparte political control was in some ways tightened even more, with the introduction of strict censorship of the press, to give only one example. As regards science, it will be for the reader to decide whether government control of the Academy – partly political and partly financial – should be considered as repressive. Certainly many nineteenth-century critics outside France would have interpreted the very existence of government regulations as repressive.[2] Yet, with a few exceptions, the system was much more liberal in practice than that which had operated under the *ancien régime*. Also one should not think of government involvement as

[1] Norbert Elias, 'Scientific establishments', *Sociology of the sciences*, 6 (1982), 3–69 (p. 10).

[2] 'The regulations of the French Institute would appear ... revolting and injurious to the feelings of an Englishman'. G. Moll, *On the alleged decline of science in England*, 1831, p. 24.

necessarily one-sided. The Academy or, strictly speaking, the First Class of the Institute in the period 1795–1815, received from the government a special authority, which it welcomed, and, as a government-sponsored agency, its activities acquired a valuable legitimation.[3] Any control that the government exercised over science was long-term, while the Academy's control over science was both long-term and short-term. Thus the government laid down general guidelines and demanded annual reports but it was the Academy which, week by week, decided the merits of particular scientific contributions.

The control exercised by the Academy over science depended very much more on a system of encouragement than of restraint. There was no censorship as such, but rather a subtle combination of recognition and reward, which no ambitious scientist could afford to ignore. Because of extreme centralisation and the prestige associated with government-sponsored organisations in France, nearly all major scientific research had to be directed to the Academy if it was to make its mark. Thus, although the Academy never had a legal monopoly over science, even high-level science, in many ways it had a virtual monopoly in practice. There was no alternative national forum of comparable importance. It was the Academy which decided whether a piece of work was to be considered as real science, quite apart from the question of whether it was good science. And although a lot of science went on outside the Academy, the latter was always enough of a magnet to attract to it most of the best science done in the nineteenth century in France and, sometimes, in other parts of the world. And, for a part of our period, much of the best science in the world was done in France. Although it is the early nineteenth century, which is sometimes referred to as the 'golden age' of French science,[4] the third quarter of the century was the period of the greatest work of Louis Pasteur. At the very end of the century the new science of radioactivity began in France and, with the work of Henri Poincaré (1854–1912) and some of his contemporaries in the Academy of Sciences, mathematics in France began to recover some of the distinction it had earlier enjoyed.

This study of the Paris Academy of Sciences is intended firstly to provide some understanding of a key institution; one might almost say that it is work of demystification, pulling aside the veil, which has obscured the activities of a major scientific society of international repute. If, as has been claimed, the Academy was (among other things) a stage, we need to know something of what went on behind the scenes. When Marcellin Berthelot, as secretary of the Academy, was asked by a newspaper to say something about activities within the Academy on the occasion of the centenary of the Institute, he replied deliberately in the vaguest generalities, saying:

[3] C. C. Gillispie, *Science and polity in France at the end of the ancien regime*, Princeton, N.J., 1980, p. 54

[4] See e.g. Robert Fox and George Weisz, *The organisation of science and technology in France, 1808–1914*, Cambridge, 1980, pp. 26, 33.

> You see, therefore, that in my opinion a member of the Academy of Sciences ... should behave with discretion and say nothing about his life as an Academician.[5]

Any personal clashes should obviously be hushed up as inimical to the image of a purely rational science. In connection with the elections to membership, there were regular frank discussions of the relative merits of candidates and their work, and understandably, these had to be held in private. To learn about all this it has been necessary to gain access to confidential minutes and study a large volume of surviving manuscript notes. Although French archives can be very rich, the Academy archives – despite intermittent valiant efforts of different generations of archivists – constitute a complex web with many holes; a visiting scholar, who recently tried to describe them,[6] only succeeded in revealing some strands of the web. There is no catalogue to the archives such as exists, for example, in the Royal Society of London and photo-copying of manuscripts is not permitted. Some introduction to the Academy or general survey of its functioning is, therefore, all the more desirable.

British and North American scholars, knowing the pattern of scientific organisation in their own countries, can sometimes be misled by making assumptions about the French, who follow a very different tradition. Some science historians have in particular failed to understand why, when a genius presented himself for election, he was not always immediately successful. The relationship between scientists and government in France has been another mystery. Finally the relation of the Academy of Sciences to the Institute of France has proved a stumbling block. It is worth understanding to what extent the actions of the scientists in this official body were constricted by institutional association with bodies representing other leading French intellectuals. The reader might be surprised how many writers in the English language, including several reputable historians and scientists, refer to the Academy of Sciences as the Académie Française, which is of course the name of a sister academy concerned with French language and literature.

But the purpose of demystification is only a beginning. One may hope to do more than simply transcribe French documentary sources or explain the administration of French higher education. The Paris Academy of Sciences is of major importance even to those with no interest in France because it was an arena for the working out of the place of the scientist in modern society. It established many precedents of world-wide influence in the performance of

[5] M. Berthelot, *Science et morale*, 1897, p. 207. For an even more extreme example of secrecy, see a letter from the secretary Flourens to Libri: 'Si il s'agit de l'Académie ... il me semble que tout discussion autant que possible, n'y doit pas laisser de traces.' Bibliothèque Nationale, N.A.F. 3269, f.334. The question of the admission of the public to ordinary meetings of the Academy is discussed in Chapter 10, Section 7.

[6] E. Stewart Saunders, 'The Archives of the Académie des Sciences', *French Historical Studies*, 10 (1977–8), 696–702.

research, its publication and its rewards. We need to understand the circumstances in which various procedures were evolved by the Academy to meet the needs of science and the state.

We should not forget that the Academy contained many of the most eminent scientists of the nineteenth century, who will be introduced in succeeding pages. In addition distinguished scientists from other countries were elected as corresponding members or foreign associates. A study of the Academy seems, therefore, at first sight to be purely the study of an elite. But if membership of the Academy represented the peak of achievement in French science, it should also be appreciated that scientists at all levels were affected by its deliberations and publications. It will be argued that, through the Academy, one may gain a better understanding of the activities and aspirations of a whole scientific community.

The Academy also presented an image of science to the general public. At its annual public meetings it was able to justify itself in terms of national and economic goals. Some consideration of the public image of science, or at least of the image that the Academy wished to project, may be not the least rewarding aspect to emerge from this study. This body represented 'official science' and thus constituted an easy target to attack by those critical of government policy. Yet in periods of rapid political change in the nineteenth century the Academy provided a certain stability and continuity lacking in government. Its direct power was limited but it had enormous influence, something which has never been explored. The present study is, therefore, a first attempt to understand it. At some later date a large research grant might fund a team of scholars to provide an exhaustive multi-volume study. For the present, a more selective and imperfect study may go some way to filling an important gap.

In a way a study of the Academy over rather more than a century is such a vast enterprise with such wide ramifications that no two scholars working independently would be likely to tell exactly the same story. Nevertheless, the author has made a conscious attempt to provide a balanced account. He has tried to take into consideration all of the many sciences represented in the Academy, as well as several which were not officially recognised. (On the other hand considerations of space have precluded the possibility of a detailed treatment of any one science. For such detailed studies of the science content the reader must look elsewhere). The author has tried to balance the perspective of the official representatives of the Academy with the viewpoint of would-be scientists who were rebuffed by that official body.

As an independent researcher, the author has been free to complement his initial position of sympathy for many of the aims of the Academy with some fundamental criticisms of its actual working. Everyone who knows something about French science has probably heard stories about the Academy of Sciences. But a study of the Academy must go beyond the anecdote and the *bon mot*. From the outset it is important to avoid two extreme portrayals of the institution, that have developed in different circles, virtual caricatures, which contradict each other but which, nevertheless, seem to be taken as accurate representations within the respective groups in which they circulate.

The first caricature has developed through uncritical admiration of the Academy, its members and all its works. Given the distinguished membership of the Academy, it has been all too easy for some to assume that the institution embodied everything that was best in French science. This is the approach taken in the past by many Academicians themselves. When called upon to make some incursion into history, they have felt that it would be an act of disloyalty to venture any criticism of the institution which had honoured them with membership. It has been a generally agreeable custom to celebrate anniversaries and centenaries with congratulations and pious platitudes rather than to dig deeper and sometimes ask embarrassing questions about the weaker aspects of the Academy, which were as real as its undoubted strengths.

But in avoiding the caricature of the Academy as the hero, we should not go to the opposite extreme and view the Academy as a villain or a fool. Knowing some weaknesses within the Academy before the 1975 reforms,[7] it is only too easy to project this back into the previous century. But the historian should beware of reading the present into the past. One example of this mistake is the branding of its entire membership as superannuated *savants*. But, if it was true until the recent reform that the majority of members of the Academy were scientists at the end of their active careers, it was not true in the early or mid-nineteenth century. It would, therefore, be an even bigger mistake to dismiss the Academy as irrelevant to the main stream of science than it would be to assume that it represented the whole of French scientific activity. Yet it was at the centre of that activity for most of the period covered by the book.

One must certainly not overlook the existence of other important institutions in nineteenth-century France. Indeed, the history of the Academy overlaps with that of several institutions of higher education and it will be necessary to mention these as a background to the Academy. If one is interested in mathematics and physical science, it is natural that one should look at the Ecole Polytechnique, founded about the same time as the National Institute. If, on the other hand, one is interested in the biological sciences, the obvious institution to consider is the newly reorganised Muséum d'Histoire Naturelle. There are others, who view nineteenth-century French science and technology through the perspective of the Ecole Normale Supérieure, the Faculties of Science, the Ecole des Mines, the Ecole des Ponts et Chaussées, the Conservatoire des Arts et Métiers, and so on. All these are important institutions in their own way, but each only relates to a limited aspect of science or technology. The Academy related (or attempted to relate) to the *whole* of science as well as to many areas of technology. For most of the scientific community its proceedings were a matter of constant interest and this applies to the provinces as well as those living in Paris, where the pace of life was quicker and the competition stronger.

[7] See e.g., *C.R.*, *280* (1975), *Vie Académique*, 21–3, also A.S., Dossier générale, No. 39.

The Academy was unique as the central national agency for defining, reporting, rewarding and generally controlling science. All of these aspects will be discussed in the course of the book. It has been remarked that 'in pre-professional sciences, prestige and authority were personal possessions, which could not readily be reproduced'.[8] In our story that prestige and authority is held by the Academy. Readers may conclude that the Academy effectively constituted a monopoly, which may not have been entirely to the benefit of science.

But was it appropriate in the nineteenth century to have a single body concerned with the whole of science? This might have been acceptable in the seventeenth century, but surely by 1800 science had become more specialised? The reason why all branches of science were united in one Academy was both historical and ideological. It was a fact of history that the different branches of science had been included in the Royal Academy of Sciences of the *ancien régime*. Despite the many faults that had been found at the time of the Revolution with that institution – that it was under royal patronage, that it was an elitist Academy, that it encouraged inequality of rank, and so forth – the association of different sciences together did not offend republican principles. On the contrary, the spirit of the *Encyclopédie* encouraged the bringing together of different branches of knowledge into a single institution. In his unpublished *Fragment of the New Atlantis* (1793),[9] Condorcet wrote:

> One could still fear the kind of rivalry which reigns between the sciences. It is in the interests of truth that they all combine, because there is not one among them that is not related to all the other parts of the scientific system by a more or less immediate dependence.[10]

Condorcet not only felt that rivalry between sciences was bad but, more positively, that co-operation was good. The First Class of the Institute encouraged collaboration without eliminating a healthy competitive spirit between the different sections.

It is perhaps unprofitable to speculate how another model of doing science, based on independent scientific societies, each representing a different specialism, would have fared in post-revolutionary France. Probably some subjects, such as the new science of chemistry, would have flourished, and physics might still have made important advances, for example in optics and the study of heat. Yet, in independent societies, these sciences would not have been able to inform each other. Also societies representing some of the weaker sciences might well have

[8] Richard Whitley, 'Changes in the social and intellectual organisation of the sciences', *Sociology of the sciences*, (1977), 147.

[9] The context is obviously the work of Francis Bacon, which had been a source of inspiration for the foundation of the Royal Society.

[10] 'Fragment of the New Atlantis or Combined Efforts of the Human Species for the Advancement of Science', in *Condorcet: selected writings*, ed. Keith M. Baker, Indianapolis, 1976, p. 190.

collapsed or withered away. Also there would not have been the same opportunity for new sciences, such as physiology, statistics and bacteriology, to emerge within the interstices of the Academy. The main objection against placing all the sciences together is that it may have discouraged specialised studies, but this is refuted by the actual history of the Academy. The main criticism that we can accept is that by bringing all the sciences into one body, it was necessary to restrict severely the representation of each science in order that the total number of members should not be too great to form a manageable single forum for science. Increasingly, therefore, throughout the nineteenth century there was need for other societies reflecting particular specialisms. Fortunately, although the existence of the Academy may have had an inhibiting effect on the growth of specialised societies, it did not preclude their existence.

Fourcroy, in a burst of revolutionary rhetoric in December 1793, had railed against 'gothic universities and aristocratic academies',[11] but universities are not necessarily 'gothic' nor are academies necessarily 'aristocratic', both adjectives of course being highly pejorative in the revolutionary context. Indeed, there is now general agreement that contemporary science owes much to the reformed or newly-founded universities of the nineteenth century, incorporating, for example, what the Germans called *Forschung* or research.[12] Might modern science not also owe something to reformed academies of the nineteenth century, notably the French Academy of Sciences, which provided many precedents for the financial support of research and its publication as well as for the judgement of merit and the career advancement of scientists?

Up to the present, little detailed research has been done on the nineteenth-century Academy. In so far as it has been referred to at all, it has often been to suggest that it was excessively bureaucratic[13] or at least not as important as the eighteenth-century Academy.[14] Taking the last point, the Academy was arguably even more important in the nineteenth century in one respect than in the eighteenth century in that it claimed to represent the whole of France and not simply Paris. Although in the eighteenth century the Paris Academy had been the most eminent of the science academies in France, the Montpellier Academy, sometimes regarded as an offshoot, had been quite distinguished and the Dijon Academy had been of more than local importance, not to mention the

[11] 'Rapport et project de décret sur l'enseignement libre des sciences et des arts', *Procès verbaux du comité d'instruction publique de la Convention*, ed. J. Guillaume (6 vols, 1891–1957), vol. 3, p. 97.

[12] See e.g. Joseph Ben-David, *The scientist's role in society*, Englewood Cliffs, N.J., 1971, chap. 7.

[13] In an article summarising the work of Ben-David on the organisation of science, Roy Porter writes: 'The Académie...des Sciences...languished as an engine of scientific innovation during the nineteenth century through administrative ossification and over-centralisation.' See Robert C. Olby *et al.* (eds), *Companion to the history of modern science*, 1989, p. 37.

[14] Roger Hahn, *The anatomy of a scientific institution. The Paris Academy of Sciences, 1666–1803*, Berkeley, Cal., 1971.

academies at Bordeaux, and elsewhere.[15] Although some of these provincial academies continued into the nineteenth century, it was only as a shadow of their former glory.[16] In the nineteenth century Paris dominated, being even more of a magnet for ambitious provincials, whether they were looking for a career in journalism, politics, art or science.

In modern times many people tend to think of universities as the most productive location for scientific research. But academies are arguably just as good as universities for producing research, as in many countries they had a better record in effective research until the twentieth century. For *reporting* research, academies are superior. Where they seem to be inferior is in transmitting information and skills to the next generation, since they do not normally include a formal class of membership committed to learning from senior members.[17] But we must not think of academies as being without an appreciative audience. The Academy had at least three audiences, none of which were necessarily passive: those present in the public seats at ordinary meetings, those reading its weekly journal, the *Comptes rendus*, and those attending the annual public meetings. The first two categories contained a fair proportion of scientists aspiring to recognition of their own work, either young men hoping one day to join the elite, or more senior figures, who may already have distinguished themselves in a prize competition and might well be candidates at the next election in their speciality.

Probably the most important and most numerous audience was that which read the *Comptes rendus*. They read it regularly and eagerly to keep up with the latest scientific research, and in doing so they acquired a view of what science was and how it should be pursued and reported. Thus, although lacking the discipline of formal attachment, they nevertheless acquired subconsciously a general view of science, its leading French practitioners and its fashions as well as a more detailed knowledge of their own speciality. What was reported in the Academy could well inspire them to undertake some new research, or perhaps encourage them to approach their current research in a different way.

Finally we might consider the value of the Academy to its members. For modern scientists the necessity of institutional support is well understood. For French *savants* after the Revolution there were several additional reasons for valuing the opportunity of membership of a National Institute. In the first place it offered protection; in the revolutionary period security was not something that intellectuals could take for granted. Secondly, it provided recognition, recognition both of the value of their studies and also, more personally, of the

[15] Daniel Roche, *Le siècle des lumières en province. Académies et académiciens provinciaux*, 1680–1789, 2 vols, 1978.

[16] Francisque Bouillier, *L'Institut et les Académies de province*, 1879.

[17] A possible exception is the Paris Royal Academy of Sciences, which had a rank, originally labelled *élève* (student).

value of the work of individual members. But apart from these considerations, which were particularly pertinent in the final years of the eighteenth century in France, there were also social reasons of more universal validity why senior men of science would appreciate the existence of a national body of science. Most scientists need an audience; they need colleagues, with whom they can discuss their ideas. Colleagues may be required as collaborators, but they are most important for the moral support they can provide and, occasionally, to offer informed criticism. When the Royal Academy of Sciences was suppressed in 1793, many of its members, feeling the need for mutual support, joined the comparatively obscure Philomatic Society. They valued the regular meetings, where they were able to associate with their peers, and they only abandoned this society when the Institute was founded, providing renewed recognition of their talents and giving them powerful motivation for future activity.

The eighteenth-century Enlightenment had presented science as an ideal, a liberating force, a demonstrably successful method of interpreting the natural world, which exemplified human progress. For some, science would never be more than a hobby but, by the 1790s, there were already signs that science might provide not only an ideal and a new cultural value but, for the fortunate few, it might provide an avenue of recognition and, as we shall see later, even a career. Membership of the First Class of the Institute would not have delighted everyone but we are talking of men whose life was science, or who wanted their life to be science. They could hardly ask for more in life than to join the inner circle of science. Membership was a great honour. The fortunate few were offered free membership of an elite society, where theoretically they would be on terms of equality with the greatest representatives of science of the age. All would value their new status. Some might value the companionship; others would value the power they could exercise. Having benefited earlier in their careers from patronage, they themselves would now be in a position to exercise patronage. Was there some exaggeration in the comment of a British visitor to Paris during the Peace of Amiens in 1802?

> These members of the Institute borrow everywhere; it is a matter of no small importance to obtain a seat in their hall, for it is the anti-chamber to wealth, fame and power.[18]

There was certainly fame and a certain amount of power, but wealth would only come indirectly and then only in moderation, as we shall see in due course.

[18] Henry Redhead Yorke, *Letters from France*, 1814, vol. 2, p. 15.

— *Chapter 1* —

SCIENCE IN FRANCE

I must ... acknowledge that literature, which formerly held the first degree in the scale of the moral riches of this nation, is likely to decline in priority and influence. The sciences have claimed and obtained in the public mind a superiority resulting from the very nature of their object, I mean utility. The title of *savant* is not more brilliant than formerly, but it is more imposing; it leads to consequence, to superior employments and.... to riches.
(F. W. Blagdon, *Paris as it was and as it is* (2 vols., 1803), vol. 1, p. 395.)

Only imagine, however, a city like Paris, where the cleverest heads of a great kingdom are grouped together in one spot, and in daily association and strife incite and stimulate each other to mutual emulation; where all that is of most value in the kingdoms of nature and art, from every part of the world, is open to inspection; and all this in a city where every bridge and square is associated with some great event of the past, and where every street-corner has a page of history to unfold. And, in addition to all this, not the Paris of a dull and stupid age, but the Paris of the nineteenth century, where for three generations such men as Molière, Voltaire and Diderot have kept up a mass of intellectual power such as can never be met with a second time in any single spot in the whole world.
(Johann Peter Eckerman, *Gespräche mit Goethe* (1827), ed. H. Houben, Wiesbaden, 1959, pp. 476–7).

1. '*French science*'

There is considerable justification for speaking of a specifically 'French science' in the nineteenth century.[1] Yet it would be understandable if the reader's first reaction to the idea of a national style of science was the feeling that true science is, or at least should be, essentially international. Certainly in the closing years of the twentieth century, when instant communication is possible with distant parts of the world and English is almost universally accepted as an international language of science in different continents, it is more difficult to think in terms of geographical boundaries. But when it is remembered that we are speaking about the science of an earlier age, a quite different objection might be made. Science or 'natural philosophy' could be thought of as the activity of a few isolated individuals and the work they did might be more influenced by their own personal characters than the country in which they happened to live.

[1] Maurice Crosland, 'Presidential Address: History of science in a national context', *B.J.H.S.*, 10 (1977), 95–113.

But although this may be true of certain early figures in the history of science and even perhaps as late as the nineteenth century in Britain where independence was often exalted as a virtue, it is hardly true of France, which in the seventeenth and eighteenth centuries had been ruled by an absolutist monarchy and where there had been much less political freedom. Science was drawn into the state machine as early as 1666 and was regulated in a way quite alien to the British (or later, American) traditions. Government regulation of many activities, including both industry and science, gave a distinctive stamp to the activity. It also became much more homogenous in a centralised state. Although the French Revolution of 1789 destroyed the monarchy, France inherited a tradition of state control and even increased it, in so far as education, for example, now became a government responsibility. Hence if it is possible in any period of history to speak of a national science, the case of France probably provides one of the best examples. This is not to say that a national pattern may not have some effect on other countries. One justification for a detailed study of the French situation is precisely that it established certain patterns, providing a model, aspects of which could be taken up elsewhere and applied in rather different social and political situations.[2]

The phrase 'French science' could be interpreted in a number of different ways. For example, it could mean:

(*a*) science reported in publications in the French language, or
(*b*) science done by French nationals, or
(*c*) science done within the geographical boundaries of France.

Although each of these alternative definitions is rather better than the preceding one, even the last definition leaves much to be desired. If taken literally it would, for example, include the experiments on iodine carried out in December 1813 by Humphry Davy on a brief visit to Paris. This would clearly be absurd, yet we need a definition which will include not only the main work done by French scientists in laboratories and on expeditions but also work done by refugees, who had made France their home, and by foreign students in France. A better definition would therefore be:

(*d*) work done within the French institutional framework.

This definition, although still not perfect, has the advantage of suggesting *why* it is possible to speak of 'French science'. The high level of organisation and centralisation of science was such that very little science done in France falls outside the institutional framework. Even those few amateurs who had bypassed

[2] Thus the Prussian Academy of Sciences founded in 1746, was modelled closely on the French model. Maupertuis was brought to Berlin as life president and he took advantage of this situation to introduce many of the regulations of the Paris Royal Academy of Sciences. The Berlin Academy even published its *Mémoires* in French. James E. McClellan, *Science reorganised. Scientific societies in the eighteenth century*, New York, 1985, p. 73.

the state educational system and who had no direct contact with the faculties of science or the *grandes écoles*, immediately came within the institutional orbit once they submitted papers to the Academy. This is because, when drafting claims to recognition by communicating with the Academy, anyone but the most naive would consider the norms of Academy communications and would edit his presentation to conform with these norms. Such considerations would, for example, minimise vague speculations in a memoir and would encourage the precise reporting of experimental evidence. Special value would be attached to quantitative evidence and, in a subject amenable to mathematical treatment, appropriate equations would be welcome. Such a presentation could not everywhere be taken for granted in the early nineteenth century. It was not the norm on the other side of the Rhine, for example, and in many other countries science was in a fairly primitive state. In some cases it was later to develop on the French model.

Having raised the question of French language as a unifying force, one does not want to reject it entirely. It did tend to give 'French science' a certain identity and it established clear boundaries. With a few exceptions, French scientists in the nineteenth century paid much more attention to what was published in French than in other languages.[3] Many papers published in English and German remained unknown and this occasionally led to the independent 'rediscovery' in France of phenomena previously described in a foreign language publication.

Although we claim to be writing about French science, it is inevitable that much of what follows will be about science in Paris. This is not only because the Academy was based in Paris but because Paris was the focus of French science and medicine in the nineteenth century. The First Class (later renamed 'Academy of Sciences') was founded in 1795 as part of a *National* Institute which, despite some early moves to the contrary, effectively drained the provinces of talent. Thus Paris dominated in two ways. In the first place it was the administrative centre; decisions on appointments and promotions in education and the civil service were made in Paris. As if this was not enough, Paris acted as a magnet. It was not only that it influenced careers from a distance. In science, as in many other fields, young people felt that they had to come to the capital to see and hear the leading figures in their field. It was only in very exceptional circumstances that very able scientists would be content with a provincial base.[4] They almost inevitably sought positions in Paris and thus were drawn into the Academy's web.

[3] Similarly British scientists obviously tended to concentrate on work published in the English language. Yet there was a clear realisation in the early nineteenth century that much valuable material had been published in French and might merit translation. In the later century German tended to replace French.

[4] For a study of science in certain provincial centres, see Mary Jo Nye, *Science in the provinces. Scientific communities and provincial leadership in France, 1860–1930*, Berkeley, Cal., 1986.

Here again there is a marked contrast with other countries. In nineteenth-century Britain the provinces (notably Scotland and the industrial midlands and north of England) continued to make important contributions to the intellectual life of the country independently of London. The British Association for the Advancement of Science, founded in York in 1831 with an explicitly provincial bias, proved an effective rival to the long-established Royal Society of London. Again in the German states local universities were to take pride in their independence from central control and engaged in healthy competition. It was not until the unification of Germany in 1871, with Berlin as the capital, that it was possible to think of a large centralised state. Similarly 'Italy' was a creation of the 1860s. France, therefore, provides an unusually early example of a centralised state with an exceptional concentration of talent, which in other countries would more naturally have been distributed between several major provincial centres. Paris was one of the oldest of the French university centres and when the university Faculties were established by Napoleon in 1808, it was understandable that the largest and most prestigious should be in the capital. Similarly, one would expect the *grandes écoles* to be in Paris. This concentration of higher education in Paris was to provide suitable career opportunities for many members of the Academy in the nineteenth century. They were often able to confine their teaching to a particular speciality corresponding to the section of the Academy, to which they belonged.

2. *Early institutions: the founding of the Paris Royal Academy of Sciences*

In order to understand nineteenth-century scientific organisations, it is necessary to go back briefly to the seventeenth century.[5] This is the period when a movement began to bring together the previously unco-ordinated activities of a few isolated natural philosophers. The main spokesman of the movement was Francis Bacon, who argued that, if natural philosophy was to be fruitful, it must be embedded in appropriate institutions which encouraged co-operation. In his *New Atlantis*, published posthumously in 1628, Bacon described a research institution with a hierarchy of research workers, which he called 'Solomon's house'. Meanwhile there had been some collective activity in the Italian states, notably the Accademia dei Lincei, founded in Rome *c.* 1609, the most famous member of which was Galileo. Yet, when the society's patron, duke Federigo Cesi died in 1630, the society was dissolved. Of course if a society had as its patron not simply a wealthy nobleman but the King, the inevitable death of a founding patron need not signal the death of a society. Thus when the Royal Society of London was founded in 1660 by Charles II and given a Royal charter in 1662,

[5] There have been extensive writings on science in the seventeenth-century Academies. Martha Ornstein [Bronfenbrenner], *The role of scientific societies in the seventeenth century*, (Columbia, 1913 Reprint: University of Chicago Press, 1975) can still be recommended. Among many books on the early Royal Society, Michael Hunter, *Science and society in Restoration England*, 1981, can be selected for special mention. For an eighteenth-century perspective see James E. McClellan, *op. cit.*

it not only had a (nominal) patron of the highest rank, which gave its activities a welcome legitimacy, but it also required a certain permanence through the institution of the monarchy.

The founding of the Royal Society in England was one of a number of factors which encouraged Colbert, comptroller general of finance to Louis XIV, to organise an Academy of Sciences under the patronage of his royal master. Just as in England there had been various groups of natural philosophers meeting unofficially earlier in the century, so in Paris the friar Mersenne (1588–1648) had organised meetings in his convent, as well as establishing an international network of correspondence. Colbert felt that such activities should be under state control and the founding of the Académie Royale des Sciences in 1666 was intended as further glorification of *le roi soleil*.[6] Indeed the sun was the symbol used in the seal of the Academy and it was ordered to meet in the King's own library and later in the Louvre. The Academy was to be very much a state institution, providing a marked contrast with the Royal Society, which had to find its own premises and pay for expenses by the subscriptions of individual members.

The senior members of the French Academy (the *pensionnaires*) could expect to receive pensions, equivalent to a salary. The more junior members, associates or 'students' (*élèves*, later re-named assistants) could hope for promotion in due course. There was state funding for experimental work and the Academy mounted a number of international expeditions, which would have been impossible without full state support. Although the members of the Academy may have been the envy of some of the members of the unendowed Royal Society, it should be remembered that, in accepting pensions, Academicians lost their independence.

The extent of state control in France is best illustrated by the formal regulations drawn up in 1699, which set out clearly the procedures members were expected to follow.[7] All had to be resident in Paris and they were expected to attend meetings every Wednesday and Saturday from 3 to 5 p.m. There were fixed places at meetings, which emphasised the hierarchy of membership: honorary members and pensioners were given pride of place. Although in several ways the structure of the Academy mirrored the hierarchical divisions of French society at that time, it also made important contributions to the advancement of science. Thus its formal organisation mentioned six specific branches of science: mathematics (*géométrie*), astronomy, mechanics, anatomy, chemistry and botany, which were the main scientific subjects recognised in the eighteenth century. The Academy held certain privileges with regard to

[6] The standard source on the Academy of Sciences under the old regime is Roger Hahn, *The anatomy of a scientific institution. The Paris Academy of Sciences, 1666–1803*, Berkeley, Cal., 1971.

[7] The text of the Royal Academy regulations of 1699 is given in Fontenelle, *Oeuvres complètes* (3 vols., Paris, 1818, Slatkine reprint, Geneva, 1968) vol. 1, pp. 40–6.

publication. In addition to its annual volume of *Mémoires*, the secretary was to compile a general annual overview of the work of members, to be published under the title of *Histoire*.

By contrast in Britain the publications of the Royal Society under the title *Philosophical Transactions*, began as the private enterprise of its polyglot secretary Oldenburg, only taken over later by the Society. In the early years the Royal Society was particularly keen to enrol members of the aristocracy as Fellows but it was soon to find that many, who had formally undertaken to pay a regular subscription, failed to do so and the Society had continual financial problems. Also it was all too easy for a gentleman with a dilettante interest in science to be elected as a Fellow, thus reducing those with a serious commitment to a small minority. For many members the Society was little more than a club, an attitude far from that of the Paris Academy. Whereas the Royal Society had no restriction in numbers, increasing from around 100 in 1700 to more than 700 in the 1840s, the Academy was always characterised by its severely restricted numbers, having no more than six members associated with any one subject. This worked very well for much of the eighteenth century and ensured that nearly everyone elected had some definite commitment. Even if the actual knowledge of his subject was sometimes not very great for an Academician elected to the lowest rank, he was expected to devote some time to the cultivation of science. For election to the senior rank of pensioner some tangible achievement such as a book, was expected (Article XIII). When in 1785 the question was raised of increasing the number of Academicians, Lavoisier argued against it, saying that he wanted to avoid the election of mediocre candidates. He claimed that 'outside the Academy of Sciences... there are only a very small number of *savants*'.[8] The Academy reached a high standard and its *Mémoires* compare very favourably with most of the *Philosophical Transactions of the Royal Society* for that period.

3. *Science, government and industry in eighteenth-century France*

Under the *ancien régime* successive government ministers made considerable use of science and men of science in the service of the state.[9] The brief ministry of Turgot (1774–6) was particularly noteworthy for giving great power to a man, previously known both as an administrator and as a *philosophe* in his own right. In the next section, we shall see how Turgot, believing in the power of science to improve old practices, enrolled Lavoisier in the *régie des poudres*, the newly nationalised gunpowder industry. Already in 1747, under the ministry of Daniel Trudaine de Montigny, a school for bridges and highways, the Ecole des Ponts

[8] Lavoisier, *Oeuvres*, vol. 4, p. 567.

[9] This section draws extensively on Charles Gillispie, *Science and polity in France at the end of the old regime*, Princeton, N.J. 1980. Other useful sources include: Harold T. Parker, 'French administrators and scientists during the old regime', in Richard Herr (ed.), *Ideas in History*, Durham, N. C., 1965, 85–109 and Rhoda Rappaport, 'Government patronage of science in eighteenth-century France', *History of Science*, 8 (1969), 119–36.

et Chaussées, had been founded, which provided good training for young men as civil engineers. They were taught a range of subjects including mathematics, mechanics, hydraulics, surveying, cartography and architecture, supplemented by projects providing good practical experience. It was thanks to the Ponts et Chaussées that by the late eighteenth century France had an enviable network of main roads and a number of excellent bridges built on scientific principles. The government was also interested in exploiting France's mineral wealth but the foundation of the Ecole des Mines had to wait till 1783. Here the basic scientific knowledge required, whether mineralogy, geology or chemistry, was in a more primitive state, but again it illustrates the desire of the government to make use of all the scientific knowledge available. A Royal military engineering school was founded at Mézières in north-east France in 1748 and provided good mathematical and practical training for officer cadets, who were normally required to come from the nobility. Thus the social selection of the officer corps, which was usual in European armies, was complemented in France, at least for engineers and the artillery, by an insistence on mathematical competence. We shall see later that occasionally mathematical ability reached such a high level that army officers were elected to the Academy of Sciences.

The French government and its agencies, notably the Bureau de Commerce, would request members of the Royal Academy of Sciences from time to time to test new industrial processes and products. The chemists Macquer and Berthollet held successively the salaried position of *inspecteur général des teintures* at the Gobelins tapestry works, where they applied their scientific knowledge to the art of dyeing. A succession of government ministers, notably the Trudaines, father and son, Bertin, Turgot and Malesherbes, believed that the application of science to agriculture and industry would increase national prosperity. Throughout the eighteenth century the government adopted a paternalistic attitude to agriculture and industry, giving pensions, grants and financial privileges but, in addition, it came increasingly to make use of science and men of science to solve its problems. When there was an agricultural crisis in 1784 the government tackled the emergency by establishing a committee of administrators and men of science, including Lavoisier.

An interesting parallel might be drawn between the control of industry in eighteenth-century France and the potential control of science. Both were helped by government finance, the attraction of foreign experts, the award of prizes and a certain amount of government direction. State control of industry in France goes back to the time of Colbert in the 1660s. Under the mercantilist system which Colbert established, French industry, far from being allowed to flourish according to the ability of individual entrepreneurs, was carefully controlled and protected by tariffs. There were detailed regulations to guarantee the quality of various grades of cloth and inspectors were appointed to enforce the regulations. The production of tapestries at the former Gobelins works was a government concern, as was the production of porcelain at Sèvres. All this was in stark contrast to the situation in England, where private enterprise flourished in the eighteenth century and successive innovations, most notably in spinning and weaving, led to an 'industrial revolution' in textiles and other industries. The French could claim with pride that the quality of their textiles was guaranteed, but their detailed regulations constituted a severe obstacle to innovation, so that during a large part of the nineteenth century French industrialists had to work hard to catch up with British advances. Of course, it is only with hindsight that one can see the British eighteenth-century model as superior to the French system. And the reasons why the industrial revolution of the eighteenth century took place in Britain rather than France are much more varied and complex than the above summary might suggest.[10] Of special interest to us are the implications for science.

The analogy between industry and science is not a perfect one but it would be true to say that in science too there was more control in France than in Britain. Yet the important thing is that the level of control was much less intense in science than in industry. The regulations of the Royal Academy of Sciences were more restrictive than those of the Royal Society of London, but they were not draconian. There was pre-publication censorship in eighteenth-century France, when there was relative freedom of publication in Britain, but this applied much more to political and religious works than to science. It was under these very moderate restrictions that science flourished in eighteenth-century France, surpassing in many ways British scientific production after the death of Newton in 1727. State funding helped science in France as it helped industry, but science greatly benefited also by acceptance as a creative activity in its own right and a cultural ornament rather than simply an avenue leading to the market place.

4. *The French Revolution and the revolutionary war*

The effect of the French Revolution of 1789 on science could be the subject of a whole book.[11] Here it is appropriate merely to make a few brief points which may, incidentally, help set the scene for the later discussion of the nineteenth-

[10] The extensive literature on the subject includes: F. Crouzet, 'England and France in the eighteenth century: A comparative analysis of the two economic growths', chap. 7 of R. M. Hartwell, *The causes of the industrial revolution in England*, 1967, pp. 139–74; C. Fohlen, 'The industrial revolution in France', in R. E. Cameron (ed.) *Essays in French economic history*, 1970, pp. 201–25; N. F. R. Crafts, 'Industrial Revolution in England and France: Some thoughts on the question, "Why was England first?"', *Journal of Economic History*, 37 (1977), 429–41.

[11] Discussions of the subject include: Henry Guerlac, 'Some aspects of science during the French Revolution, *Scientific Monthly*, 80 (1955), 93–101; R. Taton, 'The French revolution and the progress of science', *Centaurus*, 3 (1953), 73–89; Dorinda Outram, 'The ordeal of vocation: the Paris Academy of Sciences and the Terror, 1793–5', *History of Science*, 21 (1983), 251–74. See also Hahn, *op. cit.*, especially Chaps. 8 and 9. A major analysis of the whole question is expected soon from the pen of Charles Gillispie.

century Academy. The Revolution called for the abolition of privilege, which included all corporations, universities and academies. The Royal Academy of Sciences could claim that it was doing work in accordance with the new ideas, since it was deeply involved in the development of the metric system to replace the feudal system of weights and measures. This work, however, did not require the whole Academy *as an Academy*, but merely a few of its members. In the end, therefore, the scientific Academy was to disappear with the others. It was under attack for many reasons, because it was a *Royal* Academy and because it was an Academy which was an elitist institution, not only intellectually but because it included a special class of honorary members, drawn from the aristocracy. It was a privileged body in other ways; in its meetings, which were usually private, and in its publications. It did not include artisans in the membership and when they approached the Academy to ask for a recommendation for an invention, they were often treated with contempt.[12] It set itself up as a judge of good science and had not shown favour to the work of charlatans, including Marat. Marat became the implacable enemy of the Academy and, when he acquired power in the Revolution, he was happy to use it against the Academy.[13]

The suppression of the Royal Academy of Sciences in August 1793 along with the other royal academies is sometimes interpreted as evidence that the Revolution was fundamentally hostile to science. Similarly the execution of France's greatest man of science, Lavoisier, in May 1794 seems to confirm this innate hostility. But the Royal Academy of Sciences was suppressed *despite* the fact that it was concerned with science, not because of it, and Lavoisier went to the guillotine not as a great chemist or an Academician, but as a prominent member of the 'tax-farm' in a patently unjust system of taxation.[14] One obtains quite a different perspective if one looks at the revolutionary war, and particularly at the new scientific institutions established in 1795–6.

In some ways the Terror had been a response to the threat of invasion, and the war, which began in 1792, was an area in which science and scientists had a part to play. The revolutionary war marked a new development in the history of warfare, which previously had largely been between the hired retainers of rival monarchs. This war was going to involve a whole nation called to arms, as can be seen in the *levée en masse* of August 1793, when all French citizens, both men and women, were called upon to 'defend liberty'.[15] One category not explicitly mentioned in the decree was the men of science. But the potential value of this small group was recognised by at least one member of the Committee of Public Safety, Prieur de la Cote-d'Or.[16] Although much less famous

[12] See Hahn, *op. cit.*, p. 41.

[13] *Ibid.*, pp. 198, 202.

[14] Douglas McKie, *Antoine Lavoisier, scientist, economist, social reformer*, 1952, chap. 16, pp. 305–8. E. Grimaux, *Lavoisier*, 1743–94, 1888, pp. 388–92.

[15] J. M. Thomson (ed.), *French revolution documents*, 1789–94, Oxford, 1933, pp. 255–6.

[16] Georges Bouchard, *Un organisateur de la victoire: Prieur de la Cote-d'Or, membre du comité de salut public*, 1946.

than Robespierre or Lazare Carnot ('the organiser of the victory'), he played a crucial part in mobilising men of science to help in the war effort.

The only explosive and propellant known at this time was gunpowder, of which the major constituent is 'saltpetre' or 'nitre' that Lavoisier had shown to be nitrate of potash (potassium nitrate). Traditionally this had been imported from India, but when France lost possession of this sub-continent in the Seven Year War, she was obliged to look elsewhere for supplies. In 1775 the manufacture of saltpetre was put under state control and Lavoisier was one of four administrators appointed.[17] He began an experimental investigation of the subject, which ended with recommendations as to how saltpetre might be made artificially, in addition to the traditional method of extracting it from the walls of farm buildings and stables. Thus he helped introduce improvements in the *quality* of French gunpowder. With the revolutionary war the great problem was to increase the *quantity* of powder. Several scientists including Berthollet and Monge were involved. Instructions were published, telling people how to extract saltpetre from their farms and villages. Crash programmes were mounted in Paris to teach young soldiers how to extract saltpetre, make gunpowder and cast cannon. By January 1795 Fourcroy was able to report to the Convention on a vast increase in the production of arms and munitions.[18]

There were improvements in the manufacture of steel for swords and bayonets.[19] Monge drew up instructions for the casting of iron in sand to produce cannon. The army, however, mainly used bronze cannon and this required copper, previously imported from abroad. The new source of copper was to be drawn from church bells, confiscated from towns and villages throughout the land; but before the sonorous bell metal could be used, the proportion of tin had to be reduced. The chemists were able to do this by drawing on the new oxygen theory. The army needed large amounts of leather and the government gave Berthollet money to improve the traditional method of tanning. Seguin was able to reduce the time taken to tan leather from nearly two years to two weeks.

Armies in the field need good communications. The newly developed hydrogen balloon was considered a useful observation platform and in at least one battle (Fleurus, 26 June 1794) a captive balloon, used to observe the enemy's position, was crucial in the French victory. A school of ballooning was consequently established at Meudon for several years. But it was probably the 'telegraph' of which the French were proudest. In 1793 Chappe's proposal of

[17] Robert P. Multhauf, 'The French crash program for saltpeter production, 1776–94', *Technology and Culture*, 12 (1971), 163–81.

[18] 'Rapport sur les arts qui ont servi à la république', *Réimpression de l'ancien Moniteur*, 23, 139ff.

[19] A standard source for the contributions of science and scientists to the war effort is: C. Richard, *Le comité de salut public et les fabrications de guerre*, 1922, e.g. pp. 255ff on bell metal, pp. 612ff on communications and 663ff on the role of the *savants*. See also Hahn, *op. cit.*, pp. 256–62.

constructing a line of semaphore towers was put into effect between Paris and the north-east frontier. This visual telegraph depended on the establishment of a code of signals and good telescopes to view the signals. Previously messages had depended on messengers on horseback and it might take a day or longer to transmit a message. In August 1794, when the town of Quesnoy was recaptured by French troops, it took no more than one hour for the news to be transmitted from Lille to Paris. Barère, presenting this achievement to the Convention, interpreted it as an example of science serving 'liberty'.[20]

The French revolutionary experience did much to convince politicians of the *utility* of science. The many ways in which men of science were able to help the war effort increased their prestige and, when the Terror was over, and a more constructive period of French history began, the scientists were able to make powerful claims for the recognition of science as an important subject in the new system of higher education.

Up to the Revolution education had been largely in the hands of the Roman Catholic church. Although the Jesuits were expelled from France in 1762, a high level of secondary education continued to be given by other religious orders, notably the Oratorians and Benedictines.[21] With the Revolution, Church property and lands were confiscated, thus removing the financial basis of the Church's educational and charitable activities. Universities too were closed down. Education was now secularised and nationalised. Nationalisation in higher education meant that teachers were now paid by the State and lectures were free. Another innovation was student grants, paid, for example, to young men who had passed the competitive entrance examination for the Ecole Polytechnique. The important question now was no longer 'what is your family background?' or 'do you know some Latin and Greek?' but 'how good is your mathematics?'

The Ecole Polytechnique established in September 1794, was one of the most prominent of the educational institutions established after the Terror. We have seen that there had been both military engineering and civil engineering schools under the old regime but one of the purposes of the new school was to give both types of engineer a common training. Also in addition to vocational training, it was hoped to provide a good higher education in physical science. Students, having been recruited on the basis of their mathematical ability, could be taught to quite a high standard by a distinguished group of teachers. The curriculum of the school placed special emphasis on mathematics and chemistry; there were also lectures on mechanics, engineering drawing and fortification. The Ecole Polytechnique together with the Ecole Normale were the two *grandes écoles* founded by the Convention and of major importance throughout the nineteenth century. They will be discussed further in Section 9 of this chapter.

The French Revolution also marked an important change in the organisation

[20] *Richard, op cit.*, pp. 614–5.

[21] See Part 1 of R. Taton (ed.), *Enseignement et diffusion des sciences en France au XVIIIe siècle*, 1964.

of medicine.[22] When universities were abolished, Faculties of Medicine were included; yet the war produced an immediate need for medical personnel on the battlefield.[23] Men without qualifications were enrolled and hoped to learn by experience. The end of this first stage in a period of medical anarchy in the 1790s came with the foundation of schools of medicine in Paris, Montpellier and Strasbourg by the law of 14 *frimaire* year 3 (4 December 1794). Important innovations in the new Ecoles de médecine (of which the largest was to be in Paris) were that surgeons and physicians were to be given a common training with great emphasis on practice, observation and clinical examination of patients. From the very beginning students were trained on the hospital wards. The hospitals themselves were reorganised and expanded with special emphasis given to the classification and separation of different diseases. Physical examination of the living patient was complemented by regular autopsies, from which much was learned. The staff of the Paris Faculty of Medicine and the Paris hospitals contained many eminent physicians and surgeons, including Pinel, Corvisart and Laënnec. When peace came in 1815 Paris became an international centre for medical study[24] and the French innovations were gradually taken up in other countries.

One of the few scientific institutions not to be closed during the revolution was the Jardin du Roi, which cleverly transformed itself in a way consonant with the new political climate into the Muséum d'Histoire Naturelle. It has also been suggested that the study of plants was more acceptable to the disciples of Rousseau than abstract physical science and this was another reason for favourable treatment.[25] The applications of botany to agriculture were also stressed as a strategy for survival at a time when the best argument in favour of any kind of science was its utility. The law of 10 June 1793 accepted the new name of the institution and increased the number of professors who were to be collectively responsible for its running. By transferring the menagerie from Versailles, a basis for the study of the animal kingdom was added to the traditional emphasis on the vegetable kingdom. The mineral kingdom was represented through chairs in chemistry, mineralogy and the new subject of geology. The rich collections of the Muséum were further improved over the next few years.

Another museum, but with no previous history before the revolution, was the Conservatoire des Arts et Métiers, intended to provide a collection of machines,

[22] E. H. Ackerknecht, *Medicine at the Paris hospital, 1794–1848*, Baltimore, 1967, chap. 4.

[23] D. M. Vess, *Medical revolution in France, 1789–96*, Gainesville, Florida, 1975.

[24] See Ackerknecht, *op. cit.*, chap. 16; also R. C. Maulitz, 'Channel crossing: The lure of French pathology for English medical students', *Bulletin of the History of Medicine*, 55 (1981), 475–96.

[25] C. C. Gillispie, 'The *Encyclopédie* and Jacobin Philosophy of Science' and L. Pearce Williams, 'The politics of science in the French revolution' in Marshall Clagett (ed.), *Critical problems in the history of science*, Madison, Wisconsin, 1962, pp. 255–89 and pp. 291–308.

tools and drawings useful to industry. Although approved in 1794, it was not until three years later that the former Abbey of St. Martin was converted into a museum, which was also able to house the models of inventions sent to the Academy of Sciences. The Conservatoire was later to provide useful lectures relating to science and technology for artisans.

5. *Napoleonic science*

In political history the Napoleonic period marks a gradual swing back to the right, culminating in the declaration of the Empire in 1804 and the re-establishment of a class of nobility. After the coup d'état in November 1799, Bonaparte had very cleverly managed to gain support both from former Jacobins and former Royalists. As a victorious general, he pursued a policy of national glory and, on the civilian front, a part of this was the role he played as a patron of science. Literature was suppressed as potentially subversive to his increasingly authoritarian regime, which was soon to introduce censorship of the press. But science was encouraged as a modern cultural value, which could only reflect glory on the mother country. Since the Bourbon kings had been patrons of science and learning, it was all the more important that science should flourish and should be seen to flourish under the reign of the usurper to the throne of the Bourbon dynasty.

But one should not think of this as a sham. Bonaparte as a young artillery officer had shown a genuine interest in mathematics and science and had cultivated the friendship of several members of the First Class of the National Institute, including the chemist Berthollet and the mathematician Monge.[26] These scientists were to become members of the Napoleonic Senate as was the mathematician Laplace, who even briefly held the demanding post of Minister of the Interior after the coup d'état. Laplace dedicated one of the volumes of his monumental *Mécanique céleste* to the head of state and Bonaparte said that he would read it – when he had time! When Berthollet fell into debt it was Bonaparte who bailed him out and with the sinecure of a senatorial position he was able to buy a country property at Arcueil on the outskirts of Paris and make this the focus of a scientific society that flourished for a few years under the Empire.

But Bonaparte valued science not only for the prestige it carried in a society influenced by Enlightenment values, but also for its utility. Bonaparte used the First Class of the Institute as a platform from which to launch a number of prizes in science, technology and medicine. The prize for electricity is particularly famous. There was to be a grand prize of 60000 francs for a discovery equivalent to that of Volta (of the electric cell and battery) but also, and more realistically, an annual prize of 3000 francs for good work on the subject. Among the winners were Gay-Lussac, Thenard and, surprisingly, the English-

[26] Maurice Crosland, 'Napoleon Bonaparte and science', chapter 1 of *The Society of Arcueil. A view of French science at the time of Napoleon 1*, 1967.

man Humphry Davy. When the French later heard of Davy's isolation of the new elements sodium and potassium, using a giant electric battery at the Royal Institution, Napoleon authorised the necessary expenditure for the construction of a comparable battery at the Ecole Polytechnique, so that French scientists could produce equally spectacular results, a hope never actually realised.

One of the ways in which science was applied during the Napoleonic period was to look for alternative sources of sugar, given that in time of war cane sugar could no longer be imported directly from the West Indies. Nor would Napoleon allow the French to import material carried by British ships. Early experiments with sugar from grapes were far from satisfactory, although Napoleon did reward Proust with 100000 francs for his research on this problem.[27] More promising was the extraction of sugar from sugar beet, which kept the Napoleonic chemists busy for many years. At first the sugar was extracted in the laboratory with alcohol but the great problem was to extract the sugar economically and on a large scale. By 1812 this had been achieved and in 1813 Napoleon launched a national plan for the cultivation of the beet, the construction of factories and the training of chemists. All this, however, had happened too late in the Empire to be put into effect. The Napoleonic state was crumbling and the final defeat at Waterloo was not far in the future. Nevertheless the extraction of sugar from beet was taken up again not only in France but also in Britain and other countries, where nearly 200 years later it continues to supply indigenous sugar, which is chemically equivalent to cane sugar.

An important educational innovation under Napoleon was the re-establishment of universities in various parts of France with Faculties of Science, the first in the world. Previously recruits to science had sometimes learned some mathematics as part of an Arts curriculum. The nearest they had come to science was probably in Faculties of Medicine. Now science was an area of study in its own right, with professors appointed in mathematics, astronomy, physics, chemistry and the biological sciences. The level required for a licence was not particularly high; physics, for example, being taught without what we would now regard as the necessary basic mathematics. The highest degree available was the doctorate, for which candidates were required to write two short theses. It was expected that future appointments in the Faculties would be made from those who possessed the doctorate although, of course, the first generation of teachers in the Faculties of Science did not have the formal qualification. The lectures were open to the public so that, for example in Paris, genuine students sat side by side with visitors seeking free entertainment and perhaps enlightenment.[28] The number of science students enrolled was small, although in the 1820s there was a temporary influx of medical students, for whom the

[27] *Ibid.*, pp. 34–6.

[28] On the style of these lectures, see Robert Fox, 'Scientific enterprise and the patronage of research in France, 1800–70', *Minerva*, 11 (1973), 442–73.

baccalauréat in physical science became a prerequisite for a medical degree. Although the establishment of Faculties of Science was an excellent idea, their actual administration was excessively bureaucratic.

At the same time as Napoleon founded the Faculties in 1806–8, he also re-established the Ecole Normale of the Convention. But this time students were carefully selected and monitored as future loyal servants of the Napoleonic state. They were all boarders, subject to a rigid discipline and, in qualifying, they all undertook to teach in the state system for at least ten years. Later in the nineteenth century the Ecole Normale Supérieure was to develop as one of the great elite institutions, rivalling the Ecole Polytechnique.

6. *Professionalisation and the recognition of young* savants

One of the more obvious consequences of the founding of many institutions of higher education in which science was prominent was the emergence of a generation of young men who, for the first time, could be described as 'science graduates'. Many of the most distinguished of these graduates were able to gain employment in the higher educational system, which was expanding. In addition industry was coming to recognise more and more the utility of a scientific training and was offering employment to science graduates. Probably the branch of science in which this aspect of utility was the most marked was chemistry. Thus Gay-Lussac (1778–1850), one of the first generation of graduates from the Ecole Polytechnique, was not only able to acquire a teaching post at his old school and a further post at the newly-founded Paris Faculty of Science, but also to obtain lucrative employment at the Mint and at the Saint Gobain chemical works.[29] The possibilities of a *higher* education in science and full time employment at the level of research contrasts with the situation at the end of the *ancien régime*. Lavoisier, for example, had no comparable opportunities for a higher education in science, although he did study for a university degree in law. Also France's most famous chemist earned a living as a tax-collector, doing his scientific research in his spare time. An authority on the Academy under the *ancien régime* who had expected to find clear indications of professional commitment and cohesion in that body, found that 'the financial necessities of scientists required them to be pulled in different directions, and to be turned away from the calling of science'.[30]

One does not, however, want to place too great a gulf between the two generations. One of the characteristics of a profession is membership of a body exercising control over standards of performance.[31] It is true that there are a

[29] Maurice Crosland, *Gay-Lussac: Scientist and bourgeois*, 1978.

[30] Roger Hahn, 'Scientific careers in eighteenth-century France' in M. Crosland (ed.), *The emergence of science in western Europe*, 1975, pp. 127–39 (p. 135).

[31] There is a vast literature on professionalisation but two useful short analyses are: J. Ben-David, 'The profession of science and its powers', *Minerva*, **10** (1972) 362–83 and E. Shils 'The profession of science', *Advancement of science*, **24** (1968), 469–80.

number of problems about treating early science as a profession equivalent to the three traditional professions of law, medicine and the church.[32] Yet the criterion of membership of a professional body can be applied with a little adaptation to French science, taking the organisation of the Academy as including this function. Under the old regime the election of young men to a junior rank meant that, by their regular association with senior colleagues, they were able to acquire at least a tacit knowledge of what it meant to be a *savant*. Throughout their careers Academicians were encouraged to present memoirs at meetings and to benefit by questions and criticisms raised. After the Revolution the standard of entry was raised, so that a member would normally have learned what was required of a *savant* before he was elected. Nevertheless, membership of the Academy continued to exercise some control over the way that members performed their research and presented it to their colleagues. Moreover, such published memoirs tended to act as models for the whole scientific community. This was particularly the case after the *Comptes rendus* was founded in 1835, since it appeared much more frequently than the old *Mémoires* and was much more widely diffused.

Another characteristic of a profession is formal certification and in this too the Academy had a part to play. Even under the old regime, to be a member of the Academy was to be recognised as a *savant*. The trouble after the Revolution, however, was that, whereas all Academicians were *savants* or 'scientists', not all scientists were Academicians. This was a consequence of the growth of the scientific community without any corresponding growth in the size of the Academy. This clearly makes membership of the Academy unsatisfactory as a criterion of the professional man of science. Faced with the problem some modern scholars would be tempted to abandon the Academy dimension completely and fall back on categories of training and employment. An alternative approach, however, consonant with the theme of this book, is to agree that, whereas full membership of the Academy was the preserve of an elite, *some association* with the Academy could be used as the basis for recognition.

A decision of the First Class of the Institute in 1809 gives us evidence for this claim of official recognition of a wider circle. The context of this recognition was a discussion as to whether members of the public should be entitled to attend ordinary meetings. The First Class ruled that, whereas the general public should be excluded, serious auditors with some professional status should be admitted. This group included corresponding members of the First Class, members of the main European scientific academies, representatives from the Société Philomatique and the Société de Médecine and, finally, those who had merited the favourable attention of the Academy either by presenting two memoirs or by winning a prize. A list of some fifty names was appended to show who was recognised by the Academy. Among those named were: Ampère, Chevreul,

[32] Thus the privilege of a monopoly in exercising the professional function comes only later with science.

Chladni, De Candolle, Hassenfratz, Poisson and Thenard, most of whom were eventually to become members of the Academy.[33] The name of the young Arago was added to the list shortly after his return to Paris from a hazardous geodesic expedition.[34]

7. *French precedents for the organisation of modern science*

Many people tend to assume that the application of modern science to war is something characteristic of the twentieth century. We have already shown that the French revolutionary wars, which began in 1792–3, made systematic use of science and technology and of scientific personnel. Indeed, France provided many precedents for modern science. If we are focusing on the time dimension, we must acknowledge that the Royal Society of London was founded in 1660, six years before the Royal Academy of Sciences in Paris. But the most interesting question in comparing these two historic societies is how they differed permanently in character.[35] There was no question of the second foundation simply copying the first. On the other hand, in the case of many of the precedents to be discussed in this section, there is little doubt that they constituted genuine precedents; that is to say that they established practices which were taken up later in other countries a generation or even a century later.

It is important to note that the claim is not being made that France wonderfully *anticipated* modern procedures, as if it had been able to see into the future. What happened in the early nineteenth century must obviously be understood in terms of the state of science and the social context of the time. Certain practices which were developed by Academicians or which were imposed from above, such as specialisation in named subject areas, proved to be particularly fruitful. If one wanted to trace the antecedents of the practice of narrow concentration on a particular physical problem, acknowledgement would obviously be made of individual pioneers like Galileo, but for the systematic division of labour over the whole of science in an institutional context, it is the Academy of Sciences which deserves recognition. It is claimed that some procedures introduced in the Academy had a *continuous history*, taking us eventually to the present day.

If, on the other hand, subject specialisation, for example, had been abandoned after a brief trial and only taken up again in the twentieth century, then its original introduction would seem little more than an interesting historical antecedent. It was the ongoing implementation of such practices in a large and central European state of exceptional cultural influence, that led to their gradual adoption (and adaptation) in other countries. Whatever criticisms one may wish to make of the Academy, no-one can doubt its visibility at the centre of the

[33] *P.V.I.*, 4, 237 (21 August 1809).

[34] *Ibid.*, 242 (28 August 1809).

[35] The totally different financial basis of these institutions, with the Royal Society being dependent on private subscriptions, had a profound effect on their respective activities.

French capital. As we shall see in Chapter 11, its proceedings were reported throughout the scientific world. In the case of professionalisation, with which we begin, Charles Babbage in 1830 could hold up the French example as a model.[36] If it was not immediately adopted in Britain, it was because the custom of men of science receiving government salaries did not accord with British traditions. Having avoided the trauma of political and social revolution, Britain was to change more gradually. By way of contrast, on the industrial front, the British had given the lead and the French would try to follow.

There is still some disagreement among scholars as to exactly when science became a profession.[37] Part of this lack of agreement is due to the fact that professionalisation occurred at different times in different countries, Britain, for example, coming considerably later than France. It is not only that French *savants* received payment for their services long before this became the practice in England. The crucial factor is the availability of scientific education and this is complemented by employment opportunities. The new establishments of higher education, set up after the Revolution, enabled the generation of Arago and Gay-Lussac to obtain a highly sophisticated education in mathematics and the physical sciences. When they had graduated and gained a little practical experience, they were able to benefit by the employment opportunities in science in Paris in the early 1800s, which were infinitely better than under the *ancien régime*, when the best they could have aspired to might have been to launch a small private lecture course. Education had now been taken over by the state and a situation was created in which leading scientists could expect to hold a permanent salaried position, enabling them to teach at university level but leaving them time to do research. This is still not quite the modern picture, since few were paid to do research, but it is a considerable way along this path. The subject of professionalisation of science is complicated not only by the fact that it happened at different times in different countries but also because of differences between the various sciences. Thus from the perspective of the biological sciences, it is much more difficult to see a turning point around the year 1800 than it is to see one, for example, in chemistry, which was benefiting from a vigorous growth at this time and many applications. The biological sciences were to continue longer in the amateur tradition, although this was more the case in Britain than in France.

A second feature of French science was its high degree of specialisation of fields of investigation. As previously mentioned, the Royal Academy of Sciences distinguished six sciences: mathematics, astronomy, mechanics, chemistry, botany and anatomy. Yet for more than a century after this in Britain

[36] *On the decline of science in England*, 1830, pp. 32–6. c.f. Babbage's statement: 'Science in England is not a profession: its cultivators are scarcely recognised even as a class.' *The Exposition of 1851*, 2nd edn, 1851, p. 189.

[37] Thus the present author's perspective: 'The development of a professional career in science in France', *Minerva*, 13 (1975), 38–57, would be different from that of a scholar who had worked principally on British, German or American sources.

there was no corresponding specialisation. A man like Michael Faraday (1791–1867) could take pride in being a 'natural philosopher', equally at home in physics and in chemistry. Members of the Royal Society were not distinguished by subject in the nineteenth century. They were simply 'Fellows' with the freedom to take up any subject which interested them and equally, one may say, to abandon completely any scientific interests they had previously. Such specialisation as there was was focused on new scientific societies, like the Geological Society (1807).

We shall be discussing later (Chapter 4) the Academy's division into different subject specialisms. It was not only the Academy which encouraged such specialisation. Indeed in the Academy it was limited to such broad divisions as physics, zoology and the other main branches of science. In institutions of higher education it could be taken further. Thus we shall see below that by 1800 the Muséum d'Histoire Naturelle had not one but *three* chairs of zoology, dividing up the subject according to the complexity of the organism studied. This contrasted with Britain, where even the subject of zoology seemed excessively specialised. For much of the nineteenth century the largest British contribution to the biological sciences was within the broad compass of natural history. French medicine too could advance by specialisation and division of labour. In the new Paris medical school, established in 1794, there were no less than twelve chairs, each doubled with an adjunct professor. There could, therefore, be special chairs devoted respectively to obstetrics and legal medicine, for example, and two chairs dealing with pathology. In mathematics the Ecole Polytechnique failed to distinguish explicitly between professors of pure and applied mathematics, but by appointing two professors (later four), the work could be shared out. One would hardly expect the great Lagrange to teach anything other than pure mathematics and the engineer Prony was a recognised specialist in mechanics.

A third precedent was established in relation to publication. The phrase 'publish or perish' has been used to describe a feature of modern university careers, particularly in the United States. Publication is good evidence of research, and anyone in the present-day world who claimed to be doing successful research but failed to publish, would be very suspect. This modern situation can be traced back to the post-revolutionary period in France and to the interaction of several factors. First there was a growing number of professional scientists exploring the frontiers of knowledge. Secondly, there was a growth in the number of scientific journals. What is particularly worthy of note is the foundation of journals devoted to one particular science, for example, the *Annales de chimie* (1789) and the *Annales de mathématiques pures et appliquées* (1810). The new professional scientist would no longer be a person who devoted half a lifetime to the production of a large book. Rather he would be someone who wrote and published memoirs based on the research of a few weeks or months, thus producing acceleration in the communication of data and new ideas.

Most of these scientific journals would take a natural pride in the quality of their contents and would therefore implicitly set standards, even if there was still little formal refereeing. For a scientist to have a paper published in a good journal was, therefore, a mark of recognition. In the modern world to be the author of a number of research papers was to become an important qualification for professional employment or promotion. In the early nineteenth century appointment procedures were less well defined, but, at the highest level, publications were regarded as among the most relevant credentials, and this has been shown to have been the case most notably in elections to the Academy of Sciences, at least by the 1820s and 1830s.[38] From the earliest years of the Institute, candidates for election had had privately printed a summary of what they regarded as their most relevant qualifications. At first seniority was emphasised and any books by the candidate would naturally be mentioned. But increasingly these *Notices* consisted of a list of publications by the candidate, mainly in journals, with some explanation of the significance of the research involved. Later in the century the *Notices*, printed to influence the members of the Academy, came to be taken as models in higher education for application for senior positions, for example, at the Collège de France.[39]

Such practices were to become the norm for university science appointments in the twentieth century. Few people, however, appreciate the key part played by the Academy in establishing this practice. It would be easy to assume that it had developed from the German university system; yet although German professors at the beginning of the nineteenth century were often expected to be the author of a book, such a publication was not required to be at the level of research.[40] What distinguished the French Academy was its concern with research publications, to the exclusion of anything which hinted at popularisation. It was not enough for a scientist to be an author. What distinguished a true scientist was that he had explored a relatively narrow field and had advanced the frontiers of knowledge.

The use of publications as credentials was an important contribution by the Academy to the organisation of modern science, but the story goes further than this. One aspect not previously known was the informal introduction of an early form of citation analysis. Because of the evidence available it may be worth discussing. On 17 November 1806, just before an election for a vacancy in the botany section, the published minutes of the meeting record that (one of the

[38] M. Crosland, 'Scientific credentials: Records of publications in the assessment of qualifications for election to the French Académie des Sciences', *Minerva*, 19 (1981), 605–31.

[39] *Ibid*., 628.

[40] R. Stephen Turner, 'The growth of professional research in Prussia, 1818 to 1848, causes and context', *Historical studies in the physical sciences*, 3 (1971), 137–82, especially p. 170. Also R. Stephen Turner, 'University reformers and professional scholarship in Germany 1760–1806', in Lawrence Stone (ed.), *The University in Society*, Princeton, N.J., 1975, vol. 2. pp. 495–531, especially p. 522.

candidates) 'M. Palisot Beauvois gives a list of botanists who have cited (*qui ont fait mention*) his works or his discoveries'.[41] It should be noted that, in keeping with emerging professionalism, the candidate did not mention any *general* references to his work, only those by specialists in his field, that is to say, botanists.

Yet one may ask what prompted Palisot to take this unusual step, unless it was simply the action of an exceptionally ambitious (or vain) man of science? Fortunately, there is a letter in the Academy archives, dated 16 November 1806 (the day before the election), which answers this question. It seems that Palisot had had reported back to him the discussions in secret session at the end of the meeting of the previous Monday and the question had been raised if his publications were known or accepted by French or foreign botanists.[42] Palisot, therefore, wrote to the president of the First Class giving a list of authors in the biological sciences who had referred to his publications. He began by apologising for appearing to praise his own work, but a question had been raised and only he had the data to hand to provide an answer. He first cited his article on mushrooms in the *Encyclopédie méthodique*, published in 1785, which had been praised by Lamarck as editor. He then listed nine other publications, including one in the *Transactions* of the American Philosophical Society of Philadelphia. Cuvier had cited several of the observations on animals, which Palisot had seen in Africa, and Jussieu had referred several times to his work in the *Annales* of the Muséum d'Histoire Naturelle.

We must now report on the outcome of the election at the end of that same meeting. Out of a total of fifty-five votes, Palisot received thirty-one, as against the rival candidate, De Candolle, who received twenty-four. Palisot was, therefore, elected. It could never be proved absolutely that the citations had settled the matter but they probably helped overcome the advantage that De Candolle enjoyed through his connections with influential members of the Academy.[43] Palisot was older than his rival but the crucial factor was that he had invested his extra years in various expeditions and that *this field work was recognised* by the scientific community.

One might like to think that such citation now became the rule in the Academy but there is little evidence of this and it would therefore be misleading to make too much of the Palisot case. What counted in the elections was patronage, publications, and *previous endorsements of the candidate's work by the Academy*. Palisot de Beauvois was only called upon to justify himself in this way because he had been out of France for many years and the First Class wanted reassurance of the acceptance of his work by other botanists. The citation that

[41] *P.V.I.*, 3, 447.

[42] 'On a demandé si mes ouvrages étaient connus ou adoptés par les botanistes français et étrangers', A.S., dossier Palisot de Beauvois, letter from Palisot to president of First Class, 16 November 1806, 4pp.

[43] Candolle was one of the members of the Arcueil circle. M. Crosland, *The Society of Arcueil*, pp. 116–20.

one finds regularly in candidatures is not general approval by the scientific community but approval by the elite, namely, the Academy. Candidates regularly made a point of quoting from favourable reports on their previous work submitted to the Academy. A favourable citation from an official Academy commission was worth many times more than a favourable reference from an individual colleague. It seems appropriate now to mention how the Academy judged scientific work, although a fuller treatment will be given in Chapter 7.

An important feature of the French academic system, which has come to the fore in the organisation of modern science, is peer review. Strictly speaking, peer review should mean judgement by equals but in practice it often means judgement by more senior people in the same field. In nineteenth-century France the senior scientists in each discipline were regularly called upon to pass judgement on the merits of the work of their colleagues. This happened in the first place when ambitious scientists submitted their research to the Academy, hoping for a favourable report. A commission of experts in the field would be appointed and, if all went well, particularly in the early 1800s, a report would be drawn up within a few weeks or perhaps months, and the report would often provide encouragement for a young researcher.

At a slightly higher level there were prize competitions judged by the Academy, and at the highest level of all there were elections. Consideration of candidacies for membership occupied a regular part of the time of the Academy and we shall see that there was a well articulated mechanism for processing candidates and examining their credentials. Even a reader who has an instinctive distrust of academies could hardly fail to be impressed by the thoroughness and apparent objectivity of the series of processes involved in judging the merits of aspiring Academicians. Although in human affairs questions of personality and political and religious prejudice are never entirely absent, the Academy tried to ensure that personal considerations were pushed into second place. Its judgement of the merits of scientific research provided an important precedent for later universal practice.

A further important feature of the Academy was the influence it exerted through prizes and grants. The Academy of Sciences was not unique in offering prizes for outstanding work. This had been a feature of many academies under the old regime. The prize system of the Academy of Sciences, however, probably reached its peak in the first two decades of the nineteenth century, with several major contributions to mathematical physics, including those of Malus, Fourier and Fresnel. After that it would be easy to interpret the prize system as being in decline. A closer examination of the records, however, reveals that the prize system was being complemented and eventually transformed into a grant system.[44] Grants began in a modest way in the 1820s as *encouragements*,

[44] M. Crosland and A. Galvez, 'The emergence of research grants within the prize system of the French Academy of Sciences, 1795–1914', *Social studies of science*, 19 (1989), 71–100.

subsidiary to the main prize, but by the 1840s there were specific grant funds and the system developed considerably in the second half of the century, as we shall see in due course. The Academy had, therefore, begun to give grants earlier than the Royal Society, which only started this practice in 1850, and then only on a small scale.

In the late nineteenth century, when private benefactors gave increasingly large sums of money to the Academy for prizes, a major precedent was established which one might interpret as the inspiration for the Nobel prizes of the twentieth century. Alfred Nobel (1833–96), who had lived in Paris for the last two decades of his life, was well aware of these Academy prizes. A recent historian of the Nobel foundation, in discussing 'precursors to the Nobel prizes', emphasises the importance of the French Academy. She refers to the creation of many prize funds in the late nineteenth century at the hands of 'a new class of patrons of science made up of wealthy industrialists', of whom Alfred Nobel was typical. She concludes that 'the prizes he founded represent the culmination' of this movement.[45] It is significant that he chose an Academy to award his prizes, although the Academy chosen was that of his native Sweden. The Nobel prize is now regarded as the ultimate reward of the successful scientist. In the nineteenth century a prize-winner, although enjoying the prestige associated with the French Academy, was usually considered as being at a lower level than an Academician. To be a full member of the Academy was, for most French scientists, the ultimate accolade.

It might be noted that one precedent *not* claimed for the nineteenth-century Academy is that of consultancy on technical problems on behalf of the government. Such tasks had been undertaken from time to time under the old regime both by the Royal Academy and the Royal Society of London. Recent research on the Royal Society has revealed that in the early nineteenth century it was consulted at least once every few years and 'after 1840 the Royal Society almost routinely advised the government' in this way.[46]

8. *New sciences and the development of old ones*

Our discussion in the previous section related mainly to the organisation of science and its social context rather than its content. It is time we considered the science itself. If one were making claims for the outstanding quality of French science at the beginning of our period, two of the leading subjects would probably be chemistry and mathematics. It was after all Lavoisier and his colleagues who had given chemistry a completely new theory and a new language, which were gradually to be accepted in other countries.[47] In

[45] Elizabeth Crawford, *The beginnings of the Nobel Institution. The science prizes, 1901–15*, Cambridge and Paris, 1984, p. 16.

[46] Marie Boas Hall, 'Public science in Britain: The role of the Royal Society', *Isis* 72 (1981), 627–9. See also by the same author, 'Relations with government', chap. 6 of her book *All scientists now. The Royal Society in the nineteenth century*, Cambridge, 1984, pp. 162–81.

[47] M. Crosland, *Historical studies in the language of chemistry*, 2nd edn, 1978.

mathematics Lagrange and Laplace were outstanding, and there could hardly be a greater contrast between the state of mathematics in France around the year 1800 and that in Britain. Charles Babbage and others have testified to the lack of knowledge of (or even interest in) continental mathematics in Cambridge.[48] The level of mathematics taught to students had hardly advanced beyond where it had been left by Newton a century earlier. Had respect for Newton's genius been combined with nationalistic pride and taken to an extreme? It was not only that the most original mathematical work was being done in France but the general standards in mathematics reached a surprisingly high level. Throughout the country senior schoolboys with some competence in mathematics were encouraged to specialise in the subject to win entrance to the prestigious Ecole Polytechnique. Only a minority were successful but many of the unsuccessful had still attained a better than elementary knowledge of the subject.[49] Thus social and institutional factors help to explain great differences in the level of science in two neighbouring countries. Babbage had some grounds for his diatribe *On the Decline of Science in England* (1830), but he was speaking of relative decline. In several branches of science including his own (mathematics), France was well ahead of other countries, which could only catch up slowly.

There was another major contribution made to science in France, in which the Academy played a part, and that was the establishment of the discipline of physics or 'la physique'. In addition to the original six subjects recognised in 1699, Lavoisier played a prominent part in 1785 in the establishment of a further section devoted to 'la physique'. By the mid-eighteenth century experimental demonstrations relating to physics, and particularly the new science of electricity, had won over a fashionable audience in Paris. The abbé Nollet (1700–70), a member of the Royal Academy since 1739, was a particularly popular lecturer and in 1752 a special chair in 'physique expérimentale' was created for him at the Collège de Navarre. In a specially-built lecture theatre he entertained audiences of several hundreds with well thought out demonstration experiments. His *Leçons de physique*, which appeared in six volumes in the 1740s, went through many editions. At the same time as this popular science was helping to establish 'la physique' as one of the major branches of science known to the public, experimental philosophers were busy pushing back the frontiers of knowledge. The new science of electricity was probably crucial in enlarging the subject matter of the new and existing physics and in helping the Academy to decide to grant full recognition to this science.

After the French Revolution physics had advanced far enough to be considered as a subject to be taught along with chemistry in the new *écoles centrales*, so called because there was to be one in each department. In the new

[48] Harvey W. Becher, 'William Whewell and Cambridge mathematics', *Historical Studies in the Physical Sciences*, 11 (1980), 1–48 (pp. 6–9).

[49] The future novelist Stendhal was one of these.

programme of text-book production, physics and chemistry were again almost equal partners. The neo-Newtonian school of Laplace, Biot and Poisson helped to consolidate the new discipline or rather to inject a more mathematical content. Jolly experiments, giving people electric shocks with Leyden jars, may have represented a prominent part of the physics of the 1750s, but by the early 1800s a greater seriousness and rigour were required. Fresnel's wave theory of light represents the cream of the new mathematical physics, which was transmitted to Britain where it helped to establish 'physics' to replace the older native tradition of natural philosophy.

This alternative British tradition of 'natural philosophy' corresponded approximately with the subject matter of physics but was much wider in its general approach. As the name suggested, there was still a residual link with general philosophy, and there was also a theological dimension. A related British tradition of natural theology, which may be traced back to the seventeenth century through such prominent figures as Boyle and Newton and on to Hales and Priestley in the eighteenth century, ensured that the natural world was rarely discussed in isolation from a consideration of its Creator. This broad base of natural philosophy in Britain meant that it was of potential interest to a wide middle-class audience,[50] including some prominent clerics. In France it was easier to distinguish the practitioner from the dilettante.

A few words may be said about the transmission of physics (particularly mathematical physics) from France to Britain.[51] John Playfair, professor of mathematics and later of natural philosophy at the University of Edinburgh, paid due tribute to the work of Laplace and made good use of his work and that of Biot in his own text-book of natural philosophy in 1814. In Cambridge, William Whewell's treatises on mechanics (1819) and dynamics (1823) drew heavily on Poisson's *Traité de mécanique* of 1811. Poisson also made important contributions to both electricity and magnetism, which were used by British men of science, but it was the work of Ampère in the 1820s on electromagnetism which aroused particular interest. In optics Thomas Young's advocacy of the wave theory involved criticism of the French neo-Newtonians Laplace and Malus, but he was glad to welcome the contributions of Arago and particularly Fresnel. It was not until the 1830s that the full import of the Fresnel–Young wave theory was felt in Britain. Fourier's *Traité analytique de la chaleur* (1822) also took some time to be appreciated.

We must also consider the German states, whose major contributions to science belong to the mid- and late nineteenth century and, perhaps in physics, more to the early twentieth century. German scientific thinking around 1800 was strongly influenced by the new *Naturphilosophie*, which its critics accused of

[50] Thus books on natural philosophy were regularly reviewed in *The Edinburgh Review*.

[51] M. Crosland and C. Smith, 'The transmission of physics from France to Britain: 1800–40', *Historical Studies in the Physical Sciences*, 9 (1978), 1–61.

superficiality, vagueness and excessive generality. It provided an atmosphere which was the very antithesis of modern specialised laboratory science. Liebig later commented on the bad effect it had had on his own early scientific training, from which he was saved only by a prolonged visit to Paris in 1822–3 where, as a student of Gay-Lussac, he was able to imbibe the French tradition of rigorous and exact experimental science.[52] Liebig returned to Germany, where he was able to found his own research school at Giessen.

A major recent study of physics in the German states points out that *Naturphilosophie* also came under strong attack from within Germany.[53] Yet Goethe's retrograde theory of optics was taught in the 1820s at the prestigious University of Berlin by a former student of the philosopher Hegel with full government support.[54] The new analysis shows how little physics beyond an introductory level there was in Germany before the first important work of Ohm in 1826. Ohm was then a schoolteacher who asked for study leave, commenting on the domination of physics by the French, whose work he was studying, in order to see what was left for him to do.[55] Mayer at Göttingen made a close study of the optics of Malus[56] and Fischer studied the electromagnetic work of Ampère.[57] About 1828 Gustave Magnus, later professor at Berlin, went to Paris to acquire experimental skills.[58] So did Fechner, who made the acquaintance of Biot and Thenard and later translated their respective text-books into German.[59] Fechner's report of 1832 on mathematical physics was dominated by discussions of the work of Poisson, Cauchy and Laplace.[60] Thus up to about 1830 physics in both Britain and Germany was heavily derivative from the French and it was the French who had done most to establish institutions in which physics was one of the major basic disciplines.

French scientific institutions can also take credit for the recognition of several other branches of science, including geology and zoology. When in 1793 the Jardin du Roi in Paris was reorganised as the Muséum d'Histoire Naturelle, it was decided that in addition to the traditional course on mineralogy, there should be one on 'géologie'[61] and Faujas de Saint Fond, who had previously held the appointment of 'adjoint à la garde du Cabinet d'Histoire Naturelle', was appointed as professor of geology, a post he held (without much distinction, it must be said) until his death in 1819.[62] The new organisation of the Muséum arose out of a desire for reform by the staff of the former Jardin, which led to a fruitful discussion between one of the senior employees, Daubenton, whose

[52] A. W. von Hofmann, 'The life work of Liebig', *Nature*, Extra number, 6 Feb. 1880, i–xl.

[53] Christa Jungnickel and Russell McCormmach, *Intellectual mastery of nature. Theoretical physics from Ohm to Einstein*, 2 vols., Chicago, 1986.

[54] *Ibid.*, vol. 1, p. 17.

[55] *Ibid.*, p. 53.

[56] *Ibid.*, pp. 21–2.

[57] *Ibid.*, p. 30.

[58] *Ibid.*, p. 107.

[59] *Ibid.*, p. 61.

[60] *Ibid.*

[61] *Procès-verbaux du Comité d'Instruction Publique de la Convention*, ed. Guillaume (6 vols., 1891–1957), vol. 1, p. 485.

[62] Faujas never became a member of the Academy.

subject happened to be mineralogy, and the legislator Lakanal, who eventually presented the new constitution for approval to the Committee of Public Instruction.[63] The study of minerals was a well-established subject in the eighteenth century, but it was very far-sighted of Daubenton to appreciate that no less important was the related science of geology, based on a study of the structure and composition of the earth. The organisation of the First Class in 1795 was less far-seeing and simply recognised mineralogy, leaving to the nineteenth-century members of that section the gradual incorporation of the burgeoning new geological studies.

The term 'geology' gradually came into use in Britain in the late 1790s. The 1797 (third) edition of the *Encyclopaedia Britannica* made no mention of geology, although the next edition in 1810 had a long article on the subject.[64] The foundation of the Geological Society in London in 1807 did much to put the subject on the map. Yet it has been remarked that, despite important early British contributions to geology, 'the most popular general account of the Earth in Britain in the 1810s tellingly emerged from totally outside the community of English geologists – the Frenchman Cuvier's so-called *Theory of the Earth*.'[65]

A subject in which France was even more clearly in advance of Britain was zoology, which up to the late eighteenth century had generally been regarded as an integral part of natural history. Of course distinctions between the animal, vegetable and mineral kingdoms had long been commonplace but it is worthy of note that, whereas a serious study of the vegetable kingdom under the title of botany goes back at least to the sixteenth century, the study of animals had no generally accepted name. In France the change came with the revolution of 1789. The Jardin du Roi had been a centre for the study of natural history under the famous eighteenth-century intendant, Buffon, but it was only in 1793 that zoology was formally made a subject of study.[66] Moreover, in accordance with the French genius for specialisation, a decree of 10 June 1793 named not one but two professors of zoology. Etienne Geoffroy Saint-Hilaire was to be responsible for quadrupeds, cetacea, birds, reptiles and fish, while the more lowly insects, worms and 'microscopic animals' were given to Lamarck. But in the following year, when Lacépède returned to favour, the first chair was divided into two and he was given responsibility for his special subject of reptiles and fish, leaving Geoffroy with mammals and birds.[67] It was as a result of such specialisation, associated with rich collections, that the publication of definitive monographs such as Lacépède's *Histoire naturelle des Poissons* (5 vols., 1798–1803) became possible.

[63] Guillaume, *op. cit.*, p. 480.

[64] Roy Porter, *The making of geology. Earth science in Britain, 1660–1815*, 1977, p. 202.

[65] *Ibid.*, p. 209.

[66] The moving of the royal menagerie from Versailles to the Muséum at this time helped to consolidate the serious study of the animal kingdom.

[67] Paul Lemoine, 'Le Muséum National d'Histoire Naturelle, Son histoire, son état actuel', *Archives du Muséum d'Histoire Naturelle*, 6e Série, 12 (1935), 1–79 (p. 42).

Three parallel chairs of zoology in the same institution before 1800 was an important achievement and it was such specialisation which helped to make Paris the scientific capital of the world in the early nineteenth century. One might add that Cuvier was able to take advantage of the fact that the Muséum was concerned with both the study of human anatomy and animal anatomy to make that institution a world centre for the study of comparative anatomy. It is only the fact that the terms of reference of the Muséum confined it to the study of the biological sciences that prevented it in its early years from becoming as important as the Academy, which had an official mandate to study the whole of science.[68]

In Britain at that time there was no such thing as zoology, although natural history flourished and botany had found a home in the Linnean Society, founded in London in 1788. After the Napoleonic wars in 1817, Sir Stamford Raffles is said to have discussed with Sir Joseph Banks, president of the Royal Society, the establishment of 'a zoological collection which should interest the public'.[69] In a prospectus of 1825 calling for the foundation of a society to study zoology, it was said that:

> It has long been a matter of deep regret to the cultivators of Natural History that we possess no great scientific establishment for either teaching or elucidating Zoology.[70]

Raffles was influenced by the existence of the menagerie at the Muséum d'Histoire Naturelle in Paris, which he hoped to surpass, and he is held to have been the inspiration both of the London zoo and the Zoological Society, founded in 1826. Of course, it was all done in an English way with private subscriptions and due mention of natural theology. It was not the model but the *inspiration* which was French.

If one were looking for all the possible precedents established by French science, a final claim might be made for internationalism. At first sight this may seem implausible because it seems to contradict the idea of a national science, which France represented so much more clearly than many other countries. Indeed one of the ways in which one might interpret internationalism was as a kind of super nationalism. There is the example of a kind of cultural imperialism advocated by Napoleon Bonaparte when he invited Rumford to take up residence in France, saying he hoped that France would be the home of all future science.[71] In other words, by attracting foreign *savants* to Paris, he hoped it

[68] The decline of the Muséum in the later nineteenth century is described by Camille Limoges, 'The development of the Muséum d'Histoire Naturelle of Paris, *c.* 1800–1914', in R. Fox and G. Weisz (eds.), *The organisation of science and technology in France, 1808–1914*, 1980, 211–40.

[69] P. Chalmers Mitchell, *Centenary history of the Zoological Society of London*, 1929, pp. 1–2. For a recent study of the beginnings of zoology in London in the 1820s, see Adrian Desmond, 'The making of institutional zoology in London, 1822–36', *History of Science*, 23 (1985), 153–85; 223–50.

[70] P. Chalmers Mitchell, *op. cit.*, p. 10.

[71] Napoleon I, *Correspondance*, vol. 9, no. 7141.

would become the international capital of world science. Paris was virtually in that position anyway in the early 1800s even without bribing and flattering foreign men of science to live there.

But France did make a genuine and permanent contribution to international science, and it did this by establishing the metric system. There is no doubt about the importance of a universal system of measurement to replace the feudal system of weights and measures, which in France diverged wildly from any single standard. The French government delegated the work to the Royal Academy of Sciences. The Academy was anxious not to select a new unit of measurement that would be peculiarly French. Following Enlightenment ideas, which looked to nature to provide a model, it eventually decided to take as the basis for the whole system a unit of length, the metre, related to the circumference of the earth.[72] This could be presented to other nations as a truly natural and international unit.

In order to increase the international acceptability of the new metric system, the French government decided to convene a meeting in Paris in 1798 of neutral and allied powers to approve the final stages in the establishment of definitive new standards of length, mass and volume.[73] This is discussed more fully in Chapter 11. The acceptance of the metric system was slow at first in commerce, although it was soon found to be useful in science. As concern for precision increased during the century, it became more necessary to have truly international agreement on standards of measurement and in 1875 an international bureau of standards was established at Sèvres on the outskirts of Paris.[74] Thus France remained at the centre of the metric system, which was soon to become the international language of scientific measurement.

9. *Teaching and research in post-revolutionary France*

Whenever a need arose to study a new subject the French reaction was to found a school to teach it.[75] It was no accident that when the Jardin du Roi was reorganised in 1793 as the Muséum d'Histoire Naturelle, the teaching function of the institution should be emphasised. While accepting the importance of extensive collections of plants and other specimens, revolutionary ideology saw teaching as the real justification of the institution and the function for which the professors should be paid.

But a review of teaching institutions should probably begin with the Ecole

[72] M. Crosland, 'Nature and measurement in eighteenth-century France', *Studies on Voltaire and the Eighteenth Century*, 87, (1972), 277–309.

[73] *Science in France in the Revolutionary era described by Thomas Bugge*, ed. Maurice Crosland, Cambridge, Mass., 1969.

[74] G. Bigourdan, *Le système métrique des poids et mesures*, 1901.

[75] There had been a particular mania for founding new *ad hoc* schools in the 1780s. See C. C. Gillispie, *Science and polity*, p. 501. After the Revolution there was greater practical justification for the founding of new schools.

Normale, established at the beginning of 1795 to provide a crash programme for the new category of lay teachers. Education was now seen as the responsibility of the state rather than the Catholic church and this involved a massive training programme. There were many eminent professors at the short-lived Ecole Normale of the year 3. In science they included Lagrange, Laplace, Haüy, Monge and Berthollet, all former members of the Royal Academy and soon to become members of the First Class of the Institute. It was optimistically assumed that students would be familiar with the elementary aspects of each subject and so many lectures were given at quite an advanced level. It provided an interesting experiment in the history of education. In evaluating its limited success, one should consider not only the effect on the students but also on the professors themselves. Many were distinguished scholars who were being asked for the first time to explain their subject in the form of lectures. It thus added a new dimension to science. It was not that science had not previously been popularised. What was new was the large audience for science beyond the elementary level.

Men of science were also encouraged to write text-books on their respective subjects. Again text-books were not new, but they had tended to be the preserve of a few text-book writers. Now *savants* working at the frontiers of knowledge, like Haüy[76] and Biot[77], were encouraged to write text-books of physics, explaining the subject to the thousands who were expected to take advantage of the new educational opportunities. Thus, for a few years science subjects formed the basic curriculum for boys aged fourteen to sixteen in the new *écoles centrales*.[78] The actual teaching of science in these schools varied from place to place, depending on available teachers, apparatus and books. Probably the three *écoles centrales* established in Paris had the best facilities. In 1802 this revolutionary experiment came to an end and the schools became *lycées*, reverting to a traditional classical curriculum.

For the highest level of physical science in the early nineteenth century we have to turn to the Ecole Polytechnique, which included among its teachers the mathematicians Lagrange and Monge, the engineer Prony, and the chemists Guyton, Berthollet and Fourcroy. Laplace was an examiner at the school. The first generation of students included several who were later to join the teaching staff and become distinguished scientists: Gay-Lussac (entered 1797), Poisson (1798), Dulong (1801), Arago (1803), and Petit (1807).[79] Among the very first students (1794) were Biot and Malus and a little later came Fresnel (1804).

[76] R. H. Haüy, *Traité élémentaire de physique*, 1803.

[77] J. B. Biot, *Traité de physique expérimentale et mathématique*, 1816.

[78] On the écoles centrales, see H. C. Barnard, *Education and the French revolution*, Cambridge, 1969, chaps. 12 and 13.

[79] Ambroise Fourcy, *Histoire de l'Ecole Polytechnique*, 2nd edn. with introduction by Jean d'Hombres, 1987. For a modern social analysis of the school see Terry Shinn, *Savoir scientifique et pouvoir social, l'Ecole Polytechnique, 1794–1914*, 1980.

Ampère taught mathematics at the school for twenty years, although he was not a graduate of the school.

Bonaparte had spoken of the (revolutionary) army as 'a career open to talents' and one might say the same about science at that time through such institutions as the Ecole Polytechnique. Of course many graduates did not take up scientific careers but became army officers (particularly before 1815), engineers or administrators in the higher civil service. The training was a rigorous one, doubly so after 1804 when the school was militarised. But the governing body of the school, which included Berthollet and Monge, prevented Napoleon from turning it into a military academy. Its importance increased when it was decided after the first few years that all students at the Ecole des Ponts et Chaussées and the Ecole des Mines would have to study first at the Ecole Polytechnique. This helped to broaden the curriculum of the Polytechnique in contrast to the strictly vocational training given at the other schools, called *écoles d'application*. Later in the nineteenth century the complaint was often made that the training given was *too* academic. Great importance was attached to the order of merit of students in the passing out examination, as in the Cambridge Mathematical Tripos. Those coming near the top of the list were able to follow the career of their choice, while those at the bottom might be drafted into the army, at least during the Napoleonic wars.

There were also a small number of research institutions, in which any teaching function was secondary. Two of these were connected with astronomy: the Paris Observatory and the Bureau des Longitudes. The Observatory was a seventeenth-century foundation, which throughout the eighteenth century remained firmly in the hands of the Cassini family, the first member of which had been invited to Paris from Bologna in 1669 and cosseted with a generous pension and good funding for equipment. Designed by the architect Perrault, the Observatory was situated in what was then the southern outskirts of Paris. At the time of the Revolution the director was the great grandson of the original Cassini; often described as 'Cassini IV', he had made some effort to bring the Observatory up to date. An ardent Royalist, he resigned his position in 1793 and was reluctant to participate in revolutionary institutions, even the First Class of the Institute.[80] Some of the staff of the Observatory, notably Méchain and Delambre, played a prominent part in obtaining triangulation data required as the basis of the metric system. In 1805 Arago was appointed first as secretary to the Observatory, subsequently becoming assistant astronomer and finally director. The Observatory provided a convenient place for his optical research but he also insisted on giving popular lectures on astronomy. Under Arago the Observatory was later to become a centre for the collection of meteorological

[80] Jean-Dominique Cassini (1748–1845) was first elected to the Royal Academy of Sciences in 1770. He was elected a resident member of the First Class of the Institute in December 1795 but refused to accept the position. He was elected as a non-resident associate in 1798 and became a full member in July 1799.

data. After the death of Arago, Le Verrier (1811–77) was appointed Director and applied himself with renewed vigour and the help of several assistants to the collection of meteorological data. His authoritarian stance earned from his superior Victor Duruy, the Minister of Public Instruction, the comment that, as director he directed too much and that he did not recognise the authority of the Minister.[81] The episode provides an interesting case study of weakening of control, with a significant gap between government and senior scientist. After the death of Le Verrier, meteorological services were reorganised as a separate agency, part of a general trend which we shall have occasion to comment on in the final chapter.

A second research institution was the Bureau des Longitudes, established in 1795.[82] Although inspired by the British Board of Longitude and having the responsibility of publishing astronomical and navigational tables, it was always more concerned with pure science than its British counterpart. Its personnel was predominantly scientific, consisting of two mathematicians, four astronomers and a geographer, together with two former naval officers and a technician. Posts in the Bureau provided a useful source of income for many Academicians, particularly those in the astronomy section. The Bureau met once a week and in addition to its original concern with navigation, it acted as the guardian of the metric system, having formal responsibility for establishing and maintaining the physical standards. Laplace was associated with the Bureau from the beginning and he took a leading part in directing it, even asking the members to investigate his own research problems. Thus in 1822 he arranged for the Bureau to undertake experiments on the velocity of sound.[83] Under the later inspiration of Arago, the Bureau became much involved in the collection of data on terrestrial magnetism.[84] Thus both the Observatory and the Bureau, which had been founded with general goals relating to astronomy, became increasingly involved in the nineteenth century in studies which were described as 'physique du globe' and which were much more concerned with the environment of the earth.

Although we have described these two institutions as research institutions, we have mentioned that Arago gave lectures at the Observatory on popular astronomy. Also in 1834 it was decided to admit to the Observatory young men as pupil-astronomers (*élèves astronomes*) and some thought was given to their training. The emphasis on teaching was widespread. In technical education the teaching function of the Conservatoire des Arts et Métiers was extended under

[81] Jean Rohr, *Victor Duruy, Ministre de Napoleon III*, 1967, p. 109.

[82] M. G. Bigourdan, 'Le Bureau des Longitudes. Son histoire et ses travaux de l'origine á ce jour', *Annuaire du Bureau des Longitudes*, 1928, A1–A72, C1–C92; 1930, A1–A110; 1931, A1–A145; 1932, A1–A117.

[83] *A.c.p.*, **20** (1822), 210–23.

[84] John Cawood, 'Terrestrial magnetism and the development of international collaboration in the early nineteenth century', *Annals of Science*, **34** (1977), 551–87.

the Restoration.[85] An impressive programme of lecture courses was begun, the lectures being given in the evening so that workmen could attend. By 1824 the lectures on applied mechanics, industrial chemistry and business economics attracted a total of 2000 students. A further contribution to technical education was made by the Ecole Centrale des Arts et Manufactures, founded originally as a private venture in 1829 and taken over by the Ministry of Commerce in 1856. It provided a more practical curriculum than the Ecole Polytechnique and its graduates included many distinguished engineers, including Gustave Eiffel, best remembered for his tower built to commemorate the centenary of the Revolution of 1789.[86] To mark the growing importance of electricity in urban life in the late nineteenth century the Ecole Supérieure d'Electricité was founded in 1894.

The two most important *grandes écoles* continued in the late nineteenth century to be the Ecole Polytechnique and the Ecole Normale Supérieure. The former had a military atmosphere with students drawn from the upper middle classes who often became senior civil servants or army officers, while the latter appealed more to students of more modest social background. Students for both schools had achieved entry through a rigorous *concours* and always considered themselves superior to ordinary university students. By the 1860s, the Ecole Normale Supérieure had come to rival the Ecole Polytechnique as a leading centre of scientific education in France.[87] Originally specialising in classics with some mathematics, the arrival of Louis Pasteur in 1857 as administrator and director of scientific studies had greatly helped to strengthen the science side. Typically the course of study culminated in the *agrégation*, an advanced teaching degree, but bright science students were encouraged to take the doctorate. A large proportion of the graduates became lycée or university teachers and, through its graduates, the school exercised enormous influence in higher education. A further reason why the Ecole Normale Supérieure was so influential in the educational world is that it came under the Ministry of Public Instruction, whereas the Ecole Polytechnique was controlled by the Ministry of War. During the Third Republic the Ecole Normale gained a reputation as a breeding ground for left-wing politicians.[88] Particularly influential was the socialist librarian Lucien Herr; graduates included Jean Jaurès, the founder of the French Socialist Party (1902) and the mathematician, Academician and politician Paul Painlevé (1863–1933), who held the position of premier three times.

Given this bias towards teaching rather than research institutions, the role of the Academy became all the more important. Its main function was to

[85] Frederick B. Artz, *The development of technical education in France, 1500–1850*, Cambridge, Mass. 1966, pp. 216–17.

[86] Joseph Harris, *The Eiffel tower, symbol of an age*, 1976.

[87] Craig Zwerling 'The emergence of the Ecole Normale Supérieure as a centre of scientific education in the nineteenth century' in R. Fox and G. Weisz (eds.) *op cit.*, pp. 31–60. A notable exception to this generalisation would be the science of geology.

[88] Patrick H. Hutton (ed.), *Historical Dictionary of the Third French Republic*, Westport, Ct., 1986, vol. 2, p. 897.

encourage research, not only by the work of its members but by providing a platform, an audience and, we might say, a refereeing system for the presentation of research by anyone with scientific pretensions. Some might want to reproach the Academy for contributing so little to teaching. This would be unjust. Its terms of reference specifically excluded teaching and emphasised research. It is, therefore, on its research record that it should be judged, that is to say the research presented at meetings by members and that submitted by non-members. We shall see later that this covered the whole spectrum from the most brilliant to the most feeble, but in an open system the latter could not easily be excluded in advance. The general standard, however, was very high.

10. *Social opportunities and pressures: the* 'salons'

Having said something about the formal organisation of science, we must finally consider the informal and social dimension. The emphasis in this book will naturally be on institutions. Yet it would be naive to assume that only two sources of control affected the Academy: the government and the official actions of the Academy. According to the simplest analysis, where the government did not exercise power this left the Academy in full control. But to accept such an analysis would be to overlook the existence of unofficial outside forces, pressure groups, cabals, political and religious affiliations, 'old-boy networks' and various *salons*. We shall have occasion later to discuss the political and religious background and also the existence of powerful family dynasties and alliances. Here it should be sufficient to say something about various *salons* in nineteenth-century Paris, where information was exchanged and alliances developed.

First there is the problem of identifying the scientific *salon* and of learning something of what transpired on these social occasions. If at one extreme a *salon* was entirely social in nature, then it could be argued that it was irrelevant to science. Thus the Thursday evening gatherings in the early 1800s in the house of Madame Lavoisier, where music was played, may be of little interest although the dinners she held were more selective and guests included Berthollet, Cuvier and Humboldt.[89] If on the other hand the occasion was *purely* scientific, then it should be considered as a scientific society rather than a *salon*. A few *salons* are mentioned below and are known because they were held regularly over a period of several years. In addition there were, of course, a large number of social occasions and dinners, to which scientists were invited yet of which little is known. It is only by chance that we learn that when the Dutch natural philosopher Van Marum was in Paris in 1802, he was invited to the 'fortnightly supper' given by the anatomist Portal, where the other guests included Lalande and Cuvier.[90] The dinner invitations of a scientist are not the centre of attention

[89] D. I. Duveen, 'Madame Lavoisier', *Chymia*, 4 (1953), 13–29 (pp. 23, 27–8)

[90] *Martinus Van Marum, Life and Work*, ed. R. J. Forbes (6 vols., Haarlem, 1969–76) vol. 2, p. 373. No other guests are mentioned by name.

for the historian, but they can provide valuable background information which may help us to understand better the academic politics of the period.

One of the earliest and better documented of the informal groups which met regularly in the nineteenth century, was that which began to meet about 1801–2 at Berthollet's country house at Arcueil, a few miles to the south of Paris.[91] Soon after Laplace had bought a neighbouring property at Arcueil in 1806, the group described itself as the 'Society of Arcueil', but this name should not lead us to think of the group as a formal scientific society with officers and minutes of meetings. Rather it was something of a *salon*, where Berthollet and Laplace presided, sometimes with their wives, offering hospitality at the week-end to a small and select group of younger scientists, who were thus encouraged to further their scientific careers. The elite of a generation of physical scientists, notably Gay-Lussac, Biot and Arago, owed much to the Arcueil circle, which may be interpreted as a major pressure group within the First Class of the Institute, ensuring the election of its younger members. Alternatively the Society of Arcueil may be interpreted more as a research school of neo-Newtonian science.[92] The various interpretations are not mutually exclusive.

Although the Society of Arcueil went into decline with the fall of Napoleon, this did not mean its influence was dead. Visiting foreign scientists continued to visit Berthollet in retirement at Arcueil. Also friendships made there continued during the period of the Bourbon restoration and beyond to form the basis of a powerful network.[93] If we look at the names of the senior scientists chosen by the Academy to be successive annual presidents between 1822 and 1828, we find that in six out of seven years the men chosen were former members of the tiny and exclusive Arcueil group: Gay-Lussac (1822), Thenard (1823), Arago (1824), Chaptal (1825), Poisson (1826), Dulong (1828).

In 1823, the year after Berthollet's death, the young physiologist and future chemist J. B. Dumas arrived in Paris from Geneva. He was delighted to be invited regularly to the weekly receptions held by the mineralogist Alexandre Brongniart in his house at 3, rue Saint Dominique.[94] Entry to the house was probably the most important single event in Dumas' personal life as well as his subsequent career, since he was later to marry into the Brongniart family.[95] As far as scientific contacts were concerned, he had the opportunity to meet there some of the most eminent scientists of the capital, including Arago, Humboldt, and Thenard. British visitors included Faraday and Brewster.[96]

[91] Maurice Crosland, *The Society of Arcueil*, 1967.

[92] Gerald L. Geison, *Michael Foster and the Cambridge school of physiology*, Princeton, N.J., 1978, p. xiv.

[93] M. Crosland, *op cit*., chapter 9.

[94] Marcel Chaigneau, *J. B. Dumas, chimiste et homme politique*, 1984, pp. 69–71.

[95] For a discussion of how marriage into the Brongniart family helped Dumas' career and his entry into the Academy, see Chapter 5, Section 5.

[96] There is further information on the *salon* in Louis de Launay, *Une grande famille de savants. Les Brongniart*, 1940, p. 142.

But we have neglected to mention a more famous *salon* of the Restoration period, that held by Cuvier on Saturday evenings in his house attached to the Muséum d'Histoire Naturelle.[97] Many famous people attended this *salon*, some with literary reputations, including Stendhal, but more particularly guests drawn from the scientific world. Here were to be found[98] regularly his brother Frédéric and many naturalists including the zoologist Valenciennes (1794–1865), the botanists Mirbel (1776–1854) and the young Adrien de Juisseu (1797–1853), his old friend[99] Etienne Geoffroy Saint-Hilaire (1772–1844) and the chemist Chevreul (1786–1889). All of these had positions at the Muséum and became members of the Academy. In addition there was Ampère, whose son was attracted to Cuvier's daughter Clémentine, whose death in 1827 put an end to these happy social occasions. On one occasion the guests included Brongniart, director of the porcelain factory at Sèvres, Prony[100] director of the Ecole des Ponts et Chaussées, the physicist Biot and the German naturalist and explorer Humboldt.[101] Foreigners, visiting scientists and explorers were frequently to be found there. When Mr and Mrs Nathaniel Bowditch from Boston, Mass. went to Paris in 1819,[102] they benefited many times from Cuvier's hospitality, as did Charles Lyell in the 1820s, who described them as 'a great treat'.[103] Cuvier welcomed his guests graciously, assisted by his wife and step-daughter Sophie. Soon tired by petty gossip, however, Cuvier liked to pick the brains of his guests,[104] particularly foreigners, whom he expected to be fluent in French. Sometimes he would withdraw temporarily to his well-stocked library, which few were permitted to enter. Generally, however, he was happy to be at the centre of attention, dazzling his guests with his erudition. Some special friends were invited to stay late.

Here surely was a gathering which could operate at many different levels, from the political to the personal, from the scientific to the social. As at Arcueil, patronage would be dispensed and careers planned, but an important difference was that Cuvier held simultaneously an extraordinarily large number of administrative and scientific posts, which it has been remarked 'enabled him to

[97] Unfortunately, there is no single comprehensive source for this subject. The following footnotes indicate several of the main sources of information, both primary and secondary.

[98] Edmond Pilon, 'A propos d'un centenaire. Le Salon de Cuvier au jardin des plantes', *Revue des deux mondes*, 15 July 1932, 382–94 (p. 389). (A source of disappointingly limited use). See also J. Viénot, *Georges Cuvier, le Napoléon de l'intelligence, 1769–1832*, 1932, pp. 193–6, and especially Dorinda Outram, *George Cuvier. Vocation, science and authority in post-revolutionary France*, Manchester, 1984, e.g. p. 217.

[99] This friendship came to an unfortunate end with the Academy dispute of 1830 on transformism. See Chapter 3, Section 7.

[100] Madame Prony held her own salon.

[101] (Mrs) R. Lee, *Memoirs of Baron Cuvier*, 1833, p. 304.

[102] *Ibid.*, p. 5.

[103] *Life, letters and journals of Sir Charles Lyell*, 1881, vol. 1, pp. 125, 134, 136.

[104] H. E. Negrin, *Georges Cuvier, administrator and educator*, Ph.D thesis, New York University, 1977, pp. 441–4.

exercise patronage in almost every corner of the French political world'.[105] But, it is argued, to operate successfully in a world of personalised power, it was necessary to cultivate social contacts. His *salon* played a crucial part in extending this power, since he was able 'to cultivate personal contacts with every shade of opinion'. Cuvier was an unusually political animal but even the more innocent *salons* might bring influence to bear on decisions affecting scientific appointments and the Academy.

The spacious quarters and concentration of naturalists and their families made the Muséum an obvious social rendezvous. The zoologist Henri Milne-Edwards (1800–85) used to hold regular social gatherings in his house there in the mid-century. Marcellin Berthelot was later to reminisce about these social evenings and about the 'elite', both French and foreign, who used to gather there.[106] Another institutional setting was the Ecole Normale, where Henri Sainte-Claire Deville (1818–76) used to gather his students and friends on Sundays. Although experiments were sometimes performed, the guests included not only scientists but also industrialists, literary figures, philosophers and historians, thus placing science in a wider context.[107]

Usually at these salons ladies were present, reinforcing the occasion as a social one. This is explicit in a description of the Bertrand *salon* in the 1860s. Darboux, secretary of the Academy in 1907, described it in the following terms:

> I became acquainted with M. and Mme Berthelot some 45 years ago in that hospitable house on the rue de Rivoli, where M. and Mme Joseph Bertrand brought together men such as Pasteur, Foucault, Boissier, Renan, Deville and many others whom I now forget. At that time Berthelot was not yet a professor at the Collège de France.[108]

The implication here and elsewhere is that these *salons* were not simply meetings of friends of the same generation. They always included young scientists and must, therefore, be seen as important agents in the system of patronage. Of this *salon* Berthelot was later to admit that one of the regular subjects of discussion was forthcoming appointments in the academic world, a subject which he said was always discussed discreetly![109] There must have been many other *salons* involving scientists, whose stories remain to be told.[110]

It would be a mistake to conclude that the only influence of these *salons* on science was in academic politics and elections to the Academy, for example to organise a cabal to vote for *X* or conspire against *Y*. The *salons* also helped to

[105] Outram, *op. cit.*, p. 108.

[106] M. Berthelot, *Science et education*, 1901, p. 151.

[107] D. Gernez, 'Henri Sainte-Claire Deville' in *Ecole Normale, Livre du centenaire, 1795–1895*, 1896, pp. 407–25 (p. 411).

[108] *C. R.*, **144** (1907), 668.

[109] M. Berthelot, *op. cit.*, p. 131.

[110] The role of freemasons, for example, still remains to be investigated. We do however know that a few Academicians, notably J. J. Lalande, were at one time prominent freemasons.

guide the actual science done, not only by passing on news but also by helping to encourage conformity and reinforce orthodoxy. Cuvier used his *salon* to consolidate his absolute authority among French naturalists and it was not until 1830 that Geoffroy Saint-Hilaire dared even to challenge his doctrine of the fixity of species. Blainville (1777–1850) was one of the few naturalists who refused his patronage outright and, in a spirit of extreme independence, made the successful prediction:

> One day I shall take my seat at the Institute and the Muséum beside you, opposite you, and in spite of you.[111]

When elected, Blainville was more than once to break the polite conventions of conformity within the Academy. The *salon* was very much connected with conformity and with patronage, which was the key to advancement in a scientific career.[112] Of course, this was not quite the outright favouritism of the *ancien régime*. After the Revolution all men of science were expected to have basic qualifications for a scientific post, whether a diploma or research publications or both. But qualifications alone would not take a young person very far in his career. A well-placed patron made all the difference.

In the Arcueil group too, patronage was important. Among the younger members of the group was Arago (1786–1853), who grew increasingly uneasy with the intellectual dominance of Laplace. The latter had obtained his post at the Observatory for him, but he could not agree with Laplace's totally committed Newtonian approach. Under the Restoration Arago himself was to become a patron, a patron of Fresnel (1788–1827) and his wave theory of light, which came to replace the corpuscular theory of the traditional Newtonian orthodoxy. It was, however, only after 1830, when he was elected secretary of the Academy, that Arago became a really powerful figure. But even earlier, soon after 1815, there had been a split of physical scientists into two camps. The radicals, headed by Arago, were opposed by a more conservative faction who worked within a neo-Newtonian framework; they included Biot (1774–1862) and Poisson (1781–1840) and were sometimes referred to pejoratively as the 'secte des Biotistes'.

One could, if one wished, make more of the conflict between the followers of Laplace and Biot on the one hand and those of Arago on the other.[113] Indeed, going beyond the theme of patronage, this could be interpreted as an outstanding nineteenth-century example of a battle for the control of (physical) science and one which took place partly in the Academy. During the Napoleonic period the neo-Newtonians had tried to bring order to various parts of physics,

[111] Flourens, 'Eloge historique de Ducrotay de Blainville', *M.A.I.*, 27, Part 2 (1860), i–lx (xii).

[112] For a useful discussion of patronage in the nineteenth century, see Terry N. Clark, *Prophets and patrons. The French university and the emergence of the social sciences*, Cambridge, Mass., 1973.

[113] See Robert Fox, 'The rise and fall of Laplacian physics', *Historical Studies in the Physical Sciences*, 4 (1974), 89–136.

and notably to optics and the study of heat, by applying the concept of particles subject to short-range forces. But some of the 'new' physicists, such as Fourier (1768–1830) and Fresnel, saw no need to postulate particles and put forward alternative mathematical theories. The kind of control sought by the new men was in the first place intellectual mastery of the phenomena. But for the politically ambitious Arago this intellectual quest soon developed into a quest for power. In so far as the school of Laplace had represented not only a research programme but also a powerful political force within the Academy, which manifested itself in support of suitable candidates for election, it is understandable that Arago should seek to challenge this school with his own growing band of allies. His election as perpetual secretary to the Academy in 1830 was Arago's greatest political victory and one which was going to have widespread repercussions within that institution, as we shall see in due course.

— *Chapter 2* —

THE STRUCTURE OF THE ACADEMY

> Our working meetings follow each other according to a common pattern and with the regularity of astronomical events. No festival, no great event prevents the Academy from meeting once a week to receive and register research work and to discuss impersonal truths.
>
> (G. Lippman, Public meeting of 16 December 1912, *C.R.*, **155** (1912), 1277.)

> A person, who devotes his life to the study of the sciences or the arts, must surely know how to control his desires and regulate his needs.
>
> (Report by Villers on honoraria for members of the Institute, 21 May 1796, reproduced in L. Aucoc, *L'Institut de France*, 1889, p. 36.)

> Turning up every Monday at about 3 o'clock at the Institute, signing the attendance register on arrival, accepting the minutes of the previous meeting, helping at every other meeting (each fortnight) in analysing the correspondence without taking in a word of it, paying no attention to most purely scientific communications but being all ears on matter of personality, cutting short the public part of the meeting, which is mainly concerned with the former, in order to be able to prolong the secret session, where the latter can be discussed at leisure, nominating for the examination of memoirs sent by scientists from outside, commissions which will not in fact examine anything, finally between 5 and 6 o'clock going to dinner – this is what [the secretary of the Academy] calls working with the greatest activity.
>
> (*Victor Meunier, Scènes et types du monde savant*, 1889, pp. 182–3.)

1. *The foundation of the Institute*

The Revolution was a time for drafting ambitious new projects and in this period one has to be careful to distinguish between plans and their ultimate realisation. The history of the Institute is a case in point. As early as 1791 Talleyrand had an idea for a grand National Institute, located in Paris, where all the sciences, arts and literary studies would be pursued at the highest level.[1] This would replace not only the existing academies but also the universities and institutions like the college founded by François I, soon to be renamed the Collège de France, as well as that seventeenth-century foundation, the Jardin du Roi, since the members of the Institute were going to be required to teach as well as to carry out research. But Talleyrand also wanted to encourage new knowledge rather

[1] R. R. Palmer, *The Improvement of humanity. Education and the French Revolution*, Princeton, N.J., 1985, pp. 98–9.

than merely the transmission of existing knowledge. The report was discussed but never implemented.

The idea of a National Institute surfaced again in December 1792 in a bold new educational project of Condorcet, although there was some change in vocabulary.[2] Condorcet, 'the last of the *philosophes*', advocated a succession of tiers of education throughout France and, crowning the whole structure in Paris, there would be a 'National Society' of Arts and Sciences. The 'National Society' or Institute was to have half its members resident in Paris and the remainder scattered throughout France. It was to consist of four classes, of which the First represented the mathematical and 'physical' [and biological] sciences, the Second the moral and political sciences, the Third (and largest) applications of science to the industrialised practical arts, and the Fourth literature and fine arts. The Third Class would include medicine and it was from this class that most immediate benefits to society were expected to come. Members of the National Society were not required to teach, but they were expected to supervise teaching at lower levels and appoint professors in the lycées among other duties. Although Condorcet was concerned to introduce the highest standards throughout his system, he strove hard to combine this with the political principle of equality. Yet when the project was discussed in the Convention, it suffered serious criticism not only because of the great cost involved but because it was seen as elitist and the National Society was seen as an academy in disguise. It would have been more powerful than any single academy under the old regime and, like the old academies, its members were to be leading intellectuals.

For the actual founding of the National Institute we have to wait till the constructive period after the Terror, although discussion predated *thermidor*. When the subject was raised again on 23 June 1795 by Boissy d'Anglas in the Committee of Public Instruction, he presented it as a high-level teaching organisation for 'l'enseignement public'.[3] But the very next day a rival project was put forward by Daunou, who insisted that the new National Institute should be responsible for accumulating knowledge rather than for teaching.[4]

Daunou was one of the several people who shared responsibility for planning the Institute which actually came into existence, other principal architects being Lakanal and Fourcroy.[5] All elected members of the Convention, Lakanal and Daunou were former priests who had held teaching positions at provincial *collèges* before the Revolution. Fourcroy was a chemist who had resented many features of the *ancien régime* and, like the others, had welcomed the Revolution as a means of establishing a new social order. Daunou had a legalistic flair and

[2] *Procès-verbaux du comité d'instruction publique de l'Assemblée Législative*, ed. J. Guillaume, 1889, pp. 188–246. Palmer, *op. cit.*, pp. 124–31.

[3] *Procès-verbaux du comité d'instruction publique de la Convention*, ed. J. Guillaume (6 vols., 1891–1957), vol. 6, p. 335.

[4] *Ibid.*, p. 339.

[5] *Ibid.*, pp. 644–5.

was the chief architect of the political constitution of 1795, but the key figure in relations between men of science and government in the revolutionary years was Lakanal.[6] In the final year of the Royal Academy of Sciences Lavoisier had found Lakanal a useful intermediary in dealing with the government. Lakanal's value increased when he became a member of the Committee of Public Instruction. In a letter of 17 July 1793 Lavoisier explained that, unlike literature, 'most of [the sciences] cannot be pursued with success by isolated individuals.'[7] Scientists *needed* something like the Academy. Moreover, Lavoisier insisted on the utility of science. Lavoisier and Lakanal nearly succeeded in saving the Academy of Sciences by calling it a *Société libre*[8] but the Convention was determined to abolish all Academies, which it did by the decree of 8 August 1793. The next year Lavoisier himself was executed but Lakanal lived on as a member of the Committee of Public Instruction with great sympathy for science and learning.

According to article I of the draft drawn up by Daunou, the proposed Institute was to be concerned with 'extending the progress of the sciences and the arts'. In the report presented to the Convention on 19 October 1795 Daunou said:

> We have taken from Talleyrand and Condorcet the plan of a national Institute, a grand and majestic idea, the execution of which should efface in splendour all the academies of kings, as the destiny of republican France already effaces the most brilliant periods of France under the monarchy. It will be in some way an epitome of the learned world, the representative assembly of the republic of letters, the honorable goal of all ambitions of science and talent, the most magnificent recompense of great efforts and great success.[9]

But lest anyone should conclude prematurely that Daunou alone was the father of the Institute actually established, it is necessary to point out that his project followed Condorcet in advocating *four* Classes. The influence of Condorcet was also evident in the *separate* representation of pure and applied science. The Class concerned with 'the application of science to technology' was headed by medicine and surgery and included veterinary science and rural economy, mechanical arts and navigation.[10] All of these subjects were eventually to be incorporated into a single class concerned with both pure and applied science. The proposed separation of pure and applied science would not have worked well. It is Fourcroy whom we have to thank for bringing together pure and applied science and thus reducing the number of classes from four to three.[11]

Having settled the number of classes and the general areas of knowledge they

[6] *Ibid.*, pp. 831–9. This reproduces parts of the text from Lakanal's own *Exposé sommaire des travaux de Joseph Lakanal*, 1838.

[7] Quoted by R. Hahn, *The anatomy of a scientific institution. The Paris Academy of Sciences, 1666–1803*, Berkeley, Cal., 1971, p. 234

[8] *Ibid.*, p. 247.

[9] L. Aucoc, *L'Institut de France*, 1889, p. 5

[10] Guillame, *op. cit.*, vol. 6, p. 340.

[11] *Ibid.*, p. 576n.

were supposed to cover, we may return to consider the purpose of the actual Institute that was founded in 1795. We have seen that the teaching function was disavowed before its foundation, leaving it with the principal obligation to do research. Article I of the constitution of the Institute gave its purpose as 'perfecting the sciences and the arts by continuous research (*des recherches non interrompues*), by publication of discoveries, and correspondence with learned and foreign societies'. We may note that this is a very early expression of the concept of research, which was certainly not known in Britain at the time. Even in the second half of the nineteenth century the idea of research was not fully acceptable in the older universities in England[12] and, when Mark Pattison in 1876 argued in favour of it, he apologised for the limitations of the English language, saying that no one word was adequate to express the concept. Nevertheless he used the term 'research', giving in parenthesis the French word *recherches* for greater clarity.[13]

As regards the membership of the new Institute, Lakanal drew up a list of leading representatives of the ten sections representing science, two for each. No one would have disagreed that Lagrange and Laplace were ideal representatives for the mathematics section, but when it came to some of the other sections the Directory was not happy with the inclusion of a number of royalist *savants*, notably Borda in the mechanics section, Adanson and Jussieu in the botany section and Parmentier in the agriculture section. On the other hand, no one could accuse Fourcroy of right-wing sympathies, yet his name was removed so that chemistry was now represented by the two senior (and less political) chemists, Guyton and Berthollet (see Table 2). It was the task of the first two nominated to propose two more candidates in each section, and then these four proposed a further two names to bring the strength of each section up to size. In this way people like Fourcroy and Cousin, originally on Lakanal's list, joined the First Class of the Institute a few weeks' later. Scientific talent had been recognised by their peers and no harm had been done to the *savants*, except possibly to their pride if they were exceptionally sensitive. Nomination by a section did not, of course, automatically secure election by the whole Institute but generally speaking the electoral body, which in these early years represented all departments of knowledge, tended to accept that the specialists in each discipline would know who were the most talented individuals in their field.

The basic constitution of the Institute was drawn up by the Legislative body, which asked the Institute to draw up its own regulations about the conduct of meetings and then submit these regulations to it for approval. This system, together with that used to choose the original members of the Institute, shows that there was a great deal of co-operation between government and *savants* in

[12] Some leading dons, notably Jowett, opposed the idea of research.

[13] Mark Pattison, *Essays on the endowment of research*, 1876, pp. 22–3. By this time Pattison could equally well have explained the concept by reference to the German *Forschung*.

Table 2. *Foundation members of the First Class of the National Institute*

	Nomination*		
Section†	Lakanal's provisional list	Directory's definitive list	Election
Mathematics	Lagrange, Laplace	Lagrange, Laplace	Borda, Bossut, Legendre, Delambre
Mechanics	Borda, Cousin	Monge, Prony	Le Roy, Perrier, Vandermonde, F. Berthoud
Astronomy	Lemonnier, Lalande	Lalande, Méchain	Lemonnier, Pingré, Messier, Cassini
Geography		(section added in 1803)	
Physics	Brisson, Monge	Charles, Cousin	Brisson, Coulomb, Rochon, Lefèvre-Gineau
Chemistry	Darcet, Fourcroy	Guyton, Berthollet	Fourcroy, Bayen, Pelletier, Vauquelin
Mineralogy	Haüy, Gillet	Darcet, Haüy	Desmarets, Dolomieu, Duhamel, Lelièvre
Botany	Adanson, A. L. Jussieu	Lamarck, Desfontaines	Adanson, A. L. Jussieu L'Héritier, Ventenat
Agriculture	Thouin, Parmentier	Thouin, Gilbert d'Alfort	Tessier, Huzard, Cels, Parmentier
Anatomy and Zoology	Daubenton, Lacépède	Daubenton, Lacépède	Tenon, Broussonet, Cuvier, Richard
Medicine	Portal, Sabatier	Des Essartz, Sabatier	Portal, Hallé, Pelletan, Lassus

* *Source:* J. Guillaume (ed.), *Procès-verbaux du comité d'instruction publique de la Convention*, vol. 6, pp. 833, 838. (October–November 1795).
† For a more exact title of each section, see Table 4.

the establishment of the Institute. Yet the Institute was very aware that its existence depended very much on its political masters and it had to do what it was told.

2. *The First Class as a part of the Institute*

It is not possible to discuss the history of the First Class without considering it as an integral part of the National Institute. This applies particularly to the period 1795–1802, but even after the reorganisation of 1803 the First Class continued to share premises with the other classes and suffer from the same constraints. Anyone with a knowledge of modern science will appreciate that it might have special claims for certain facilities but in the historical context of the foundation of the Institute, the First Class was merely one of three divisions, albeit the largest and, in the eyes of some, the most important. The esteem in which science was held fell sharply with the Restoration, when the Académie Française resumed its former position of seniority. The only permanent privilege which the First Class or the Academy of Sciences enjoyed over its neighbours was that it had *two* secretaries in recognition of its size and the wide spectrum of knowledge which it embraced. Even nearer to our own times there may be occasional murmurs from the other Academies that this constitutes an unforgivable infringement of the republican principles of equality and fraternity.

If the First Class corresponded to the range of subjects taught in the Faculty of Natural Sciences of a modern university, the Second Class corresponded to the Faculty of Social Sciences (see Table 3). It included several traditional subjects, such as ethics and history but, not content with the classical tradition and a Christian inheritance, studies were added which were supposed to be more scientific.[14] One new subject was geography but this was now intended to go beyond cartography and provide an integration of topographical studies and climate in a comprehensive 'science of man'. Of the six sections of the Second Class, three were relatively new fields: 'social science and legislation' extending some of the earlier ideas of Montesquieu, 'political economy' and, finally, the new territory of the *idéalogues*, led by Cabanis and Déstutt de Tracy, 'analysis of sensations and ideas'. It was in the latter section that most of the younger members of the Second Class were to be found, in contrast with the history section, which had the highest average age. This is partly because the history section was largely filled with former members of the Academy of Inscriptions.

The Third Class corresponded largely to a combination of the former Académie Française and the former Académie de Peinture et de Sculpture. It was divided into eight sections, each of six members, representing respectively grammar, classical languages, poetry, antiquities and monuments, painting, sculpture, architecture, music and declamation. Given the different number of sections in each class, the First Class had sixty resident members, compared with thirty-six for the Second Class and forty-eight for the Third Class.

At the Restoration the classes of the Institute were again called Academies but they continued to share a common budget and the same buildings. Occasionally the Academy of Sciences would collaborate with another Academy to appoint a commission to examine a problem which was of mutual concern. Thus photography was of interest to the Academy of Fine Arts as well as to the Academy of Sciences. After the 1830 Revolution a further Academy was added, that of Moral and Political Sciences, since the new regime felt itself to be sufficiently liberal to encourage the study of political theory at the highest level. It was in this liberal atmosphere that the Academy of Sciences was able to begin its historic new journal, the *Comptes rendus*. However the foundation of a fifth major academy in 1832 had imposed new demands on government coffers and

[14] Martin S. Staum, 'The Class of Moral and Political Sciences, 1795–1803', *French Historical Studies*, 11 (1980), 371–97.

Table 3. *The Paris Academies and the different classes of the Institute*

	Ancien régime Academies	National Institute 1795	National Institute 1803	National Institute 1816–
Mathematical and natural sciences	Académie Royale des Sciences (f. 1666)	Première Classe (Sciences mathématiques et physiques) (10 sections) 60 members	Première Classe (Sciences mathématiques et physiques) (11 sections) 65 members	Académie [Royale] des Sciences de l'Institut (11 sections) 75 members
Social sciences	—	Deuxième Classe (Sciences morales et politiques) 36 members	—	[1832 Académie des Sciences morales et politiques (5 sections) 35 members]
Literature	a. Académie Française (f. 1635)	Troisième Classe	Deuxième Classe (Littérature)	Académie Française (no sections)
Art, music	b. Académie Royale de Peinture et de Sculpture (f. 1648)	(Littérature et beaux-arts)		
	c. Académie Royale de Musique (f. 1669)	48 members	Quatrième Classe (Beaux-arts)	Académie des Beaux-Arts (5 sections) 50 members
	d. Académie Royale d'Architecture (f. 1671)			
Classical studies, archaeology	e. Académie des Inscriptions et Belles Lettres (f. 1716)		Troisième Classe (Histoire et littérature ancienne)	Académie des Inscriptions et Belles Lettres

we shall see in Chapter 8 how the Academy of Sciences had to look elsewhere for the necessary finances to launch this journal.

One of the features of the Academy of Sciences not shared by other Academies was the potential novelty of the work to be reported and a comparative urgency to publish. By its very nature, science works near the frontiers of knowledge. The other Academies could take a more leisurely view of their responsibilities. Besides, it could be argued that there was also a certain artificiality in their work. In the eighteenth-century Académie des Inscriptions et Belles Lettres members were called upon *in rotation* to deliver a memoir. In the nineteenth century this Academy had a regulation that all ordinary members were obliged to deliver a memoir once a year, whether or not, one might add, they had anything original to say. When the Académie des Sciences Morales et Politiques was re-established in 1832, the president would arrange in advance for a memoir to be read as the principal business of each meeting. If the author wished it to be published in the *Mémoires* of the Academy, he was obliged to read it twice to his colleagues, who were encouraged to comment on it. This prevented some unconsidered trifle from embarrassing the *Mémoires*, but it was none the less artificial for that.

Yet some of the other Academies took science as their model. In 1832 Guizot in a report to Louis Philippe to justify the re-establishment of the Académie des Sciences Morales et Politiques said that it was only recently that these disciplines had regained 'a truly scientific character', which justified a separate Academy.[15] Again a recent review of the Académie des Inscriptions claimed that it was to be considered as a laboratory rather than a museum.[16]

In the Academy of Sciences there was much less uniformity in the contributions of members. Active Academicians were continually at the rostrum, others only occasionally. Academicians would not present a paper merely to confirm that they were still alive. They would address their colleagues when they had solved a problem or had discovered something new. It is true that with hindsight we might judge many of these discoveries trivial and even, occasionally, false, but at least it could be said that most of it was *original* work, pushing back the frontiers of knowledge and even, from time to time, opening up completely new areas.

While considering divisions of knowledge in an institutional context, it would be appropriate to mention one inherent weakness in the Academy throughout its history. Just as the First Class had been hindered in the first few years by constant association with members of the Second and Third Classes who had no knowledge or interest in science, it could be argued that the First Class (or Academy) itself was permanently handicapped by including the representation of such a wide spread of interests. To take an extreme case, it was not that non-mathematicians, say, actually *hindered* the mathematicians in the presentation of their memoirs, but in so far as they found it difficult if not impossible to

[15] *Institut de France, Catalogue de l'Exposition*, avril/mai 1983, p. 218.

[16] *Ibid.*, p. 51.

understand, they might become bored and inattentive. Some visitors commented on the fact that an Academician presenting a memoir could not rely on the undivided attention of all his colleagues. There would be occasions during a meeting when other Academicians would be looking at their notes and some would even converse in low tones with their neighbours, which left a bad impression on visitors who had not expected such lapses from formal behaviour in an official body.

The Academy of Sciences had to share the same building with the other Academies. It also had to share a common government budget, which was discussed by a committee on which all the Academies were represented. As a means of enhancing good relations between the different Academies, distinguished members were sometimes elected to sister Academies. Thus certain members of the Academy of Sciences credited with literary talent were elected to the Académie Française, for example. Among those so honoured were Cuvier, Biot, Flourens, Dumas and Pasteur. Cuvier and Biot also became members of the Academy of Inscriptions and Belles Lettres. However, because of the stricter criteria for membership of the Academy of Sciences, it was not normally the practice to honour non-scientists from sister Academies in this way.

3. *Change and continuity*

There were a number of obvious differences between the old Academy and the First Class of the Institute. Most obviously the new organisation brought the men of science into regular contact with other *savants*. Whether this amounted to a dilution for the purposes of serious science or an expansion of horizons for the specialised men of science is a matter of possible debate but it was an exemplification of the basic principle of Diderot's *Encyclopédie*, that knowledge is one. As the *Encyclopédie* had been intended to bring together in a series of volumes all the knowledge of the eighteenth century, so the Institute was intended to assemble under one roof the wisdom of the age right across the spectrum of intellectual and artistic talent. Secondly, the grades indicating differences of social rank or seniority were abolished. Thus there were no longer honorary members, a class previously reserved for persons of superior social rank, nor was a distinction made any longer between the grades of pensioner, associate and assistant – all were theoretically equal as resident members. Thirdly, the resident members in Paris were to be balanced by an equal number from the provinces. (As we shall see later, this admirable idea was difficult to put into practice.) Fourthly, the traditional position of permanent secretary was abolished, since this post was seen as providing too much power for one man. Instead secretaries were to be elected every twelve months, or since there were two secretaries, election for each post in turn was held at six-monthly intervals. Similarly the president was elected for a period of six months. It is significant that the president was to be elected by the members, whereas previously he had been nominated by the King.[17] Even more important, new members were elected by the (whole) Institute, their election being subject to government approval. This contrasted with the previous system, by which the Academy forwarded three names, of which the King selected one. The involvement of the whole Institute rather than the relevant Class in elections is a further point which will be discussed below (Section 11).

But although there were these striking differences between the old Academy and the First Class, there was also considerable continuity. Of all the Classes of the Institute, the First Class probably corresponded most closely to the previous Royal Academy. No one could deny that there was a new organisation and a new ideology, but sceptics might be forgiven if they concluded that what they were seeing was the former Royal Academy dressed up in new clothes.[18] First there was the question of membership. There was a remarkable degree of continuity in the membership of the two bodies. Of course we must not forget the Terror, in which Lavoisier was the most famous of ten deaths of former members of the Academy of Sciences, arising from the political situation.[19] There were also a few natural deaths, such as Vicq d'Azyr, the former secretary of the Société de Médecine. Despite these losses, we find that of the sixty original members of the First Class, two thirds had been members of the former Royal Academy. In other words, after the Terror France was not foolish enough to turn its back on the talent surviving from the *ancien régime*. Even the ultra royalist Dominique Cassini was nominated[20] and several elderly former Academicians were elected in recognition of past achievements rather than in the expectation of continuing active contributions to their respective subjects. A specific claim of continuity was made in 1805 in an official Academy publication which spoke with some exaggeration of 'the link between the two institutions, animated by the same spirit and composed of the same members'.[21]

One of the consequences of the continuity was the preponderance of older men among the original membership. In 1795 the average age of election was fifty-three years two months, whereas by 1803 it had fallen to forty-five years one month.[22] The higher original figure was due to the presence of several septuagenarians and octogenarians, whose authority belonged entirely to the *ancien régime*, and whose early deaths made way for men of a new generation. One wonders what the astronomer Le Monnier (1715–99), born in the final year

[17] Actually the Royal Academy of Sciences had won this right in December 1792, seven months before it was suppressed.

[18] The revolutionary period witnessed many changes of nomenclature. Here the term 'academy' was being avoided as redolent of the *ancien régime*. Similarly the word 'médecine' was avoided and often replaced by 'santé'.

[19] Dorinda Outram, 'The ordeal of vocation: The Paris Academy of Sciences and the Terror, 1793–5', *History of Science*, 21 (1983), 251–73.

[20] Although he resigned on 18 January 1796, to be re-elected as a full member in 1799.

[21] *Mémoires présentées à l'Institut...par divers savans*, 1 (1805) xv.

[22] A. Potiquet, *L'Institut National de France*, 1871, p. xv.

of the reign of Louis XIV and originally elected to the Royal Academy as early as 1736, would have made of the Revolution and the new Institute. A similar comment may be made about several other very elderly Academicians, such as J. B. Le Roy (1720–1800). Neither of those named survived into the nineteenth century. An analysis of ages of members of the Institute in 1795 shows little difference between the different Classes. Thus, despite revolutionary fervour for innovation, the original Institute had clearly adopted the policy of associating itself with the wisdom and talent of senior figures from the *ancien régime*. It had not been rash enough to assume that inexperienced young men or even those of middle age could bear alone the burden of representing knowledge.

If the continuity of membership is the most striking feature of resemblance between the two bodies, almost equally striking is the continuity in the conception of science. Table 4 shows the names of the sections in the Academy at the end of the old regime (column 3) and those adopted in 1795 (column 4). (For convenience previous and subsequent titles of the sections are also shown.) Of the ten sections of 1795, eight correspond exactly to sections in the Royal Academy. The section of rural economy and veterinary science is partly foreshadowed in the previous section of botany *and agriculture*, but a section devoted to medicine and surgery is an entirely new concept. Both new sections, however, represent the new concern with applied science.[23] Leaving this dimension to one side for the moment, one sees the continuity in the designation of the main branches of pure science. There is a hint in the name that the section of 'mathematics' was intended to embrace a subject much wider than 'geometry', the term inherited from the seventeenth century. Given that physics had been recognised as a new subject in 1785, this leaves only 'vegetable physics' (i.e. plant physiology) and zoology as having acquired the status of new subjects, or rather half subjects, in the new Institute. Considering that shortly before the foundation of the Institute the reconstituted Muséum d'Histoire Naturelle had recognised the new subject of geology with a chair, it may seem a pity retrospectively that, in preference to this, the more traditional subject of mineralogy was chosen as the focus of a section. But to introduce the growing importance of geology in the early nineteenth century is to use hindsight, a luxury which should not be indulged in too often.

Even the number of members in each section (six) was inherited from the former Royal Academy. Again when on 21 *prairial* year 4 (9 June 1796) the First Class discussed what subject should be proposed for the mathematics prize of that year, it could think of no better subject than that previously suggested in 1793 by the Royal Academy before the dissolution.[24] It is perhaps understandable that there should in some ways be more continuity between two

[23] The emphasis on applied science is discussed in dealing with the respective sections in Chapter 4.

[24] *P.V.I.*, 1, 57.

Table 4. *The differentiation of disciplines within the Academy*

1666	1699 *Académie Royale des Sciences* (6 sections)	1785 *Académie Royale des Sciences* (8 sections)	1795 *Première Classe de l'Institut National* (10 sections)	1803 *Première Classe de l'Institut National* (11 sections)	
(No specialist sections)	Géométrie	Géométrie	Mathématiques	Géométrie	Division des sciences mathématiques
	Mécanique	Mécanique	Arts mécaniques	Mécanique	
	Astronomie	Astronomie	Astronomie	Astronomie	
			—	Géographie et Navigation	
		Physique générale	Physique expérimentale	Physique générale	
	Chimie	Chimie et Métallurgie	Chimie	Chimie	Division des sciences physiques
		Histoire naturelle et Minéralogie	Histoire naturelle et Minéralogie	Minéralogie	
	Botanique	Botanique et Agriculture	Botanique et Physique végétale	Botanique	
			Economie rurale et Art vétérinaire	Economie rurale et Art vétérinaire	
	Anatomie	Anatomie	Anatomie et Zoologie	Anatomie et Zoologie	
			Médecine et Chirurgie	Médecine et Chirurgie	

institutions belonging to the 1790s than, say, between the original Royal Academy of 1666 and that of 1785.

There was also a great deal of continuity in the procedures of the First Class, where memoirs were read as before. As before, there was the hope that the best work would be published in the official *Mémoires*. Where there had been Academy practices with clear political implications these were understandably modified, but since most of the procedures of the Royal Academy had been concerned with receiving and approving scientific work, these procedures were allowed to continue. Thus, for example, the First Class continued the practice of the *ancien régime* of receiving sealed notes, safeguarding priority for new scientific discoveries without revealing what they were.

In publication the First Class hoped to continue the annual publications of its *Mémoires* but was handicapped by a rule that all the Classes of the Institute should be treated equally in matters of publication. This idea of equality was taken to the extreme that the First Class could publish no more pages of text than any of the other Classes. In this extreme example of equality by levelling down, no allowance was made for the fact that the First Class had more members. The fundamental issue was, of course, that scientific research demanded facilities for the publication of new knowledge, which was not required to the same extent in the infant social sciences or in the humanities, and the political masters of the Institute consistently failed to recognise this.

The First Class did, however, manage to make the case that, if its members were to do even simple experiments, a budget allocation would be necessary. Thus by 1800 it had a budget of 7500f. whereas the Second and Third Classes considered that by economising they could manage with expenses of only 650f. and 850f. respectively.[25] This brings us to consider the whole question of finance. Once the financial background of the Institute, and in particular salaries, has been established, we will be in a better position to appreciate the obligations of members. Finance often provides a powerful means of control in any organisation.

4. *Finance*

We must consider at two levels the question of financial support of the Institute. There was first the question of payment of members. It was felt that all members should accept some modest payment, not exactly a salary but rather an honorarium, described originally as an *indemnité* and later as a *traitement*. We shall review two alternative approaches, one of the Directory, which wanted to give salaries at least comparable with those offered under the old regime, and the alternative more severe approach of the finance committee, which asked what was the minimum that should be paid to *savants* who would probably have other sources of income. We may note that in 1790 a handful of members of the Royal Academy had received a basic stipend of as much as 3000 livres but others had

[25] A.I., Box 6A1, Dossier Commission des Fonds, an 7.

received as little as 500 livres.[26] To suggest a very approximate modern equivalent, the *livre* (soon to be re-named the *franc*) was worth rather more than £2 sterling or about 3 U.S. dollars.

The other financial problem was the general support of the Institute, running costs, publications, prizes and possibly the support of experiments. We shall have to see whether science could make any special claims. Although in some ways science seemed to be favoured in the Institute, in other ways it was merely one of a number of accepted activities of intellectuals in the late eighteenth century and should not have been allowed to infringe the shibboleth of equality. Nor, at a time when many rival claims were put forward for government expenditure, was it clear that the Institute deserved special treatment and the *savants* did not yet feel confident enough to demand it.

We shall have to examine the historical background in order to establish what was a reasonable 'going rate' in the period 1794–6. In a report to the Convention in October 1794 on the financial aid which the government was to give to *savants*, men of letters and artists, Grégoire had first to convince the assembly of the practical utility of these activities. For example, when speaking of science, he said:

> It is to chemistry that we owe the beauty and fastness of our dyes. It is this science which teaches the art of transforming sand into a transparent mass, which provides us with glass. Without the help of optics a man with poor sight would not have a pair of glasses to help him. Geometry helps to make highways to your frontiers, bridges spanning rivers and ships on the high seas...But the first person to study gases hardly thought that one day balloons would rise in the air and that balloons would help to beat our enemies.[27]

Having shown the relevance of science and other intellectual activities to the problems of the day, Grégoire then came to the delicate question of the scale of awards or pensions:

> *Savants* do not demand riches. The precious modicum of Horace will always be their motto.[28] Yet we should observe that often positions which require the greatest talents, are the worst rewarded. A shipping clerk often has a salary of 3000 livres, 6000 even, whilst a professor may only obtain his post after twenty years of preliminary study, and he can only usefully do his job by means of continuous work and is, therefore, not able to aspire to a second modest salary; although he is often overburdened with scientific duties, he only has a very small income and is permanently in a state not far from

[26] Eight (senior) members were given pensions of 3000 livres, eight members 1800 livres, eight members 1200 livres, and sixteen (junior) members 500 livres. Aucoc, *op. cit.*, p. cxcix.

[27] 'Rapport sur les encouragements et récompenses à accorder aux savants, aux gens de lettres et aux artistes', *Réimpression de l'Ancien Moniteur*, **22**, 183.

[28] I must thank my colleague Dr Graeme Anderson for identifying a passage in Horace (*Epistles*, I, i, 43) where he uses the expression *exiguus census* to express this modest income.

> penury. Now who would in future be willing to take up such a career, if, after a painful journey, the goal should appear only a sad prospect?[29]

The same theme (and almost the same example) was taken up by Fourcroy later the same year when he was arguing that professors at the new école de santé in Paris should not have financial worries to distract them from their teaching and research:

> It is therefore necessary that their salaries should be sufficient for their needs and that they should not be obliged to look for other employment as a means of completing their livelihood. Men, who have spent twenty years of their lives in study in order to acquire profound knowledge and be able to transmit it to others, should be treated by the country which employs them in such a way that they are not tormented by domestic anxieties. By the exercise of their useful talents they should be able to draw on resources sufficient for their own maintenance and that of their families.[30]

Teachers had traditionally received low salaries under the old regime. In so far as education was largely in the hands of the Catholic church through various celibate teaching orders and depended on a vocation, teachers did not even expect to receive a proper salary. When the Revolution secularised education it created the need for secular teachers who would, at the very least, earn enough to keep their families. The social status of teaching, especially in higher education, rose enormously in the new society but it took several years for the new standards to emerge.

But to return to Grégoire: after his report an annual sum of 300 000 livres was approved for pensions and on 26 *brumaire* (16 November 1794) the Convention decreed that the minimum sum to be awarded would be 1000 livres and the maximum 3000 livres.[31] Candidates had been divided into three classes for benefits, the largest group being those who received 2000 livres.[32] Again at a later distribution,[33] fifty-three people were given 2000 livres, as opposed to twenty-seven receiving 3000 livres and thirty-eight having 1500 livres.

The information about pensions has been cited because pensions helped to establish a norm. Although the Institute had been founded at the end of 1795, it was not until July 1796 that the law was passed on the honorarium which was to be given to each member of the Institute. The Minister of the Interior had proposed 2000 francs but the Council of 500 considered that 1500 francs 'was sufficient'.[34]

This reduction was decided after a submission by Villers on behalf of the commission on expenditure. Since this was the crucial speech on which the

[29] *Réimpression de l'Ancien Moniteur*, 22, 192.

[30] 'Rapport et project de décret sur l'établissement d'une Ecole centrale de santé à Paris', *Réimpression de l'Ancien Moniteur* (Paris, 1794), 22, 665.

[31] Guillaume, *op. cit.*, vol. 5, p. 221.

[32] *Ibid.*, pp. 359–60.

[33] *Ibid.*, vol. 6, pp. 628–31.

[34] L. Aucoc, *op. cit*., p. 37. The livre was renamed the 'franc' about this time.

decision was based, it is worth some detailed discussion. Villers seemed to think that there was a danger of members of the Institute living in opulence:

> A person who devotes his life to the study of the sciences or the arts must surely know how to control his desires and regulate his needs. The worries involved in acquiring or keeping riches is difficult to reconcile with the love of literature and philosophy. The *savant*, whose only pleasures are in the charm of study, cannot find happiness in a life of dissipation.[35]

He considered a figure should be arrived at which avoided the two extremes of luxury(!) and penury. If nothing were offered, the *savant* might be tempted to go abroad and work for the enemies of the Republic. Nor should he have to turn to non-scholarly occupations in order to earn a living. Science must not be the prerogative of the wealthy, as in former times. On the other hand mere membership of the Institute was itself a great honour. Nor were the duties very onerous; the short time ('les moments') required could be reconciled easily with business interests or the exercise of a 'lucrative profession'. Here was a curious interpretation from someone who had little understanding of the years of training necessary to produce scholars, nor of the full obligations of members of the Institute. Villers (1747–1807) had started life as a Capucin friar but in the Council of 500 he spoke often on questions of commerce, manufacture and finance. Unlike Lakanal and some other members of the Convention, he had no special sympathy for education or the intellectual life. On the contrary, he seems to represent a strain of anti-intellectualism. He was keen that no member of the Institute should be allowed to refuse the honorarium (otherwise he would become an honorary member) and he approved of *droits de présence* (attendance money), to give members a strong financial incentive to attend as many meetings as possible. He was a hard task master!

The delay in deciding upon the question of payment of members allowed members of the Institute to express their own opinions on the subject. In March 1796 Baudin had found it necessary to urge the necessity of payment. He had argued that members of the Institute should have some independence, both financial and in the expression of opinions.

> He presumes that someone, who devotes all his time and his talents to move back the limits of human knowledge, and who spends his life in study, will be content with an honorable modicum ('*honorable mediocrité*), which places him above basic needs.[36]

He was taking up again the idea of a modicum being sufficient for *savants*, a concept which Grégoire had used eighteen months earlier.

The 'honorable modicum' was fixed by the law of 29 *messidor* year 4 (17 July 1796) at the value of 750 mirigrams of wheat (so stated to deal with the

[35] *Ibid.*, p. 36.

[36] A.I. 3A1, f.59 (15 *ventose* an 4 = 5 March 1796).

problem of severe inflation), which soon became 1500 francs.[37] However, as a reminder of the importance of regular attendance at meetings, one fifth of the amount was withheld to form a fund to be paid out to members of the Institute in proportion to their attendance and called '*droits de présence*'.

Really the members of the Institute had the worst of both worlds. By accepting an honorarium, however modest, they accepted obligations to the state. It is the *principle* of payment which counts, rather than the amount, as when a token rent is paid or, in law, token damages are awarded. Yet the sum involved was so miserly that many members of the Institute with other means of support might well have felt they could dispense with the honorarium, if only they had been allowed to do so. The cash value of the honorarium was comparable with that of a basic grant for a modern university student. In other words it could at most provide basic subsistence. It was on their salaries from other employment that the members of the Institute really lived and kept their families.

Coming now to the financial provision for the Institute as a whole, it took several years before the legislature could accept the Institute as a body with a basic minimum budget. The budget for the Institute for the republican year 7 (September 1798–September 1799) was approximately 400 000f. Accepting the need for economy, the Institute had itself proposed a total budget for the following year of 272 333f., an effective reduction of 35%.[38] In a spirit of abnegation and extreme patriotism the Institute took pride in making the minimum demands on public money and when the government had previously offered more in error, the *savants* had hastened to point out the mistake. However such sacrifices, far from being appreciated, sometimes made the politicians suspect that further economies were possible. It was in these circumstances that in 1799 the Institute felt obliged to emphasise how frugal were its own demands. Thus having had 120 000f. in the year 7 for travellers in the three Classes, it was asking for nothing in the following year. It reduced the budget for prize money from 15 000f. to 12 000f. Building work for the library was postponed and it reduced the amount to be spent on books to 3000f. with the following comment:

> Although the Institute lacks many books, both French and foreign, which would be very useful for its work, it thought that it should reduce the expenses under this heading by half this year, in order to comply with the

[37] The Institute itself suggested in August 1796 that honoraria should be paid at three levels: 2100 livres to those nominated by the Directory, 1500 livres to the next two members of each section to be elected and 900 livres to those elected last. 'Arrêté de l'Institut Nationale pour la répartition des Indemnités accordées à ses membres. 19 *thermidor* an 4 (6 August 1796).' E. Maindron, *L'Académie des Sciences*, 1888, pp. 183–5. Such blatant inequality was not, however, politically acceptable and was quickly replaced by a system in which everyone received the same.

[38] The exact previous allocation was 414 333f. A.S., Boite: Commission [centrale] administrative, 1795–1831.

> interests of economy, which circumstances render necessary; but it is still necessary to keep up with discoveries.[39]

The Institute was also prepared to reduce its allocation for apparatus and publications from 18 000f. to 9000f., nearly all for the First Class, which had to do experiments which were said to be important and urgent. Although they had an allowance for experiments, members were hesitant about making claims on the fund, so a commission of the First Class in 1799 decided that each section should meet to establish what funds its members might require and then representatives from each section would negotiate to balance supply and demand.[40] The commission also suggested that each section would accept responsibility for the experimental work done with the money, thus imposing a further layer of control.

Although foreign visitors like Thomas Bugge[41] were very impressed by the money set aside by the French government for science, the reality was sometimes less impressive. Even the modest salaries of members of the Institute were often in arrears for the first few years and in March 1800 we find the Institute making representations to Bonaparte as the new head of state for payment due to them.[42] There was considerable improvement, however, under the Consulate and Empire in the payment of the honoraria of members but it was not until 1811 that these honoraria, as well as the trifling salaries and wages of librarians and porters, were granted automatically without the necessity of presenting a basic budget to the Minister of the Interior. On the other hand, the same decree of 11 January 1811 stated that no additional expenses could be introduced into the budget unless a special case had been made to the Minister. The records reveal that the Minister of the Interior at that time, Montalivet, was particularly severe in his financial control of the Institute and hence of the science budget.[43] Three thousand francs for a statutory prize for physics in 1812 was refused on the grounds that it would not actually be awarded until the following year. He also suggested that the Institute should be most exacting in its prize competitions. Nor would Montalivet allow the First Class the 3000f. for the annual offer for the prize on galvanism until he was sure that it would be awarded. If we add that he proposed a reduction on the allowance for heating and lighting and insisted on monthly rather than annual accounts, one may regard this period as one of the most stringent financial control. In the light of such rigid controls we can sympathise with the deliberate vagueness of the

[39] Institut Nationale. *Observations sur l'article du rapport du cit. Daubermesnil concernant les dépenses du matériel du ministère de l'intérieur pour l'an VIII*, etc.

[40] P.V.I., 1, 535–6, 11 *ventose* an 7 (1 March 1799).

[41] *Science in France in the Revolutionary Era described by Thomas Bugge*, ed. M. Crosland, Cambridge, Mass. 1969, chap. 5, pp. 152ff.

[42] A.S., *Commission administrative, 1795–1831*, dossier an VI – XIV, letter of 1 *germinal* an 8 (22 March 1800).

[43] *Ibid.*, letter from Minister of 18 May 1811 and 20 July 1811.

Academy's system of accounting under the Restoration, vagueness which was accompanied by the beginning of greater financial freedom.

5. *The sections: specialisation without fragmentation?*

Probably the single most important feature of the organisation of the Academy was its compartmentalisation, its division into sections (see Table 4). In this it contrasts with many other societies like the Royal Society, which was largely organised without regard to specialisms. Indeed recognition of certain subject specialisms by the Academy constituted a distinctive feature of French science. Moreover, as explained previously, it was not a nineteenth-century innovation but dates back to the organisation in 1699 of the old Royal Academy, which had recognised six areas or 'subjects', to which a seventh, physics, was added in 1785. The listing of these subjects had several major effects. First it defined exactly what was meant by the term 'sciences' in the title of the Academy. They were clearly not something as general as the German 'Wissenschaft' and they were more specific than a portmanteau 'Naturwissenschaft'. With the exception of chemistry, the sciences officially recognised had all been profitably studied in the ancient world and their study had been resumed in the universities of medieval Europe. By the seventeenth century chemistry had begun to shake off its legacy of alchemy, and seemed worthy of recognition and encouragement. Anyone elected to the eighteenth-century Academy, therefore, had a double identity. He belonged not only to the whole Academy but more particularly to one section. As the sciences became more specialised in the nineteenth century, this double identity became more important.

It would have been invidious to suggest that any one branch of science was more important than any other. Hence all the sections were made up of an equal number of members and this number was maintained. Thus for a purely social reason (the avoidance of jealousy) the recognised sciences were formally implied to be of equal importance and this formal equality was maintained despite the unequal growth of the different sciences. Any possible imbalance under the old regime had not been too marked, but the growth of chemistry at the end of the eighteenth century and the opening of whole new fields, particularly organic chemistry in the nineteenth century, received no recognition in any increase in numbers within the chemistry section. It might be fancifully suggested that the number six was perhaps almost as sacred to the Academicians as the number ten had been to the Pythagoreans. Where a *savant* had qualifications relevant to more than one branch of science, the section he entered might depend on which had the first vacancy, or which was the easier section to enter. The growth of chemistry in the nineteenth century meant that it became one of the hardest sections to join. Yet because of the relation of chemistry to physics on the theoretical front and the many applications of chemistry on the practical front (for example, to agriculture), the sheer force and number of well-qualified chemists produced a situation in which the formal obstacles were overcome and several chemists were able to enter other sections. Although we will argue that in general the structure and size of the Academy imposed a constraint on the pursuit of science, we have in the case of chemistry an exceptionally powerful force or concurrence of forces, which partly overcame the constraint.

By the late eighteenth century it was clear to many people that knowledge of the natural world was growing so quickly that the days of the generalist, the person who could keep in touch with developments in different and unrelated branches of science, were numbered. The science of the future would increasingly be characterised by specialisation. In view of this, the foundation of an Academy (or 'First Class' of the National Institute), bringing together all the main branches of science, might seem to be going against the trend. Specialisation was not simply a council of despair in view of the growth of knowledge, it was actually good for science in so far as it encouraged investigators to tackle problems in greater depth.

But specialisation has its disadvantages; it implies the wearing of blinkers. To get results, it requires intense concentration in a very limited area, with disregard for what is going on outside, even in related fields. It was one of the strengths of the Academy that, while encouraging a finite number of specialisms, it encouraged specialists to see their work only as a part of the whole spectrum of the sciences. Association in the same institution with other specialists encouraged interdisciplinary work. A member of the mechanics section might well have made a contribution to applied mathematics. In presenting his research to the Academy, he would be conscious not only of other applied mathematicians in his own section but also of other colleagues who might, on the one hand, be pure mathematicians or, on the other, members of the physics section. Every specialist, therefore, had to consider not only fellow specialists but a penumbra of related disciplines. Even more important was potential feedback; representatives of these disciplines might well come up with new ideas which could help him in his own work. The sterility of extreme specialisation was, therefore, avoided. One of the most impressive examples of interdisciplinarity came from the application of mathematics to the solution of physical problems. French physical science owed much to the mathematicians, whom it would have been counter-productive to isolate in a specialist institution. Laplace provides the best example of a mathematician who could faithfully apply his talents to the solution of a variety of astronomical and physical problems.

One price to be paid for the broad spectrum of science represented in the Academy would be that specialists at each end of the spectrum, the pure mathematicians and the medical scientists, could hardly expect to gain very much from each other's memoirs. But some tedium in this extreme case is an acceptable price to pay for the stimulus of crossing disciplinary lines. In any case Louis Pasteur, whose career began with studying the angles of crystals and ended as the universally acclaimed author of the germ theory, illustrates that it is, after all, possible to span the immense distance from geometry to medicine.

Another example of interdisciplinary influence in the Academy between sciences not normally considered as having any obvious connection involves the sciences of astronomy and physiology, with photography as a connecting link. In 1874 the astronomer Janssen, recently elected to the Academy, explained to his colleagues the use of what he called 'un revolver photographique'.[44] This was a camera which could be fitted to a telescope to provide a rapid succession of exposures of any eclipse or similar phenomenon. (Janssen's current interest was in recording the transit of Venus.) Two years later Janssen spoke of extending the use of the same apparatus to study rapid changes of movement, such as the flight of birds. The physiologist Marey, soon to be elected to the Academy in the medicine section, wrote to Janssen asking for details of the apparatus. Marey had already published a book on animal locomotion and he was interested in extending his studies from horses and human beings to birds. Marey was able to construct a camera with a shutter speed 700 times faster than that used by Janssen and in 1882 Marey was able to announce to the Academy that he had been successful in producing a succession of photographs with a time difference of a fraction of a second, which enabled scientists to analyse the details of wing movements of birds.[45] Thus a technique used in one branch of science was applied to a totally unconnected science, something made possible by the broad spectrum of the Academy's interests and the fact that, since the time of Daguerre (*c.* 1839), it took photography seriously. Scientific and technological developments in photography were a subject of regular interest in the Academy. Yet it is not too difficult to imagine a nineteenth-century learned society in which photography would be dismissed as an amusing toy, whose development was more a matter of commerce than of science.

We need to explain how the existence of sections affected the working of the Academy. A study of the sections enables us to see the Academy from a different perspective – as a federation of semi-independent specialist interests. Although to the outside world successful candidates were elected to the Academy, the new Academician would be very conscious that, unless he was an *Académicien libre*, he had been elected to one specific specialist section and his first loyalties must be to that section and to that subject. In due course he would probably become the *doyen* of the section and speak on its behalf. This would happen, for example, in elections, when the *doyen* would have the major responsibility for scrutinising the qualifications of candidates and drawing up a list in recommended order of merit. Again the responsibility of the Academy for recommending persons to fill chairs in higher education establishments fell, in practice, on the corresponding section. Where the interests in question transcended an individual section, several sections would combine, but only those relevant to the institution. Thus when there were vacancies in the Bureau des Longitudes in the mid-nineteenth

[44] *C.R.*, **79** (1874), 6–7.

[45] *C.R.*, **94** (1882), 684–5, 823, 1013–20. Marey also expresses his indebtedness to the American, Muybridge. Georges Sadoul, *Histoire générale du cinéma*, vol 1, *L'invention du cinéma, 1832–97*, 1948, p. 77.

century, it was the three sections immediately concerned, those of mathematics, astronomy and geography/navigation, which were summoned to elect a Commission to make recommendations.

Again the award of prizes in specific subjects would usually correspond to one particular section. Of course, some subjects were more favoured with prizes than others and Lacaze-Duthiers took advantage of his tenure of the presidency of the Academy in 1893 to complain that zoology was one of the few subjects for which there was no specific prize.[46] In fact annual presidents were elected not only with reference to the personality and seniority of Academicians, but also with an eye to equitable representation of the whole range of *subjects* represented in the Academy. The physician Bouchard, thanking his colleagues for his election for the year 1908–9, made the subject association explicit when he said:

> He thanks you for himself, whose qualifications for the position you have judged with too much indulgence; he thanks you on behalf of medicine a subject which he has served with love and respect; he also thanks you on behalf of the body of practising physicians, to which it is his glory to belong and which you had the intention of honouring in his person.[47]

The fact that most of the major sciences were represented by sections in the Academy probably had an inhibiting effect on the establishment of specialist societies. To support this hypothesis, we may quote the date of foundation of specialist scientific societies in France and compare these with the (normally much earlier) dates of foundation of the corresponding societies in Britain, where the generalist approach to science as 'natural philosophy' or natural history originally produced a reaction:

France[48]		*Britain*	
1830	Société géologique	1807	Geological Society
1854	Société botanique de France[49]	1778	Linnean Society
1857	Société chimique de France	1841	Chemical Society
1873	Société française de physique	1874	Physical Society
1876	Société zoologique de France	1826	Zoological Society

It is interesting that one of the first specialist societies to be founded in France, the Société géologique,[50] catered for a science not only of fairly recent emergence but one not explicitly represented in the Academy. Members of the mineralogy

[46] *C.R.*, 117 (1893), 875 (18 December 1893).

[47] *C.R.*, 147 (1908), 1095.

[48] Source: Robert Fox, 'The savant confronts his peers: scientific societies in France, 1815–1914', in R. Fox and G. Weisz (eds.), *The organisation of science and technology in France*, Cambridge, 1980, pp. 241–82.

[49] Various *local* Linnean societies had been founded earlier, e.g. in Lyons in 1822 (*ibid.*, p. 246). For a discussion of the significance of the date of the foundation of the Botanical Society (1854), see Chapter 6.

[50] See Lapparent, 'Rapport d'ensemble sur les travaux de la Société géologique de France', *Bulletin de la Société géologique de France*, 1 April 1880, xix–lv.

section of the Academy were happy to join it. There were other societies, like the Société chimique, which was begun in a much more modest way by three young chemical assistants as a means of learning about the subject. Only after the first year did the character of the society change, when the founders asked the Academician Dumas to become president.[51] In fact there was a special need for a national society for chemistry since the stupendous growth of the subject; the number of practitioners in the nineteenth century had long since outgrown the ability of the Academy to provide adequate representation, even for its more senior members.

6. *A restricted number of members*

There had been more flexibility over numbers of members in the Royal Academy than in the Academy of the nineteenth century. There had also been less direct control by the Royal Academy over its membership.[52] For most of the eighteenth century the Royal Academy had had to accept occasional ministerial preferences over their own choice. For example in 1768, when the young Lavoisier had been the Academy's first choice, the Minister decided to appoint the Academy's second choice, the more senior Antoine Gabriel Jars, as a reward for services to the state. Lavoisier was, however, accepted as a supernumerary member at the same time, a questionable category which was never permitted in the nineteenth century.

After 1795 the official number of members in the respective sections was strictly adhered to. The number six for every section (except geography) remained unchanged throughout the century (see Table 5), despite the different rates of development of individual sciences. As we have suggested above, one section which was particularly restricted in this way because of its rapid growth, was chemistry. In the foundation year of 1795 chemistry was virtually synonymous with mineral (or inorganic) chemistry, although a few years previously Lavoisier had allowed that there might be a few dozen substances of vegetable and animal origin to be included as part of the subject matter of the newly reorganised science.[53] The number soon grew to hundreds, then to thousands and by the end of the century was of the order of a hundred thousand. These substances were not only numerous, they posed special problems and they came to have important applications. Thus organic chemistry emerged as very much a nineteenth-century science, beyond the wildest dreams of the founders of the National Institute.[54] Even without additional branches of chemistry, such as biochemistry and physical chemistry, the science had grown so much that, by the second half of the nineteenth century, few impartial witnesses would have said that the chemistry section should have no more members than, say,

[51] *Centenaire de la Société chimique de France*, 1957, p. 7, Fox, *op. cit.*, pp. 269–70.

[52] Hahn, *op. cit.*, pp. 81–2, 133–4.

[53] A. L. Lavoisier, *Traité élémentaire de chimie*, 1789, vol. 1, pp. 127, 161.

[54] Marcellin Berthelot emphasised the possibilities of *synthetic* organic chemistry, in which the chemist actually *created* the subject matter of his research, a procedure with limitless possibilities.

Table 5. *The limited growth of the Academy of Sciences*

		Ancien régime*			Nineteenth century				
	Title of membership	1699	1716	1785	1795	1803	1816	1866	1908–13
In Paris	Honoraires	10	12	12	—	—	—	—	—
	Pensionnaires/membres résidants	20	20	24	60	63	63	66	66
	Associés	12	12	24	—	—	—	—	—
	Associés libres	—	4	12	—	—	10	10	10
	(Elèves) Adjoints	20	12	—	—	—	—	—	—
	Secretaire(s) perpetuel(s)	(1)	(1)	(1)	—	2	2	2	2
	Tresorier perpetuel	(1)	(1)	(1)	—	—	—	—	—
	Total resident members	**62**	**60**	**72**	**60**	**65**	**75**	**78**	**78**
In remainder of France	Membres non residants	—	—	—	—	—	—	—	6
	(Associés non residants) Correspondants†	c. 70	c. 40	c. 100	60	100‡	100	100	116
Abroad	Associés étrangers	8	8	8	8	8	8	8	12

* There is some inevitable simplification in this part of the table. In particular some minor reorganisations are omitted, as are the existence of supernumerary members.

† Figures for the *ancien régime* are approximate and are taken from J. E. McClellan, 'The Académie Royale des Sciences, 1699–1793: A statistical portrait', *Isis*, 72 (1981), 541–67 (p. 551).

‡ After 1803 a fair proportion of corresponding members were elected from outside France.

the botany section. Botany had been a major science in the eighteenth century and had advanced particularly through classification and systematic nomenclature. By the nineteenth century botany was more static, although there were of course always new species to record and specimens to collect. The suggestion[55] made in 1869 that the chemistry section should be increased from six to twelve seemed to some a very reasonable modification of the original organisation. But such changes were easier to propose than to implement.

The effect of this restriction in numbers on a burgeoning science like chemistry was to exclude some notable chemists, the most famous being Gerhardt and Laurent. If one examines the situation more closely, however, one finds that the restriction in numbers had the effect not so much of total exclusion as delay. It is true that in their respective provincial bases Gerhardt and Laurent had pursued very independent approaches to chemistry, not always approved by their senior colleagues in Paris. But their other weakness was their mortality. Gerhardt died at Strasbourg in 1856, only a few days before his fortieth birthday. The Academy had only just managed to elect him as a correspondent a few months previously. Meanwhile Laurent at Bordeaux had been elected a corresponding member in 1845 but, impatient with provincial life, had soon moved to Paris hoping for great recognition. Yet this never came, as he died in 1853 at the age of forty-six. Also an examination of the chemistry section shows that there were no vacancies in that section in the late 1840s, when both Gerhardt and Laurent might have been considered at the right stage of their career to be serious candidates. Indeed there was no vacancy in the chemistry section after the election of Balard in 1844 till that of Fremy in 1857. Again a further ten years elapsed before another vacancy in the chemistry section arose, filled by Wurtz in 1867.

Although the most obvious effect of such a situation was to exclude many able chemists from the Academy, a less obvious effect was that chemists posed their candidatures for other sections and with such success that Marcou in 1869 could claim that 'the Academy of Sciences was being transformed into an Academy of Chemistry'.[56] He was writing immediately after J. B. Dumas had been elected permanent secretary, thus freeing a further place in the chemistry section. But already the pharmacist Bussy had been elected as *Académicien libre* in 1850. On the strength of his important crystallographic work, Pasteur, classified by many people as a chemist, had been successful in 1862 in entering the section of mineralogy, where he joined another chemist, Henri Sainte-Claire Deville, elected the previous year. Pasteur had previously reported a reaction of fear of chemistry within the mineralogy section: 'They say that chemistry wants to invade everywhere'.[57]

But it was the section of agriculture which had really opened its doors to the

[55] J. Marcou, *De la science en France*, 1869, fasc. 2, p. 119.

[56] *Ibid.*, p. 118.

[57] 'Ils disent que la chimie veut tout envahir'. Pasteur to his mother, 10 March 1857, *Correspondance*, vol. 1, p. 422.

chemists. After the precedent established by admitting the agricultural chemist Boussingault in 1839, it had been possible to introduce the industrial chemists Payen in 1842 and Peligot in 1852, followed finally by the son of the chemist Louis-Jacques Thenard, Paul Thenard, in 1864. Here then was an arguably academically weak subject, agriculture, giving way to pressure from a strong subject, chemistry, although any strengths and weaknesses have to be considered politically as well as cognitively. Berthelot, unsuccessful candidate in the fiercely fought elections in the chemistry section in the 1860s, finally secured a place in 1873 in the physics section. He had after all made contributions to physical chemistry. Thus, by cunning as well as wide-ranging academic ability, the chemists were able to double their numbers in the Academy by the 1860s and they remained in a strong position for the rest of the century.

It was in the 1860s that Henri Sainte-Claire Deville circulated a proposal designed both to relieve the pressure on the chemistry section and to give recognition to a number of major advances in physics since the foundation of the Institute.[58] He suggested the formation of a whole new section or division of physical and chemical sciences with a total of twenty-four members, thus doubling the official representation of these sciences. The occasion for the proposal was a government suggestion for the expansion of the (half) section of geography to include military sciences.[59] He thought that the Academy should accept the Minister's proposal on the condition that physics and chemistry were given the special recognition he advocated. He pointed out, however, that the Academy would have to give almost unanimous support to such a proposal before the Minister would give it serious consideration. He was disappointed not to receive the united support he would have required.

There were two ways of making more room in the Academy. One was to increase the number of members. Some modest increase would have been generally welcome, so that it would have been easier to become a member yet not so easy that it was not still a great honour. Much more difficult to decide, even if there had been the opportunity, would have been how the increase was to be effected. If we disregard financial implications, it would have been quite easy administratively to increase the number in *all sections* from six to, say, ten, but was this what was required? We have seen that the sciences developed at different rates during the nineteenth century and there would have been unending squabbles within the Academy if its members had been asked to decide which sections should be allowed to expand, since expansion in one area would imply relative contraction in other areas. There would probably have been a marked reduction in scientific output by members of the Academy for many years if they had been asked to reorganise their institution. Given this prospect,

[58] *Réflexions sur la création d'une section à l'Académie des Sciences*, n.d. (1863), Institut, 4° N. S. Br. 48/P.

[59] For further details see Chapter 12.

there was something to be said for equality represented by the number six and something also for the status quo, which meant that members did not waste too much time planning a scientific utopia.

But an alternative method of opening up the Academy would have been to introduce a retiring age. Many would have agreed in principle but, on discussing details, one would probably have found a retiring age suggested some years in advance of the age of the proposer. A forty-year old scientist can contemplate retirement at sixty but a scientist of fifty-five would probably prefer sixty-five. There might be general agreement on retirement by, say, seventy-five but only a few Academicians were over seventy-five and it would therefore have caused some unpleasantness without much corresponding benefit for younger men. A proposal that Academicians should retire at sixty to become honorary or emeritus Academicians[60] was not only the proposal of a younger man who assumed that all scientific creativity had ceased by then but, by allowing reasonable pensions to retired Academicians, called for a large increase in the Academy's budget. The nineteenth-century Academy seldom felt itself to be in a position to ask for anything but the most modest increase. Once again, therefore, what might have seemed a reasonable and constructive proposal for reform was found to be impractical, all the more so in a society where seniority was widely respected.

7. *The conduct of meetings*

For the first few years, when the Republican calendar was in force, the First Class would meet twice every *décade* or week of ten days. Once the Gregorian calendar had been restored, it met on Monday afternoons. The scene in the mid-nineteenth century was described by a British visitor in the following terms:

> Every Monday the Académie des Sciences opens its sitting at three o'clock in the presence of a crowded assembly. The desk at which the Perpetual Secretaries are seated, with the President and Vice-President, is literally piled with letters, memoirs, books, papers and documents of every description addressed to the Academy and the larger proportion of them by persons who are almost entirely unknown.[61]

The last comment would suggest that the visitor had expected to hear great names in the Academy and was surprised to find that obscure people from every quarter were in the habit of directing their work to the Academy.

It may be useful to describe the way business was conducted at meetings. After reading the minutes of the previous meeting, it was usual for the secretaries to report on correspondence. Yet, since this part of the proceedings was often in danger of dominating the meeting, later in the nineteenth century it was transferred to the end, with the secretaries being permitted merely to note the arrival of a letter or other communications; the contents of the letter might

[60] Marcou, *op. cit.*, p. 178.

[61] Anon., 'The Institute of France', *Quarterly Review*, 93 (1853), 315–48 (p. 341).

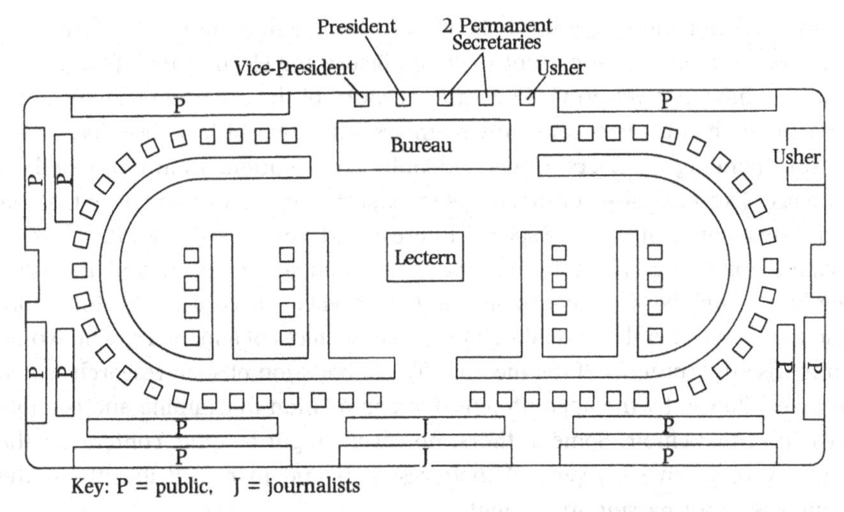

Fig. 1. Plan of the meeting room of the Academy of Sciences in the late nineteenth century. Each Academician had a designated place. Members of a particular section would not sit together. (Based on E. Maindron, *L'Académie des Sciences*, 1888.)

be briefly summarised rather than read out in full. In early meetings the titles of books or memoirs presented to the First Class would be read out. After the foundation of the *Comptes rendus* in 1835 it saved time if a list of such items could simply be reported in an appendix to the published account of the meeting.

The main part of the meeting was obviously the presentation of original research by both members and non-members. It would be nice to think that when an Academician had a memoir ready for presentation he simply read it out at the next meeting. The First Class hoped that this simple approach would work but in practice it broke down. The Academy never forgot the embarrassment of June 1809, when two successive meetings had to be adjourned after a few minutes because there was no substantial business.[62] It became a matter of prime importance, of honour even, for at least one Academician to read a paper at every meeting. This could only be ensured by asking members who wished to read memoirs to write their names on a list in advance of the meeting. Unfortunately this became effectively a waiting list. The honour of the Academy was saved but at the price of the informality and spontaneity that might have characterised a smaller society which was less in the public gaze. The waiting list had the effect of making a memoir by an Academician something of an occasion, even a privilege; it could not just be taken for granted but had to be planned in advance. Considering that the contributions of non-members did not reflect directly on the honour of the Academy, it was sometimes easier for them to submit a memoir but, if there was already plenty of business on the agenda, they might have to wait for a later meeting. There was, however, another well-publicised waiting list, that of candidates for election. When a vacancy occurred,

[62] *P.V.I.*, **4**, p. 227–9.

aspirants would want to bring their work to the attention of the Academy and would often queue up to present memoirs before the election took place.

But we have not yet provided a strict account of the order of business. After the memoirs by members came any memoirs which might have been presented by corresponding members. Understandably, contributions from non-members came last. Exceptionally non-members were permitted to read an entire memoir to the Academy if it were of special interest and not excessively long.[63] Some advanced planning might be necessary for such an occasion and the non-member would be very conscious of the privilege accorded to him. More generally a non-member would obtain the good offices of a member, who would summarise the contents of the memoir. The submission of such research would often be followed by the appointment of a commission to examine and possibly report on the memoir. Some authors, however, might be quite content for the Academy to receive a report of their research and to record its title in the minutes without passing judgement.

In the eighteenth-century Academy of Sciences, suitable allowance had been made for holidays. There was not only an extended summer vacation from early September to early November but also at least a week's holiday at the respective Christian festivals of Easter, Whitsun and Christmas.[64] The 1795 constitution, however, perhaps reacting against what it saw as the laxity of the *ancien régime*, made no allowance for extended breaks in the pattern of meetings. In 1801 we find members of the Institute asking unsuccessfully for official consideration to be given to the question of holidays.[65] Again after the 1848 Revolution the issue of holidays was raised but to no avail.[66]

The government might have imagined the Institute as a permanent machine. In theory it functioned for fifty-two weeks of the year under whatever government was in power, being closed only briefly for national holidays. In practice the pattern of activity was more variable, following the cycle of the academic year. In so far as many members of the Academy held positions in higher education, they had teaching and administrative duties in Paris during the university terms but hardly anything during the summer months. This provided an opportunity to escape for a few weeks from the over-crowded capital into the countryside and many Academicians left Paris, often returning to their roots in the provinces. We should remember too that France was for most of the nineteenth century a predominantly agricultural country. Academicians sometimes still had land in their home villages that they needed to attend to at harvest time. In the agriculture section of the Academy they even had a professional excuse for absence for at least several weeks in the autumn. This meant that attendance at the Academy tended to be slack in the late summer

[63] Such memoirs were later described in the *Comptes rendus* as 'Mémoires lus'.

[64] *Règlement ordonné par le Roi pour l'Académie des Sciences*, 1699, Art. 18.

[65] A.I., Registre 3A4, p. 54 (5 *thermidor* an 9)

[66] A.S., *Comité secret, 1845–56*, p. 145 (24 September 1849).

and early autumn and elections and other major corporate decisions sometimes had to be postponed until the late autumn.[67] It also meant that memoirs were less abundant in the summer, whereas in the winter the Academy was sometimes overwhelmed with material.[68]

8. *Corresponding members*

In order to justify the claim that this was a truly National Institute, the resident members in Paris were balanced by an equal number of 'non-resident associates' in the provinces. Accordingly, in February 1796 the Institute proceeded to the election of these provincial 'associates'. Though many may seem to be rather obscure figures, a few associates have a recognised place in the history of science. In mathematics they included Montucla, the author of an important history of mathematics and Arbogast, a contributor to the metric system and the author of text-books of mathematics. In physics they included another prolific author, Sigaud de Lafond, who was interested in the popularisation of science, and Etienne Montgolfier, younger brother of Joseph Montgolfier, who had together made history in 1783 by sending a large hot air balloon 2000 metres into the air. Etienne Montgolfier ran the family business near Lyons, while his brother later came to Paris and was elected as a full member of the First Class. The chemistry section was able (at least temporarily) to call on Chaptal, soon to come to Paris, and Seguin, former collaborator of Lavoisier, most recently known for his process of tanning leather, which had been invaluable to the revolutionary armies. In mineralogy the representatives were not exactly household names but they had all published at least one book on their subject. Looking at the location of these non-resident associates, one can see the existence of certain centres of scientific activity including Strasbourg, Toulouse, Montpellier and Marseilles. At least three successive directors of the Marseilles observatory were to be recognised by the First Class for their astronomical talents.

Despite the existence of a few fairly eminent non-resident associates, it soon became clear that the ideal of a provincial counterweight to the concentration of talent in Paris was an illusion. Even if one disregarded quality and looked only at quantity, everyone could see that these associates were no more than token members of the Institute. They took no part in its meetings and had very little contact with the parent body. At the earliest convenient opportunity, therefore, it was decided to revert to the practice of the *ancien régime* and describe provincial members of the Institute as corresponding members.

In the eighteenth-century Academy of Sciences correspondents had been officially assigned to specific Academicians in Paris to keep them informed of developments in their own region. With the growth of newspapers and scientific journals, this source of private information was hardly appropriate in most of the

[67] A.S., e.g. *Comité secret, 1857–69*, p. 313 (22 August 1864), p. 383.

[68] A.S., e.g. *Comité secret, 1837–44*, p. 118 (5 December 1842).

branches of science recognised by the nineteenth-century Academy. There is no question therefore of non-resident associates taking up the pen as soon as they received their new title. It was a largely honorific title but it was also something of a demotion for men who had previously been, in theory, the equal of their Parisian colleagues. At least one associate publicly complained about his demotion.[69] One of the few nineteenth-century corresponding members to interpret his title literally and assiduously by frequent contributions to the Academy was the Italian Jesuit, Angelo Secchi (1818–78), elected as a correspondent in the astronomy section in 1857.[70]

But to introduce an Italian as a corresponding member is to admit that the replacement of non-resident associates in 1803 with correspondents involved more than a change of nomenclature. Having abandoned the idea of equal representation of subjects in the provinces of France, the rank of correspondent was open equally to men of science living beyond the boundaries of France. Among the early correspondents elected in 1803–4 we therefore find such eminent names as van Swinden, van Marum, Blagden, Landriani and Gauss. In 1808 the addition of the names of Jenner and Watt was a further proof of the international remit of the First Class, transcending problems of war. In some sections, such as agriculture, the great majority of correspondents continued to be Frenchmen from the provinces but other subjects, such as astronomy and chemistry, were determined to be more international. In 1808 it was stipulated that a minimum of a quarter of the correspondents for each section should be French.[71] In practice the proportion was usually much higher than this.

The other problem that needs to be mentioned is the uncharacteristic recklessness with which the First Class elected correspondents, giving some sections many more 'members' *extra muros* than others. Some Academicians argued with considerable justification that an observational science such as astronomy needed many more correspondents than an experimental science like physics. In a rationalisation of correspondents in 1808 it was, therefore, agreed that astronomy could have as many as sixteen correspondents, compared with six for mathematics.[72] But already chemistry, for example, had as many as a dozen correspondents and the section of botany had nearly as many. A fixed (but different) number was therefore agreed for each section, which allowed the

[69] *P.V.I.*, 2, 41 (Saucerotte, 21 *germinal* an 11 = 11 April 1803).

[70] A.S., dossier Secchi.

[71] *P.V.I.*, 4, 75–6 (6 June 1808)

[72] *P.V.I.*, 4, (15 February 1808). At a later meeting in 1808 (*ibid.*, p. 76) it was hinted that correspondents might be considered on probation, so that each year a check could be made on whether they had published anything or communicated something to the First Class. If not, they could be dropped from the official list of correspondents. This resolution was quickly forgotten. It is surprising that it was even accepted provisionally, since it conflicts with the idea of permanence in the other positions in the First Class. Perhaps it reveals that the resident members could be guilty of applying double standards. Correspondents were never able to speak up for themselves at meetings.

ninety-six existing corresponding members to be increased to the round number of 100.

But if the First Class was generous in bestowing the title of correspondent on a comparatively large number of provincial men of science, the honour conferred was strictly limited. The reorganisation of 1803 had explicitly stated (Art. 7) that correspondents could not call themselves members of the Institute. A further regulation stated that if they took up residence in Paris they would lose their title of correspondent. In 1808 it was pointed out that, even for publication, the correspondents fell between two stools. Not being full members, they were not permitted to publish in the *Mémoires*; yet since they were associated with the First Class they could not publish either in the *Mémoires des Savants Etrangers*, reserved for non-members.[73] It was agreed that outstanding memoirs by correspondents would in future be accepted for publication in the ordinary *Mémoires*.

Yet one should not be left wondering why men of science outside Paris bothered to be elected as correspondents. For many it provided considerable local status and in a provincial town far from Paris a correspondent might be treated as if he was a member of the Institute, whatever the regulations might say. For the more ambitious and the more mobile it could also be seen as a stepping stone to full membership, although the gulf was greater than many appreciated.

Another early idea, intended partly as a countermeasure to extreme centralisation, was to appoint travellers (*voyageurs*), essentially agricultural experts, to travel round France and abroad over a period of several years and report back to the Institute. The idea drew on the experience of the Jardin du Roi under the old regime, which had made regular use of travellers to bring back new specimens of plants from different parts of the world.[74] For the Muséum this continued into the nineteenth century as a useful activity, but it was soon found to be quite impractical for the Institute. According to the original constitution of the Institute, which went into the matter in some detail, it could nominate as many as twenty individuals a year to travel for a period of three years to make observations relevant to agriculture.[75] They would keep a regular diary and send their observations every few months to the Institute. In order to qualify, travellers had to be at least twenty-five years of age, be the owner of a country estate or a well-established farm or the son of either, and also have a good knowledge of mathematics, political economy and natural history (especially botany or mineralogy). The Institute itself was expected to nominate six of its own members to undertake similar travels.

[73] Law of 3 *brumaire* an 3 (25 October 1795), Titre V, arts 1–4.

[74] Yves Laissus, 'Les voyageurs naturalistes du Jardin du roi et du Muséum d'Histoire Naturelle: essai de portrait-robot', *Revue d'Histoire des Sciences*, **34** (1981), 259–317. For French government interest in the eighteenth century in *voyageurs*, see Rhoda Rappaport, 'Government patronage of science in eighteenth-century France', *History of Science*, **8** (1969), 119–136 (pp. 129ff).

[75] A.S., Carton 36, file: 'Voyageurs'; A. I. 3A3, f.151 (5 *fructidor* an 8).

An insurmountable problem was to find suitably qualified observers who would be willing to abandon their families and friends and their careers for several years for the uncertainties of travel far from home. One of the few travellers appointed was the naturalist and agriculturalist Broussonet (1761–1807), a member of the First Class, who went to Africa and received an allowance of 3000 francs a year from the Institute. The Institute was in the curious position of having a budget, which in one year amounted to 120 000 francs, which it could not use for the intended purpose. It is a pity that in its early years it did not feel confident enough in its relations with the government to propose alternative uses for these funds. With hindsight one can think of many facilities that such funds could have bought in Paris, although the money could not, of course, have been siphoned off purely for the benefit of the First Class. Instead the Institute, conscious of the many calls for economy in government expenditure, told the Minister that the money could be saved. If it hoped that, at some later stage, this saving would be held up to its credit, the Institute was to be disappointed.

9. *Buildings*

Under the old regime the Royal Academy had met in its early days in the Bibliothèque du roi but there were also vague plans to build additional facilities including a laboratory.[76] The early practical fulfilment of these plans, however, was the building of the Observatory (1672), which was to have a parallel and largely separate history from the Academy.[77] In the eighteenth century the Academy met in the Louvre in a suite of rooms on the first floor. Adjacent to the main meeting room were smaller rooms housing natural history collections and cupboards with anatomical specimens and a few mechanical appliances.[78] In addition the Academy gradually accumulated a collection of machines and models submitted to it for approval. In 1785 the desirability of additional accommodation for these machines was recognised by the King, who designated particular rooms to house them. With the Revolution, however, the situation changed. A patent law was introduced. The Academy was no longer required to give judgements on inventions and the existing collections were transferred to the newly-founded Conservatoire des Arts et Métiers, established in the buildings of the abbey of St Martin.[79] These included sixty hydraulic machines, many clocks and agricultural implements and a variety of models of machines.

When the National Institute was founded in 1795 it first met in the Louvre,

[76] R. Hahn, *op. cit.*, p. 16.

[77] Gillispie points out that, in the early years, the Observatory had been thought of as an adjunct to the Academy. It had been considered not simply as a place where astronomical observations were made but as a focus for practical science of all kinds. The plan to include a chemical laboratory or anatomical theatre did not, however, work out well (C. C. Gillispie, *Science and polity under the ancien régime*, Princeton, N.J., 1980, p. 100).

[78] E. Maindron, *L'Académie des Sciences*, 1888, pp. 34–5.

[79] M. Crosland (ed.), *Science in France in the Revolutionary Era described by Thomas Bugge*, 1969, pp. 148–51.

now being transformed from a royal palace into a museum and art gallery.[80] This necessitated an extensive programme of renovation and repair. Alternative temporary accommodation was therefore sought for the Institute and the choice fell on the former Collège des Quatre Nations, founded by Cardinal Mazarin, opposite the Louvre on the left bank of the Seine. Becoming state property in 1792, it had served successively as a prison, an *école centrale* and an Ecole des beaux arts. The chapel under the gilded dome was now to be used for the most solemn and public meetings of the Institute, the place of the altar being taken over by the president and secretaries of the respective Classes. The Institute was soon to find that what had begun as a temporary measure was to become its permanent home.

From the beginning of the use by the Institute in 1805, space was at a premium, all the more so as a part of the group of buildings continued to be occupied for some time by the Ecole des beaux arts. Nor was there any change of use contemplated for the neighbouring Bibliothèque Mazarine, although this did serve to provide space for meetings until a separate meeting room was built in 1832–5. Yet the proximity of this meeting room to the noisy street was a source of constant distraction and the Institute took advantage of the crumbling state of another wing of the building to urge the Minister of the Public Instruction in 1838 to demolish the wing and construct a larger locale for meetings. The new meeting rooms on the second floor were completed in 1845. The room given to the Académie des Sciences was certainly quieter than the previous meeting place, but the windows allowed in only a moderate amount of daylight. There was a permanent problem with lighting and heating. Until 1875 artificial illumination was provided by candles, which were subsequently replaced by gas lighting. Primitive central heating columns in the four corners of the room supported busts of the scientists Laplace and Cuvier, the artist Gros and the sculptor Droz; a reminder that the room was shared by three of the five Academies: the Académie des Sciences meeting on Monday afternoon, the Académie des inscriptions et belles lettres meeting on Friday afternoon and the Académie des beaux arts meeting on Saturday afternoon.

On the upper parts of the wall were arranged a series of portraits, including those of d'Alembert, Buffon, Coulomb, Lavoisier and Lagrange, collectively representing the sciences. These were to be found side by side with portraits of prominent figures of the Enlightenment, notably Montesquieu, Voltaire and Rousseau. Among the busts of former members of the Academy of Sciences were (in addition to the two previously mentioned) Bonaparte, Lagrange, Monge, Haüy and A. L. de Jussieu, all representative of a heroic age of science in the early years of the Institute. A second adjoining smaller room served for the Académie Française, meeting on Thursdays and the youngest of the Academies, founded in 1832, the Académie des sciences morales et politiques, meeting on Saturdays. On the other side of the meeting room of the Académie des Sciences

[80] *Ibid.*, p. 173.

was a large ante-room, sometimes called the *salle des pas perdus*. On the floor below were rooms for the secretaries of the respective Academies and committee rooms for the meetings of official commissions.

Although many members of the Royal Academy would have liked a laboratory in the Louvre, we have seen that their material support consisted of collections. With the foundation of new institutions after the Revolution, there were now more appropriate buildings to house such collections: the Conservatoire des Arts et Métiers for the machines and models, and the Muséum for the skeletons of several large animals. The only available space in the Collège des Quatres Nations was in the attics. The lack of space was the principal consideration in parting the First Class from its collections, although for many Academicians it must have been a relief, since most of the objects really represented the science and technology of a previous age. Also we should remember that the First Class, although the largest, was merely one of four classes of the Institute and any request for special favours for science would have met with little sympathy from the representatives of literature, history and fine art.

10. *Access to apparatus and laboratories*

In the early days of the Royal Academy it had been thought that meetings might consist partly of collaborative work by members, but very soon meetings were organised more around *reports* of work done earlier and often quite outside the institution. The meetings became occasions for receiving these reports and possibly criticising them or relating them to other knowledge. Of course there might be demonstration experiments at meetings but they only represented repetition or verification of earlier work. The actual process of discovery was quite separate from later verification. Also, as experimental science became more complex, it became increasingly difficult to bring everything under one roof.

Nevertheless, it was arguably a weakness in the Academy that it seemed to be detached from the actual production of new knowledge. For example, the Academy buildings did not include laboratories where research was actually carried out. Critics may not appreciate that it was quite impossible for any one building in central Paris to provide physical space to undertake the whole spectrum of scientific activities of the Academy, which ranged from mathematics to medicine, from anatomy to agriculture, from physics to zoology. Not only were there no laboratories, there was hardly room for a few cupboards containing instruments; and there was certainly no room for a full collection of specimens needed for subjects such as mineralogy or comparative anatomy. Even the Muséum d'Histoire Naturelle, originally built on a large site on the edge of the city, could only provide for a limited range of sciences, and even then work performed in the institution had to be supplemented by field work and expeditions.

This does not mean that Academicians were not tempted from time to time to try and organise a modest central store of apparatus and material. There was some demand for pieces of apparatus which Academicians did not themselves possess and which might be necessary for particular experimental studies. An additional justification for such a collection was that Academicians appointed to commissions to examine memoirs submitted to the Academy might wish to repeat experimental work described in the memoir. Thus in 1801 we find regulations drawn up about special collections for different interests within the Institute, with special mention being made of physics and chemistry apparatus.[81] Even at this stage there was some anticipation that the condition of apparatus might deteriorate. There was, however, an overriding problem in the cramped quarters of the Institute. Even if the money could be found to buy apparatus, space was always at a premium and in 1806–7 the idea of a central collection was abandoned in the face of practical problems. Existing apparatus was donated to other scientific institutions in Paris.

Yet by 1834 these difficulties had been forgotten and a further collection of instruments was established, with the physicist Antoine César Becquerel appointed as honorary curator.[82] A list of regulations was drawn up, allowing apparatus to be borrowed for a period of up to three months, although in special cases this period might be extended to one year. Yet Academicians did not always observe the regulations, instruments were damaged and the idea of a collection of instruments being run like a lending library was finally abandoned in 1864. Lack of space was given as the main reason for abandoning the idea. The Academy had never claimed to provide laboratory space. At the most it intermittently offered Academicians the loan of apparatus, which they were expected to use elsewhere. We must now consider, therefore, what laboratories were available.

Academicians normally held at least one teaching post in Paris, where there might be laboratory facilities which they could use. In discussing the origins of professional science, it should be pointed out that although France made a major contribution *before* the German universities, a weakness in the simplistic claim that France provided the model for modern professional science is that in France scientists were usually paid to *teach* in establishments of higher education rather than, as later in Germany, to do research. Thus an Academician who was a professor at the Ecole Polytechnique, the Ecole Normale or the Paris Faculty of Science, if he did research in the laboratories of his educational institution, could be considered not as doing it primarily for the institution but rather *for the Academy*.[83] It was to the Academy that it would normally be directed; it would be judged and published by the Academy.

[81] Arrêté de l'Institut relatif à ses collections, 5 *fructidor* an 9 (23 August 1801). Aucoc, *op. cit.* p. 59.

[82] A.S., dossier: Appareils scientifiques; also *P.V.I.*, 10, 555.

[83] The foundation under Pasteur of the *Annales scientifiques de l'Ecole Normale Supérieure* in 1864 was intended to focus loyalty on that institution, just as the earlier *Journal de l'Ecole Polytechnique* had been intended to strengthen the esprit de corps of that school. Nevertheless, such publications can be seen as almost peripheral, compared with the central importance of the *Comptes rendus* of the Academy.

The main conclusion we might draw about the failure of the Academy to provide a location for research is that, far from breaking the chain of productivity, it merely led to production taking place within a range of complementary institutions of higher education or research, or in private laboratories. If one tries to imagine the thousands of square metres of space, the hundreds of thousands of francs (an inconceivable expense in a strictly limited budget) which would have been necessary to satisfy the minimal research interests of some sixty senior scientists representing some fifteen[84] widely different disciplines, one might turn with relief to an alternative scenario, in which more realistic support is provided by a network of educational or commercial enterprises established in or near the French capital city. There were a variety of supporting institutions, ranging from the Collège de France to the Muséum d'Histoire Naturelle, from the Ecole Polytechnique to the Paris hospitals, where research could be carried out and *was carried out* by Academicians. Thus, instead of a situation of the most extreme centralisation, we find a network of scientific activity and mutual support covering the whole Paris region.

In the case of chemistry in particular, there were several laboratories with varying facilities. By modern standards French government investment in laboratories in the early nineteenth century was small, and scientists sometimes made use of private laboratories. A recent study of the research school of J. B. Dumas helps us to understand the general situation. As a young man, Dumas had a laboratory at the Ecole Polytechnique, where he was successively *répétiteur* and professor from 1824 to 1838.[85] It was here that he did important research in the 1830s in organic chemistry, which was duly reported to the Academy. When he helped to found the Ecole Centrale des Arts et Manufactures, he drew up plans for another small research laboratory. From 1838 to 1848, however, he made greatest use of his private laboratory near the Muséum d'Histoire Naturelle. At the Ecole Normale Henri Sainte-Claire Deville established a laboratory where he carried out important research in inorganic and physical chemistry. In 1855 a research laboratory was finally authorised for the Paris Faculty of Sciences. Further laboratories were associated with the Mint and with commercial enterprises including the Gobelins dyeworks and the Sèvres porcelain factory. All these institutions at one time or another counted Academicians among their senior scientific staff.

These laboratories were available not only to carry out original research but, as has been suggested, they could be used to verify work reported to the Academy; often commissions of the Academy would meet in a laboratory associated with the employment of one of its members. The Academy, far from

[84] Although there were only eleven sections, this figure allows for sections which included two different disciplines.

[85] Leo J. Klosterman, 'A research school of chemistry in the nineteenth century: Jean-Baptiste Dumas and his research students', *Annals of Science*, **42** (1985), 1–80 (pp. 7–9).

being an organism cut off from other institutions, might be considered more like a huge octopus with tentacles stretching all over the Paris academic scene. We shall see later[86] that this metaphor applies not only to laboratories but also to paid positions in higher education in the Paris region. Although officially these institutions were independent, or sometimes under different Ministries, in practice they were all part of a network with, arguably, the Academy at the centre.

11. *The new constitution of* 1803

It is helpful to describe the 1795–6 constitution of the Institute as a provisional, even an experimental, one. The inspiration of the *Encyclopédie* brought together all branches of knowledge in a single institution, something which was fine in theory but in practice presented many problems. It was not enough that the members all belonged to an intellectual or artistic elite. In practice they were interested in quite diverse things, and the three classes provided much more coherent groups of related subjects, so that the members wanted to associate only with the members of their own class. It was also quite ridiculous in elections to expect a poet or a painter to vote for an astronomer, or a mathematician to vote for a dramatist or a historian. There were monthly meetings of the whole Institute as well as quarterly public meetings, which were probably felt by most members of the Institute to be largely a waste of time.

A new constitution in 1803, which replaced all these meetings by a single annual meeting of the whole Institute, must have been a relief. But these administrative changes of 1803 included other changes which had a direct political motive and often these alone are reported. In other words it is not always appreciated that the new constitution of the Academy of 1803 was on the whole an *improvement* on the earlier organisation since it could draw on the experience of several years of unwanted association and poor efficiency. (The latter arose largely from constantly changing secretaries, who were to be replaced by *permanent* secretaries, as in the former Royal Academy of Sciences.) If we are principally interested in the First Class, the political changes are less important since they were focused entirely on the Second Class. Bonaparte as First Consul decided that he could not tolerate state support for the political theorists of the Second Class, who constituted a challenge to his increasingly authoritarian rule, and he therefore took the extreme step of abolishing the Second Class and dividing the members between the other classes.[87] It was in this way, as we have said, that a new (half) section of geography was added to the First Class in 1803.

The new constitution was drawn up on 3 and 8 *pluviose* year 11 (23 and 28 January 1803) and transmitted to the Institute by the Minister of the Interior,

[86] Chapter 6, Section 12.

[87] It was in this way that the Idéologues shifted their research interests to innocuous historical subjects – Martin S. Staum, *op. cit.*, p. 393.

Chaptal. The First Class held an extraordinary meeting on 31 January to hear the new organisation.[88] As shown in Table 3, the Institute was henceforth to consist of four classes instead of three:

First Class:	Physical and mathematical sciences (as before).
Second Class:	French language and literature.
Third Class:	Ancient history and literature.
Fourth Class:	Fine Arts.

Thus the controversial new social sciences had disappeared and in the new Third Class historians were directed towards Latin, Greek and oriental texts from the distant past, which were far removed from sensitive areas of modern history. Here then was a silencing of potential opposition but it was done subtly by administrative reorganisation, not by a political purge, which had been the fate in 1802 of the political assembly known as the Tribunate.

Under the Empire Napoleon was to go further. Not content with eliminating opposition, he expected the Institute to play a positive part in his cultural politics. Writing to the Minister of the Interior in 1807, he said:

> The Institute is a splendid weapon for the Minister; if he uses it properly, it will do all that the Government requires...The Institute will raise no objection to our demands. It is obliged by the constitution to do anything the Minister asks of it.[89]

It should be pointed out, however, that the context of this letter was the use of the Institute 'to prevent the corruption of taste and language by vices of thought and expression', in other words the censorship of politically subversive works of literature. The suppression of great literature under Napoleon, notably the expulsion of Madame de Stael and the restrictions placed on Chateaubriand, as well as the refusal to countenance the study of the social sciences within the framework of the Institute, are in marked contrast to the encouragement given to the natural sciences, which flourished in this period both in the First Class and outside.[90]

Elections were henceforth to be the business only of the class in which the vacancy occurred. There was to be an administrative commission, consisting of representatives of each of the classes, to deal with general matters affecting the whole Institute and particularly with the budget. Once the money had been divided between the different classes, expenditure was the sole responsibility of the class concerned. On this commission three of the classes had one representative each but the First Class had two representatives, just as it had two

[88] *P.V.I.*, 2, 619–22.

[89] 'Remarks on the Minister of the Interior's reports about the encouragement of literature', Finkenstein, 19 April 1807, J. M. Thompson (ed.), *Napoleon's Letters*, pp. 178–9.

[90] See for example Maurice Crosland, *The Society of Arcueil. A view of French Science at the time of Napoleon 1*, 1963.

Table 6. *Structure of the Academy for most of the nineteenth century*

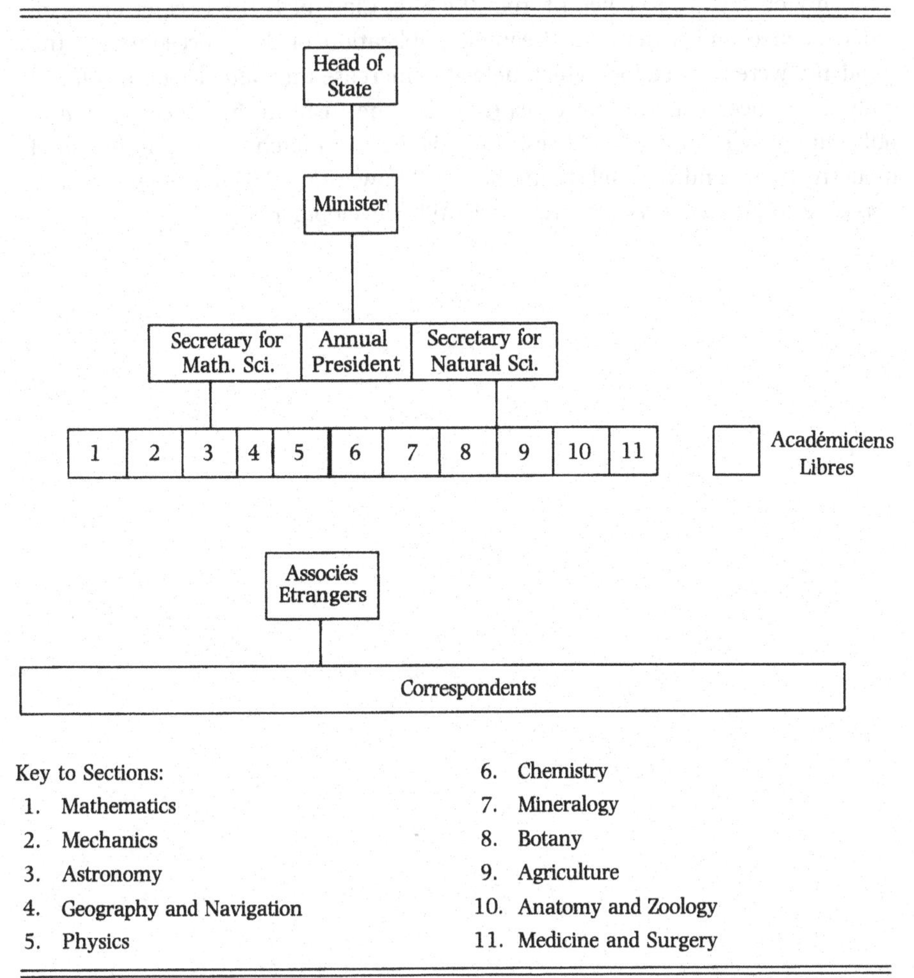

Key to Sections:

1. Mathematics
2. Mechanics
3. Astronomy
4. Geography and Navigation
5. Physics
6. Chemistry
7. Mineralogy
8. Botany
9. Agriculture
10. Anatomy and Zoology
11. Medicine and Surgery

salaried permanent secretaries. Thus although we have to remember that one of the constraints on the activities of the First Class was that it was only *one* of the classes in a broadly based Institute, and consequently was always in a minority in any discussion about special claims for science, the Bonapartist constitution gave it a double representation.

This chapter has emphasised the administration and regulations of the Academy. A general schema of the structure of the Academy is given in Table 6. The regulations continued in force throughout the nineteenth century and, whereas some undoubtedly helped, others might be seen increasingly as a strait-jacket. Indeed, this might be an appropriate description if there was not good evidence that the Academy was capable of change and growth. Different aspects of change and development will be described in subsequent chapters but we might take this opportunity of mentioning two important areas which made the Academy of, say, 1850 different from the First Class of 1800. Greater financial independence and a new and frequent publication of the proceedings of the Academy were both factors which helped to increase the influence of the official body of science. The original constitution of the Institute had specified *yearly* publication by each of the classes. But this did not match the requirements of modern science and we shall see in Chapter 8 how from 1835 the *Comptes rendus* was able to provide *weekly* news of scientific developments.

— *Chapter 3* —

THE FUNCTIONING OF THE ACADEMY: SOME POSSIBLE ROLES

> A crowd of young people come regularly on Mondays to hear memoirs which are read at the Academy; it is a means of keeping in touch with science.
>
> (*Journal des Débats*, 19 September 1832.)

> At the [Royal] Academy [of Sciences] one can see many people in a short time
>
> (Letter from Jean Trembley-Colladon, Paris, 28 November 1786.)

> The Academy has among its members judges, who know how to examine [scientific work] prudently and conscientiously; they are able to reach a decision, to give their approval or to express reservations. Guardian of scientific traditions, ...the Academy appears like a superior tribunal.
>
> (Armand Gautier, *C.R.*, 153 (1911), 1274.)

1. *The many roles of the Academy*

A good scientific institution fulfils a number of roles. In the first place its very existence should encourage or even *provoke* the production of new knowledge. Secondly, it should provide some *facilities* for the production of that knowledge, possibly by giving access to a laboratory or by loaning scientific equipment or by giving grants, although we should beware of too high (and essentially twentieth-century) expectations in these areas. Thirdly, it should *receive* the knowledge produced. This reception of knowledge is possibly the most fundamental aspect of a scientific society. The presentation to members of a well-organised piece of research at one of its regular meetings is surely at the heart of any scientific society. If there is discussion and even criticism, so much the better. But the existence of meetings, at which papers are presented, constitutes the minimum functioning of a society. In many cases the work of the society would stop at that point, but a few societies would have some means of publication, usually a journal. Publication, even in the most obscure journal, would give a certain status to the work. This, therefore, brings us to the fourth aspect of a society, the ability to *sanction* research. The more important the society, obviously the more important its acceptance of work and, even better, its approval or recommendation.

The Academy contributed to all of these aspects and very powerfully indeed in the case of the last two functions. Its weakest aspect was probably its failure to provide immediate facilities for research, although it helped indirectly. We have considered the question of laboratories in the previous chapter. In later chapters we will discuss at length other aspects mentioned above; for example, the Academy's role in provoking new knowledge, which was helped by its visibility in the nation's capital under government patronage. It was recognised nationally as an elite group, representing one important aspect of national culture. The Academy's role in provoking new knowledge was greatly helped by the prize system, which is described in Chapter 7. The role of the Academy in receiving knowledge is dealt with in the same chapter, where the institution is considered as a registration bureau. This leads on naturally to the final aspect mentioned above, that of sanctioning the research. One major aspect of this is publication, which is the subject of Chapter 8.

The Academy also performed to some extent the role of providing a model. The control of science by the Academy included the influence it exerted on how science should be done. This was a legacy from an earlier idea of an academy as establishing *rules* for the pursuit of a discipline. Thus under the *ancien régime* the Academy of Painting dictated how painting should be done, and the Académie Française not only decided what was good literature but was also charged with compiling an official dictionary of the French language.[1] In the Academy of Sciences there was no formal dictionary, but there is no doubt that certain linguistic rules were followed in its memoirs. Thus by 1800 a chemical memoir would be expected to use the language of the new French school of chemistry.[2] By the end of the century all measurements were expected to be reported in metric units.[3]

The rules, although seldom explicit, extended to methodology. In a memoir on an observational science it would be expected that the observations reported should be systematic and repeated several times where possible. In physics and chemistry one would expect reports of experiments carefully carried out using standard instruments and pure materials. Such experimental work should be described in enough detail for the experiment to be repeated by independent workers. The Academy would not look kindly on the submission of a speculative paper by a young scientist which was not grounded in careful observation and/or experiment. One of its sister academies, the Académie Française, in 1829 rashly denounced Romanticism, which, it said, 'puts in disorder all our rules, insults our masterpieces ... '[4] It is arguable whether science has 'masterpieces' in the same way as literature, but it certainly has standards, models and even rules,

[1] See e.g. Paul Mesnard, *Histoire de l'Académie Française depuis sa fondation jusqu'en 1830*, 1857. There are several general surveys of the activities of the respective academies which later became part of the Institute, e.g. *Institut de France. Catalogue de l'exposition, avril/mai, 1983*.

[2] Maurice Crosland, *Historical studies in the language of chemistry*, 2nd edn, 1978, pp. 177ff.

[3] See Chapter 12, Section 4.

[4] Roland N. Stromberg, *An intellectual history of modern Europe*, 2nd edn, Englewood Cliffs, N.J., 1975, p. 225

and in France these were largely in the hands of the Academy of Sciences. Apart from the work of the Academicians themselves, the models were provided by the Academy reports, which singled out some of the memoirs submitted for special praise, and by Academy prizes. In the language of T. S. Kuhn,[5] the Academy helped establish paradigms and it was not too difficult for the scientific community to produce 'normal science' conforming to these paradigms. But, by the same token, it was more difficult for the Academy to deal with a 'scientific revolution', particularly if it came from the outside, Often, however, new ideas emerged within the Academy, where they would be discussed while further experimental evidence was sought to support new ideas. Admittedly, new ideas often took a long time to be accepted. But in the end this might have been preferable to a hypothetical institution at the opposite extreme, captivated by every passing fashion.

We can speak of the Academy as providing a model, but in practice a large part of this role would devolve on individual Academicians, whose work would be regarded as exemplifying good science. That is why it was particularly important not to elect anyone who might deviate from the conventions of good science. In elections, emphasis was accordingly placed on a good track record rather than on future potential. In case of doubt it was preferable to elect someone safe rather than an erratic genius. This helps to explain why the list of Academicians contains, in addition to many famous names, a number of mediocrities. If the Academy had taken a few more risks, it might have had a more disturbed history. Given its high visibility in the capital city, this is a situation it understandably wished to avoid.

But if we are to understand at a deeper level the actual working of the Academy, it may be helpful to consider its various roles, using this term in its widest interpretation to include a number of analogies with other institutions such as the family or the church. We may refer to a sociologist's discussion of a parallel between the world of the scientist and that of the priest. Both priests and scientists may be considered as professionals, 'whose social function is to administer a specific fund of symbolic representations, which can serve people as a means of orientation',[6] which is arguably the central social function of knowledge.[7] Both groups have shown a tendency to monopolise the means of orientation as a source of power, which has led to a number of well-known (though imperfectly understood) conflicts between 'science' and 'religion'.[8] But the view that scientists, and more particularly the members of the Academy,

[5] *Structure of scientific revolutions*, Chicago, 1962.

[6] Norbert Elias, 'Scientific establishments', *Sociology of the sciences*, 6 (1982), 3–69 (p. 43).

[7] *Ibid.*, p. 37.

[8] The crude nineteenth-century conflict thesis typified by A. D. White (*A history of the warfare of science with theology in Christendom*, 1896, republished by Dover, 1960) has been justly criticised in the Open University course 'Science and belief: from Copernicus to Darwin'. For a sophisticated modern interpretation see the work of John H. Brooke.

were a new clergy should not be regarded merely as the fantasy of a sociologist or a modern historian. It is a parallel which was often drawn in the 1790s. It is interesting that a similar point had been made in England in the seventeenth century in relation to natural philosophers such as Robert Boyle.[9]

Again, the idea of the Academy as a family may not only be inferred indirectly from the historical evidence, it was sometimes made explicit. There are other roles to be considered, however, which by the nature of things would never be spelled out. If we claim below that the Academy sometimes acted as a gerontocracy, we can hardly expect to find this word in the proceedings of the Academy. What we find is considerable documentary evidence of special respect given to the elderly (always male in this case), which supports our hypothesis. Similarly when we speak of a bureaucracy, this is a useful way of reminding ourselves of the complex administrative machinery which underpinned the efficient working of the Academy. One might assume that that institution was under the immediate control of Academicians, but they could not carry out all the necessary administrative tasks on a day-to-day basis. There had to be a few salaried officials, whose full-time undivided loyalty to the Academy ensured its regular and efficient functioning.

The previous chapter dealt with the structure of the Academy. It provided some necessary basic information, supplemented by some interpretation. But it left many questions unanswered, for example, the question of authority, relations between members, discussion of scientific research, and many other issues. We must now, therefore, turn from structure to function or, if a metaphor from the life sciences is permissible, from the anatomy of a scientific institution to its physiology. One function of the Academy was to present science as a corporate rather than as a private activity. By focusing on a number of aspects of the Academy, we hope to give the reader a better idea of the actual working of the institution, an interpretation which will be further amplified in later chapters.

2. *Priesthood and authority*

In seeking to understand the many roles of the Academy, we have argued that it might be appropriate to include a religious analogy. We shall provide evidence to show that politicians made some use of this analogy in discussions about the early Institute and the re-establishment of higher education. There were in fact at least two major periods when traditional Christianity was under attack and science appeared in some ways as an alternative: the early years of the First Republic and the period beginning in the 1860s and leading into the Third Republic. In the first case the representatives of science and learning were accused of presumption. In insisting on special authority for science, they were violating political ideals of equality. By the 1870s on the other hand, traditional religious belief had been weakened by increasingly bold claims, both from

[9] Steven Shapin and Simon Schaffer, *Leviathan and the air-pump: Hobbes, Boyle and the experimental life*, Princeton, N. J., 1985, for example pp. 310, 319.

scientists and Biblical scholars regarding themselves as scientific, so that science itself and scientism came to be adopted by many, especially by extreme republicans, almost as a new religion. The main counter attack this time had to wait till the 1890s and will be considered later. We must first focus on the 1790s.

In a popular novel of 1791 called *La chaumière indienne*, Bernardin de Saint-Pierre, one of the most influential of the disciples of Jean-Jacques Rousseau, launched a scathing attack on academies in general and their members by comparing their attitude with the insolent pride of a Brahmin priest.[10] The high priest claimed that the truth was only open to Brahmins, since they alone could read the ancient books. Similarly, members of the Royal Academy of Sciences claimed that they alone could read the book of nature. They thought that the very title of Academician guaranteed the purity of their judgement. Saint-Pierre was careful to dissociate himself from other disciples of Rousseau who were opposed to all learning. It was not science and literature that he was attacking, but learned societies which, because of their pride and prejudice, often acted as an obstacle to progress. He contrasted the alleged arrogance of priests with the supposed wisdom of simple men who communed directly with nature.

Opposition to the idea of a priestly caste also arose in political debates on the reform of education. The idea of Condorcet of a national institute at the apex of the system of higher education aroused considerable opposition. When it was discussed in the Committee of Public Instruction, the lawyer Durand Mallaine objected that it would constitute 'a formidable corporation'. He continued:

> After having shaken off the yoke of tyrants and rid ourselves of priestly domination... it is strange that, on the pretext of science and enlightenment, it is proposed to the nation to confer, at its own expense, a permanent and particular status on one class of its citizens. And what citizens? Those most able to dominate public opinion by guiding it, for there is a superstition about what are called men of learning (*savants*), as there used to be about kings and priests.[11]

Durand Mallaine raised a stir in the meeting by publicly avowing his own Catholic faith, but similar sentiments were shared by the Protestant Jeanbon Saint-André, who asked:

> Is it after overturning the hierarchy of priests that you are going to create a new hierarchy of the learned?[12]

[10] This paragraph is based entirely on R. Hahn, *The anatomy of a scientific institution. The Paris Academy of Sciences, 1666–1803*, Berkeley, Cal., 1971, pp. 183–4.

[11] J. Guillaume, *Procès-verbaux du Comité d'Instruction publique de la Convention* (6 vols., 1891–1957), vol. 1., p. 126, quoted by R. R. Palmer, *The Improvement of Humanity*, Princeton, N.J., 1985, pp. 131–2.

[12] *Ibid.*, vol. i, p. 273. The speech was written in December 1792 in opposition to Condorcet's plan. It was not delivered but printed in 1793 in opposition to the Lakanal plan.

This, he said, would be

> a true corporation, *a new clergy*, armed by yourselves with dangerous influence, a group of masters, a guild.[13]

The idea of Academicians and professors as a new kind of priesthood was not restricted to those excluded from academic circles. In December 1793, a few months after the suppression of the Academies, Fourcroy, a former member of the Royal Academy of Sciences, spoke out against any 'corporation of *savants*'.[14] To create permanent state positions, he said, was to give men 'certificates of immortality'; it was to reinstitute sheltered positions rather like those of canons in a cathedral. He feared the establishment of a kind of priesthood. If, on the other hand, the Convention refused to establish privileged educational institutions, it would have 'won a further victory over fanaticism and doctoral superstition'.[15] The French republic, he suggested, should be a unity and it should not be split up into a republic of letters, a republic of science, and so on. He could not help seeing permanent educational appointments as necessarily involving intrigue and cabals, and this would happen even if there were open competition or elections.

It was easy for politicians to think of Academicians and professors as potential threats to the authority of 'the people', as represented by themselves. But we must also consider the viewpoint of the Academicians. Far from seeking power at this time, most simply wanted protection. Many had been living in hiding during the Terror and they looked on their membership of the Institute to give them protection against undue political pressures. A basic freedom in the Institute must be freedom of speech. It was actually a member of the Second Class, Naigeon, who made this point publicly at a meeting of the Institute on 9 *ventose* year 4 (28 February 1796). The minutes record that Naigeon made some observations,

> on the necessity for the progress of science and learning that all members of the Institute should enjoy unrestricted liberty to think, speak, write and publish on all matters with which they are concerned.[16]

Thus, although some politicians had said that intellectuals should have no further privileges beyond their talents,[17] here was a claim that members of the

[13] *Ibid.*, vol. i, p. 278 (my italics).

[14] J. Guillaume, *op. cit.*, vol. 3, pp. 97–102. Already in 1792, Fourcroy had spoken out in the Academy, asking it to expel members opposed to the Revolution (W. A. Smeaton, *Fourcroy, chemist and revolutionary*, Cambridge, 1962, pp. 46–7).

[15] 'Vous avez fait une nouvelle conquête sur le fanatisme et la superstition doctorale', Guillaume, *op. cit.*, vol. 3, pp. 100–1. A plausible interpretation is that Fourcroy had a historical appreciation of the history of European universities and their close links in the early days with organised religion.

[16] A. I., Registre 3A1, p. 59 (9 *ventose* year 4).

[17] Thus Jean Bon Saint André in notes intended to be presented to the Convention wrote: 'La république n'est pas obligé de faire des savants; de quel droit demanderait-il pour eux un privilège? Leur privilège c'est leur génie,' J. Guillaume, *op. cit.*, vol. 1, p. 275.

Institute required a special privilege of freedom of expression. In practice this affected literature and the emerging social sciences rather than natural science. In theory, however, all members of the Institute were to become part of a privileged elite. They claimed rights over and above those of ordinary citizens and these rights related to the area of knowledge, which had been officially recognised by the State. Natural science was one of the cornerstones of that knowledge.

After the first few years of the Institute the idea of Academicians as a new priesthood seems hardly appropriate. Some scientists became legislators and civil servants but many were simply technicians. But we need to consider not only the status of science but also that of traditional religion, which came increasingly under attack. In the second half of the nineteenth century there developed an explicit conflict between science and religion which was to end with the former claiming much of the ground of the latter, particularly the claims of truth and authority. Berthelot was to refer repeatedly in his speeches to the authority, not of scientific data or scientific theories, but of 'scientific *truths*'.[18] By implication, if not explicitly, leading scientists became the high priests of this new religion. They were listened to with the same respect previously accorded to leading theologians.

Meanwhile there had been important developments in the writing of history, which now claimed to be 'scientific'. German scholars turned these new sceptical techniques on the Gospels. In France a turning point was the publication in 1863 of the *Vie de Jésus* by Renan, a close friend of Berthelot. Renan combined erudition with great literary skill to banish completely the supernatural dimension of the traditional Christian story. Many basic Christian claims were challenged. But the 1860s was also the period when increasing claims were made for science. In 1866, a French translation of Darwin's *Origin of Species* appeared with a preface denying the Christian doctrine of creation.[19] Some people began to think of science as the modern religion, displacing Christianity.

This movement was given extra force in France by the spread of positivism. Auguste Comte (1798–1857) had argued that theology corresponded to a primitive interpretation of the world, which had already been challenged for several centuries by philosophers who represented a second stage of development, that of metaphysics. The final triumphant stage in the evolution of human thought was that of positivism, which drew on discoveries of science.[20] It was only taking earlier writers like Condorcet slightly further to claim that natural science represents a model for all knowledge.

There was a sense then in which science seemed to represent a new religion

[18] *Science et éducation*, 1901, e.g. p. 26 (my italics).

[19] *De l'origine des espèces*, trans. Clémence Royer, 1866.

[20] *Cours de philosophie positive*, 6 vols., 1830–42.

and its practitioners a new priesthood. Whereas in the early nineteenth century the French had turned again for illumination to the Catholic Church and its claims for authority, they now looked elsewhere. In particular they preferred the authority of science. This is the message in a famous speech of Sainte Beuve to the Senate in 1868:

> There is, gentlemen, another great diocese [outside the Church], with no fixed boundaries, extending over the whole of France, over the whole world... which continually increases its membership and its power... which embraces minds in various stages of emancipation, but all in agreement on one point – that above all else we must be freed from an absolute authority and a blind submission – which counts in their thousands deists and adherents of spiritualistic philosophies, disciples of natural religion, pantheists, positivists, realists, sceptics and seekers of every kind, the devotees of common sense and the followers of pure science.[21]

If one passes on to the Third Republic, one finds people like Berthelot trying to make a religion of science. We may also imagine a visitor, straying into a meeting of the Academy on a winter afternoon, who might be forgiven if he thought he was attending a religious service. In the dim light of flickering candles[22] he might see shadowy figures intoning an esoteric ritual. The statues around the room might be mistaken for those of the saints. In a way they were, since they were representations of outstanding *savants* of an earlier age, providing role models for later men of science and other intellectuals, and through them giving inspiration and guidance to the public at large.

3. *A gerontocracy*

To what extent was seniority important in the Academy? Was the Academy controlled by a group of old men? At first sight it might seem that the First Class of the Institute, like the revolutionary army, provided exceptional opportunities for young men. Certainly the new educational institutions of the Revolution provided opportunities for the generation of Biot (1774–1862) and Gay-Lussac (1778–1850), and Arago (1786–1853) was elected to the First Class in 1809 at the incredibly early age of 23.[23] Yet, although young men had an important part to play, particularly in the early years, we should not forget the preponderance of older men. The First Class in its early years had a majority of members who had already been members of the Academy of the old regime. Older *savants*, like Laplace and Berthollet, exercised great influence.

Moreover, it was inevitable that, in a system of election depending on dead men's shoes, seniority should be important. Each member belonged to a section, and on the death of one of his colleagues he moved up the hierarchy, which was

[21] Quoted by John McManners, *Church and state in France, 1870–1914*, 1972, pp. 16–18, on which much of the above summary is based.

[22] The Academy chamber continued to be lit by candles till 1875.

[23] Another scientist to be elected at an early age was Cuvier at 26.

headed by the *doyen* of the section, the senior member in terms of years of service. The *doyen* was called upon from time to time (for example, in election nominations) to represent the collective view of the section. In the election of 1822 for a secretary for the mathematical sciences the nominating committee decided not to recommend an explicit order of preference but to list the three candidates, Fourier, Biot and Arago, in order of age. Arago took the hint and withdrew (to return eight years later) and in the election Fourier, the senior candidate, had a clear majority.[24]

Unlike most positions in higher education the Academy made no provision for retirement, which could have constituted a stepping off point on the ladder of seniority. The Academy contained, therefore, not only many old men but a few exceptionally old men, who were treated with particular respect and accorded special privileges. Thus Flourens, speaking of the zoologist Duméril, who had died recently at the age of eighty-seven:

> The Academy, filled with respect for this patriarch of science, has devoted three volumes of its *Mémoires* to the publication of his *Ichthyologie analytique* and his *Entomologie*.[25]

Similarly Chevreul, who celebrated his hundredth birthday in 1886, was allowed in the last thirty years of his life to publish at length in the Academy's historic *Mémoires* and was also dispensed from the strict interpretation of the regulations limiting the number of pages allowed to members in the *Comptes rendus*. When finally an exasperated Academician asked how it was that Chevreul alone was able to have his papers published in their entirety by the Academy and on what condition the regulation could be ignored, Dumas as secretary is said to have replied that the condition was that one had reached the age of 100![26] Scientists like Dumas, Claude Bernard, Pasteur and Berthelot regarded Chevreul with increasing veneration.

Again we may take the case of Biot, a member for nearly sixty years, who, in his old age, was said to be the despair of the librarian of the Academy as he made a habit of removing its books to his home. After meetings of the Academy a group of members was often to be found in the library gathered round him,

> to hear anecdotes told by the venerable man of science, who had lived under all the political regimes and appeared as the last representative of a heroic age.[27]

Considering what a 'golden age' had been constituted by the closing years of the eighteenth century and the opening years of the nineteenth century, it is all the

[24] *P.V.I.*, 7, 386 (11 November 1822).

[25] *M.A.I.*, 35, (1866), xxii. Duméril's *Entomologie analytique* occupied the entire volume 31 of the *Mémoires*, consisting of 1339 4to pages.

[26] Albert B. Costa, *Michel Eugène Chevreul: Pioneer of organic chemistry*, Madison, Wisconsin, 1962, p. 18.

[27] E. Picard, *Eloges et discours académiques*, 1931, p. 285.

more understandable that by the mid-nineteenth century, when French science no longer clearly led the world, many Academicians, who had known the great men of science of the earlier age, should be treated almost with reverence.

The Academy also attributed special honour to the dimension of time to non-members. Candidates for election were expected to be defeated at their first candidature. When they applied for a later vacancy they had the advantage of a little seniority over any new contestant. Eventually they would hope to rise to the top of the list of candidates, drawn up in order of merit. Thus there was seniority among aspirants to the Academy as well as among its members. An able young man ambitious to enter the Academy was encouraged to bide his time, adding years as well as further research and publications to his credentials.

From a modern perspective it would be only too easy to deride the Academy as consisting uniquely of a group of old men whose active research was mainly in the past. Activity is not always easy to judge, unless it is simply measured by publications. But Academicians would probably have claimed that, although their independent original contributions may have declined, their judgemental status had in fact increased. One does not after all accuse a lawyer of laziness on account of his few clients after he has been elevated to the rank of judge. But this still leaves the question of age of election.

In 1871 it was calculated that the average age of election of members of the Academy had been forty-seven and the average age of death seventy-one, providing an average span of about twenty-four years as an Academician.[28] It is unfortunately true that, as competition to enter the Academy increased through the century, the average age of election steadily rose, from about forty around 1815 to nearly fifty in 1900. This average disguises the fact that in mathematics Cauchy could be nominated in 1816 at the age of twenty-six and Poincaré elected in 1887 at the age of thirty-two, whereas others had to wait till their sixties (or exceptionally seventies) for election. A candidate aged thirty-five, whose abilities untypically related to several different sections of the Academy, commented that, if he stood for a vacancy in the mathematical sciences, he would have to compete with 'a multitude of able young men' and he felt that he would stand a better chance among the natural sciences.[29]

If one takes the respective sections of mathematics and medicine as providing the two extremes, mathematics being a subject where intellectual maturity comes relatively early, if it comes at all, and medicine being a subject where the vast amount of knowledge necessary, together with appropriate experience, means that a peak of ability is not likely to come before early middle age, we find an interesting variation. In mathematics the average age of election in the first half of the century[30] was 40.0, rising to 43.9 in the second half of the century.

[28] A. Potiquet, *L'Institut national de France*, 1871, p. xv.

[29] Letter of Bravais to Libri, 11 February 1847, Bibliothèque Nationale, F.R. NA 3267 f.171.

[30] To start from 1801 rather than 1795 reduces (but does not eliminate entirely) the carry-over effect of membership from the old regime.

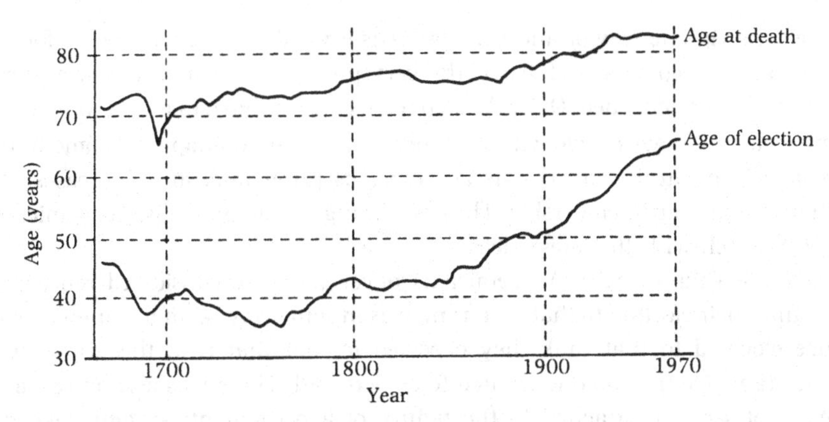

Fig. 2. The age of election of Academicians. (Adapted from A. Kastler, 'Evolution de l'age moyen des membres de l'Académie des Sciences depuis la foundation de l'Académie', *Comptes-rendus*, cclxxvi, (12 Feb., 1973), Vie académique, Part 1, p. 65.)

In medicine the average age was 54.3 in the period 1801–50 and 58.2 in the period 1851–1900. This shows that the slight rise in the average age of election during the nineteenth century is completely overshadowed by an average difference of fourteen years in the age of election between the sections.

Nevertheless, some general quantitative statements about the whole Academy may be of value if they are treated with caution. In 1815 the Academy had eight members under the age of forty but by 1850 this number had dropped to four.[31] When, by the second half of the *twentieth century*, the sixties became the *average age* of election, the problem was really serious (see Fig. 2). Fortunately this extreme situation, attributable to increasing human longevity in the twentieth century, was not reached in the period covered by this book. Also we would want to argue that the age of election was only one of several factors governing the dominant place of older men in the Academy.

The main basis of the claim that the Academy consisted largely of old men depended as much on longevity as the age of election. We have said that there was no retirement age and indeed the concept of retirement was quite foreign to the Academy. It had inherited from the *ancien régime* the idea of the old man as the wise man, a concept still accepted in several parts of the world and perhaps with greater empirical support in many areas of conduct than the recent western idea of the young person as the model and guide. The wisdom of the old is, nevertheless, a concept with limited value in the changing field of science, as opposed to a more traditional study, like law. It might, therefore, be regretted that, with great talent outside waiting for admission, the Academy did not impose a retiring age. A minority of Academicians would certainly have done well to accept retirement, had it been offered. But the majority of Academicians

[31] 'L'Académie des Sciences' (1867) in M. Berthelot, *Science et philosophie*, n.d., p. 213.

were active in one way or another. Whereas it would have been so easy for the Academy to become a backwater, the very prestige of belonging to it ensured that this never happened. Older Academicians did not have to go out looking for new ideas. Younger people with new ideas were only too happy to come to the prestigious institution to seek its approval, its prizes and, in a few cases, the ultimate honour of membership. Thus by serving on commissions, Academicians kept in touch with the latest ideas.

One good reason why Academicians never seriously considered retirement was that their election to that high rank was in some ways akin to ennoblement. Once elevated to that rank they expected to maintain it to the grave; it is understandable that no one wanted to be demoted. There was also an enviable feeling of security attached to the tenure of a permanent position. Like the judiciary, Academicians should not be subject to recall or early retirement on the grounds that their views were not acceptable to the government. An Academy consisting of distinguished figures who could not be dismissed on the grounds of unsound opinions or old age held a valuable position of independence. Thus any Academician who threatened to resign[32] was indirectly undermining one of the basic freedoms of the Academy.

Undoubtedly the high average age of members increased the conservative tendency of the Academy. In so far as membership was a reward for past work rather than a calculated risk of future promise, even the youngest new member had to have gone some considerable way along his career path, and most Academicians would show a natural tendency to look back rather than forward. There is, of course, much to be said for upholding traditions, but in so doing the Academy, by its very constitution, was prevented from being as open to future developments and changes as might have been desirable.

4. *A club or a family?*

Among the qualifications looked for among candidates for election to the Academy, collegiality was not one of the most important but it was a consideration. In contrast to election to, say, a Fellowship of a nineteenth-century Cambridge or Oxford college, the electors did not have to consider the prospect of living with the successful candidate. They had to share a meeting once a week, together with a few committees, and only an exceptionally disagreeable character would disqualify himself from this. On the other hand, membership of the Academy imposed a certain sociability on its members. Before and after every meeting there was the opportunity for social intercourse with one's peers. There would also be at least one or two occasions annually that

[32] A very rare example of a resignation involved the *Académicien libre*, Jaubert (1798–1874), elected in 1858. After the Franco-Prussian war he proposed a re-organisation of the Institute and, when this was rejected, he resigned in 1872, dying two years later. In at least one other case the threat of resignation was withdrawn after pressure from the Academy.

were primarily social. On committees eminent men in different scientific disciplines would be brought together and would hope to reach a consensus.

For most of the week many of these men were absolute rulers in their own domain, usually one of the institutions of higher education in Paris. Leading scientists, like Claude Bernard at the Collège de France or Louis Pasteur at the Ecole Normale, would have existed simply as unchallenged authorities within their respective institutions without some central institution like the Academy to give them a wider perspective. They might well have enjoyed their great power over their junior colleagues and their students, who would normally obey orders without question, but such power is ultimately bad for the scientific enterprise. In the Academy they mixed with men who claimed to be their peers, even if not all of them were of equal scientific eminence. Although outright criticism was the exception rather than the rule in the Academy meetings, the possibility of informal comment and criticism off the record is good for even the most brilliant scientific genius. This was the Academy acting not so much as a judge but rather as a club; a club which had full control over its own membership. If formality often stifled the possibility of much informal exchange of views, that is a pity. But in principle men who, after a painstaking process of selection, have reached a great height, tend to see in a special way not only 'outsiders' but also, and more constructively, their colleagues. Sharing privileges and a community of interest, they could think of themselves not only as being equal members of a college but even, perhaps, of a new family. The metaphor of the family was particularly appropriate in the nineteenth century, much of which corresponded to the Victorian period in British history.

There were several occasions on which the whole Institute presented itself as a large and presumably happy family. At the public meeting of 1831 the president claimed that all members, whether *savants*, writers or artists, 'form one single family'[33] and at a similar occasion two years later, the president spoke of the Institute as a 'large academic family'.[34] The justification for this metaphor was to bring together the members of the different Academies, who performed most of their work in separate meetings. To a certain extent it expressed an aspiration rather than a reality. Yet for each Academy considered separately it came very much nearer to the truth.

The Academy of Sciences, for example, was described by Emile Blanchard as a family in a speech of 1886, when he spoke of 'notre famille intellectuelle'.[35] The Academy made a show for important anniversaries, such as that celebrating fifty years of the life of J. B. Dumas as a member[36] or the one hundredth birthday of the phenomenal Chevreul.[37] In the case of a scandal, the Academy closed its ranks and, like a family, could usually rely on the loyalty of its members. Indeed

33 *Séance publique des 4 Académies*, 30 April 1831.

34 *Séance publique des 5 Académies*, 2 May 1833.

35 *C.R.*, **103** (1886), 443.

36 *C.R.*, **95** (1882), 1077–81.

37 *C.R.*, **103** (1886), 443.

one aspect of the Academy as a family was mentioned by Alphonse de Candolle as a criticism of French science; the desire not to offend colleagues, which greatly inhibited the expression of originality.[38]

The idea of the Academy of Sciences as an extended family is illustrated by the genuine concern for the welfare of its members when they were absent through illness. If a member was known to be seriously ill or to have had an accident, concern was expressed and often someone would be asked to visit the sick Academician and report back at the next meeting. Thus at the meeting of 6 June 1808 Biot reported that Gay-Lussac had had an accident; there had been an explosion in his laboratory and caustic potash had been splashed into his eyes. Biot was charged to express the concern of the First Class. At the next meeting Biot reported that his friend's eyes were better. On 27 June he reported that Gay-Lussac's health was improving, and finally on 11 July the absent scientist was able to attend the meeting and took the opportunity of thanking his colleagues for their expression of concern.[39] Frequently concern was expressed in the Academy for the welfare of the widows of former Academicians.

The Academy would also come together on the death of a colleague to accompany the body to the grave. Soon after the foundation of the Institute, a committee established to consider the funerals of its members stressed the deep obligation for members to be present at this solemn occasion affecting their 'brother', a family term chosen deliberately in preference to the more impersonal 'colleague'.[40] When Cuvier died in 1832 the Academy received a letter of condolence from the Royal Institution in London,[41] and when Berthelot died in 1907 individuals and institutions from all over the world sent their condolences to the Academy[42] as one might write for a family bereavement. On the death of a member, a large number of his colleagues from the Academy would make a point of attending the funeral and there would be a speech on behalf of the Academy as well as one from the principal institutions of higher education with which he had been associated. In 1828 the Academy asked the Minister of the Interior for the necessary authorisation to pay the funeral expenses of the penniless former member of the agriculture section, Bosc d'Antic.[43]

The death of an Academician had a significance rather less than that of a monarch, but far greater than that of a retired professor. The death of a king brought the reign to an end but, unlike an Academician, his successor would normally be known in advance. The holder of a prestigious chair in a university or *grande école* would normally step down at retirement age. His subsequent death would, therefore, be strictly irrelevant to his professional succession. The death of an Academician was always more important since, in theory,

[38] Alphonse de Candolle, *Histoire des Sciences*, 2nd edn, 1885, p. 434.

[39] *P.V.I.*, **4**, 74, 77, 79, 81.

[40] Report of Baudin at Séance générale of Institute, 5 *frimaire* an 7 (25 November 1798).

[41] *P.V.I.*, **10**, 92.

[42] 'Les adresses de condoléances affluèrent à l'Académie de tous les centres de la vie intellectuelle à la surface du globe', *C.R.*, **145** (1907), 965.

[43] *P.V.I.*, **9**, 90.

Academicians died while holding office. Thus they were nearly always in close touch with their colleagues until, perhaps, their final illness. They were part of a close-knit community or brotherhood, at least as much as their intermittent contact at the centre of a bustling metropolis allowed.

We have argued that Academicians felt themselves much more than formal colleagues; they were almost brothers,[44] at least if they belonged to the same generation. Older Academicians might be considered as father figures, which would be particularly apt in the case of a patron. Berthollet had written to the young Gay-Lussac at the beginning of his career: 'I wish to be your father in scientific matters'[45] and this metaphor is quite appropriate, considering the friendship, advice and patronage offered to the younger man. The whole teacher–pupil relationship may be considered as an extension of the family. This is recognised explicitly in the German language, where the supervisor of a successful postgraduate research student is sometimes still described as the 'Doktorvater' of the new 'Doktor', and where the relationship may continue throughout life as that of an academic godfather. In the electoral system of the Academy the favour conferred on a former student may be considered as an extension of nepotism.

5. *A bureaucracy*

An organisation, particularly a private one, may depend very much on the personality of its leader or patron. If the charisma fails, or if the leader is absent or his interests change, serious problems are likely to develop. Alternatively an organisation, particularly a public one, once established can be run as a machine, with precisely specified procedures and clerks recording scrupulously every detail and carefully filing the relevant documents. There was a good tradition in the old regime of such bureaucracy in various government departments. Clerks did not necessarily have to understand the documents they handled, only where they fitted into the machine. In the case of science, the understanding and evaluation of documents was the task of members of the First Class, notably the secretary, and of commissions appointed to examine specific issues. We shall later be considering the position of the secretary but at the moment what is relevant is that his power to act always lay within certain defined limits or constraints.

In the earliest days meetings of natural philosophers had developed in the most informal atmosphere and in several countries, even in the nineteenth century, such meetings would still be fairly informal. This would tend to be the case particularly with purely local societies, where the members were well known to each other. It may be instructive to consider the opposite extreme, that

[44] In 1773 the President of the Royal Society when conferring the Copley medal on Joseph Priestley, described him as 'our worthy brother', John Pringle, *Six Discourses*, 1783, p. 5.

[45] D. F. J. Arago, 'Eloge de Gay-Lussac', *Notices biographiques*, vol 3, 2nd edn, 1865, p. 7.

of rigid formality. Obviously individual personality counts for less but this does not mean that the individual could not make his mark. Also what might appear stifling to some foreign visitors could well have been regarded as normal by many French *savants*. Rather than being crushed by the machine, they would accept it and use it to make their own contribution. After all the whole justification of the bureaucracy was to help the cause of science. Also we should remember that formality would be largely confined to the official part of the meeting, held under public gaze. Before and after meetings Academicians would communicate with each other informally, but this is an aspect to be considered separately.

In order to provide a better idea what it could mean at its most extreme for science to be part of a large bureaucratic machine, we will reproduce a description of the Institute in the early years provided by Henry Redhead Yorke:

> There is an orderly register with five columns in front, with the following titles, viz:
>
> Dates of the receipt of papers.
> Numbers of their order.
> Designation of the classes.
> Titles of the Mémoires and Objects of Letters.
> Dates of the decisions of the Institute.
>
> By means of this arrangement, and an index, any paper which is wanted may be found in an instant.
>
> When the monthly bureau have opened the letters and packets addressed to the Institute, the secretary writes on every piece, the despatch to the class to which the object relates. He sends these pieces afterwards to be registered, in order that they may be placed by the *commis* in the hands of the secretary of the class, to which they have been transferred. The letters and parcels, addressed immediately to the classes, are opened at the commencement of the sitting. Immediately after they have been announced, and the nomination of the commissioners, if such be requisite, the secretary sends them to the *commis* to be enregistered, and the latter returns them immediately to whom they pertain, conformably to the marginal note which the secretary of the class had written. Besides the minute books of the three classes, and that of the general assemblies, there are four registers, one of which is for the labours of each class, and one for the general and public assemblies, in which the minutes with the reports and official letters are inserted.
>
> The *commis* remains in the secretary's office during all the sittings of the Institute. Before each sitting, he places on the bureau the papers relating to the class which have reached the secretary's office, and he collects afterwards those which are to remain there. At the three first sittings of the month, he places on the bureau of each class, the list of the reports, which remain to be made, and the names of the commissioners appointed to make them. It contains firstly, a list of the places which have become vacant, secondly, a list of the mémoires read by the members, and another of works presented, and which have been esteemed worthy to be published among those of the Institute; thirdly, a list of all the commissions respecting the

> internal administration, prizes, &c.; fourthly, a list of the papers intended to compose a volume; fifthly, a list of the mémoires received for prizes, and another of the names of those who have deposited at the Secretary's office, sealed packets containing discoveries, and which are not to be opened, except at their own request.[46]

Yorke was obviously struck by the efficiency of the system at a time when the First Class had to share most of its activities with the other classes. After 1803 the First Class had much greater independence, but the general point about organisation and the keeping of records would still apply. From the above account it is difficult to recognise that the subject matter of many of these communications was knowledge of the natural world. It might almost have been an office for the registration of trade and commerce or the collection of taxes.

We might refer to another aspect of the bureaucracy, the attendance register. Academicians were, of course, not only adults but senior figures in the scientific world. Yet from one point of view it might seem that they were treated rather like children with a meticulous record being kept of their attendance at meetings. The attendance record had minor financial implications in so far as a part of the Academicians' honorarium was temporarily withheld and then paid out in proportion to attendance. Although this might appear as yet another aspect of control, it is probably better to accept the attendance register as no more than a standard part of the bureaucracy. It is after all a recognised part of modern committee procedure to make a note of attendance.

6. *Miscellaneous roles*

The Academy could be considered in a number of other ways. It could, for example, be regarded simply as a hall of fame. Certainly membership did confer fame, but members are not to be considered as a lot of stuffed dummies. If it is legitimate, therefore, to regard the Academy as a hall of fame, this was no more than one of a dozen different roles which it filled. Again the Academy could be regarded as no more than a stage, a platform for the presentation of science. There was certainly something theatrical about the formal public meetings, '*sous la coupole*' (under the dome of the former church). Even ordinary meetings were sometimes enlivened with demonstrations, but to consider the Academy as no more than a show (the view of several casual visitors) would be too superficial an impression, even grossly misleading.

The Academy was not a theatre. Ordinary meetings were held for the benefit of Academicians with members of the public present almost on sufferance. Memoirs were often read into the record with a view to publication in the *Comptes rendus* rather than declaimed to an appreciative audience. It was only

[46] H. R. Yorke, *Letters from France*, 1814, pp. 52–4. This account is a free translation of the regulations of 1796, reproduced in Aucoc, *L'Institut de France*, 1889, pp. 40–3.

at the annual public meeting that eloquence was appreciated. At ordinary meetings no attempt was made to popularise science or to play to the gallery. If the public found some of the proceedings baffling or partly inaudible – *tant pis*! Academicians could not be expected at one and the same time to fulfil their professional obligations to their colleagues and to act as entertainers or educators.

We may put aside another possible role of the Academy, that of providing formal education. The Academy was *not* (despite one possible literal interpretation of its name) a school. Even under the *ancien régime* its educational function was implied rather than stated in its three-tiered membership. Junior members had often been elected with only a superficial knowledge of science which, however, improved by association with senior members. In other words a kind of apprenticeship system was concealed within the constitution of the Royal Academy. It was only for the more senior grades of membership that actual qualifications were asked for, for example, authorship of a book. The government which established the Institute in 1795, however, also established a separate system of higher education in science. It was only after benefiting from this education that *savants* would in future be in a position to aspire to membership of the Academy. The political motives which insisted on equality between members, therefore, had the additional effect of divesting the Academy of its previous semi-educational role.

It would be much more appropriate to consider the Academy as filling the role of a court of law. The judicial function of the Academy is important but it will only briefly be mentioned here because the idea will be further developed in a later chapter. Then we might consider the Academy as an informal meeting place which in some ways developed almost like a royal court under some bygone monarchy. Finally, it might be useful to consider whether the Academy had anything in common with a political assembly.

In some ways the Academy acted almost as a court of law, pronouncing judgement after examining evidence, accepting some work as valid contributions to science and rejecting others, also incidentally bestowing different degrees of praise and honour on the work of lower-ranking scientists. In so far as the Academy acted as a court, it was a higher court, since once it had pronounced on a subject, there was no higher court of appeal to which dissatisfied parties could turn. Although the original Academy of 1666 had not been established with a judicial function in mind, it has been shown[47] how the Academy, jealous of its good name, had soon begun to sit in judgement on the work of members who wished to publish their own research as a part of the work of the Academy. It had also acted as a censor for scientific books, fitting in with the whole system of the censorship of books under the *ancien régime*. Finally it had been asked to pass judgement on the inventions of artisans, a duty it accepted, although with

[47] R. Hahn, *op. cit.*

less enthusiasm than the other roles mentioned. There is no doubt, therefore, that in different ways the Academy of the *ancien régime* had accepted the role of the judge. But after the Revolution, when censorship was abolished and a patent system for invention superseded the previous system of protection, much of the wider judicial role of the Academy became redundant.

Yet there were important new ways, that had emerged in the eighteenth century, in which the Academy was called upon to pass judgement, and these continued and indeed blossomed in the nineteenth century. The first was the judgement of the work of scientists who were not members of the Academy. This was work submitted voluntarily by scientists who sought a seal of approval. The Academy's official report was considered of great importance and numerous reports were drawn up in the first half of the nineteenth century. But as the system of prizes began to expand, the custom of drawing up lengthy reports fell into decline. There was probably a causal relationship between the two. As the Academy began to award more and more prizes in the second half of the century, each demanding considerable time for judgement, there was less time available for drawing up reports on work not submitted for prizes. Thus the judicial function of the Academy continued to be important but in a different way.

Although the Academy accepted its role as a judge, it might have seemed to some outsiders to be much more than this. To someone without much scientific training and whose work was criticised, it might well have seemed that the Academy constituted not only a judge and jury but a prosecuting counsel as well. There would be no counsel for the defence. On the other hand, someone with scientific training who had been criticised by an Academy commission would have done well to take notice of the criticisms in doing later work. Judgement in this way had an educational value.

The ante-room to the Academy had a number of important functions which should not be overlooked. It was an excellent rendezvous for anyone who wanted to waylay and have a few words with an Academician, either before or after the meeting. The Academicians themselves used it for informal contact and so did a number of more junior scientists, hoping to solicit some favour. With its comfortable chairs and side-tables providing almost the atmosphere of a *salon*, the *salle des pas perdus* positively invited intrigue and the many private conversations there would be supplemented by others involving endless pacing of patrons and protégés together, which gave the room its distinctive name. With a virtual guarantee that nearly all leading figures in French science would be in a particular building in Paris every Monday afternoon, this was too good an opportunity to pass over for an ambitious young man. In modern times it is not unknown for junior scientists to attend scientific conferences in foreign countries so that they can become known to their seniors.[48] The Academy of

[48] John Ziman, *Public Knowledge*, Cambridge, 1968, pp. 131–2.

Sciences presented a better opportunity since it was a permanent fixture at the very heart of the capital city and there was little danger that the 'big men' would be absent.

Complementing the formal communications in the meeting, there would also be general gossip between members before and after the meeting. Academicians could hardly expect a full report on the research of their colleagues since such reports belonged to the formal meeting, but they might be favoured by a private progress report or an indication that a particular line of research was going well or badly. They would regularly hear about the fortunes of colleagues and their health. In an extreme case, news of failing health might indicate a future vacancy, something not to be ignored in a system which depended so much on dead men's shoes. Since scientific positions were not advertised, one was well advised always to look out for opportunities for employment for oneself or protégés. The ante-room provided an admirable location for personal solicitation and recommendation from a group of men of exceptional power and influence in French academic life. Sometimes business could be done with little more than a nod or a wink. The informality and use of oral rather than written communication leaves this area of the Academy's influence very poorly documented. At the most the historian finds occasional reference in the correspondence of scientists about informal contacts.

There can be little doubt, however, that the weekly Academy meetings provided an excellent forum for deals and for the exercise of patronage as well as more innocent chit-chat. It also provided an excellent opportunity to carry out business such as consultation in the running of a scientific journal. In the Napoleonic period, when the *Annales de chimie* had a large editorial board, many of whom were members of the First Class, it made good sense to use the Institute as a regular meeting place to discuss policy and individual papers. In December 1807 the editorial board of the *Annales* formally agreed to meet on the first and third Monday of every month at 2 p.m. at the Institute.[49] Thus, by coming slightly earlier than the time of the meeting at the First Class, they had a regular opportunity to discuss their journal. No doubt other scientific journals also made good use of the expert advice readily available at the Monday meetings.

Associated with the informal side of the Academy was the exercise of patronage. Would it be too fanciful to draw a parallel between it and a court of the old regime, a court like Versailles but on a much more modest scale? Of course, there was no King. In the informal gathering before and after meetings it would be the two permanent secretaries, and perhaps one or two powerful patrons, who would be the focus of different groups. The greatest distinction of course was that between Academicians on the one hand and, on the other, those seeking favour from Academicians. This group of hangers on, as at any court,

[49] The present author is preparing a general history of the *Annales de chimie*, which was founded in 1789 and soon became a key journal not only of chemistry but (from 1816) of physics also.

came to see and be seen. Occasionally they came with the formal purpose of presenting a memoir but, if they lived in Paris, they would come more often simply to hear the latest scientific news.

Some would keep their ears and eyes open for prospects of employment or preferment. There might, for example, be an Academician with a post at one of the establishments of higher education in Paris who was looking for a *suppléant* to take over on a temporary basis some of his teaching. A modest salary would be negotiated, although the young man appointed might be less concerned with the salary than the prospects. In any permanent vacancy which might arise, the experience of having acted as a *suppléant* could be a valuable recommendation. It constituted a foot in the door and doors did not often open to those who stayed at home and hoped that fortune would somehow smile on them. Many spectators came in search of patronage or, if they were already known to an Academician, they might seek to develop the acquaintance, to speak informally of their work and to ask for advice. Was there any likelihood that that year could be one of the many occasions when the Academy had a surplus from the Montyon legacy which might, for example, be available as a research grant? It paid a young man to know what was going on. To those more advanced in their careers there might be the possibility of a prestigious prize. It helped to know in advance what ideas were in favour. To the most ambitious there remained the greatest prize of all, membership of the Academy. This required many years of careful planning as well as luck. It paid not only to invest time in research and publication but also in lobbying. Could one make a favourable impression to prepare the way for the next vacancy? There were many who played the role of courtier.

Yet this was no longer the *ancien régime*. Patronage might help but it had to be supported by solid evidence of ability, usually demonstrated by a good record of publication. But in claiming that this was the beginning of the modern world as far as scientific careers were concerned, one cannot entirely neglect the social milieu. Given a choice between two men of equal scientific ability, the courtier would probably win.

Finally, the Academy may be compared in some respects to a political assembly, although this comparison would have to be with a second chamber or an upper house,[50] such as the reformed[51] House of Lords in the British parliament, rather than the House of Commons, which is subject to regular dissolution and re-election. There was never any question of the Academy being a representative college of French professional scientists, which would have required a large and frequently changing assembly. Rather on the model of the

[50] Certainly the Academy of Medicine was considered by political analogy as an upper house. F. Schiller, *Paul Broca*, 1979, p. 62.

[51] In Britain senior politicians and others who have rendered public services are now regularly nominated to the House of Lords, where they are members for life, thus supplementing the traditional hereditary basis of membership, which has no acceptable parallel in science.

former Royal Academy, it was intended to be a permanent working body of manageable size.[52] Academicians were elected on the basis of their scientific or technical expertise and they were elected for life. Consisting as it did of senior representatives of all the major established branches of science, the Academy was regularly called upon to legislate, or rather advise, on matters scientific in at least two distinct ways. First, it could investigate the authenticity of scientific claims, and secondly, it could advise the government on specific applications of science, a question dealt with in Chapter 9.

There was some discussion at meetings but hardly in the same way as a political assembly. The claims made in scientific memoirs are seldom as controversial as the claims of politicians. They are often presented simply as matters of fact for the record and sometimes the most heated discussion would relate not so much to content as to claims of priority. Discussion was understandably most free when the Academy met in secret session. Such meetings were not usually concerned with the content of science but rather with the more confidential aspects of the administration of science and particularly elections. It was probably in elections that the Academy came closest to a political assembly. One cannot ignore the lobbying which often preceded an election. Moreover, an election was sometimes less for an individual as for a general principle, even a party. Thus the failure of Charles Darwin to be elected as a correspondent, despite several candidacies before 1878, can be interpreted as a rejection of his evolutionary ideas.[53] On the other hand, there were elections in which personalities sometimes appeared more important than theories or policies. Certainly the public often liked to think in terms of personalities and did not draw a sharp distinction between academic elections and political elections.

7. *Discussion*

The outline given above of various roles filled by the Academy should have helped the reader to understand more clearly what the organisation stood for, and to begin to understand how it functioned. We will conclude this chapter by a consideration of two issues of importance to an understanding of the functioning of the Academy. One is not mentioned at all in the official regulations; it is the question of discussion and debate within the Academy. The second, that of the appointment of secretaries, is certainly specified clearly in the

[52] George Weisz has made a valuable study of the French Academy of Medicine, founded in 1820, which, with a membership of some 200, was extremely cumbersome. Drastic steps were taken to reduce membership by one half. See his paper 'Constructing the medical elite in France: The creation of the Royal Academy of Medicine, 1814–20', *Medical History*, 30 (1986), 419–43 (pp. 439, 442).

[53] Between 1870 and 1878 Darwin's name appeared six times as a candidate as corresponding member of the zoology section. P. Corsi, 'Recent studies of French reactions to Darwin', in David Kohn (ed.) *The Darwinian heritage*, Princeton, N.J., 1985, pp. 698–711 (p. 699). See also Harry Paul, *From knowledge to power. The rise of the science empire in France, 1860–1937*, Cambridge, 1985, pp. 70, 126.

regulations, but that leaves open the whole question how this worked out in practice. It is by studying the actual working of the Academy rather than simply reading the regulations that one appreciates the great power possessed by the two permanent secretaries. But first, we will consider the question of discussion.

Science does not lend itself to discussion and debate as easily as a subject like politics, in which there is much less hard evidence and where opinions and feelings play a correspondingly greater role. Hence, although there might occasionally have been some ideological issue related to science which could be debated, or a conflict between two alternative theories, the ordinary run of the mill science was simply presented to the Academy without much discussion. In a few cases, where special claims were being made, the Academy was called upon to verify the claims. But even then confirmation could hardly be obtained in the course of a meeting. It required the appointment of a commission, which would work behind the scenes, perhaps repeating the experiment in question, before the commission reported back to the full Academy.

Given the nature of science and the respect in which scientific knowledge was held, the organisation of the Academy was geared to establishing a consensus and it deliberately avoided encouraging an adversarial stance, which might have resulted in a division of opinion. Unlike a political assembly, it was not constructed as a forum for debate. Whereas a political chamber can usefully give rise to the expression of a whole range of political opinions, a scientific society is expected to arrive at a consensus. Its authority is undermined if the leading members of the society cannot agree among themselves on the conclusions to be drawn from the evidence presented. The Academy preferred to deal with facts rather than opinions. In practice this meant concentrating on experimental evidence and avoiding making pronouncements on matters of theory.

On the other hand, there could be no justification for the practice of receiving all memoirs in silence. In the early nineteenth-century Royal Society it was considered a breach of manners to comment publicly on a memoir, since it was held that such discussion 'led to the loss of personal dignity',[54] but this was in a society when social values were often rated more highly than intellectual ones. In the Academy, on the other hand, it was quite in order to make some observations on a memoir that had just been read. The format of a formal meeting was not conducive to a detailed question and answer session. It did, however, allow for some intervention, such as the expression of mild criticism or even occasionally the claim that the work reported had been done previously by someone else. If there was any discussion, it was always strictly between Academicians and could never involve members of the public.

There were many occasions when a memoir would have benefited by criticism from a small group of colleagues in an informal setting rather than in front of the whole Academy. Representing as it did a diversity of specialist interests, the

[54] Marie Boas Hall, *All scientists now. The Royal Society in the nineteenth century*, 1984, p. 69. Only in the later nineteenth century was discussion encouraged.

Academy was too broadly based for really constructive discussion of many specialist issues relating to only one branch of science. It was also too public, and any presentation was too definitive for someone unsure of his ground. There was, of course, nothing to prevent scientists sounding out their more knowledgeable friends before presenting a memoir to the Academy, and there were a number of junior scientific societies where young men could acquire experience in the presentation of memoirs. One of these was the Société Philomatique, which had similar scientific interests to those of the Academy, but with a more junior membership. In order to maintain the relatively youthful character of the society a regulation was later introduced that no one could remain an ordinary member for more than ten to fifteen years.[55] After this period members were usually promoted to the status of honorary members. Of the twenty-three honorary members listed in 1825, only two never became members of the Academy. One can understand why the Société Philomatique was sometimes described as the antechamber to the Academy.[56]

For medical students there was the Société médicale d'émulation de Paris, a society found useful by Magendie[57] before he was ready for consideration for election to the Academy, just as Fresnel had found the Société Philomatique an invaluable proving ground when he first came to Paris.[58] There was also the small and select Society of Arcueil, which thrived during the Napoleonic period.[59] Berthollet and Laplace invited a group of protégés to their country house just outside Paris, usually at the week-end. It was the practice to discuss memoirs, which were to be presented on the following Monday at the First Class. Although one can imagine the benefits to be derived from the lively discussion of scientific ideas on a summer afternoon in a country setting, there is also evidence that the younger members were constrained by the imposition of a rigid orthodoxy, and of being subject to a greater pressure than was ever exerted in the Academy.[60] However organised, such satellite societies could not replace the official body of science, although they could most usefully complement it.

Whereas the Academy view might have been that too much outspoken criticism tended to undermine the dignity of the institution, one should also consider the public reaction. For the general public much of science is frankly boring, but the potential boredom would be relieved if, instead of having reports of a factual nature, a question was introduced which was open to different interpretations. The audience might be tempted to take sides, and this feeling of

[55] 'Règlement de la Société Philomatique', *Annuaire des sociétés savantes de la France et de l'étranger*, 1846, pp. 340–4.

[56] J. R. Mandelbaum, *La Société Philomatique de Paris de 1788 à 1835*, Ph.D., 1980, Ecole des Hautes Etudes en Sciences Sociales, University Microfilms 8270086, pp. 71, 76, 130.

[57] J. M. D. Olmsted, *François Magendie*, New York, 1944, p. 20.

[58] Fresnel, *Oeuvres*, 1866, vol. 2, p. 846.

[59] M. Crosland, *The Society of Arcueil. A view of French science at the time of Napoleon 1*, 1967.

[60] A good example is Berthollet's refusal to allow Gay-Lussac to consider chlorine as an element. M. Crosland, *Gay-Lussac, scientist and bourgeois*, 1978, pp. 48–9. Laplace was even more influential in imposing a neo-Newtonian orthodoxy on physicists.

participation could apply almost as much to readers of reports in newspapers of meetings as to the few who had actually been present. One can understand why journalists would have liked there to be more discussion.

The amount of discussion varied, but it seldom reached the level of a formal debate. Believing in the importance of consensus, the Academy had no wish to turn itself into a debating society. Yet occasionally powerful feelings came to the surface and a vigorous discussion resulted. One such case was the meeting of 1 February 1847, which was occupied almost entirely by a defence of the new technique of ether anaesthesia by Velpeau and a sustained criticism by Magendie.[61] Another example of an important debate was that which took place in May/June 1877 between Wurtz and Berthelot on the subject of atomism. As a positivist, Berthelot refused to go beyond chemical equivalents and he attacked the acceptance of atoms and the use of atomic weights as sheer speculation.[62] Most commonly discussion arose out of the presentation of a memoir, when there might be some criticism at the end, perhaps followed by more sustained criticisms in a further memoir. Alternatively, an Academy report could be used to defend a new approach or attack an established one. Both of these approaches were used in the famous debate between Etienne Geoffroy Saint-Hilaire and Cuvier in 1830.

The dominance of Cuvier in the Academy helped to ensure that the doctrine of fixity of species was generally accepted in the early nineteenth century. It is true that Lamarck had publicly advocated a theory of transformism[63] but his standing in Parisian scientific circles was low, and in 1829 he died. At the Academy meeting of 8 February 1830, the Academy considered nominating a successor for his chair at the Muséum and Geoffroy Saint-Hilaire, as the doyen of the zoology section, made some suggestions. Geoffroy had long been a supporter of transformism, and although he had presented his theory of the unity of living things to an audience at the Muséum,[64] he had lacked an opportunity of presenting this theory to the Academy. At the same meeting of 8 February 1830 a memoir submitted by two young zoologists gave him the opportunity.[65] Geoffroy was appointed together with Latreille to examine the memoir by Meyraux and Laurencet, entitled 'Quelques considérations sur l'organisation des mollusques'. Geoffroy alone took on responsibility for the report and had it ready by the following week.[66] What excited him about the

[61] *C.R.*, **24** (1847), 129–52. Both Velpeau and Magendie each made two long speeches. Roux and Lallemand also contributed to the debate.

[62] *C.R.*, **84** (1877), 1189–95, 1264–76.

[63] *Philosophie zoologique* (1809), trans. as *Zoological philosophy*, New York, 1963.

[64] *Sur le principe de l'unité de composition organique, discours servant d'introduction aux leçons professées au Jardin du roi par M. Geoffroy Saint-Hilaire*, 1828.

[65] *P.V.I.*, **9**, 402–6, See also Toby Anita Appel, 'The Cuvier–Geoffroy debate and the structure of nineteenth century French zoology. Ph.D. thesis, 1974, Princeton University, University Microfilms 76–20–788, pp. 278–96. Also Jean Piveteau, 'Le débat entre Cuvier et Geoffroy Saint-Hilaire sur l'unité de plan et de composition', *Revue d'Histoire des Sciences*, **3** (1950), 343–63.

[66] *P.V.I.*, **9**, 402–6.

memoir was that it prepared a way to bridge the gap between two separate classes in Cuvier's classification: vertebrates and molluscs. Going much further in his report than the authors themselves, Geoffroy claimed the work as evidence to support his own theories and 'the principle of connections'.

In so far as Geoffroy commended the observations described, the Academy could have simply accepted the report. But since Geoffroy had dared to bring out into the open theoretical issues of the unity of organic composition, Cuvier could not let this pass. Rather than make an immediate objection, however, Cuvier decided to give his side of the argument in a memoir which was read at the meeting held the following week, 22 February. He denied that molluscs and vertebrates were built on the same plan and he attacked the use of such vague expressions as 'unity of composition'. He then went on to attack the original memoir of Meyraux and Laurencet, taking advantage of the fact that he had a much better knowledge of molluscs than Geoffroy. In showing his evident superiority in this field he further offended Geoffroy, who made a brief reply but promised a further memoir for the following meeting. On 1 March, Geoffroy accordingly read a memoir 'Des caractères de la doctrine de l'unité d'organisation, appelée théorie des analogues'. The original memoir was now forgotten as Geoffroy attempted to resolve a 'vital question in philosophy'. He tried at length, but with great difficulty, to give a precise definition of his concept of organic unity.

Over the next two weeks the Academy was allowed to turn its attention to other matters. Geoffroy wrote to say that he was ill and could not read the memoir he had prepared.[67] Cuvier, who had intended to reply to Geoffroy at the meeting of 8 March, announced that he would keep it until Geoffroy was present. The following week Geoffroy was back, but Cuvier was absent. It was, therefore, at the meeting of 22 March that Cuvier read his prepared reply, to be followed immediately by a memoir from Geoffroy. It was now clear that not only were they arguing from different premises, but they were arguing past each other. Yet on 29 March the Academy was embarrassed by a public argument between them on who should be allowed to speak first.[68] Geoffroy was allowed to present his memoir, nominally on the hyoid, but really about the philosophy of biology. Cuvier contented himself with a few remarks at the end of the paper; remarks which he continued, at greater length, at the meeting of 5 April.[69] Cuvier certainly brought considerable detailed zoological evidence to support his ideas, whereas Geoffroy relied too much on his intuition and even defended intuition as a means of arriving at the truth.

The debate having enjoyed considerable press coverage, the public benches were filled to overflowing. The large and noisy audience helped to turn meetings into theatrical events rather than occasions for the discussion of serious issues. Geoffroy himself, who had earlier suggested that the issue would be decided by

[67] *P.V.I.*, 9, 414.
[68] *Ibid.*, 426.
[69] *Ibid.*, 427.

the public,[70] now deplored the effect the public were having on the debate.[71] He defended himself against charges of irreligion and generally regretted the outcome of the debate, which had gained him little and had lost him the friendship of Cuvier. He concluded that the best way to advance an argument was by publication, and he distributed to members of the Academy a pamphlet 'On the necessity of printed writings to replace by this mode of publication verbal communications in controversial questions'. Meanwhile he hurriedly wrote a book entitled: *Principes de philosophie zoologique, discutés en mars 1830 au sein de l'Académie royale des Sciences*, which was published by the end of May in a limited edition.

Geoffroy had learned that the Academy was not a suitable forum for philosophical debate, although it might have been an appropriate body to dispute matters of fact. Perhaps he himself had been guilty of an abuse of the report system of the Academy. Probably a more constructive debate could have been carried out in a smaller and less public body. Considering that this is probably the most famous of all the nineteenth-century Academy debates, being witnessed by, among others, the German poet Goethe, it confirms our impressions about the unsuitability of the Academy as a forum for debate. Transformism was a subject on which evidence from so many different areas could be introduced, making the resolution of the basic issue particularly difficult. In the end the protagonists were probably more influenced by feelings than by the evidence produced.

It has been noted that Geoffroy's behaviour from 1830 on tended to alienate him more and more from the other members of the Academy.[72] He made constant demands on the Academy's time, yet he could not endure even the slightest comment or criticism. Cuvier made a point of avoiding confrontation, preferring to accumulate quietly evidence against transformism. But if Geoffroy had misused the report system, Cuvier, as secretary, was even more guilty of misusing the *éloge* system. It is customary to speak well of the dead, but Cuvier decided to use his *éloge* of the recently deceased Lamarck to launch a scathing final attack on transformism and, by implication, not only on Lamarck but on Geoffroy also. He spoke, for example, of some biological scientists,

> Believing themselves able to dispense with experiment and calculation, they have laboriously constructed vast edifices on imaginary bases, similar to those enchanted palaces of our old novels ('*romans*') that one can cause to vanish by breaking the talisman on which their experience depends.[73]

[70] 'C'est pour l'inventeur...s'adresser à la sagacité et aux lumières du véritable juge en toutes choses, le public', *ibid.*, 403 (15 February).

[71] Cuvier as secretary might have been even more concerned about the reputation of the Academy as well as his own reputation.

[72] Appel, *op. cit.*, p. 309.

[73] Cuvier, *Recueil des éloges historiques*, nouvelle édition (3 vols., Paris, 1861), vol. 3, pp. 179–210 (p. 180). The *éloge* was due to be read at the meeting of June 1831 but Cuvier's colleagues tried to persuade him to modify the wording, which he refused to do. The unaltered *éloge* was actually read by Sylvestre on 26 November 1832 after Cuvier's sudden death earlier that year.

However one may interpret the Cuvier–Geoffroy debate, it is clear that it was a contest which was, in principle, equally balanced. The two contestants were both senior scientists. But if we turn briefly to the equally famous debate of the 1860s between Pasteur and Pouchet, what has struck people most forcibly is that the contestants were *not* equally matched. Pasteur, a member of the Academy since 1862, is considered a member of the establishment, whereas Pouchet was no more than a corresponding member from the provinces. In a very influential paper published in 1974, a conspiracy theory was therefore advanced in order to explain why Pouchet lost the debate.[74]

If we are to consider factors external to science, there is one great point of similarity between the debates of 1830 and the 1860s. It seems that Cuvier supported the fixity of species not only because of a mass of evidence in its favour, but because he believed that the alternative theory of transformism was politically subversive. Given the political unrest of 1830, which by July had culminated in the overthrow of the King, Cuvier felt particularly strongly that it was his duty to oppose any ideas which might weaken the social fabric. In the 1860s revolution did not seem imminent. Nevertheless, the Second Empire supported the Catholic church and any atheistic ideas would have been considered as politically subversive. But the theory supported by Pouchet, that of spontaneous generation, had clear atheistic implications since, if true, it would mean that life could be created without coming from other life or ultimately from God. It is arguable, however, that although some intellectuals might have thought in these terms, most of the scientists in the Academy concerned themselves exclusively with the scientific dimension and concentrated entirely on *experimental* evidence.[75]

But if there is one interesting similarity between the debates of Cuvier and Geoffroy on the one hand and Pasteur and Pouchet on the other, there are also important differences. In the first case the 'debate' consisted largely of memoirs read by senior Academicians in front of the whole Academy. In the second case the 'debate' consisted of the examination of experimental evidence by a commission appointed by the Academy. In view of what we have said about the difficulty of useful discussion in a large body meeting in public, it might seem that the weighing of evidence by a commission was an ideal solution. Yet it raised the question of impartiality and, whatever evidence might be produced by the historian, some people would always tend to believe that when a body of professionals is asked to adjudicate between one of its members and someone outside, they are most likely to favour their own colleague.

[74] J. Farley and G. L. Geison, 'Science, politics and spontaneous generation in nineteenth-century France: the Pasteur–Pouchet debate', *Bulletin of the History of Medicine*, **48** (1974), 161–98.

[75] This still leaves open the question as we shall see shortly, of *which* experimental evidence is to be counted, as there was some genuine experimental evidence favouring Pouchet, depending on the fact that the hay infusion he used probably contained heat-resistant bacilli endospores.

Already in awarding the Jecker prize for 1861 to Pasteur, the Academy commission had praised his entry for 'the precision and clarity, which characterise the work of M. Pasteur'.[76] By the time the commission was appointed to adjudicate between Pasteur and Pouchet, the former had been elected to the Academy, and the new commission again looked favourably on Pasteur's experiments. However, Pouchet was not interested in repeating Pasteur's experiments but rather in extending the scope of the enquiry which, the commission said, could only result in 'vague and badly determined data, a new source of doubt and discussion',[77] although they did agree to a postponement till the summer, when Pouchet considered conditions were more favourable to spontaneous generation. The members of the commission: Flourens, Milne-Edwards, Claude Bernard and Brongniart have been characterised as establishment figures[78] and it certainly could not be claimed that they were completely impartial. Pouchet himself felt driven into a corner, his critics having become his judges.[79] It certainly would have been difficult by 1865 to find leading members of the Academy who favoured spontaneous generation. It has recently been argued that the claim that the Academy report depended more on political than scientific considerations can hardly be substantiated.[80] The judges were very much guided by experimental evidence, but the choice of *which* experiments to consider (Pasteur's yeast or Pouchet's hay infusions) turns out with hindsight to have been crucial.

8. *The role of the permanent secretaries*

A chapter intended to help explain the working of the Academy cannot be concluded without saying something about the crucial role of the secretaries. It would be easy to make the mistake of assuming that power in the Academy lay in the hands of the president. But the presidency was a purely honorific position, held for only one year. The secretaries from 1803 were *permanent* secretaries; that is, they were elected for life and they were each given a salary of 6000 francs, several times the honorarium of an ordinary Academician, in compensation for their onerous duties. As a commentator remarked:

> The president is only a man of straw, whilst the secretaries direct everything and are omnipotent; they are the true masters of the house.[81]

In the eighteenth-century Academy permanent secretaries such as Fontenelle

[76] *Oeuvres*, vol. 2, p. 634.

[77] 'Rapport sur les expériences relatives à la génération spontanée', *ibid.*, 637–47 (p. 639).

[78] John Farley and Gerald Geison, *op. cit.*, p. 181.

[79] 'nos adversaires devenus nos juges', *C.R.*, **58** (1864), 191.

[80] See N. Roll-Hansen, 'Experimental method and spontaneous generation: the controversy between Pasteur and Pouchet, 1859–64', *Journal of the History of Medicine*, **34** (1979), 273–92. Also see Antonio Galvez, 'The role of the French Academy of Sciences in the clarification of the issue of spontaneous generation in the mid-nineteenth century', *Annals of Science*, **45** (1988), 345–65.

[81] J. Marcou, *De la science en France*, 1869, fasc. 2, p. 141.

had built up such a powerful reputation that in the newly formed Institute it was decided to abolish the post of permanent secretary and replace it by annual appointments. The inefficiency of the system led to a reversal of policy and the new constitution of 1803 called for two permanent secretaries, one for the mathematical sciences and the other for the natural sciences.

In fact the practice of changing the secretary annually had lapsed several years before this. Delambre was elected secretary for the second half of the republican year 8 (1800) and continued through the year 9 and served the first half of year 10, thus bringing his term of office up to 1802. Cuvier too had served as secretary for the non-mathematical sciences in the years 8 and 9. The election held on 31 January 1803 simply confirmed these two in office.

The power of any one Academician was limited; it might, for example, be largely confined to the section of which he was a senior member. The Academy was no man's fief unlike, say, the private Society of Arcueil, which was firmly under the control of Berthollet in matters chemical and of Laplace in physics. In the Academy not even a secretary ever had quite the power that Sir Joseph Banks (1743–1820) held as President of the Royal Society for forty years. Yet perhaps some secretaries occasionally came close to this: secretaries such as Cuvier, who held the post for thirty years, and the very active Arago, who held his position throughout the 1830s and 1840s (see Table 7). After Arago's death in 1853 there was no single dominant figure in the Academy for a few years, although Dumas was becoming increasingly important in scientific politics. His election as secretary (natural sciences) in 1868 merely confirmed his dominant position. After Dumas' death in 1884 five years were to elapse before Berthelot emerged as his successor and the centre of scientific authority in France for the remainder of the century. In theory each of the permanent secretaries mentioned was balanced by a second secretary responsible for the other half of the Academy,[82] but in practice the dominant personalities of such men as Cuvier, Arago, Dumas and Berthelot meant that each in turn was effectively King.

The two secretaries provided a link between the scientific input of the Academy and its bureaucratic machinery. They would organise committees of the Academy, going so far as to propose names of members even if theoretically these were elected by the whole assembly. Although most of the week the secretaries would be responsible for organisation behind the scenes, at the weekly Monday meetings they would be centre stage. They would organise the proceedings, taking it in turns to summarise the correspondence, giving emphasis to matters they considered important and passing quickly over matters considered minor. This affected particularly non-Academicians. Any Aca-

[82] This avoided the situation in some other academies, such as the Académie Française, where a single secretary reigned supreme. Thus Louis Simon Auger (1772–1829) helped set the Academy against Romanticism; he committed suicide a few months before the election of Lamartine in 1829.

Table 7. *Permanent secretaries of the Academy*

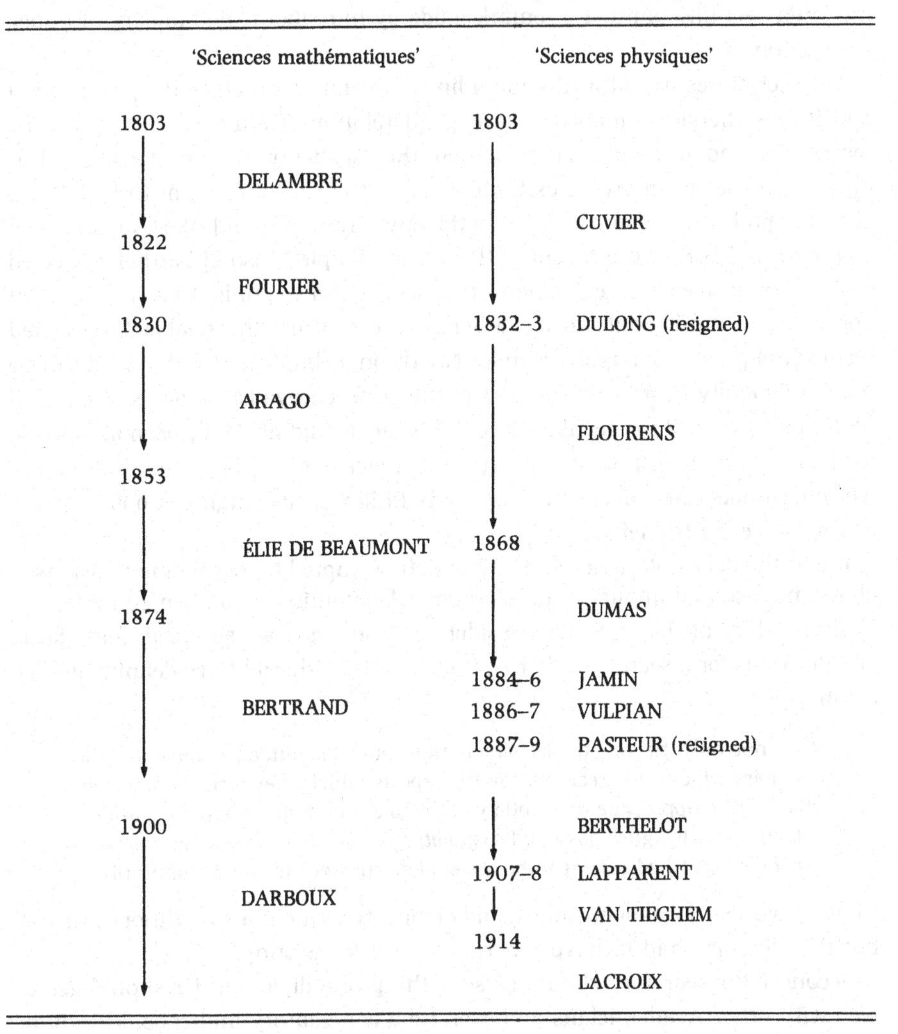

	'Sciences mathématiques'		'Sciences physiques'
1803		1803	
	DELAMBRE		
			CUVIER
1822			
	FOURIER		
1830		1832–3	DULONG (resigned)
	ARAGO		
			FLOURENS
1853			
	ÉLIE DE BEAUMONT	1868	
1874			DUMAS
		1884–6	JAMIN
	BERTRAND	1886–7	VULPIAN
		1887–9	PASTEUR (resigned)
1900			BERTHELOT
		1907–8	LAPPARENT
	DARBOUX		
			VAN TIEGHEM
		1914	
			LACROIX

demician who wanted to make a serious contribution to the proceedings would have to arrange it with the secretaries. A secretary in more modern times might have been referred to as 'Mr Fix-it'. With the advent of the *Comptes rendus* in 1835 the power of the secretaries increased considerably. Instead of having responsibility for the (private) minutes of the meeting, they had responsibility for a published version of events. Anything not recognised by the secretary would be omitted and it would be as if it had not happened. In the eighteenth century the secretary had been responsible for writing an annual *Histoire*, explaining briefly the work of the Academy and assessing its significance. In the nineteenth century the secretaries continued in a different way to be the official historians of the Academy. Blainville, for example, could accuse Arago of censorship when the latter tactfully omitted a remark made by him about the Ministry of Public Instruction.[83]

The secretaries would be the main line of communication to the government and it was therefore important that good relations should exist between the secretaries and the minister to whom the Academy was responsible. It is interesting that in many cases the politics of the secretary seem to have been very acceptable to the government of the day. One thinks of J. B. Dumas, elected secretary in 1868 at the height of the Second Empire, also of Berthelot, elected in 1889, who was so acceptable to the early Third Republic that he was even appointed to ministerial office. Earlier in the century the pliable Cuvier had seemed the perfect Bonapartist under Napoleon 1, but after his downfall Cuvier had no difficulty in playing the part of the perfect Royalist. Arago's election in 1830, on the eve of the revolution of that year, would not have been acceptable to the tottering government of Charles X; a fact which made him all the more welcome under the government of Louis Philippe, although his politics were much further to the left.

It was the death of Cuvier in 1832 which prompted the Academy to ask itself about the essential qualities of a secretary. A commission, appointed under the chairmanship of Lacroix to consider the succession, specified four basic qualifications for a secretary.[84] First, it was said, he should have definite literary talents:

> It is not only by the works which they publish, but even more by the memoirs which they read and by the reports which they write continually that we can appreciate the subtlety of the talent of drafting and composition which our colleagues possess. I say *subtlety* because it is no longer possible for a distinguished savant to be a complete stranger to the art of writing.

In other words all Academicians could obviously write in a straightforward way but the secretary had to have a particular gift for writing.

Secondly the secretary should be something of a diplomat. He should get on well with other Academicians and exercise a conciliatory influence. Thirdly he had to be able to present the work of each member of the Academy in a kindly and impartial way to the public. Finally he should be capable of representing the Academy in its relationship with the government with 'dignity', a remark which might well have implied that the secretary should be one of the senior

[83] H. D. de Blainville, 'Observations à l'occasion de la séance du lundi 10 juillet 1843 de l'Académie des Sciences', Institut, *Mémoires*, **14**, 4to AA35 (21 pp.) privately printed.

[84] *P.V.I.*, **10**, 82 (2 July 1832). Cuvier had already indicated that he would like to be succeeded by his former pupil Flourens (aged 38 in 1832) but the Academy understandably decided that such nepotism could not be passively accepted. Geoffroy Saint-Hilaire was also a candidate but the person finally chosen was Dulong, a compromise candidate who, however, resigned after one year. Flourens subsequently became secretary, a post he held for 34 years.

members. Seniority was certainly a quality possessed by the ailing Louis Pasteur when, at the age of sixty-five he foolishly allowed his name to go forward as secretary. Yet he already had an international reputation and the Academy was no doubt hoping to shine in his reflected light. Pasteur resigned after hardly more than a year and was replaced by the very different figure of Berthelot.

Given a mixture of different personalities and political persuasions, one quality which all the secretaries possessed in common was a considerable literary talent. This was the first quality listed in the report of the 1832 commission quoted above, but that report hardly did justice to the role of the secretary as the representative of the Academy to the general public. It was particularly in the composition of *éloges* of deceased members that the literary art of the secretary achieved its highest form of expression. Since the *éloges* were written to be delivered at public meetings of the Institute and their content represents the image of science which the secretary wanted to place in the public mind, a discussion of these is postponed to Chapter 10, which deals with the public dimension.

— *Chapter 4* —

SCIENCE DIVIDED: THE SECTIONS

The main feature of the constitution of the Academy of Sciences consists of being composed exclusively of specialists, devoted entirely to the separate cultivation of the different sub-divisions of natural philosophy.

(August Comte, *Correspondence générale*, 1836, p. 264.)

...eleven permanent divisions or little academies, each sovereign within its own sphere...

(M. Berthelot, 'L'Académie des Sciences' (1867), *Science et philosophie*, 1886, p. 209.)

In an assembly of professors and *savants*, when some financial grant is made to a laboratory, at once all the other colleagues protest noisily. I do not think that an example to the contrary can be cited. Suppose that 3000 francs or so is given to a laboratory of Inorganic Chemistry. Well! What about Organic Chemistry? And Physiology? And Botany? And Zoology? And Physics? And Mechanics? And Geology? Jealous claims like these are after all legitimate. The zoologist feels he has a mission to defend Zoology; the botanist, to defend Botany. They have faith in their science; they do not wish it treated as a negligible quantity.

(Charles Richet, *The natural history of a savant*, 1923, trans. 1927, chap. 2, p. 20.)

1. *Defining science*

The Academy of Sciences and its predecessor, the First Class of the National Institute, constituted a powerful force in defining science in the post-revolutionary era, which in the traditional division of French history constitutes the period marking the beginning of the modern world. Yet in some ways the First Class perpetuated ideas clearly established in the former Royal Academy. In the earlier eighteenth century 'science' had been what was studied in the Royal Academy of Sciences and when, for example, Mesmer's 'animal magnetism' failed to live up to the criteria of rigour and objectivity applied by the commission of the Academy, it was rejected.[1] In so far as it violated the norms of eighteenth-century medicine, it was 'quack' medicine, but it was also bogus science. This is not so much a question of standards as a decision of what constitutes real science. It was not genuine science done badly, but simply non-science. As the general status of science grew in the eighteenth and nineteenth centuries it was

[1] Robert Darnton, *Mesmerism and the end of the enlightenment in France*, Cambridge, Mass., 1968.

a matter of considerable importance for any new approach to the study of the natural world to be accepted as science. In the classification of knowledge, therefore, it is crucial to consider where a dividing line is drawn. Since the Academy was a state-sponsored institution, any decision by the Academy about any new branch of knowledge would have official status. In a centralised state, moreover, the Academy in Paris was speaking for the whole of France and its influence spread far beyond the geographical boundaries of that country.

The visibility of the Academy of Sciences produced a situation in which the Academy exemplified science. More than that, it defined the meaning of science. It did this in a way that the Royal Society of London never did. First because it contained the key word in its title, unlike its British counterpart, and secondly because of its official status, which was not precisely replicated in the British situation. One does not want to minimise the great importance of the Royal Society in the institutionalisation of science but the rhetoric of the early Society was always associated with experiment and occasionally with 'natural philosophy' rather than with science. It is an anachronism to speak of early Fellows of the Royal Society as 'scientists', a term not coined till the 1830s after the foundation of the British Association for the Advancement of Science. This title made an important claim about knowledge. The Latin *scientia* had meant knowledge in the widest sense (cf. *Wissenschaft*), so that theology had claimed in the medieval world to be 'the queen of the sciences'. In the increasingly secularised world of the eighteenth and nineteenth centuries theology was devalued in favour of knowledge of the natural world, and the acquisition of the term 'science' to denote exclusively this kind of knowledge was something of a landmark in the history of science in the English-speaking world. Our claim, however, is that this issue had been anticipated in France before 1800 and had been given additional force by events arising out of the Revolution of 1789.

The original division of the Academy into named sections in 1699 had been introduced as a means of ministerial control, which marked the final abandonment of earlier Baconian ideas of engagement in general collective work. The division into sections was merely one of a long and detailed list of regulations intended to cover every aspect of the work of the Academy. Academicians were even instructed where they should sit during meetings![2] Both before and after the French Revolution the Academy generally accepted the divisions into sections, which they had been given, and helped to impose them on the scientific community. In its origins the designation of sections was rather like the division of a ministry into departments; a bureaucratic device to share out work and responsibility and possibly to increase productivity. Instead of being an amorphous mass of men with a common interest in natural philosophy, the Academy came to represent a federation of subject groups. Lines

[2] R. Hahn, *The anatomy of a scientific institution. The Paris Academy of Sciences, 1666–1803*, Berkeley, Cal., 1971, pp. 19–20; Fontenelle, *Oeuvres complètes* (3 vols., Paris, 1818, Slatkine reprint, Geneva, 1968), vol. 1, pp. 40–6.

Table 8. *Links between respective sections of the Academy and some major institutions of higher education in the Paris Region*

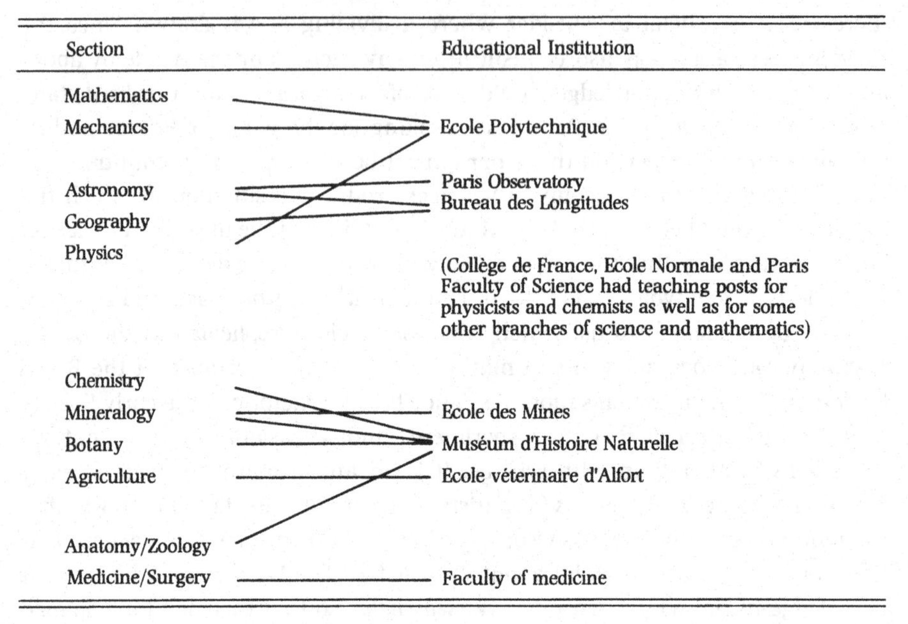

of activity were thus well marked out and the whole organisation was subject to order and control.

Although the French Academy represented a model very different from that of many other scientific societies of the time (notably the Royal Society of London) and one which might have seemed unnecessarily restrictive, the division into sections did have notable advantages. Indeed it was the only way in which one could have at the same time a society embracing the whole range of mathematical and natural sciences and yet encouraging specialisation. In the late seventeenth century it had been possible for men of science to do important original work in several disciplines. A century later the growth of knowledge made specialisation almost a necessity; the Academy not only permitted this, it encouraged it. As Table 8 suggests, there was a close correspondence between concepts of science fostered within the Academy and those adopted in French higher education. Thus the structure of the Academy provided a reinforcement of a career structure for mathematicians, astronomers, chemists and specialists in most other branches of science. In the nineteenth century it was the norm for the French scientist to be a specialist.

In its subdivisions the First Class recognised a basic distinction between mathematical sciences and natural sciences. Astronomy and physics were seen as mathematical, whereas chemistry and mineralogy were seen as more experimental. Although chemistry had recently begun to make use of a quantitative approach, this involved no more than basic arithmetic. For example, according to Lavoisier's principle of conservation of matter, in any self-contained chemical reaction the sum of the weights (or masses) of the products should be equal to the total weight of the reactants. Although crystallography was beginning with Haüy to make good use of geometry, it was still largely a classificatory science. Hence chemistry and mineralogy were still associated with the respective study of plants and animals (i.e. botany and zoology) rather than with physics. In the early years of the nineteenth century, chemists like Berthollet had still to make a case for the alliance of physics and chemistry. Fortunately in the working of the First Class the distinction between the two broad divisions of mathematical and natural science was never regarded as absolute and served mainly as a convenient division of responsibilities between *two* permanent secretaries, one appointed for each of these divisions, on the principle that no one person after about 1800 could be expected to encompass in his understanding the whole spectrum of mathematical and natural science.

The main way in which the constitution of the First Class affected science was by the recognition of ten named branches. Some of these branches, such as astronomy and botany, had been recognised for centuries and in the Academy at least since the beginning of the eighteenth century. Others, such as physics, were newer, with a brief history before the Revolution, mainly emphasising the experimental aspects. In the chemistry section the first two members to be nominated by the Directory were Guyton and Berthollet. These were not only two scientists who had served their country well in the war effort, but also two of the principal associates of Lavoisier. 'Chemistry' in the First Class, therefore, came to mean Lavoisierian chemistry, and supporters of the phlogiston theory could not expect to be elected as full members of the Class.

'Mechanics' was an ambiguous title for a section since it could represent equally the severely practical artisan or the (applied) mathematician. The nomination of the mathematician Monge and the engineer Prony marked the beginning or rather the resumption of a tradition in which greatest honour was attached to mathematical ability, albeit in a practical context. One of the most curious descriptions of the sections was that of 'Géometrie', adopted from 1803 for the (pure) mathematics section. In the original Academy of 1666 geometry was practically synonymous with mathematics but by the end of the eighteenth century mathematics had developed so much that it hardly took any account of geometry. The section of Astronomy, classified by the Academy as one of the mathematical sciences, continued to recruit men of high mathematical ability, so helping to perpetuate the French tradition of excellence in theoretical astronomy although, as we shall see, observational astronomy was not neglected either.

As regards the life sciences, zoology shared a section with anatomy, thus giving an important place to comparative anatomy, to which Cuvier made major contributions. Thereafter Cuvier's reputation reinforced the influence of the formal organisation of the section in encouraging further study of comparative anatomy. On the other hand the double title of the section created a number of conflicts and raised the question of whether, because of the existence of a separate section for medicine, 'anatomy' meant non-human anatomy. The botany section was more fortunate in not having to share with another branch of science. In 1785 the growing interest in scientific agriculture had led to the addition of agriculture to the botany section of the Academy of the *ancien régime*. In the Institute, however, there was a separate section for agriculture. Thus instead of being concerned with pure and applied botany, the section could luxuriate in the pure science. It was only a pity that more was not happening in nineteenth-century botany to justify a whole section, given the severe restriction in the number of places for representatives of other sciences. Only the new science of palaeontology, through its branch of palaeobotany, was able to find a natural home in this section and do important new work.

It was in its failure to anticipate the unequal development of the recognised sciences, and slowness in recognising new sciences, that the nineteenth-century Academy revealed areas of major potential weakness. As regards the development of the different sciences we can, with the benefit of a modern perspective, see physics and chemistry as two sciences which were to grow almost beyond recognition in the course of the nineteenth century. Chemistry took off first with the search for new elements, electrochemistry and the enormous expansion of organic chemistry. In physics the discovery of polarised light, the wave theory of light, the discovery of a connection between magnetism and electricity and the enunciation of the principle of conservation of energy all contributed to the recognition of physics as one of the truly basic sciences with great scope for development. If physics did not flourish as much in nineteenth-century France as chemistry did, this may be because the education system tended to syphon off any young man with mathematical ability and train him as a mathematician – a class of high prestige in the lycées – leaving to physics men of lesser mathematical ability or men whose primary interests were experimental. In other words any influence the Academy had in shaping and defining 'physics' was undermined by the educational system. Chemistry had much greater influence but was seriously handicapped by the restrictions in number, although this provided a justification for the foundation of the Société chimique de Paris (1857), which by the 1860s was beginning to play a major part in the organisation of French chemistry. The Société de physique was not founded until 1873, a further indication perhaps that, at least until the discovery of radio-activity, physics was less thriving in nineteenth-century France than chemistry.

We shall argue that by means of the sections the Academy was able to give a special meaning to most of the main branches of science. Of course there were new branches of knowledge which emerged in the nineteenth century and were never properly incorporated within the Academy. Anthropology was one such discipline. It has been plausibly argued that 'anthropology' had only the

Table 9. *Estimate of comparative scientific eminence of respective sections based on listing in Dictionary of Scientific Biography*

Section	Number of members, 1795–1914	Number listed in *D.S.B.*	Percentage in *D.S.B.*	Adjusted Number in *D.S.B.*	Adjusted Percentage in *D.S.B.*	Ranking Order
Mathematics	32	28	88	26	81	1
Mechanics	38	22	58	22	58	6
Astronomy	33	20	61	20	61	5
Geography	25	4	16	4	16	11
Physics	40	29	73	26	65	4
Chemistry	34	23	68	27	79	2 =
Mineralogy	42	34	81	33	79	2 =
Botany	32	16	50	16	50	8
Agriculture	34	13	39	11	33	9
Anatomy/ Zoology	35	20	57	20	57	7
Medicine/ Surgery (including Physiology)	44	12	27	14	27	10

vaguest of meanings until the foundation in 1859 of the Société d'Anthropologie and the rise in the reputation of its principal founder, Paul Broca.[3] If a comparatively small private society can have such influence on the definition of a discipline, one might expect a large official organisation to have an even greater influence on defining the disciplines which it recognised.

2. *A comparison of the sections and a few anomalies*

In Chapter 2 the division of the Academy into sections was explained. It is of some interest to estimate the relative scientific eminence of the different sections. A ready approximation is obtained by seeing how many members from each section have, with the perspective of the elapse of at least a century, satisfied the judgement of history by inclusion in the modern standard international multi-volume *Dictionary of Scientific Biography*.[4] If we were primarily interested in individuals, we would take into account the relative lengths of different biographies, remembering that it was a prominent feature of editorial policy to give entries varying lengths according to an estimate of the relative eminence of the subject. But a more basic question relates to the relative eminence of different branches of French science, represented so clearly and explicitly in the different sections.

[3] Elizabeth A. Williams, 'Anthropological institutions in nineteenth-century France', *Isis*, **76** (1985), 331–48.

[4] C. C. Gillispie, ed., *Dictionary of scientific biography*, 16 vols., New York, 1970–80 (denoted throughout as *D.S.B.*).

Although the figures in Table 9 should not be given any absolute significance, they can be used as evidence to support the view that it was particularly in mathematics, chemistry and 'mineralogy' that French science excelled in the nineteenth century, although this does not imply a uniform eminence for over a hundred years. In mathematics, for example, it was particularly at the beginning of the nineteenth century, and again at the end, that the most eminent French mathematicians flourished. We must also remember that the sections often encompassed a number of sub-disciplines, and that in some of these the French contribution may not have been especially strong. The claim is simply that *in general* the French showed considerable distinction in these subjects. Comparing sections within the Academy, it is understandable that the more practical sections relating to agriculture, medicine and surgery, and geography[5] should have had the lowest proportion of eminent scientists, some having overall no more than a third of those attached to the most scientifically distinguished sections. It may be worth emphasising that this is the judgement of history and not necessarily that of the nineteenth century. In theory all sections were equal in prestige, although in practice one cannot imagine an ordinary member of the agriculture section, for example, thinking himself the intellectual equal of one of the great mathematicians.

But before the Table is used to compare all of the subjects recognised by the Academy, a further word of caution is necessary. In a simple reading, it is assumed that in all cases a scientist is elected to the section which best corresponds to his posthumous fame. In well over 90% of cases Academicians were elected to the section representing their specialist competence, but in a few instances some adjustment of the data is required to allow for scientists who belonged to a section different from that in which they established their main reputations. The anomaly arises sometimes from a change of interests after election or, more commonly, from the Academy using a vacancy in another section to secure the election of an outstanding candidate.

Thus, looking through the membership of the Academy, one is occasionally struck by an unexpected affiliation. A man like Ampère (1775–1836), remembered today principally for his contributions to electricity, was actually a member of the mathematics section; Gay-Lussac (1778–1850), known primarily as a chemist, was formally a member of the physics section. In both these cases the apparent anomaly may be largely explained by later developments. The most famous work of Ampère was done on electromagnetism in the 1820s, when he had been a member of the Academy for at least six years. He had started his career as a mathematics teacher in Lyons. In 1802 he had begun work on a major memoir on probability, and on the strength of this he had been appointed as *répétiteur* in mathematics at the Ecole Polytechnique. In

[5] The section of geography and navigation, of course, suffered from the additional disadvantage that for most of the nineteenth century it consisted of only three members instead of the usual six.

the Napoleonic period, therefore, Ampére was most closely identified with mathematics. To refer also to Ampère's early contributions to chemistry and the structure of matter, where he independently formulated a principle similar to 'Avogadro's Hypothesis', would merely complicate the issue further. But the French system of higher scientific education in general, and the structure of the Academy in particular, encouraged scientists to specialise and Ampère was an exception in several ways, extending beyond the disciplinary context.

The case of Gay-Lussac is also easily explained. At the time of his election to the physics section in 1806, he was equally interested in physics and chemistry and had done research reflecting the respective concerns of his Arcueil patrons, Laplace and Berthollet. There was no vacancy in the chemistry section in the early 1800s and he was probably well advised to apply for the first vacancy relevant to his interests. He was not to know that three years later Fourcroy was to die, thus creating the first chemistry vacancy for more than a decade. Gay-Lussac's early reputation was based on research such as the law of thermal expansion of gases, commonly known as 'Charles' Law', as well as his more widely publicised balloon ascent; but by the end of the Napoleonic period his work on chlorine, iodine and cyanogen had demonstrated that his main concerns were really chemical. The existence of a chemist in the physics section may possibly have contributed to the development of physical chemistry. On the other hand it may have tended to weaken the physics section especially when, after the death of Lefèvre-Gineau in 1829, Gay-Lussac became the doyen of the section, thus exercising disproportionate influence on the election of future physicists to the section.

The only similar case of a chemist being elected to the physics section was that of Berthelot (1827–1907), elected in 1873. This has both a negative and a positive aspect. On the one hand the chemistry section was one of the most difficult to enter because of the large number of well-qualified chemists. Berthelot, who had made something of a reputation for himself in organic chemistry,[6] had, nevertheless, been unsuccessful in the 1860s in the relevant elections.[7] However, the Franco-Prussian war in 1870 widened Berthelot's interests. He headed a Commission under the Ministry of Education for the application of physics and chemistry to national defence.[8] In particular his study of explosives helped to direct his research interests from organic to physical chemistry. The prominent part played by Berthelot in applying science for the defence of Paris during the siege of 1870–1 won him some respect from his

[6] See e.g. *Chimie organique fondée sur la synthèse*, 2 vols., 1860. The focus of this book may be contrasted with his later *Essai de mécanique chimique fondée sur la thermochimie*, 2 vols., 1879.

[7] When the organic chemist Wurtz, ten years Berthelot's senior, was elected to the chemistry section in 1867, Berthelot had received only three votes compared to Wurtz's forty-six votes, not a very encouraging situation for him. *C.R.*, **65** (1867), 101.

[8] Maurice Crosland, 'Science and the Franco-Prussian War', *Social Studies of Science*, 6 (1976), 185–214 (pp. 193–4); art. 'Berthelot', in *D.S.B.*, **2**, 63–72.

fellow countrymen. His favourable political profile in the Third Republic, together with his new but undoubtedly genuine physical interests, earned him his place in the physics section when a vacancy arose in 1873. He became a prominent enough member of the Academy to be proposed and elected to the position of permanent secretary in 1889, thus bringing to an end what other members of the physics section probably regarded as an undesirable anomaly.

Among other anomalies were the election of Biot (1774–1862) to the mathematics section in 1800, despite his principal interests in physics and astronomy, and of his younger contemporary Poisson (1781–1840) to the physics section in 1812, despite the fact that Poisson was a mathematician with no interest in or aptitude for experimental physics. Biot's election can be related to his mathematical training at the Ecole Polytechnique but the fact remains that both scientists felt a little uncomfortable in their respective sections and even proposed a private exchange. The rigid rules of the Academy, however, did not permit such informal arrangements. An election was potentially a commitment to a specified place for life; it could not be regarded as a credit for later transfer, as sometimes happened with university chairs. As far as statistics are concerned, these two cancel each other out and this is equally the case for the mathematician Liouville among the astronomers, which compensates for the astronomer Delambre, who found himself among the mathematicians. It would not be superfluous, however, to repeat that all the cases we have detailed above were *exceptions*. In the vast majority of cases, scientists were elected to the appropriate section.

Nevertheless, when the Academy elected someone who, *at the time of election*, seemed to belong more appropriately to another section, it was in a sense rebelling against the bureaucratisation of science imposed by the government. It was bestowing its prerogative of favour on the able candidate, despite the fact that his previous career did not exactly fit the title of the section when the vacancy occurred. A case where a section was used by the Academy in a way which was at variance with the original government intentions, was that of the *Académiciens libres*, founded by royal decree in 1816. Because of the strong political dimension and because of the different status of the section, which was quite unlike the other specialist sections, a discussion of it is postponed until Chapter 12.

3. *Interpretation of the titles of the respective sections*

In the revised constitution of the Institute of 1803[9] the duties of the Second and Third Classes were spelled out. Thus the new Second Class (corresponding to the former Académie Française) was charged, among other duties, with compiling a dictionary of the French language. No parallel duties were prescribed for the First Class. Instead the *names of the sections* were listed. Thus the very title of a section was intended to indicate the duties of its members.

[9] *P.V.I.*, 2, 619–25.

While considering briefly the sections in turn, it is worthwhile trying to understand what was meant by the respective titles of individual sections and, in order to do this, it seems desirable to focus on the people chosen to represent that science. The first question we intend to answer, therefore, is: who were the men elected as members of that section and what did they stand for? Since there were more than thirty members in most sections in the period 1795–1914, only some of the more important members can be considered. More detailed coverage, however, is given to the original half-dozen members of each section on the grounds that they reveal most clearly the original conception of the subject when the Institute was founded. Sometimes it reveals a new conception of a discipline, but there are other cases where traditional conceptions continued from the *ancien régime*. This was particularly the case when older men, who had been members of the previous Royal Academy of Sciences, were elected as members of the First Class. This had the effect of providing some continuity but also means that, instead of a sharp break in the conception of a subject between 1793 and 1795, new ideas emerged more slowly and only with the next generation. We are particularly concerned to study the change in the conceptions of a subject during the nineteenth century and we often find that the very practical emphasis of 1795 was replaced by a greater respect for theory.

What the following pages do *not* represent is a balanced survey of developments in the different sciences in the nineteenth century, such as one might expect to find in a general history of science. Only indirectly does it provide a survey of French science in general. Rather what follows is a number of snapshots of work in different branches of science specifically related to the members of the respective sections. In some cases this may be the main opportunity to learn briefly of the respective contributions of the Academicians to science but, although it is inevitable that there should be many names and dates, we must not lose sight of our main purpose, which is, by means of a few brush strokes, to depict at least an outline of the respective sciences represented in the Academy. Because there were eleven sections and even more disciplines represented, we must be brief. It is part of our thesis that the whole was greater than the sum of the parts, but to make this point one needs first to know something of the parts.

4. *The mathematics section*[10]

That the mathematics section should continue to be called 'Géometrie' is evidence of the power of tradition transcending revolutions, since although

[10] Among relevant histories of mathematics we may mention Carl B. Boyer, *A history of mathematics*, New York, 1968, e.g. chapter 22: 'Mathematicians of the French Revolution'. Morris Kline, *Mathematical thought from ancient times* (New York, 1972) also contains useful material but pays less attention to the national patterns. See also the later chapters of John Fauvel and Jeremy Gray (eds.), *The history of mathematics: a reader*, 1987. Of great potential value is Ivor Grattan-Guinness's magisterial *Convolutions in French mathematics, 1800–1840*, 3 vols., Basel and Berlin, 1990.

'geometry' had been a major part of mathematics at the time of the foundation of the Royal Academy in 1666, it had been reduced considerably in relative importance by the late eighteenth century. Indeed one of the leading members of the section, Lagrange, could boast in the introduction to his *Mécanique analytique* (1788) that he had used not a single figure in the book. Yet the French continued during the nineteenth century to refer to mathematicians as *géomètres*, a usage which had apparently been introduced originally to distinguish a *real* mathematician from an ordinary numerate person.[11] We shall find that although the majority of outstanding pure mathematicians were elected to this section, many French mathematicians made important contributions to both pure and applied mathematics and there may, therefore, have been equal justification for electing them to the mechanics section when a vacancy arose.

The fact that Lagrange (1736–1813) and Laplace (1749–1827) were the first two mathematicians to be nominated by the Directory provided a brilliant start not only for the section but for the whole of the First Class.[12] Lagrange had been born in Turin of partial French descent. It was only at the age of fifty that he came to Paris and became a naturalised Frenchman. His first thirty years in Turin and a second period of twenty years in Berlin had been the most fruitful in original work in calculus, mechanics and the theory of numbers[13]. His years in Paris were really a period of consolidation, when he composed great treatises summarising his mathematical views. Thus, although his work is usually described as a part of French science, he was really more of a multi-national figure. Some of his earlier work had been inspired by prize competitions set by the former Royal Academy of Sciences and we might interpret his arrival in Paris, just before the Revolution, as the logical culmination of a distinguished career. When he died his body was buried in the Panthéon among the heroes of French history.

Laplace too was one of the great mathematicians of all time but his interests and influence were wider, embracing as they did not only pure mathematics but applied mathematics, physics and astronomy.[14] As a young man Laplace had worked on a number of mathematical topics, including recurrent series, partial differential equations and determinants. In the 1780s Laplace began to work on integral solutions to partial differential equations and today Laplace's name is probably best known by mathematicians for the 'Laplace transform' method of solving differential and integral equations.[15] Yet outside the field of professional mathematics it is Laplace's work on probability which deserves to be remembered. Laplace had written on the subject even before he became a

[11] Grattan-Guinness *op. cit.*, vol. i, p. 57.

[12] These two figures would continue to be cited throughout the nineteenth century as exemplifying the highest standards of 'French science'.

[13] Jean Itard, art. 'Lagrange', *D.S.B.*, 7, 559–73.

[14] C. C. Gillispie *et al.*, art. 'Laplace', *D.S.B.*, 15, 273ff.

[15] I. Grattan-Guinness, *op. cit.*, pp. 382–7.

member of the Royal Academy of Sciences in 1773 and he was still developing the subject in the last years of his life. His major contribution to the subject was presented to the First Class in 1812 and developed as a book, *Théorie analytique des probabilités*, which went through several editions. Later editions are prefaced by his more popular *Essai philosophique sur les probabilités* (1814), which was a development of lectures that he gave at the Ecole Normale in 1795. His discussion in the latter work considers a number of applications of probability theory, beginning with the long-established interest in games of chance and extending to the moral sciences, medicine and law.

Perhaps Laplace's greatest work was in physical astronomy. He made early studies of universal gravitation and the distribution and orbits of comets, going on later to study secular inequalities, the moons of Jupiter and lunar theory. He continued these studies in his monumental *Traité de mécanique céleste* (5 vols., 1799–1825), in which he presented astronomy as a series of problems in mechanics. His major achievement in celestial mechanics was his proof of the stability of the solar system, showing that certain variations in planetary orbits, which had worried Newton, were in fact self-correcting. More widely read was his semi-popular *Exposition du système du monde* (2 vols., 1796), which is probably best known for the speculation on the nebular origin of the solar system.

But if Laplace's work had a profound effect on astronomy, it had even greater influence on physics. Already in the 1780s he had collaborated with Lavoisier in a memoir on heat. In the early 1800s he was to apply the idea of Newtonian short-range forces to a variety of phenomena including refraction and capillary actions. The Newtonian focus of his work was reinforced by his concern to arrive at a more accurate formula for the velocity of sound than that left by his illustrious predecessor, one which depended on a determination of the specific heats of gases. In all these problems he leaned heavily on young experimentalists to obtain the necessary data.[16]

Even from this brief summary it can be seen that Laplace was a key figure not only in the mathematics section but in the First Class generally. In the early 1800s no other figure in the mathematical sciences was so influential. It is such a figure which gives the lie to the assumption that the future of science lay entirely in specialisation. His multi-disciplinary interests transcended the sectional divisions.

One of the features of the mathematics section was that the first six members had all been members of the former Royal Academy. Borda (1733–99), a former naval commander, was very much a figure of the old regime.[17] In so far as his most important work was his contributions to fluid mechanics, the study of navigation instruments, geodesy and the determination of weights and measures, he could as well have been a member of the mechanics section. The

[16] M. Crosland, *The Society of Arcueil*. R. Fox, art. 'Laplace', Part IV, *Op. cit.*, pp. 356ff.

[17] C. S. Gillmor, art. 'Borda' *D.S.B.*, **2**, 299–300.

same might be said of Bossut (1730–1814), a former professor and examiner at the Ecole de Génie at Mézières, where Monge too had been a teacher. From 1780 he occupied a special chair of hydrodynamics in Paris and was known as the author of several standard text-books. Even better known as the author of a text-book was Legendre (1752–1833), whose main subjects of research were celestial mechanics, the theory of elliptic functions and number theory. An excellent disciple of Lagrange, his work has been characterised as transitional, belonging clearly neither to the eighteenth nor the nineteenth centuries.[18] He possessed boundless, and possible naive, confidence in the powers of abstract mathematics. Just as Legendre was the author of a standard text-book on geometry, so was Lacroix (1765–1843). Less original than most of his colleagues, Lacroix has a place in the history of mathematics through his text-books, notably one on the calculus (1st edition, 1802), which was later translated into English and German.

There were a few apparent anomalies in the early elections. The inclusion of the astronomer Delambre (1749–1822) as the last of the original six members to be elected in 1795 was partly an act of kindness to a former colleague, who had not been elected to the astronomy section. Although Biot (1774–1862) is remembered more as a physicist, his election in 1803 to the mathematics section can be seen partly in relation to his mathematical training at the Ecole Polytechnique but even more to the patronage of Laplace, whose ideas he was to support consistently within the Academy. We have also explained why Ampère was a member of the section.

Many members of the mathematics section had been students at the Ecole Polytechnique, where in the early years projective geometry constituted a prominent part of the curriculum. Yet it soon lost its prominence as the influence of Monge declined.[19] One of the few representatives of the geometrical tradition in the mathematics section of the Academy in the first half of the nineteenth century was Poinsot (1777–1859), a fervent disciple of Monge.[20] It was in mechanics that Poinsot applied most effectively his geometrical talents, for example in a text-book on statics intended for candidates for the Ecole Polytechnique. He did little pure geometry and indeed he was such a perfectionist that his publication record is disappointing. He did, however, help to establish a chair for advanced geometry at the University of Paris in 1846, intended for Chasles. Michel Chasles (1793–1880) was a product of the Ecole Polytechnique, who devoted his whole life singlemindedly to the pursuit of geometry.[21] He believed that the methods of synthetic geometry were superior to those of

[18] Jean Itard, art. 'Legendre', *D.S.B.*, 8, 135–43 (p. 142).

[19] John Fauvel and Jeremy Gray (eds.), *op. cit.*, pp. 541, 548.

[20] René Taton, art. 'Poinsot', *D.S.B.*, 11, 61–2. Another important advocate of geometry was Poncelet (1788–1867) who, as a member of the mechanics section of the Academy, is mentioned in the following section.

[21] Elaine Koppelman, art. 'Chasles', *D.S.B.*, 3, 212–15.

algebraic analysis. After appointment to the Paris chair he wrote two important text-books, on advanced geometry and on conic sections. He was unfortunate not to be elected to the Academy until 1851. In the 1830s the Academy had conferred its favour on the young Genevan Sturm (1803–55), who had mixed in the social circles of Arago and Gay-Lussac[22] and who certainly had an extraordinary talent for applying mathematics to the solution of physical problems. For the vacancies in the 1840s the Academy had passed over Chasles in favour of Lamé (1795–1870) and Jacques Binet (1786–1856). Lamé represented interests in both pure and applied mathematics,[23] and Binet had been a highly recommended candidate for so long that many felt that his election was long overdue.

The leader of the French school of mathematics from the death of Cauchy[24] in 1857 to the rise of Henri Poincaré (1854–1912) and his very talented colleagues was Charles Hermite (1822–1901).[25] Even as a student at the Polytechnique, Hermite had shown a special interest in Abelian functions and elliptical functions. He continued to work on these subjects throughout his long life, but also did distinguished work in a few other areas including the theory of numbers. Perhaps he will be remembered most for his demonstration that the number *e* is transendental.

It has been said of the brilliant mathematician Camille Jordan (1838–1921) that his place in the tradition of French mathematics lay exactly half way between Hermite and Poincaré.[26] Like them he was something of a universal mathematician, who published papers in nearly all branches of nineteenth-century mathematics. His early fame depended on his work in algebra and he later came to hold an international reputation for his work in group theory. He also developed some of the ideas opened up earlier by Galois.

Whereas the majority of members of the mathematics section had studied at the Ecole Polytechnique, Darboux (1842–1917), although accepted for that school, opted for the Ecole Normale Supérieure.[27] As a mathematician his main contribution was to geometry and he established a brilliant school of infinitesimal geometry. He was also a good administrator and in 1900 was elected as secretary of the Academy.

It is high time that we said something about the work of Henri Poincaré, a precocious figure who dominated all the main branches of mathematics in the late nineteenth century. His genius and persistent hard work produced more than 500 memoirs, which were later collected into thirty volumes. It has been

[22] Pierre Speciali, art. 'Sturm', *D.S.B.*, 13, 126–32.

[23] Samuel L. Greitzer, art. 'Lamé', *D.S.B.*, 7, 601–2.

[24] Cauchy, as a member of the mechanics section of the Academy, is discussed in the next section.

[25] Hans Freudenthal, art. 'Hermite', *D.S.B.*, 6, 306–9. P. Appell and P. Montel, 'Les mathématiques', in *La science française* (2nd edn, 2 vols., 1933), vol. 1, pp. 79–91.

[26] J. Dieudonné, art. 'Jordan', *D.S.B.*, 7, 167–9.

[27] Dirk J. Struik, art. 'Darboux', *D.S.B.*, 3, 559–60.

said that mathematics in the early nineteenth century began under the shadow of Gauss and ended under the domination of the comparable genius of Poincaré.[28] Poincaré was a student at the Ecole Polytechnique before following an engineering course at the Ecole des Mines. Most of his life, however, he was a professor at the University of Paris. He became famous before he was thirty by his discovery of what he called 'fuchsian' or 'kleinean' functions after the respective German mathematicians. He was elected to the Academy at the age of thirty-three. He wrote a series of papers which inaugurated modern methods of algebraic topology, creating new concepts and tools. He is also a major figure in the history of the theory of differential equations. He worked in algebraic geometry, number theory and algebra as well as celestial mechanics. He was also called upon to work on the foundations of mathematics and criticised some of the ideas of Russell, Peano and Hilbert. He was aware of the latest developments in nearly every branch of mathematics and might himself be described as a universal mathematician. If he was rather conservative in his work in physics, his mathematics was the very opposite. As his reputation grew, Poincaré was asked to speak on many aspects of mathematics and science and his books *La science et l'hypothèse* (1906) and *Science et méthode* (1908) are important contributions to the philosophy of science. He died in 1912 at the age of fifty-eight.

Nevertheless, French mathematics did not suddenly come to an end. In the éloge of Painlevé (1863–1933), it was said that he had demonstrated that 'continuing [the work of] Henri Poincaré was not beyond human capacity'.[29] Painlevé contributed to many fields. As a mathematician he had received the *grand prix des sciences mathématiques* (1890), the prix Bordin (1894) and the prix Poncelet (1896) before being elected to the Academy in 1900. He took a practical interest in the new field of aviation and when war was declared in 1914 it was he who created the Service des Inventions pour les Besoins de la Défense Nationale, which became a ministry in 1915. He himself became Minister of war in 1917 and after the war became an active participant in the League of Nations.

The mathematics section, therefore, represented a very wide and impressive coverage of the subject. In analysis we have surveyed the field from the work of Lagrange to that of Poincaré. The scope of mathematics was extended, as in the development of the calculus, but in the centralised and authoritarian French system nothing so revolutionary emerged as, for example, non-Euclidean geometry. Because of the practical applications of descriptive geometry, however, we find important advances on this front, not only in the mathematics section, but also in the mechanics section (notably by Monge and Poncelet) and it is to this section that we must now turn.

[28] J. Dieudonné, art. 'Poincaré', *D.S.B.*, 11, 51–61. See also Carl B. Boyer, *op. cit.*, pp. 650–4.

[29] Lucienne Félix, art. 'Painlevé', *D.S.B.*, 10, 274–6.

5. *The mechanics section*

The mechanics section was one of the less homogeneous in the Academy. Everyone has a good idea what a mathematician is, but 'mechanics' could imply a number of different interests, ranging from applied mathematics to different kinds of engineering; indeed it has recently been shown that a large proportion of the members of this section had some connection with engineering.[30] Engineering could have a military as well as a civilian connotation and, while mentioning the military context, we may be reminded that the Academy's most famous member of all time, General Napoleon Bonaparte (1769–1821), was elected in 1797, ostensibly on his expertise in applied mathematics as an artillery officer, to take the vacant place of Lazare Carnot (1753–1823), exiled by the government for political reasons.[31]

At the foundation of the Institute an explicitly practical dimension had been given to the section by calling it 'arts mécaniques', which might be freely translated as 'mechanical technology'. The two foundation members nominated by the Directory were Monge and Prony, both of whom combined practice and theory to a high degree. Gaspard Monge (1746–1818) had first entered the Royal Academy as a mathematician but, in the reorganisation of 1785, had accepted categorisation under physics. He had taught both mathematics and physics at the school of military engineering at Mézières and acquired greatest fame as the founder of 'descriptive geometry'. This subject shows how to project orthogonally a three-dimensional object onto a horizontal and a vertical plane, a technique which is useful in architecture, fortification, carpentry and stone-cutting, all skills relevant to the students of the newly founded Ecole Polytechnique. From 1794 his name is linked with this school, which he administered and where he taught most ably. The name of Marie Riche de Prony (1755–1839) is equally inseparable from another great school, the Ecole des Ponts et Chaussées. He was more of an engineer than Monge and the author of many books including *Nouvelle architecture hydraulique* (1790) and *Mécanique philosophique* (1800). Like Prony, the mathematician Vandermonde (1735–96) was a friend of Monge, but his death a few weeks after election prevented him playing more than the most minor part in this story.

Most worthy of comment is that out of the original six members of the mechanics section, two were clock-makers, Le Roy and Berthoud, thus reflecting the practical ideology of 1795 and, perhaps, a more traditional or old-fashioned view of science which might have been well understood by that early curator of experiments to the Royal Society, Robert Hooke (1635–1703). Jean-Baptiste Le

[30] Eda Kranakis, 'Social determinants of engineering practice: A comparative view of France and America in the nineteenth century', *Social Studies of Science*, 19 (1989), 5–70 (p. 24).

[31] The First Class was particularly conscious of its wise choice when Bonaparte became head of state. It was not to be expected, however, that he would attend more than a few meetings. In 1815 he formally submitted his resignation.

Roy (1720–1800), son of the famous clock-maker, was a man with a primary interest in instrumentation. A member with greater skill as a craftsman was Ferdinand Berthoud (1727–1807), born at Neuchâtel (Switzerland) but moving later to France. Berthoud was a clock-maker and held the post of Mechanic (in chief) to the Navy. He did not have a deep understanding of principles, but he was a superb craftsman and he made his mark by constructing a number of chronometers in the 1760s[32] and had published a book entitled *Traité des horloges marines*.

The practical representation in the section was briefly reasserted in the reorganisation of 1816 by the inclusion of Abraham-Louis Breguet (1747–1823), another former inhabitant of Neuchâtel. Breguet had made an international reputation for himself under the *ancien régime* as a clock-maker and it has been claimed that, from the 1780s onwards, with his series of self-winding watches, he led the way in almost every branch of horology.[33] He was not actually elected by the Academy but nominated in 1816 by the King, who obviously wanted to confer a public symbol of recognition on a man of such a high reputation dating back to the *ancien régime*. The Academy could hardly expect much new work from Breguet, then in his late sixties, but it could benefit from the reflected glory of a famous name which in the nineteenth century became a byword internationally for precision timing and dependability.

In the world of professional science of the nineteenth century, clock-makers could no longer expect to be dignified by election to the Institute and the best instrument makers were hardly able to stand up to the competition from brilliant mathematicians.[34] When Breguet's grandson, the instrument-maker Louis François Clément Breguet (1804–83), was elected to the Academy in 1874, it was in the miscellaneous category of *Académicien libre*.

When Berthoud died in 1807, the mechanics section produced a long list of possible successors, mainly practical men, including Sané, ('Inspecteur général du Génie maritime'), Molard ('Administrateur du Musée des Machines') and Girard ('Ingénieur en chef des Ponts et Chaussées').[35] All three were to be elected in due course. The positions held by these men are a reminder that engineering in France had developed in a way totally different from Britain. The French engineer, far from being a largely self-educated entrepreneur, was a higher grade civil servant with mathematical training.[36] It was the latter aspect which was emphasised when engineers presented themselves for election. Thus the

[32] David S. Landes, *Revolution in time. Clocks and the making of the modern world*, Cambridge, Mass., 1983, pp. 167–70.

[33] *Ibid.*, p. 225.

[34] An exception was Gambey (1787–1847), maker of precision instruments, who was elected to the mechanics section in 1837.

[35] *P.V.I.*, 3, 560 (3 August 1807). Girard was actually elected to the physics section in 1815, but not without adverse comment from physicists about associating with a man who constructed canals.

[36] Terry Shinn, 'From "corps" to "profession": the emergence and definition of industrial engineering in modern France' in R. Fox and G. Weisz (eds.), *The organisation of science and technology in France*, Cambridge, 1980, pp. 183–208.

engineer and bridge builder Navier was elected in 1824 not, fortunately, on the strength of his most famous suspension bridge (which actually collapsed) but of his many publications. Another bridge builder, the engineer Clapeyron, was elected in 1848 because of his mathematical prowess and the fact that he had also taught a course at the Ecole des Ponts et Chaussées. Other civil engineers elected include Saint-Venant (1797–1886) and Maurice Levy (1838–1910) in 1883 from the Corps des Ponts et Chaussées, and Charles Combes (1801–72) in 1847 and Reseal (1828–96) in 1873 from the Corps des Mines. In the 1880s the section even elected an electrical engineer, Deprez (1843–1918).

But we must not forget military engineering which in France, under the *ancien régime*, had provided a parallel career to civil engineering. Poncelet (1788–1867), a student of Monge, graduated just in time to take part in Napoleon's Russian campaign. He survived and after 1814 he worked mainly on fortifications. Yet when he presented himself for election in 1834 he prepared a summary of his work under three quite different headings: *Géometrie* (including his important work on homologous figures), *Mécanique appliquée* and *Machines et instruments divers*, thus emphasising his mathematical memoirs at the expense of more practical achievements.[37] Two other military engineers, Piobert (1793–1871) and Morin (1795–1880) were elected to the mechanics section in the 1840s. Although neither are remembered today as great scientists, their work was considered sufficiently highly by the army for each to end their careers as *général de division*. Another pupil of Monge was the naval engineer, Dupin (1784–1873),[38] who made useful contributions to mechanics, geometry and also to public affairs.

Yet despite the important practical element, as the nineteenth century progressed the section tended increasingly to interpret its title as meaning rational mechanics or applied mathematics. The name of Coriolis (1792–1843) is known to modern students of physics ('Coriolis force'). As a professor at the Ecole Polytechnique and the author of several works on mechanics, he helped to clarify the basic concept of work in physics (1821). He was elected to the Academy in 1836 to succeed Navier.

The most famous mathematician in the section was Cauchy (1789–1857), who was brought into the section in 1816 by the new Royalist government to replace Monge, who had been too closely identified with the Napoleonic regime. Cauchy would have been more at home in the mathematics section since, although his important work on elasticity can be classified as 'mechanics', it tends to be overshadowed by his many major contributions to the calculus, algebra, error theory and differential equations. He was the effective founder of the theory of the complex variable. He is remembered not only as a great

[37] *Notice analytique sur les travaux de M. Poncelet*, 36 pp. n.d. Yet Poncelet's important work on the improvement of the efficiency of the undershot water wheel by the use of curved paddles had won him an Academy prize in 1825.

[38] Dirk J. Struik, art. 'Dupin', *D.S.B.*, 4, 257–8.

mathematician but as a prolific publisher of mathematical memoirs. He seems to have found the weekly meetings of the Academy a particularly valuable stimulus to making regular original contributions to mathematics.

Cauchy was succeeded in 1858 by Clapeyron (1799–1864), who had a long-standing interest in the theory and design of steam engines. It was Clapeyron who in 1834 had rediscovered and publicised the key concepts of his fellow countryman, Sadi Carnot (1796–1832), whose *Réflexions sur la puissance motrice du feu* had fallen still-born from the press ten years earlier. Clapeyron, therefore, has a firm place in the history of thermodynamics.

If the mechanics section had mathematics as its neighbour on one side, we might consider that it had physics on the other side.[39] Thus, after failing several times to be elected to the physics section, the experimental physicist Foucault (1819–68), of pendulum fame, was finally elected in 1865 to the mechanics section, illustrating once more how wide was the connotation of 'mechanics'. Yet because the section represented much of the same range of interests as the Ecole Polytechnique, the school's graduates became very prominent in this particular section, which on average had a higher proportion of *polytechniciens* than any other section (see Chapter 6, Section 10).

6. *The astronomy section*

The first two members nominated to the astronomy section were the practical astronomer Lalande and his protégé Méchain. Joseph-Jérôme Lefrançois de Lalande (1732–1807) had made a reputation for himself under the *ancien régime* both for his many publications and because of his love of publicity. He had been the editor of the astronomical almanac *Connaissance des temps*, professor of astronomy at the Collège Royale and the author of important text-books. Méchain (1744–1804), the discoverer of many comets, had also taken his turn as editor of the *Connaissance des temps*. He had a further reputation in geodesy, collaborating with Delambre from 1790 in carrying out the fundamental measurements on which the metric system was to be based. Lalande felt obliged to elect his former teacher Le Monnier (1715–99), although he was then aged eighty and paralysed. The fourth member of the section, Pingré (1711–96) was another eighty-year-old, whose reputation belongs entirely to the *ancien régime*. Then there was Messier (1730–1817), an observational astronomer associated with a long list of comets and nebulae, and Jean Dominique Cassini (1748–1845), who had been displaced in 1793 as Director of the Paris Observatory and at first declined to accept his election to the Academy. The astronomy section seems to have taken to extremes respect for seniority and service under the old order.[40] A second feature of 1795 is the emphasis on a good record of practical observations.

[39] Strictly speaking, astronomy was listed immediately after mechanics in the official order of the sections.

[40] Only Jeaurat (1724–1803), an astronomer who had made his career in the Royal Academy in the mathematics section, was omitted, to be elected in 1796 on the death of Pingré.

Although François Arago (1786–1853, elected 1809) and his brother-in-law Mathieu (1783–1875, elected 1817), continued the practical tradition, the section moved steadily towards theoretical astronomy and mathematics. In the eighteenth century one thinks of the expeditions to determine the shape of the earth associated with such Royal Academicians as La Condamine and Maupertuis, and also expeditions in connection with the transit of Venus.[41] In the Revolutionary period the geodesic work associated with the metric system gave a practical direction to the work of astronomers but this was only continued into the early Napoleonic period. It is true that the late nineteenth century was to witness another transit of Venus, for which the Academy sent Mouchez (1821–92, elected 1875), but if one wanted to choose an outstanding observational astronomer of the late nineteenth century the choice would probably fall on Janssen (1824–1907, elected 1873), who shares with the Englishman Lockyer the distinction of opening up solar astronomy through his many observations and publications. Janssen's work was taken further by Deslandres (1853–1948, elected 1902), who invented the spectroheliograph. Certainly in the late nineteenth century France occupied a leading position in astronomical photography, and the atlas of the moon by Loewy (1833–1907, elected 1873) and Puiseux (1855–1928, elected 1919), was standard in the field.

In the mid-nineteenth century, however, theoretical astronomy had clearly been gaining ground over practical astronomy and was seen to be in competition with it. Thus in the election of 1842–3, it was suggested that two columns should be drawn up to represent the best candidates in each of these areas.[42] Arago even felt it necessary to remark that places should be reserved for those representing observational astronomy! Only three years previously the Academy had elected the mathematician Liouville (1809–82) to a vacancy in the section. Liouville was the editor of a new journal, the *Journal de mathématiques pures et appliquées*, started in 1836. He was no astronomer in the ordinary sense but excelled in both pure and applied mathematics.

The view that astronomy was a kind of applied mathematics would have been favoured by the influential Le Verrier (1811–77), elected in 1846 while he was engaged in calculating the position of an unknown planet (Neptune) from perturbations of the motions of Uranus. Although his forte was on the theoretical side, Le Verrier was appointed Director of the Paris Observatory in 1854 on the death of Arago. He continued meteorological observations at the Observatory but, as far as observational astronomy goes, Le Verrier can only be associated with decline. His contemporary and rival Delaunay (1816–72) was also a theoretician, best known for his work on lunar theory.

A neighbouring institution which might have encouraged a more practical

[41] H. Woolf, *The transits of Venus. A study of eighteenth-century science*, Princeton, N. J., 1959.

[42] A. S., *Comité secret, 1837–44*, p. 117 (28 November 1842).

approach was the Bureau des Longitudes, founded in 1795 on the model of the British Board of Longitude to provide useful astronomical data for a maritime power.[43] It met weekly at the Observatory (1804–74), or later in the actual buildings of the Institute, where it still meets today. It consisted of two mathematicians, four astronomers, two naval experts, a geographer and a technician. One can see that astronomy was the predominant subject represented and there was a permanent and major overlap between membership of the Academy section of astronomy and the Bureau. In practice it often meant that an astronomer, once elected to the Academy, might expect before very long a second and more substantial source of income, amounting to 8000 francs, compared with the Academy honorarium of 1500 francs. The Bureau had the responsibility of publishing the annual *Connaissance des temps*; it also had overall responsibility for dealing with remaining problems in the establishment of the metric system. In the early years it was given specific tasks to perform, for example the exact comparison of the metric standards of length and mass, but activity soon diminished in this area. Indeed one suspects election to many posts in the Bureau as becoming little more than a sinecure, although there was a revival of activity in the 1860s.

7. *The section of geography and navigation*

For most of our period this section represented an anomaly, since it consisted of only three instead of the usual six members. There was no such section in 1795 and it was only added to the First Class in 1803 because of Bonaparte's decision to dissolve the Second Class, which had included such potentially subversive subjects as political science. The geography section was divided into two groups of three, the first of which could plausibly be related to science and was transferred to the First Class, and the other to the Third Class, which was concerned with classical literature and history.

Several of the Revolutionary legislators, such as Condorcet, had an image of geography which would integrate studies of topography and climate with the study of mankind. Such a study was seen as part of the Class of Moral and Political Sciences. But, being too near to the 'ideology' hateful to Bonaparte, geography in 1803 was given a more practical dimension by association with navigation, which was well represented by the famous explorer Bougainville (1729–1811) and the former sea captain and hydrographer Fleurieu (1738–1810). This left the representation of the subject of geography entirely in the hands of Nicolas Buache (1741–1815). He was the nephew of Philippe Buache (1700–73), who had had a special place created for him in the Royal Academy of Sciences as a geographer and who had also held the title of 'Premier

[43] M. G. Bigourdan, 'Le Bureau des Longitudes. Son histoire et ses travaux de l'origine (1795) à ce jour', *Annuaire du Bureau des Longitudes*, 1928, A1–A72; 1929, C1–C92; 1930, A1–A110; 1931, A1–A145; 1932, A1–117. Maurice Crosland, *The Society of Arcueil*, 1967, pp. 209–13.

géographe du Roi'. To this title his nephew had succeeded, so making Nicolas Buache the leading French geographer of his generation and an obvious choice as a member of the Institute. He was, however, less original than his uncle, whose ideas he developed. He held the post of Hydrographer in chief at the Depôt de la marine in Paris.

Hydrography was one of the best represented interests in the section, since more than a third of the members described themselves as hydrographers. The subject was well established in the eighteenth century, when it was taught in the naval colleges. However, it is necessary to point out that whereas the twentieth-century meaning of hydrography is restricted mainly to underwater topography and the study of ocean currents, the eighteenth-century meaning was broader, embracing all the sciences and technology of the seagoing man.[44] In the nineteenth century hydrography was approaching its modern meaning and there were many French naval expeditions to different parts of the world. The most distinguished hydrographers might receive a posting to Paris in the later part of their careers and some would aspire to become members of the Academy. One hydrographer, Rossel (1765–1829, elected 1811), even reached the rank of admiral (*contre-amiral*), but very many were members of the Bureau des Longitudes. The section also included several naval engineers, including Dupuy de Lôme (1816–85, elected 1866) and Emile Bertin (1840–1924, elected 1903). Bertin suggested limiting the damage to ships in war by combining armour plate with a cellular compartment. He headed the technical department of naval construction of the Naval Ministry.

Clearly physical geography was the kind of geography considered most appropriate to the First Class of the Institute, where it could be related, for example, to geology. Mapping[45] was regarded as one of the most important activities of the geographer and, through geodesy, there was a link with astronomy. Although an independent school of geography was established after the Revolution, it was soon militarised and in 1832 the corps of geographers was absorbed within the general staff. We therefore find senior army officers presenting themselves as geography candidates and in the late century two generals, directors of the geographical service of the army, were elected: François Perrier (1833–88, elected 1880) and Bassot (1841–1917, elected 1893). Perrier, a graduate of the Ecole Polytechnique, revitalised geodesy in the War Ministry and organised a complete revision of the geodesic and astronomical measurements carried out at the beginning of the century by Delambre and Méchain. The geography section also honoured great explorers, notably the astronomer Abbadie (1810–97, elected 1867), who had spent many years in

[44] François Russo, 'L'hydrographie en France aux XVIIe et XVIIIe siècles, in René Taton (ed.), *Enseignement et diffusion des sciences en France au XVIIIe siècle*, 1964, pp. 419–40.

[45] Josef Konvitz, *Cartography in France, 1660–1848*, Chicago, 1987.

Ethiopia, and Grandidier (1836–1921, elected 1885), whose name is inseparably linked with that of Madagascar.

The most important single event in the history of this section after its foundation was the decision of the government to increase the size from three to six members from 1866. Because of the lateness of this decision and its strong political overtones, its discussion is postponed until Chapter 12.

8. *The physics section*

We are used to thinking of physics as one of the senior branches of science which provides a model for other less well-established sciences. This modern image of physics must be abandoned if we are to understand the position of the subject in the early years of the Institute. Although the etymology of the word 'physics' takes us back to the ancient Greeks, *la physique* was a subject only recently recognised in eighteenth-century France, notably by the Royal Academy reorganisation of 1785, and still to be accepted in Britain.[46] The French in particular had contributed the term *physique expérimentale*, the title of the chair given to the popular lecturer the abbé Nollet, who was able to demonstrate with brilliant effect newly discovered properties of electricity as well as more standard experiments in mechanics, optics, sound and magnetism. Nollet, a member of the Royal Academy, had died in 1770 but the first member of the physics section to be nominated in 1795, Charles (1746–1823), bought up Nollet's apparatus for teaching purposes and can be seen as belonging to the same tradition. The fact that Charles was very much a representative of the old world of science is confirmed by the absence of any significant publications by him. The time which some authors devote to writing or polishing their writings, he is said to have spent polishing his instruments!

A further member of the section, Brisson (1723–1806), was really a naturalist but Nollet had encouraged him to extend his interests to elementary physics and in the 1780s he had begun to publish popular treatises on *la physique*. The best known name among the early members was probably Coulomb (1736–1806), who had already made most of the transition from military engineer to man of science.[47] He is particularly remembered for his torsion balance and his work on electricity and magnetism. Also in the original section was Cousin (1739–1800), who had more properly been classified as a mathematician under the *ancien régime*, and the experimentalist Lefèvre-Gineau (1751–1829), who was to play a prominent part in establishing the experimental basis of the metric system, notably in connection with the standard of mass.

In the opening years of the nineteenth century, without any unifying concept

[46] Maurice Crosland and Crosbie Smith, 'The transmission of physics from France to Britain: 1800–1840', *Historical Studies in the Physical Sciences*, 9, (1978), 1–61.

[47] C. Stewart Gillmor, *Coulomb and the evolution of physics and engineering in eighteenth-century France*, Princeton, N.J., 1971.

such as energy there was still no very clear idea of what studies should be regarded as 'physics'. The section had originally been entitled *physique expérimentale* in 1795. In 1803 the title was changed to *physique générale*, which lessened the emphasis on experiment and implied a more mathematical approach to the subject in keeping with a section within the division of mathematical sciences. The election of two young graduates of the Ecole Polytechnique, Gay-Lussac (1778–1850) in 1806 and Poisson (1781–1840) in 1812, suggested that the science should combine an experimental and a mathematical approach. The later elections of Fourier (1768–1830) in 1817 and Dulong (1785–1838) in 1823, confirmed these two complementary traditions. Probably of greater significance for the course of French physics in the nineteenth century were the elections of Malus (1775–1812) in 1810 and Fresnel (1788–1827) in 1823. The former, a disciple of Laplace, supported the theory of light as particles, while the latter was a pioneer of the wave theory. Yet the important thing about them both is that they represented the highest level of a new French interest in optics which was to continue throughout the nineteenth century. Both Malus and Fresnel were graduates of the Ecole Polytechnique and they combined a high level of ability in mathematics with experimental skill in a way which only found a feeble echo later in the century.

Fresnel was succeeded by Savart (1791–1841, elected 1827), whose work on vibrations, although relating to accoustics, was not totally unrelated to optics. Babinet (1794–1872, elected 1840) was to carry on a noble tradition by taking farther both the goniometer of Malus and the theory of Fresnel. A very powerful figure in the physics section was Fizeau (1819–96, elected 1860) and best known for his work on the velocity of light, a classic experiment which was later repeated by another worker in experimental optics, Cornu (1841–1902, elected 1878). Also working on optics were Jamin (1818–86, elected 1868), Mascart (1837–1908, elected 1884), the author of a three-volume treatise on optics, and Gouy (1854–1926, elected 1913). The early work of Pouillet (1790–1868, elected 1837) on solar radiation was taken further by Violle (1841–1923, elected 1897).

Given the fact that in Britain and the German states physics began to be transformed in the 1840s by the discovery of the principle of conservation of energy, one might expect a similar movement in France. Although the mathematical principles of heat conduction had been established by the brilliant work of Fourier (elected 1817), and the idea of a reversible heat cycle was the work of another Frenchman, the excessively modest Sadi Carnot (1796–1832), who in any case died too young to be considered for election to the Academy, it is disappointing to have to record the failure of other Frenchmen to build on this work.[48] Thermodynamics was discovered by the chemists in France before its fundamental importance was appreciated by the establishment physicists.

[48] An exception may be made in the case of Clapeyron, a member of the mechanics section.

The physical properties of gases had been a matter of special interest to Gay-Lussac, who may be considered as the real author of the law of thermal expansion of gases, often known as 'Charles' law' after his fellow-Academician.[49] Gay-Lussac's interest in the behaviour of vapours as ideal gases was taken further by Cagniard de la Tour (1777–1859, elected 1851), who discovered the concept of critical temperature. This work was later developed by Amagat (1841–1915), who had presented dozens of papers to the Academy before being elected, first as a correspondent (1890) and then as a full member of the physics section in 1902.[50]

We have left till last any mention of the Becquerel family, three successive generations of whom were members of the physics section. Although the first scientific interests of Antoine-César Becquerel (1788–1878) were in mineralogy, it was his subsequent interest in electricity which had helped gain him a place in the physics section (1829). His second son, Alexandre-Edmond (1820–91), joined the physics section in 1863 after doing research on magnetism and electricity. In the 1850s he did pioneering work on luminescence, which was taken further by his son Henri Becquerel (1852–1902, elected in 1889). The younger Becquerel's study of the fluorescent crystals of a compound of uranium (potassium and uranyl sulphate) in 1896 led to his discovery of radioactivity, which had the effect of revitalising French physics. Becquerel acted as intermediary for the presentation to the Academy of the papers of Marie Curie. Although Marie never became a member of the Academy, her physicist husband Pierre Curie (1859–1906) was elected a member in 1905, only a year before his tragic early death.

9. *The chemistry section*

A slightly anomalous feature of the chemistry section for the nineteenth century was that it was separated from physics (classified as a 'mathematical science') and labelled as a natural science (actually '*science physique*'). This meant that it was officially grouped together with mineralogy, botany and zoology, which in the *ancien régime* had been very much preoccupied with classification. To think of chemistry at a natural history stage is to think very much of pre-Lavoisier chemistry. Part of the 'chemical revolution' of Lavoisier had been built on the idea of the subject as a physical science. Although chemistry had to wait another few years for Dalton (1808) to propose atoms with specific weights, Lavoisier had made one of the bases of the new chemistry the principle of conservation of mass. Phlogiston had been eliminated largely by insistence on quantitative factors. There was no question in 1795 of advanced mathematics, but Lavoisier had definitely removed chemistry from the clutches of natural history (except at the Muséum) and had reorganised it as a *physical* science. It may seem a pity that

[49] Charles, however, had not measured the expansion under constant pressure; nor had he bothered to publish his work.

[50] As a reminder of the contributions of French Academicians to the theory and practice of the liquefaction of gases, we may also mention Cailletet (1832–1913), elected as *Académicien libre* in 1884.

purely to maintain a balance in the number of sections chemistry was included in the second group. In practice, however, it made little difference to the advancement of science. In the Academy its main effect was probably to increase the overall importance of chemistry, making it a leading science in the second division rather than an ancilliary science in the first division. By the 1860s, however, there was considerable justification for the view expressed that physics and chemistry alone should constitute a whole division of the Academy, since both subjects had shown so much development and they were not unrelated.[51]

But we must return to the situation in 1795. After the 'chemical revolution' associated with the name of Lavoisier, one might expect the composition of the chemistry section to reflect the 'new chemistry'. In a sense it did, but in a more important sense it did not, at least not immediately. It is true that the two chemists nominated by the Directory to represent chemistry, Guyton and Berthollet, were former associates of Lavoisier, and that they then proceeded to nominate their former colleague Fourcroy, but in 1794 the names of Guyton and Berthollet were probably more closely associated with gunpowder and saltpetre than with Lavoisier. In other words, part of the reason for the preference given to Guyton and Berthollet was based on the civic considerations as much as scientific ones. If the section had simply been a 'Lavoisier school', the next person for election would probably have been Chaptal. He had, however, become ineligible for consideration as a resident member by moving back to his native Midi, although fortunately only temporarily. Back in Paris a few years later he succeeded to a vacancy in 1798.

Meanwhile it is noteworthy that the remaining three elections to bring the section up to the statutory number of six, and the first replacement election (1797), were all won by pharmacists. Bayen (1725–98) had been a military pharmacist for thirty years and was elected first on grounds of seniority. His most important chemical research was that carried out in 1774–5 on precipitates of mercury. He had come to doubt the phlogiston theory but had never challenged it in the way Lavoisier did, and his work was vastly inferior to the unfortunate leader of the French chemists, whose untimely execution in 1794 had deprived France of one of her leading *savants*. Then there was Bertrand Pelletier (1761–97), who had learned pharmacy from his father. Like his master, Jean D'Arcet, Pelletier only accepted Lavoisier's theory after some hesitation. The other pharmacist Vauquelin (1763–1829) was a protégé of Fourcroy and also a late convert to Lavoisier's chemistry. He was appointed director of the Ecole de Pharmacie on its foundation in 1803. On Pelletier's early death in 1797 he was succeeded by a fourth pharmacist, Deyeux (1745–1837). We might mention that a further pharmacist, Antoine Baumé (1728–1804), who had been a senior member of the chemistry section in the Royal Academy of Sciences, was reduced to being a corresponding member of the new Academy.

[51] J. Marcou, *De la science en France*, 1869, fasc. 2, p. 103.

A supporter of the phlogiston theory, Baumé could not accept the new chemistry of Lavoisier and his colleagues.

The election of so many pharmacists must be more than a coincidence. It represents in the first place the high scientific level of French pharmacy at the end of the *ancien régime*. A high standard of *scientific* competence was to continue in the nineteenth century to be a characteristic feature of the best of French pharmaceutical training, distinguishing it, for example, from the British apothecaries, who were pulled in one direction to be shop-keepers and in the other to become general medical practitioners. But apart from reflecting a ready-made pool of talent, the pharmacy constituent in the Academy reflected the continuation of a seventeenth- and eighteenth-century tradition in chemistry, one which was strongly supported by Fourcroy but would have been played down if Lavoisier had still been alive.

Yet if one sees quite a strong representation of pharmacists or former pharmacists in the chemistry section in the first half of the nineteenth century (for example Proust, elected 1816, Serullas 1829, Robiquet 1833, Balard 1844), it should be said that these men were elected *despite* their pharmacy background rather than because of it. Proust, a relic from the *ancien régime*, had made a separate reputation for himself in chemistry, taking him far from his beginnings as a pharmacy apprentice. Balard was clearly elected for his strictly chemical work on bromine; the fact that bromides had a later part to play in pharmacy is purely coincidental. But perhaps the most useful lesson is to be learned from men like Serullas and Robiquet, who moved early from pharmacy to the rapidly expanding field of organic chemistry. Serullas had been a pharmacist and professor of pharmacy before being nominated as professor of chemistry at the Muséum d'Histoire Naturelle. He did important work on the investigation of compounds of the newly discovered elements iodine and bromine. Robiquet combined early training in pharmacy with the benefits of the patronage of Fourcroy and Vauquelin. He developed their interests in the 'principles' which could be extracted from natural products, developing this into a kind of organic analysis.

Organic chemistry was essentially a nineteenth-century science. Lavoisier's 'chemical revolution' had focused principally on inorganic or mineral chemistry, although he had also suggested the basic principles of the analysis of vegetable and animal materials. So, although a few chemists in the generation after Lavoisier made a reputation for themselves in inorganic chemistry by the discovery of new elements, the greatest innovations lay in organic chemistry. Given that most organic compounds contained carbon, hydrogen and oxygen and sometimes nitrogen, it was necessary to determine quantitatively the proportion of each of these elements. But the atoms could be arranged in different ways. Organic chemists tried desperately to make sense of chemical reactions involving organic compounds by postulating that they contained groups of elements or 'radicals'. Later attempts at classification produced various 'type' theories. The number of organic substances known rose from hundreds to thousands and some framework was necessary to make sense of experiments. In Germany Justus von Liebig (1803–73), a former student of Gay-Lussac, became the leading organic chemist. His opposite number in France was J. B. Dumas (elected 1832), who had begun as a physiologist before deciding that a career in chemistry would provide him with a more secure future.

One of the first generation of organic chemists in the Academy, elected in 1826, was the long-lived Chevreul (1786–1889), a former student of Vauquelin. He had spent more than a decade studying natural fats and the chemistry of soap, later specialising in the chemistry of natural dyes. Artificial dyes, such as those made from aniline, had to wait till the 1850s and 1860s, when they opened a whole new chapter in the history of organic chemistry. Organic chemistry changed so much around this time that a man of Chevreul's generation had difficulty in keeping up to date, all the more so because, as he approached old age, his interests widened to include pseudo-science and philosophy of science. The representation of organic chemistry in the section increased with the election of Wurtz and Cahours in the 1860s. When the next vacancy occurred in 1876 with the death of Balard, the chemistry section found it almost impossible to distinguish the merits of the main contestants, even after several long meetings.[52] The organic chemists with Chevreul as their doyen had greatest difficulty in judging the work of younger organic chemists, who reported results with new organic compounds which could not be bought commercially and which it was very difficult to prepare in the laboratory. The older chemists even admitted that they did not understand all the terms used, so rapidly was the subject expanding.

That is not to say that inorganic chemistry was dead. Mendeleef's Periodic Table of 1869 had helped to restore the morale of that branch of science by classifying the elements in groups according to their respective atomic weights. Inorganic chemists were often successful in elections, sometimes alternating with organic chemists. Perhaps worthy of special note is Moissan (elected 1891), a more unusual case now of someone trained originally in pharmacy. Moissan had the distinction of finally isolating one of the most reactive of elements, fluorine (1886), which belongs to the same family as chlorine. He received the Nobel prize for chemistry in 1906.

10. *The mineralogy section*

The mineralogy section provides a good illustration of how the Academy was able to adapt an eighteenth-century label to a nineteenth-century situation, since there is probably no better example of a branch of science changing and developing beyond the apparent confines of its foundation in 1795. The collection, classification and display of specimens of minerals and other natural

[52] A. S., dossier Henri Debray, MS. 'Presentation de la liste des candidats à la place vacante dans la section de chimie par le décès de M. Balard', (7 pp.).

objects in cabinets was one of the ways in which a fashion for science developed in the eighteenth century. But there was not only a social but an economic aspect. French governments became increasingly interested in the exploitation of their coal and other mineral resources. Thus in the 1750s Gabriel Jars was sent on a long expedition to study the mines of Saxony and central Europe; he was eventually rewarded with a place in the Royal Academy. A number of important works on minerals in Swedish and German were translated into French. Already in 1745 a chair of mineralogy had been established at the Jardin du Roi and in 1783 the French government founded the Ecole des Mines, partly inspired by the *Bergakademie* of Freiberg. In 1772 Romé de l'Isle published the first edition of his book on crystallography, which was to provide a valuable precedent for the definitive work of the abbé Haüy.

All the first members of the section had good experience of mineralogy under the *ancien régime*. The senior member, D'Arcet (1725–1801) had a life-long interest in minerals and their chemical analysis. In the early period of the revolutionary war he had played a prominent part in applying chemistry to the extraction of metals for the war effort. His interest in applied science was shared by his exact contemporary Desmarest (1725–1815), who had been inspector general for manufactures. It had been while inspecting industries in the Auvergne in the 1760s that Desmarest had become interested in the geological formation of the region and built a scientific reputation on his hypothesis of the volcanic origins of basalt. His colleague Dolomieu (1750–1801) had travelled even more extensively, including the Alps, with part of which his name is still associated. If we mention that two remaining members of the original section had contributed to mineralogy, mainly by taking part in surveys (Duhamel, 1730–1816), or through administration (Lelièvre, 1752–1835), this will serve to throw into relief the *practical* contribution of the section.

This leaves till last René-Just Haüy (1743–1822), probably the most original and influential member of the original section. He, uncharacteristically, did no field work but used the large collections in Paris to build up a theory of the mathematical relationship of crystal forms, which entitles him to be regarded as one of the founders of the science of crystallography. Delafosse (1796–1878), who became his assistant in 1817, gave his theory of crystal structure a more mathematical character, but was unfortunate in not achieving election to the Academy till 1857. He was the teacher of Pasteur (1822–95, elected 1862), who has so many other claims to fame that we are in danger of overlooking the fact that it was as a crystallographer that he had entered the Academy in the section of mineralogy. Pasteur's study of the relationship between the crystalline forms of tartrates and the optical properties of their solutions was soon recognised as a brilliant piece of research. It provided further confirmation that nature sometimes spoke in the language of crystals, the science of which had undergone something of a revival of the mid-century thanks to the work of Delafosse and Bravais (1811–63). Bravais' theory of crystal lattices cannot, however, be introduced here to characterise the mineralogy section since the polymathic Bravais was to enter the Academy in the 'miscellaneous' section of *Académiciens libres*.

But if one development of mineralogy was in the application of geometry to the study of crystals, the main development was towards the study of the earth and its history. Already in the eighteenth century Buffon had raised questions about the age of the earth, and the construction of canals (particularly in England) had revealed new evidence of different layers of rock beneath the Earth's surface. In the organisation of the Muséum d'Histoire Naturelle in 1793 a chair of 'géologie' had been established and given to Faujas de Saint-Fond (d. 1819) but he was not considered distinguished enough to enter the Academy. During the Napoleonic wars French scientists were not able to take part in expeditions to other continents but some made important studies on home ground. One of the most notable of these was the study by Alexandre Brongniart and Cuvier of the Paris basin from 1804 to 1808, when they presented their results to the First Class.[53] Brongniart, who was able to make good use of his training as a mining engineer, showed that above the chalk was a complex series of strata, probably formed by slow deposition. This suggested the elapse of a considerable period of time since the chalk period. Brongniart also used fossils for the correlation of strata and in 1815 published a geological map. With his election in 1815 to the Academy we may consider that geology had arrived. Several members of the section elected in the next few years combined interests in geology and mineralogy: Brochant de Villiers (1772–1840, elected 1816); also a former student of Werner, Cordier (1777–1861, elected 1822) and Beudant (1787–1850, elected 1824).

In 1825 Elie de Beaumont (1798–1874) was elected to the section. He had spent many of the preceding years gathering data from the eastern half of France for a national geological map and was appointed to chairs in geology at the Ecole des Mines and the Collège de France. In 1829 he had presented his first ideas on tectonics to the Academy. He was also one of the founders of the Société Géologique de France (1830).[54] It is clear, therefore, that the section did not allow the formal title to exclude the new and expanding subject of geology. On the other hand it did not want simply to transfer itself into a section of geology. Candidates for election sometimes came to be presented under one of two headings: mineralogy *or* geology.[55] In two successive elections in 1857 the

[53] Interestingly this geological memoir was presented under the title: 'Essai sur la *géographie minéralogique* des environs de Paris' (my italics). An enlarged version of the memoir was published in 1811 in a book, in which Cuvier appeared as the principal author, although Brongniart had done most of the work.

[54] Yet one biographer (Arthur Birembaut, *D.S.B.*, **4**, 350) considers that he was really a cold mathematician at heart, which would help to explain why he ended his days in the Academy as secretary of the *division of mathematical sciences*, rather than of the division to which mineralogy belonged.

[55] *P.V.I.*, **8**, 639 (24 December 1827), where in an election for a correspondent Conybeare is described as a 'geologist', as opposed to 'mineralogists' properly so-called, who were considered separately. A.S., *Comité secret, 1857–69*, p. 14 (23 February 1857).

crystallographer Delafosse was first elected (16 March) as a 'mineralogist', to be followed on 27 April by Archiac (1802–69) as the leading candidate under the heading 'geology'.

Yet the development of the subject called for more than a two-fold division. Mineralogy was already associated with mining, sometimes with crystallography, with (inorganic) chemistry, and occasionally with metallurgy. But geology too was giving rise to new specialisms such as palaeontology. Palaeontology, or the systematic study of the fossil remains of animals and plants of past ages, gradually emerged as a specialist branch of geology in the nineteenth century. Already in the Napoleonic period Brongniart had made good use of fossils for the detailed correlation of strata but although he displayed considerable skill in the identification of the fossils, he was doing no more than using them as a tool. He is judged, therefore, to have made a major contribution to stratigraphy rather than to palaeontology. The subject did not take off until the 1850s. In 1853 one of the chairs of botany at the Muséum was transformed into a chair of palaeontology, despite the opposition of a number of the professors who saw this as having radical implications. The chair was given to Alcide d'Orbigny, who had recently published two major books on the subject. D'Orbigny made two attempts in the 1850s to enter the mineralogy section of the Academy but was defeated by candidates representing better-established disciplines. Meanwhile his young brother-in-law Gaudry (1827–1908) was receiving substantial grants from the Academy to travel to Greece to undertake a prolonged study of fossil remains of certain quadrupeds,[56] showing that the Academy had more ways than one of recognising a new discipline. By 1862 Gaudry was able to publish a treatise describing his research and drawing evolutionary conclusions. In the circles of the Academy this tended to win him more enemies than friends but he was to gain support elsewhere, including that of Duruy, Minister of Education, who appointed him to a chair in palaeontology at the Sorbonne in 1868. His eventual election to the Academy in 1882 marked not only the final recognition of palaeontology by the Academy but also a growing general support for evolutionary ideas. In 1903 Gaudry was briefly joined by a second palaeontologist, Munier-Chalmas (1843–1903), who unfortunately died a few months after election.

11. *The botany section*

In contrast to the mineralogy section, we cannot pretend that botany was one of the more brilliant sections of the Academy. In the nineteenth century French botany was soon overtaken by the high level of work in the German states. Nor can this be blamed on the organisation of the Institute or the Muséum; the

[56] Gaudry received the exceptionally large sums of 6000f in 1855 and a further 8000f in 1860. M. Crosland and A. Galvez, 'The emergence of research grants', *Social Studies of Science*, 19 (1989), 71–100 (p. 83).

former provided six places for the subject and the latter had two professors of botany. Yet with the many universities in Germany and a new independent status given to botany, separating it from its traditional links with medicine, there were many more posts in botany in the German states than in any other country of the world.[57] Also describing the contributions of the botany section of the Academy, we are penalised by being obliged to omit the famous botanist Augustin Pyramus de Candolle (1778–1841), whose residence in Montpellier and Geneva precluded his ever becoming a full member of the section.[58] Equally, neither the great explorer Alexander von Humboldt (1769–1859) nor Bonpland (1773–1858), his co-author of a pioneering work on the geography of plants (1807), were elected to the botany section although the Academy, which insisted on French nationality and resident status for full membership, was able to elect each of them in other capacities.[59]

At least the section could boast retrospectively of having had as its first member Lamarck (1744–1829), nominated by the Directory on the strength of his *Flore française* (1st edition, 1779; 2nd edition, 1795), and his major botanical contributions to the *Encyclopédie méthodique*. He was not, however, always appreciated by his contemporaries and thus, in the establishment of the Muséum, where he was given a professorship, it was not one of botany but of 'insects, worms and microscopic animals'. Turning this assignment to advantage, Lamarck soon began to speculate about transformism in the animal as well as in the vegetable kingdom, but his wide-ranging interests (which included unorthodox chemistry and meteorology) only served to alienate him further from his colleagues, who thought increasingly of scientists as specialists. It was only later that his ideas on transformism were appreciated in France as a reputable theory of evolution, foreshadowing the work of Charles Darwin.

Three of the other original members of the section were Adanson, A. L. de Jussieu and Desfontaines. Michel Adanson (1717–1806) was elected purely on the basis of his seniority. He had begun his career many years earlier by studying at the Jardin du Roi in Paris and then going to Senegal for six years. Finding so many tropical plants which did not fit into the current schemes of classification he determined to establish a natural system of classification, which he published in his *Famille des plantes* (1763–4). He had little to contribute to the post-Revolutionary world and spent the last years of his life fighting poverty. It fell to his fellow-Academician Antoine Laurent de Jussieu (1748–1836), who had adopted Adanson's principles in his *Genera Plantarum* of 1789, to establish natural classification as the basis of modern systematic botany. Jussieu might have had a less original mind than Adanson, but he had the advantage of the

[57] A. G. Morton, *History of botanical science*, 1981, pp. 364, 405.

[58] De Candolle was elected correspondent in 1810 and *associé étranger* in 1826.

[59] Alexander von Humboldt was elected correspondent in 1804 and *associé étranger* in 1810. Aimé Bonpland was elected correspondent in 1817.

enormous number of new plants which poured into the herbaria at the Muséum d'Histoire Naturelle where he lived and held the post of professor of 'Botany in the countryside', as opposed to his fellow-Academician, Desfontaines (1750–1833), who held the chair of 'Botany at the Museum'.

The title of this latter chair was transformed by his successor, Adolphe Brongniart (1801–76, elected 1834) into 'Botany and vegetable physiology'. Brongniart's main reputation, however, was in palaeobotany, of which he was virtually the founder. From his first memoir on the classification and distribution of fossil plants in 1822 to 1849, his main work was in this area, which combined his deep interest in botany with some of the geological interests of his father, a member of the mineralogy section. We have neglected to mention some other botanists of distinction belonging to the early mid-nineteenth century, notably Mirbel (1776–1854, elected 1808), who collected 24 000 plants as well as many birds and insects during a six-year study in Brazil. Also worthy of mention is Louis-René Tulasne (1815–85, elected 1854), one of two brothers who, on receiving a large legacy, decided to devote their lives to the study of botany. Their main published work was a three-volume monograph on fungi completed in 1865.

Although the botany section was not as pioneering in its research as many of the other sections, it did make some contributions to more modern fields, such as morphology, plant pathology, ecology and even genetics. Van Tieghem (1839–1914, elected 1877), for example, first put the study of vascular anatomy upon its modern scientific basis. He created a new approach to the anatomy of plants, in which he emphasised the principles of plant symmetry and made special studies of the pistil, ovule, the composition of the seed and root. His son-in-law Bonnier (1853–1922, elected 1897) was a pioneer in the study of the effect of environment on the growth of plants, having demonstrated the changes that took place in the same plant when it was transferred from an alpine to a lowland habitat and vice-versa. We must not forget the unfortunate Naudin (1815–99, elected 1863), living his later years in the South of France wracked by illness. In the 1850s, however, he had begun a systematic study of hybridisation, using mainly *Datura* species. The first generation of hybrids seemed relatively homogeneous but he went on to study the diversity of second generation hybrids. Unlike his contemporary Gregor Mendel, however, he failed to collect statistical data and today we speak of 'Mendel's law' rather than 'Naudin's law'. A later botanist, Mangin (1852–1937, elected 1909), made a special study of plant diseases; he also pioneered the use of dyes for microscopic investigation.

What is missing however from most of this account is the introduction of completely new methods or approaches. One can list many minor advances but only a few major ones.

12. *The agriculture section*

The first two members nominated by the government to this section were André Thouin (1747–1824), a former member of the botany section of the Royal Academy, and F. H. Gilbert (1757–1800), often known as 'Gilbert d'Alfort' because of his attachment to the veterinary school at Alfort, near Charenton, a few miles to the east of Paris. These two appointments made clear the two aspects of agriculture that were intended to be represented in the section, one relating to plants ('*économie rurale*') and the other relating to animals ('*art vétérinaire*'). The full official title of the section was 'Economie rurale et art vétérinaire', but even within the Academy this was found to be too cumbersome for everyday use so that even semi-official documents often speak of 'the agriculture section', and we shall follow this usage.

Election to the section brought in Cels (1740–1806), who had specialised in the naturalisation of exotic plants, Tessier (1741–1837), who had studied certain animal diseases, J. B. Huzard (1755–1838), who became director of the Alfort veterinary school, and finally the pharmacist Parmentier (1737–1813), a keen utilitarian who had been an advocate of the nutritive properties of the potato and whose name still lives on in that connection. Other early members of the section included Silvestre (1762–1851), who was secretary of the Société d'agriculture de la Seine for forty-four years, and Bosc d'Antic (1759–1828), who edited a standard text-book of agriculture[60] with contributions from the entire membership of the Academy section. We may call attention to two features of the agriculture section, one specific and one more general; first the representation of the Alfort veterinary school and second the question of the range of sciences relevant to agriculture. These will be discussed in turn.

France was a predominantly agricultural country even in the nineteenth century, when Britain was becoming increasingly industrialised. But even in the eighteenth century the French government had realised that as so much national wealth depended on agriculture it was desirable to apply science to improve this area. It was in an attempt to improve the standard of farm livestock and horses (vital for transport and in time of war) that the government established a veterinary school at Lyons in 1761 and another at Alfort, near Paris, in 1766.[61] The latter taught not only agriculture but scientific subjects such as comparative anatomy and chemistry. Despite transport difficulties, modest salaries and the lower social status of the students in comparison with the Paris schools, the Alfort school was able to attract some quite distinguished professors from the metropolis, both under the *ancien régime* and in the

[60] *Nouveau cours complèt d'agriculture*, 1st edn, 1809; 13 vols., 2nd edn, 1821–3, 16 vols.

[61] E. Leclainche, *Histoire de la médecine vétérinaire*, Toulouse, 1936. For an analysis of French higher agricultural education in the later nineteenth century, see Harry Paul, *From knowledge to power. The rise of the science empire in France, 1860–1939*, Cambridge, 1985, Chap. 5.

nineteenth century. In the nineteenth century there was usually at least one representative of the Alfort veterinary school in the agriculture section of the Academy.[62] Henry Bouley (1814–85, elected 1868) was a professor at Alfort and Auguste Chauveau (1827–1917, elected 1886) was a graduate of the school.

The other question to be considered is what range of sciences could properly be found in the section. The late eighteenth and early nineteenth centuries were a time when chemistry was beginning to demonstrate its utility in its application to agriculture, especially in fertilisers, soil analysis and animal nutrition. The first major agricultural chemist to be elected to the section was J. B. Boussingault (1802–87, elected 1839), who owned an experiment farm and had carried out experiments on crop rotation. Three years later a second chemist, Anselme Payen (1795–1871) was elected to the section. Payen was an industrial chemist who had done some work on sugar beet, which could be regarded as agricultural research. The claims of Peligot (1811–90, elected 1852) to membership of the section were less clear. In fact, it was only after three unsuccessful attempts to enter the chemistry section that he had realised that he would have a better chance of entering the Academy if he presented himself as an agricultural chemist. Even the entomologist Audouin (1797–1841) was able to enter the section in 1838 on the grounds that he had studied insects harmful to agriculture.

The incursion of extraneous sciences also happened in another way. In 1828 the Academy rejected outright the recommendations of the agriculture section and elected a protégé of Cuvier, the physiologist Flourens (1794–1867). The precedent having been created, a second physiologist Dutrochet (1776–1847), was elected to the section in 1831. Dutrochet was a distinguished plant physiologist who had been unsuccessful in a previous candidature in the botany section. Although it could be argued that physiology was a fundamental science of some relevance to the welfare of animals and the production of crops, neither Flourens nor Dutrochet tried seriously to justify their membership of the section of agriculture.

An agricultural background being an advantage in any election to the section, one could in theory have equally well have been a peasant[63] or a landowner. The section had its full share of members of lowly social origins. Thus Thouin, following his father's footsteps had, under the *ancien régime*, been a gardener at the Jardin du Roi and Chauveau was the son of a blacksmith. In different generations each demonstrated the ability to relate the trade of his upbringing to science. At the other end of the social scale we may mention

[62] Paul Elliott, 'Vivisection and the emergence of experimental physiology in nineteenth-century France', in N. Rupke (ed.), *Vivisection in historical perspective*, 1987, pp. 48–77.

[63] All candidates in the republican year 4 for the position of *associé non résident* attached to the agriculture section are described as 'cultivateurs' (*P.V.I.*, 1, 14).

Silvestre, who was made a baron by Louis XVIII, Morel-Vindé (1759–1842), who was successively baron and viscount, and Gasparin (1783–1862), who was a member of the Chamber of Peers in 1834 and even briefly became Minister of the Interior (1836–7).

Yet the section tended to be looked down upon by the other sections both socially and intellectually. The members of the medicine section certainly considered themselves superior to veterinary surgeons and the scientists were sometimes snobbish about a close association with the soil. It was a great day for the section when the veterinarian Henry Bouley was elected in 1885 to the largely honorary post of annual president of the Academy. Pasteur, himself a man who never forgot his humble origins and a scientist who had made major contributions to veterinary science as well as to medicine, made a speech at a banquet given in Bouley's honour.[64] But the standing of the section did not depend entirely on intellectual snobbery. In a survey carried out in the mid-century on the relative productivity of the different sections in terms of number of memoirs presented over a period of several months to the Academy, the sections of agriculture and botany come last. They stood out as sections contributing significantly less than their share; only a quarter of the output of the more prolific sections of mathematics, mechanics and chemistry.[65] On the other hand, it could be argued that Pasteur's epoch-making work in microbiology helped to breathe new life into the section in the late nineteenth century, when successive Directors of the Institut Pasteur, the microbiologists Emile Duclaux (1840–1904) and Emile Roux (1853–1933) were elected as members.

13. *The section of anatomy and zoology*

The founding of the Institute provided an opportunity to abandon the old idea of 'natural history' and establish a series of specialist sections concerned respectively with the vegetable, mineral and animal kingdoms.[66] Of these the last had previously received the least recognition. But rather than take the enormous step of designating a whole section as concerned with the new science of zoology, the legislature introduced a compromise which involved more continuity with the old regime. The section was called 'anatomy and zoology', thus perpetuating a branch of science of the greatest importance in the sixteenth and seventeenth centuries and formally recognised in the Royal Academy since 1699. But the anatomy which had been recognised to be so important was

[64] 'Bouley est élu à la présidence de l'Académie des Sciences', in Louis Nicol, *L'épopée Pastorienne et la médecine vétérinaire*, 1974, pp. 613–17.

[65] Quatrefages, 'L'Académie des sciences et ses travaux', *Revue des deux mondes*, [5], 10 (1845), 973–1010.

[66] Although there was the *opportunity* in 1795 to abandon the idea of natural history entirely, the name lived on temporarily in the section called 'natural history and mineralogy'. The inappropriateness of this title was soon felt and in 1803 the section was retitled simply 'mineralogy', thus discarding forever the phrase 'natural history' from the official vocabulary of the Academy.

human anatomy, a standard subject in the medical curriculum. Since the time of Vesalius, anatomy had been closely associated with surgery. But the First Class had a separate section for medicine and surgery and there was, therefore, some ambiguity about whether the anatomy in the section of that name should relate to the anatomy of animals in general but excluding man.

The ambiguity is reflected in the early membership of the section. Daubenton (1716–1800), nominated by the Directory, was a naturalist and former collaborator of Buffon, outside any medical context, but he helped to elect Tenon (1724–1816), who was a surgeon and anatomist. Also elected in December 1795 was 'Georges' Cuvier (1769–1832) the youngest member of the section.[67] Cuvier was to do important work in comparative anatomy but he was also a zoologist. The senior zoologist in the section was Lacépède (1756–1825), who was France's leading expert on reptiles and fish; later, under Napoleon, he was to combine his scientific career with a government post as grand chancellor of the Legion of Honour. Cuvier's friend the zoologist Etienne Geoffroy Saint-Hilaire (1772–1844) was less politically astute and was not elected until 1807.[68]

It was Geoffroy who, in 1821, was to protest that the Academy gave insufficient recognition to zoology, since the comparable science of botany (which he regarded as virtually complete) had a whole section to itself, while zoology had to content itself with half a section.[69] The occasion for the protest was the presentation of a list of candidates for a vacancy in the section.[70] Geoffroy had presented a list of zoologists but he had to admit that only the entomologist Latreille supported him, the other zoologist Lacépède refusing to be involved. Another faction within the section, consisting of Duméril and Pinel, wanted to place on record the rights of candidates concerned with human anatomy. They accordingly produced two parallel lists. Geoffroy replied with a memoir pointing out that there was a separate section in the Academy for members of the medical profession. Moreover, the financial and honorific resources available to medical doctors outside the Academy were far greater than those available to zoologists. Zoologists were confronted with daunting tasks, given the enormous number of species to be studied, yet they tended to live in isolation. They needed protection and financial support. On the other hand he admitted that part of the problem went back to the foundation of the Institute in 1795 when, for example, no provision had been made for physiology. He was prepared to accept that half the membership of the section should be devoted to

[67] For useful hints of the politics behind this election, not unrelated to appointments at the Muséum, see D. Outram, *Georges Cuvier*, 1984, pp. 54–5.

[68] Geoffroy had been a candidate in 1803 but had received only 23 votes as against 30 votes for the physician Pinel. This election had prompted a debate as to whether the Academy should be looking for a human anatomist or a zoologist. *P.V.I.*, 2, 640.

[69] 'La zoologie a-t-elle dans l'Académie des Sciences une représentation suffisante?' *Revue encyclopédique*, 13 (1822), 501–11.

[70] *P.V.I.*, 7, 210 (23 July 1821). See also John Lesch, *Science and medicine in France. The emergence of experimental physiology, 1790–1855*, Cambridge, Mass., 1984, pp. 118—20.

'anatomy' but this still left open the question of whether this should include physiology.[71] Geoffroy's principal concern was to safeguard the position of zoology.

The name of Pinel (1745–1826) has been mentioned, and it must be said that although he was an immensely important figure in the history of medicine, nosology and particularly psychiatry, his position as a member of this section represents an anomaly. It can only be explained on the basis of the ambiguity of the term 'anatomy', which allowed a distinguished senior clinician into a section whose normal understanding of anatomy related more to comparative anatomy. A considerable number of members, beginning with the illustrious Cuvier, were not *either* zoologists *or* (comparative) anatomists but *both*, as if the very title of the section exercised a strong influence on their research. A second figure who stands out as unusual in the section was Quatrefages (1810–92), who was elected in 1852 as a zoologist but showed unusual flexibility in moving later through a study of human anatomy to anthropology (see Chapter 12).

Another important zoologist was Blainville (1777–1850) although, like Cuvier, he was also an outstanding comparative anatomist. Originally Cuvier's assistant, he quarrelled with him but still managed to be elected to the Academy in 1825. Blainville was a practising Catholic, whose religious instincts guided his science most prominently in his vision of the unity of God's creation.[72] He saw God's plan for the ordering of animals and plants as part of a *série* or hierarchical chain of being, which he defended against Cuvier's static model of creation. Among the next generation of zoologists was Henri Milne-Edwards (1800–95), partly of British descent, who made a special study of crustacea, drawing on field work he had carried out in the Mediterranean off the coasts of Provence, Italy, Algeria and Sicily.[73] He became one of the founders of marine biology. He was also known as an author, not only of a standard text-book but of a fourteen-volume encyclopaedia of zoology. In addition he spent twenty-four years bringing out a vast study of comparative anatomy. His research revealed links between living animals and fossils but he was very cautious about propounding evolutionary ideas.

One of his protégés was the zoologist Lacaze-Duthiers (1821–1901) who, like Blainville, had studied for the degree of doctor of medicine before deciding that his life's work lay in the study of the animal kingdom. After occupying the newly created chair of zoology at the Faculty of Science at Lille for ten years, he was able to come to Paris where he held successively chairs at the Muséum and the Faculty of Sciences, which allowed him to pose his candidature for the

[71] The physiologist Magendie, who had been included in the alternative list of July 1821 as an 'anatomist', entered the Academy in November 1821 in the section of medicine and surgery, thus creating an important precedent.

[72] William Coleman, art. 'Blainville', *D.S.B.*, 2, 186–8.

[73] M. Berthelot, 'Notice historique sur Henri Milne-Edwards', *M.A.I.*, **47** (1904), i–xxxvii.

Academy.[74] He made important contributions to zoological education in France and, as a marine biologist, founded two of the earliest marine zoological laboratories at Roscoff in Brittany and Banyuls on the Mediterranean. A spirited defence of zoology as a science based on observation came from Coste (1807–73), professor of comparative embryology at the Collège de France, in a reply to Claude Bernard, who had emphasised the role of experimentation in the life sciences. Coste, speaking to the Academy in 1868, pointed out that careful observational work on the life-cycle of fish, such as salmon, had important economic consequences.[75]

Members elected to the section towards the end of the century included Edmond Perrier (1844–1921) and Ranvier (1835–1922). Perrier continued work in marine biology in the tradition of Henri Milne-Edwards and Lacaze-Duthiers. He declared his acceptance of the theory of evolution in 1879 and soon became one of the principal defenders of the theory in France, being, however, more of a Lamarckian than a Darwinian.[76] The other Academician Ranvier could be classified as an anatomist but his speciality was histology. He was France's leading histologist in the later nineteenth century, combining the best of the German microscopical tradition with the best of French physiology as a former student of Claude Bernard.[77] He did important work on the peripheral nervous system, excelling in experimental techniques but avoiding discussion of the theoretical implications of his work. This strand of positivism is seen more clearly in another histologist, Robin (1821–85), who as a 'free-thinker' and partisan of the extreme left, allowed politics to take over from scientific research in his later life.[78]

14. *The section of medicine and surgery*

Although there had been an anatomy section in the Academy of the old regime, the incorporation of representatives of the professions of medicine and surgery[79] as a whole section was one of the innovations of the Revolution. Already in 1790 Vicq d'Azyr, on behalf of the Société Royale de Médecine, had proposed the formal establishment of an Academy of Medicine which, he said,

> needs to be placed near the Academy of Sciences, to be so to speak, the constant witness of its researches, to associate itself with the spirit and to follow the same path to arrive at new results.[80]

[74] Toby A. Appel, art. 'Lacaze-Duthiers', *D.S.B.*, 7, 545–6.

[75] *C.R.*, 66, (1868), 1278–84. Paul Elliott, 'Vivisection and the emergence of experimental physiology in France', in N. Rupke (ed.), '*Vivisection in historical perspective*, 1987, pp. 48–77 (72).

[76] C. Limoges, art. 'Perrier', *D.S.B.*, 10, 522–3.

[77] Toby A. Appel, art. 'Ranvier', *D.S.B.*, 11, 295–7.

[78] M. D. Grmek, art. 'Robin', *D.S.B.*, 11, 491–2.

[79] The bringing together of medicine and surgery at the time of the Revolution, as, for example, in a common educational programme, does not seem to have had any marked effect in the Academy.

[80] *Nouveau plan de constitution pour la médecine en France*, présentée à l'Assemblée Nationale par la Société Royale de Médecine, 1790, p. 150.

In other words, it was clear to some leading physicians that medicine had much to gain by association with the rigours of scientific method as well as with some of the actual content of science. Vicq d'Azyr's protégé, Fourcroy was to launch, in 1791, a new journal based on this philosophy: *La médecine éclairé par les sciences physiques*.[81] Many members of the Royal Academy of Sciences had been associated with the medical profession and it should not be forgotten that Berthollet and Fourcroy, normally classified as chemists, each had an M.D. degree. In the mid-nineteenth century there were about eighteen members of the Academy with an M.D. degree, of whom only a minority were members of the medical section.[82]

Yet many of the scientists in the First Class may not have been quite so enthusiastic about what they were to gain by close association with medicine. Indeed there were probably some mathematical scientists who deplored the inclusion of medicine as a whole section on a par with one of their own sections. Medicine had long been an established profession when science had still been struggling for recognition. The standards and interests of the medical profession, in so far as they were centred on the treatment of patients, were very different from the norms of both mathematical and experimental science. On the other hand there was a growing feeling among doctors that medicine should become more 'scientific' and it was, of course, those leading members of the profession who took science most seriously who were considered for election. In fact the existence of a medical section widened the spectrum of science covered by the Academy, giving greater social weight and prominence to the biological sciences which might otherwise have been overawed by the well-established mathematical sciences together with the new science of chemistry. The medical dimension also helped to give greater meaning to the activities of the nineteenth-century Academy in the eyes of the public, although this could work to the disadvantage of the Academy as well as to its advantage.

The six original members of the section of medicine and surgery were not an exceptionally distinguished group. Only one is included in the *Dictionary of Scientific Biography*, Portal (1742–1832), who published important works on pulmonary phthisis and on the liver, but it would only be just to mention a second name for his distinguished contributions to medicine and particularly hygiene, that of Hallé (1745–1822), who is also remembered as a fervent supporter of vaccination. The two members nominated by the Directory, Des Essartz (1729–1811) and Raphael Sabatier (1732–1811), were probably chosen mainly for their seniority, being aged respectively sixty-six and sixty-

[81] 4 vols., 8vo. 1791–2. See W. A. Smeaton, *Fourcroy, chemist and revolutionary*, 1962, p. 235. It should be noted, however, that it was largely the applications of the science of chemistry which Fourcroy had in mind.

[82] H. Meding (*Paris médical*, 1852–3, vol. 2, p. 370) gives the following list of M.Ds in the Academy of Sciences: Andral, Brongniart, Bussy, Civiale, Coste, Dumas, Dumeril, Flourens, Isidore Geoffroy Saint-Hilaire, Adrien de Jussieu, Lallemand, Magendie, Milne-Edwards, Rayer, Roux, Serres, Héricart de Thury, Velpeau.

three in 1795. The former had been a *docteur-régent* in the Paris Faculty of Medicine and it is worthy of note that when Cuvier delivered his *éloge*, he could find nothing to say about his contributions after the Revolution.[83] Sabatier on the other hand was brought in to represent surgery. The next two to be elected were Portal and Hallé, both eminently worthy but, since neither represented surgery the final two to be elected, Pelletan (1747–1829) and Lassus (1741–1807) had to be surgeons. Representation of surgery was always a problem in the Academy. In the revolutionary war it was the surgeons rather than the physicians who had distinguished themselves in the battlefield with emergency treatment of the wounded.[84] Larrey (1766–1842), Napoleon's *chirugien en chef des armées*, certainly deserved some honorable recognition when the war was over and for him the Academy may have provided a niche of distinction. But however skillful the amputations and other operations carried out by the surgeons, it was not very clear that their craft was yet related to science in the way that medicine claimed to be.

It is possible to argue that this potential weakness provided an opportunity for new medical sciences, notably physiology, to be introduced into the section.[85] Although the professional ethos of the section naturally placed great emphasis on *clinical* considerations, the Academy as a whole had little interest in the clinical dimension and tended to judge candidates for election on the basis of their *scientific* contributions. It was such considerations that allowed the physiologist Magendie (1783–1855) to join the section in 1821, despite the strong reservations of the section, which regarded him as an outsider. His entry, however, paved the way for other distinguished physiologists to enter that section later in the century, notably his student Claude Bernard (1813–78, elected 1854) Jules Marey (1830–1904, elected 1878), Paul Bert (1833–86, elected 1881), Brown-Séquard (1817–93, elected 1886) and Arsène Arsonval (1851–1940, elected 1894). This adaptation of the interpretation of the title of a section to the state of contemporary knowledge shows the organisation of the Academy at its best.

It could also be argued that the establishment of an Académie Royale de Médecine in 1820 provided the necessary focus for the professional interests of medicine, leaving the Academy of Sciences freer to encourage the application of science to medicine. On this interpretation it was no coincidence that the first physiologist to be elected to the section entered it only a year after the foundation of the specialist medical academy. The developing science of physiology permanently altered the composition of the medical section of the

[83] 'Eloge historique de Des Essarts, lu le 6 janvier 1812', Cuvier, *Eloges*, 2nd edn, 3 vols., 1861, vol. 1, pp. 339–50.

[84] D. M. Vess, *Medical revolution in France, 1789–96*, Gainesville, Florida, 1975.

[85] For detailed discussions of the emergence of physiology in nineteenth-century France and the role of the Academy, see John E. Lesch, *op. cit.*, and Paul Elliott, 'Vivisection and the emergence of experimental physiology in nineteenth-century France', in N. Rupke (ed.), *Vivisection in historical perspective*, 1987, pp. 48–77.

Academy of Sciences and from the time of Claude Bernard onwards it was generally accepted that the section should consist of two physicians, two surgeons and two physiologists.[86]

The suspicion of clinicians by scientist Academicians was based on a number of considerations. It was not simply that they were men who had made their reputations at the bedside and in the hospital wards rather than in the laboratory. There was also the suspicion that, in advancing their candidatures, the clinicians had ulterior motives. For the scientists membership was an honour, the ultimate honour, even if it also helped in some cases to acquire a (further) teaching position. For the physician or surgeon on the other hand, membership of the Academy was something like an honorary appointment in a famous hospital, to be displayed before potential patients; something which would increase their practice and boost their income considerably. Some of the members of this section, therefore, had much higher incomes than their colleagues who depended principally on their positions in higher education. The medics also tended to have higher social aspirations and we know that Pasteur, for example, never felt at his ease in gatherings of senior clinicians, both because he lacked medical qualifications himself and because of his modest social background.

The Academy was also heavily involved in medicine through the Montyon legacy, which in the late 1820s was causing the Academy to give disproportionately large sums of money as prizes or *encouragements* to members of the medical profession, while continuing to reward scientists with very limited resources. Enterprising young medical doctors could, by carefully cultivating the Academy of Sciences, achieve quite large financial rewards.[87] The story of how an agreement was worked out between the medical section and the non-medical majority, to the benefit of the latter is told in Chapter 7 (Section 9).

But to return to the membership. The physician Corvisart (1755–1821) was elected at the height of the First Empire in 1811 when he was first physician to Napoleon, and the First Class would have understood what was expected of it. Yet there is no doubt that Corvisart, as a famous cardiologist and the translator of Auenbrugger's work on percussion, was fully deserving of the honour. Another very successful and exceptionally wealthy clinician to be elected was the surgeon Dupuytren (1777–1835, elected 1825). Later members of the section who were known not only for their talents but also for their firm radical political stance were Vulpian (1826–87, elected 1876) and Charcot (1825–93, elected 1883). We must mention the Academy's election in 1901 of Laveran (1845–1922), who had devoted twenty years to establishing the role of the mosquito as a vector in malaria. As a former army doctor, he had not received the support or recognition he deserved for his detailed observations and experiments and he had felt compelled to resign on a modest pension. In 1907

[86] A.S., *Comité secret, 1903–12*, p. 439.

[87] For the case of Civiale, who won the Montyon prize of 1827 of 10000 francs, see *P. V.I.*, **8**, index.

he was awarded a Nobel prize for physiology and medicine. Another Nobel prize winner, the physiologist Charles Richet (1850–1935), was only elected to the Academy in 1914, the year after he had won the international prize. Finally another surgeon may be mentioned. Lucas-Championnière (1843–1913) was an early disciple of Lister's method of antisepsis but only achieved recognition from the Academy in 1912. There was always room for the immediate disciples of Pasteur but the French medical profession found greater difficulty in accepting the full implications of the germ theory in clinical medicine.

We will conclude this section by mentioning a dispute arising from the medical application of X-rays. In 1909 there was a series of long discussions in the Academy in secret session on the status of radiography.[88] While agreeing that the use of X-rays as possible therapy should be the exclusive preserve of medical practitioners, the Academy insisted that scientists who took X-ray photographs for diagnostic purposes should not be considered as practising medicine illegally. We see here the Academy involved in the delicate question of representing the interests of the professional scientist without wishing to antagonise the medical establishment.

[88] A.S., *Comité secret, 1902–12*, p. 264ff.

— *Chapter 5* —

THE ACADEMICIANS

The Institute of France may be far more truly said to contain the essence of French science than the Royal Society of London can be said to contain that of Britain.

(R. Chenevix, *Edinburgh Review*, **34** (1820), 411.)

For twenty years...I must plead guilty to living only to deserve the approval of the Academy.

(Pasteur, in letter of September 1866, *Correspondance*, vol. 2, p. 281.)

The more academies are justly famous, the greater the wish to belong to them, and the greater the efforts that people make to reach this goal turns to the advantage of science and to the glory of the human spirit.

(Arago, *Eloge* of L. Carnot (1837), *M.A.I.*, **22** (1850), cxvi–cxvii.)

1. *An intellectual elite*

There is no doubt that the Academy constituted an intellectual elite. In many ways it was also a social elite but this is a different issue and it is best to deal with each aspect in turn. That members constituted an intellectual elite can be demonstrated in several different ways. We could for example, examine a number of case histories of the careers of Academicians. Of course, the First Class of 1795–6 consisted of a substantial core of men who had been members of the previous Royal Academy. They had come up through the old system of promotion from assistant to associate and finally to pensioner. Most had learned a great deal of their craft, so to speak, by 'apprenticeship' as junior members of the Royal Academy. As these men died in the late 1790s and early 1800s they were replaced by a new generation, many of whom had benefited by the new institutions of higher education founded after the Revolution, such as the Ecole Polytechnique and (after 1808) the Faculties of Science. The standard of entry to the *grandes écoles* was high and they provided a very competitive system, not only on entry but through annual examinations and, most important of all, in the final passing-out grade. The standard of entry to the Faculties was lower but the system of graded examinations for *baccalauréat*, *license* and *doctorat* served to eliminate all but the most able. Of the new generation of Academicians, nearly all were products of the new educational system.

It had been argued that this rising generation constituted the first group of professional scientists, since not only were there widespread opportunities for the first time for a higher education in science but there was also related employment available for the best and most committed, either in higher education or by the direct application of science. By providing a livelihood based on their knowledge of science, the system reinforced their commitment to the subject. Sometimes the employment itself would provide access to a laboratory or some other means of making original contributions to science. Above all, such employment would enable the person concerned to see himself as a full member of a scientific community with its well understood hierarchy. If someone held a teaching position in the provinces, the possibility of an equivalent post in (or even nearer to) Paris would be interpreted as a promotion. If the post was in a *lycée*, then teaching in a Faculty would normally be at a higher level, and teaching at a *grande école* at a higher level still. But well above all these institutions was the Academy. Ambitious members of the scientific community knew that if they carried out research and published it, this work might qualify them eventually for consideration by the Academy. As Pasteur wrote in 1866:

> For twenty years...I must plead guilty to living only to deserve the approval of the Academy.[1]

Thus in considering the career of any individual, the original process of selection and refinement produced by the examination system was succeeded by a second phase, in which research was all-important. And only those who produced the best research were considered seriously by the Academy. It therefore follows that those who were elected constituted an undoubted scientific elite, although there were one or two periods of extreme republicanism when the Academy might have been anxious not to admit any possible political accusation of elitism.

An alternative method of illustrating that the Academy constituted a scientific elite would be to look for 'great names'. Whom would one select to represent the intellectual brilliance of the Academy over the period 1795–1914? In mathematics one would unquestionably pick out Lagrange and Laplace in the first generation and Poincaré almost a century later, but mathematicians would probably want to add other names such as Legendre, Charles Hermite and Camille Jordan. We have mentioned that Ampère, whose name was later used for the unit of electric current, was actually a member of the mathematics section. Similarly the great mathematician Cauchy was in the mechanics section, which had previously included such great names as Monge, Prony and Lazare Carnot. Later elections brought in Poncelet, Coriolis and Clapeyron. The astronomy section included Lalande, Arago, Leverrier and Janssen. The physics section was distinguished by several generations of eminent physicists. Coulomb belongs largely to the *ancien régime* but the generation of Fourier, Gay-Lussac,[2]

[1] Pasteur Vallery-Radot (ed.), *Correspondance de Pasteur*, (4 vols., 1940–51), vol. 2, p. 281.

[2] Although Gay-Lussac is better known for his contributions to chemistry, it was as a member of the physics section that he was elected to the Academy.

Malus, Poisson, Fresnel and Dulong is a reminder of the eminence of French mathematical and experimental physics in the first third of the nineteenth century.

Chemistry was probably even more distinguished and included several of the generation of Lavoisier: Berthollet, Guyton de Morveau, Fourcroy and Chaptal. Later generations provided Thenard, Chevreul, Dumas, Regnault, Wurtz and Moissan; and one should not forget Berthelot, although he was officially a member of the physics section. In the mineralogy section Haüy and Pasteur were probably the biggest names, and in botany A. L. de Jussieu and Lamarck. In the agriculture section Boussingault was an important agricultural chemist and Flourens and Dutrochet were physiologists who found themselves in that section. It was more usual to elect physiologists, however, to the medical section, where we find Magendie and his famous student Claude Bernard. Representative of the best of French clinical medicine were Corvisart, Dupuytren and Charcot, the teacher of Freud. Finally in anatomy and zoology one cannot omit the names of two well remembered for their mutual opposition on the question of transformism: Cuvier and Geoffroy Saint-Hilaire, to whom one might add Lacépède, Milne-Edwards and Quatrefages. But after listing some fifty of the more famous men of science of the Academy, we have omitted several dozen who were arguably also outstanding figures of the annals of science. On the basis of some of these names most modern scientists would be reassured that the Academy did indeed contain an intellectual elite.

The presence, and later the memory, of these scientists had an enormous influence on the prestige of the Academy among French scientists and helps to explain the extraordinary efforts made by many of them to qualify for admission. To be associated with a great figure like Laplace or Cuvier or, later in the century, to be a colleague of such men as Louis Pasteur or Henri Poincaré was the greatest possible honour for any man of science. The Academy had begun by honouring these outstanding scientists but in the end it was equally they who honoured the Academy. It is striking how few outstanding French scientists were *not* members of the Academy unless they died young. The record of the Academy of Sciences compares very favourably with that of the Académie Française, which failed to elect Molière in the seventeenth century, Rousseau and Beaumarchais in the eighteenth century and Stendhal, Balzac and Zola in the nineteenth century.

2. *A social elite*

To become a member of the Academy was not only to reach the highest rank within science but also to achieve a certain status in French society. The status of the professional classes generally had risen in the early nineteenth century and they had increasing influence on public affairs. They had taken over positions and power formerly held by the nobility, and leaders of the movement might well be considered as a new intellectual 'aristocracy'. An Academician would be described in society by the phrase: 'Membre de l'Institut'. This was a title of which one could be as proud as any nobleman of his title under the old regime. Indeed as a member of the Academy one would have a few members of the remaining aristocracy as colleagues. There would be social mixing too with high-ranking navy and army officers and with senior civil servants with scientific qualifications. Many member of the Academy were members (often in the higher ranks) of the Legion of Honour. By the late nineteenth century many new members of the Academy would, by the time of their election, be already sufficiently distinguished to be members of that order.[3]

But first we must go back to the beginning of the century for a few words about the holders of titles of nobility within the Academy. Although the Revolution had driven out the old nobility, Napoleon had created a new nobility. At the Restoration of the Bourbons, therefore, we find a partly revitalised nobility claiming special privileges. In the Academy the newly created section of *Académiciens libres* was intended to reintroduce the values of the old regime and it is here that we find several dukes. Among the ordinary members both Laplace and Dutrochet attained the title of marquis. Going back slightly in time to the period of Napoleon Bonaparte, we may note that Laplace, Lagrange, Monge, Berthollet and Fourcroy all became counts, as did the veteran circumnavigator Bougainville. The title of baron was conferred on Fourier (a former prefect), Roussin and Cagniard de Latour before election to the Academy, as it also was on three distinguished members of the medical section, Boyer, Dupuytren and Larrey. Several others were made barons while they were members of the Academy and in this category we may include Guyton de Morveau, Thenard, Prony, Cauchy, Dupin, Ramond and Sylvestre, as well as a further three members of the medical section, Cloquet, Percy and Portal. Also the agriculturist Morel-Vindé, formerly a baron, eventually became a viscount.

Secondly, as evidence of superior social status we may consider a number of high-ranking naval and army officers, particularly in the section concerned with geography and navigation. There was rear-admiral Roussin, elected in 1830, rear-admiral Paris, elected in 1863 and vice-admiral Jurien de la Gravière, elected in 1866. In the astronomy section there was also rear-admiral Mouchez, elected 1875. François Perrier, elected 1880, attained the rank of brigadier-general by virtue of his position in the geographical service of the army, but most of the senior army officers were to be found in the mechanics section: the military engineer Poncelet, brigadier-general, elected 1834; also Piobert, elected 1840 and Morin, elected 1843, artillery officers who both attained the final rank of lieutenant-general.

Much larger than the number of high-ranking officers were a number of Academicians holding positions in the civil service and eventually reaching the very senior supervisory grade of *inspecteur général*. Education was only one of

[3] The registration form for new members did not ask whether they were members of the Legion of Honour. It assumed this and asked them to state their rank, e.g. *chevalier*, *grand officier*, etc.

the many government departments concerned. For example, we find several holders of the post of *Inspecteur général de l'Instruction publique* (Coulomb, Georges Cuvier, Beudant), *Inspecteur général de l'Université* (Frédéric Cuvier, Geoffroy Saint-Hilaire), with parallel titles for higher education (Le Verrier), the national library (Libri), veterinary schools (Huzard, Chauvaux), fishing (Coste), the Mint (Vauquelin) and bridges and highways (Prony, Bresse). The post of *Inspecteur général des Mines* was held by a succession of Academicians from the foundation of the Institute: Duhamel, Lelièvre, Brochant de Villiers and Reseal. Like the other civil servants they had started in the lower ranks (for example, engineer) and had gradually won promotion through merit and seniority. Altogether there were at least forty Academicians who reached the rank of *Inspecteur général*, usually obtaining their final promotions after their election to the Academy.[4]

Less permanent than the positions for senior civil servants were the various government positions held from time to time by various members of the Academy. Although some of their political careers had begun as members of the Chamber of Deputies, we will pass over this aspect for the moment and concentrate on positions giving more prestige or power. If we begin with the upper chamber, we may note that among the members of the Chamber of Peers in the 1830s an 1840s were Poinsot, Roussin, Poisson, Thenard and Gay-Lussac. When the Senate was reconstituted in the second half of the century we find among its members: Dupin, Poinsot, Dupuy de Lôme, Le Verrier, Roussin, Dumas, Claude Bernard and Berthelot. But over and above these prestigious positions there were several occasions when Academicians held governmental office, positions of power which they must have found very flattering. The mathematician Monge had been Minister of the Navy from 1792 to 1793. Lazare Carnot, best known in French history as 'the organiser of the victory' for the vital part he played in the revolutionary wars, was a leading member of the Directory until a coup in 1797 forced him to flee from France, returning only after the advent to power of Bonaparte. He served Napoleon as Minister of the Interior in 1815 during the 'Hundred Days'. Less successful in that post had been Laplace, who held it for only a few weeks (11 November–25 December 1799). Laplace was to prove of much greater use to Napoleon as Chancellor of the Senate, where he presided over many acts consolidating the role of his master. In contrast, the chemist Chaptal proved an extremely able Minister of the Interior and carried out the heavy demands of this office successfully for four years.

Passing on to the reign of Louis Philippe, we may note that Dupin briefly held the Ministry of Marine and the Colonies in 1834 in the so-called 'Ministry of three days', a reminder of the transitory nature of nineteenth-century French cabinets rather than a reflection on the efficiency of Dupin. The same post was

[4] Of the 40 listed by Franquet de Franqueville (*Le premier siècle de L'Institut de France*, 1895), 28 attained the rank after election to the Academy and 12 before.

held by his Academy colleague Admiral Roussin in 1843. In the provisional government of 1848, François Arago was successively Minister of the Navy and Minister of War but these positions together lasted less than three months. In the succeeding government of the Second Republic J. B. Dumas was Minister of Agriculture and Commerce from October 1849 to January 1851 and seems to have done the job well. Under the Third Republic we find Hervé Faye as Minister of Public Instruction in 1877 in a government lasting no more than a month. Paul Bert held the same post in Gambetta's Ministry of 1881–2, which lasted only ten weeks. This should not lead us to suppose that Bert was a failure. He was a passionate believer in a purely secular education and he used his influence first as a deputy, then as the chairman of parliamentary committees, and finally as a minister, to establish the laws originally proposed by Jules Ferry in favour of free compulsory elementary education under lay schoolteachers. The portfolio of Public Instruction was entrusted to another ardent Republican within the Academy, Marcellin Berthelot in a government of 1886–7, which lasted six months. To give an eminent scientist with appropriate political convictions the brief of education was perhaps no more surprising than to give the Ministry of War to an army general. What is more surprising is that Berthelot should have been offered the portfolio of foreign affairs in the government of Léon Bourgeois in 1895–6, not a very successful appointment.

After reporting such a failure, we must remind the reader of several cases where scientists carried out their duties well, as with Chaptal and Dumas. The most important point for our purposes, however, is that a handful of France's most prominent scientists, which in the nineteenth century meant members of the Academy, were repeatedly selected as suitable material for ministerial office. This happened under successive Republics and under the two Empires. Only the monarchies of Louis XVIII and Charles X seem to have had no use for scientists in government.

So far we have argued that the prestige of a few highly visible members of the Academy reflected glory on all its members. The fact that some Academicians had rather humble social origins was forgotten. They were all now part of a social elite. The black and green uniform worn from 1801 was introduced partly with the intention of denoting a social rank. The superior social position of the Academician was also formally recognised in law. Universal male franchise was not established in France until 1848, the vote being restricted before then to men paying more than a certain amount of tax on property they owned. This system restricted voting rights to a mere quarter of a million out of a total population of more than thirty million. In the 1830s the tax threshold was 200 francs but this was reduced to 100 francs for members of the professional classes with certain qualifications. One of these categories was membership of the Institute.[5] Thus French society gave a certain recognition to the intellectual,

[5] Even then there was an appreciable number of members of the Academy who had no vote, since their income fell below the required minimum.

which had no equivalent in nineteenth-century Britain unless it was perhaps the granting of special voting rights for graduates of Cambridge and Oxford. There was a sense too in which French society after the Revolution had become a meritocracy. Although the nobility was given a role to play in the period 1816–30, French society had changed permanently in a way which did not affect Britain until much later. With selection by competition, appropriate examinations and elections giving positions to people who in more traditional societies would have obtained them only through birth or family connections, the *professeur* or the member of the Institute had a superior status, certainly much higher than the university professor in England[6] or the ordinary Fellow of the Royal Society.

The majority of members of the Academy lived among the bustle of the centre of Paris, mostly but not exclusively in the *quartier latin*. Many lived in private apartments but if they held positions in higher educational establishments in Paris, these sometimes offered the possibility of residential accommodation. This was an advantage not to be neglected, since many of the apartments must have been grander than their modest means would otherwise have allowed. Although other duties were attached, the cause of science tended to benefit by the existence of a permanent residential network of scientists in central Paris. Thus from the *Almanac* of 1840 we see that the astronomers Arago, Bouvard and Mathieu lived at the Observatory; the physicists Biot and Savart at the Collège de France; the mathematicians, Poisson and Libri gave their addresses as the Sorbonne; the chemists, D'Arcet and Pelouze the Mint, Coriolis the Ecole Polytechnique, and Pouillet the Conservatoire des Arts et Métiers. No fewer than ten Academicians and their families lived at the Muséum d'Histoire Naturelle or in the houses in the rue Cuvier immediately adjoining it: Audouin, Blainville, Adolphe Brongniart, Chevreul, Cordier, Dumas, Flourens, Gay-Lussac, Etienne Geoffroy Saint-Hilaire, and Adrien de Jussieu.

3. *The limits of a purely quantitative approach*

It is desirable to have a general picture of the sort of people who were elected to the Academy. It might be useful to have information also about the age at which they were elected and there are several reasons for wanting to know, for example, how many members were former graduates of the Ecole Polytechnique. Such basic quantitative information would seem to be both relevant and objective. Yet one should not become intoxicated by the ability to produce a few figures and allow 'number crunching' to get out of hand. There would be dangers in trying to construct a comprehensive but simplistic quantitative collective social biography or 'prosopography'[7], which attempted to categorise

[6] But not necessarily in Scotland, which has quite a different educational history.

[7] For discussions of prosopography see: S. Shapin and A. Thackray, 'Prosopography as a research tool in history of science: The British scientific community, 1700–1900', *History of Science*, **12** (1974), 1–28. L. Pyenson, '"Who the guys were": Prosopography in the history of science', *History of Science*, **15** (1977), 155–88.

members of the Academy exhaustively and mechanically. In the case of both religion and politics Academicians had an interest in *not* publicising their views. The historian might appreciate the difficulty of extracting reliable information in many cases but any scholar would do well to consider the dubious validity of many of the categories that might be employed. In seeking to display data quantitatively it is only too easy to present an estimate as a measurement. In any case data on membership of the Academy should not be seen as an end in itself but simply as a means to an end.

Individuals cannot always be reduced to numbers and categories nor even to simple social classes. It matters little that the Academicians Fourcroy and Lamarck were technically the sons of noblemen. The nobility in eighteenth-century France comprised a wide range of wealth and power from some of the most powerful in the land to others who suffered serious economic hardship, and who would willingly have changed their financial situation with many a wealthy shop keeper. The most relevant social information about both Fourcroy and Lamarck is that they both started in life as impoverished students. Each managed to secure an academic education at some personal sacrifice and, when the Revolution came, Fourcroy welcomed it unreservedly as providing opportunities that had previously been denied him.

Again it might well be possible to find evidence that, say, 74.3% of Academicians were of 'middle class' origin. But such a claim would require an agreed definition of social class covering both the late eighteenth and the whole of the nineteenth century, a period of enormous social change. Also a precise percentage, painstakingly calculated, can give an illusion of perfect understanding.[8] Even if it is not positively misleading it may be less enlightening than the impression which the researcher acquires, and can pass on, that the great majority of Academicians were of middle-class origin, broadly defined. The more traditional method of the historian, providing a vignette of a range of representative figures, may be more valuable than barren numbers and abstract categorisations.

The statistic that, say, 90% of Academicians had been baptised as nominal members of the Roman Catholic Church is less valuable than some knowledge of their adherence to that religion in adult life. Our later discussion of religion is more concerned with the *quality* of religious affiliation than the quantity. In fact, despite any counter-evidence from baptismal figures, it will be argued that the ethos of the Academy was predominantly secular and even anti-clerical, a characteristic which it would be almost impossible to quantify usefully. Similarly, the existence of a small religious faction is indicated best by the study of relevant documentary sources. An exhaustive quantitative study of

[8] Biographical information about Academicians covers the whole range from plentiful to slight. In many cases where the only information available on family background is, say, a title of nobility or ambiguous description of father's employment, this could give a quite misleading social categorisation.

membership of the Academy might in the end tell us less than a selective study, supplemented by a look at a few scientists who had special difficulty in being elected. Also a view of the Academy from the inside can be usefully complemented by considering a few critics who viewed the Academy from the outside.

A basic problem with the prosopographical approach is that it focuses far too much attention on the members of an institution and then tends to assume that the history of the institution is little more than the combined history of the individual members. We would claim, on the other hand, that the importance of the Academy was very much more than the sum of the contributions of the individual Academicians. We would not wish to divert attention from the work of the more prominent members, but any study of membership needs to be complemented by a whole range of other perspectives, including studies of procedures, publications and, more generally, power. Our main concern is not the individual scientists who became members of the Academy. Rather it is those scientists, both non-members and members, working *towards*, *for*, *in* or *through* the Academy. There is no doubt that the Academy greatly helped to shape scientific careers in the nineteenth century and it did this at least as much for non-members (particularly serious aspirants) as for members.

Most prosopographical studies tend to treat individuals, whether politicians or scientists, as equal units. But individuals come in different shapes and sizes of character and commitment, no two being exactly alike. It is not only that human beings are not simple homogenous units but that the quality of their membership and adherence to an institution can vary enormously. In the Academy there were a number of key figures, including Laplace, Cuvier, Arago, Dumas and Berthelot, all of whom were members of the Academy for many decades and exerted extraordinary influence. Successive permanent secretaries naturally wielded more influence than ordinary members. For outstanding activity and achievement in the context of the Academy one might wish to name a small number of outstanding members, including Cauchy and Pasteur. The half dozen individuals named above are among those who loom disproportionately large in the history of the nineteenth-century Academy and they cannot be simply pigeon-holed and dismissed as representing less than 2% of the total nineteenth-century membership, although numerically this is all they amounted to.

At the other end of the scale there were a few figures from the *ancien régime*, elected to the First Class of the Institute at the foundation but dying shortly after. To mention two extreme cases, the mathematician Vandermonde (1735–96) was a member for less than three weeks and the astronomer Pingré (1711–96) lived on for only a few months after election and so hardly enters our story. Then there were a few people like the chemist Proust (1754–1826), who was elected (1816) in recognition of his services to science but who preferred to live out his old age in the provinces rather than come to Paris and attend meetings.

Although the mineralogist Desmarest (1725–1815) was formally a member of the Academy, he seemed to Cuvier at the time of his death at the age of ninety, 'like a monument from another century',[9] having made all his major contributions to geology and technology before the Revolution. In the mineralogy section even the modest Haüy (1743–1822) overshadowed him. Similarly we give rather less prominence to those Academicians elected just before 1914, who barely enter the main story.[10]

To sum up, it would not be sensible to ignore quantitative data where it is readily available or where it can provide useful objective information but, for the reasons given above, we have refrained from embracing the quantitative method unreservedly. There are many cases where categorisation would be arbitrary and quantification would be misleading. There are many other cases where information is simply not available. It would, for example, be of interest to report the detailed financial situation of Academicians. At present there is only a limited amount of reliable information on this subject. Some studies of individual Academicians have taken pains to ascertain income from a variety of different sources[11] but until many more scholarly biographies of Academicians have been undertaken it will not be possible to do more than estimate income from the more obvious sources, such as salaries for those in higher education.[12] Like many other aspects of the life and work of members of the Academy, there is room in the future for further studies.

4. *The social origins of Academicians*

Despite what we have said about the arbitrariness of some attempts to divide people into different social classes, it is a matter of interest to know something of the social background of members of the Academy. Was the new possibility of a career in science an avenue for social advancement? Could the phrase used by Napoleon Bonaparte to describe the army – 'a career open to talents' – be applied equally to those who embarked successfully on a scientific career? The Revolution certainly provided new educational opportunities and career prospects in science. The world of science stressed talent and achievement rather than birth and we might, therefore, expect considerable diversity in the social

[9] Cuvier, *Eloges*, vol. 2, p. 162.

[10] An exception is made in the case of provincial members, who were admitted to full membership for the first time in 1913, since their admission marked a major policy change.

[11] E.g. Maurice Crosland, *Gay-Lussac; Scientist and bourgeois*, 1978, pp. 228–31. For useful information about the fortunes left by former members of the Académie de médecine see George Weisz, 'The medical élite in France in the early nineteenth century', *Minerva*, **25** (1987), 150–70 (pp. 160–3). For the fortunes of other professionals see Adeline Daumard, *La bourgeoisie parisienne de 1815 à 1848*, 1963, pp. 79–90, and *Les fortunes françaises au XIXe siècle*, 1973.

[12] For information about the professional salaries of several mathematicians benefiting from the system of *cumul* see I. Grattan-Guinness, *Convolutions in French mathematics* (3 vols., Basel and Berlin, 1990), vol. 2. p. 1273.

origins of leading scientists, whose eminence was recognised by election to the Academy.

Although a large proportion of nineteenth-century Academicians were of broadly middle-class origins, a few were from the working classes and this is true both for the older generation, who began their careers before the revolution, and those who came later. In this section, we shall deliberately draw attention to a number of Academicians who were *not* born in comfortable circumstances. Many started life in very modest circumstances, if not in grinding poverty, and their later membership of an intellectual elite like the Academy must be considered a considerable social advance. The astronomers Bouvard (1767–1843) and Bigourdan (1851–1932) were of peasant stock, as was the chemist Thenard (1777–1830). The parents of Claude Bernard (1813–78) were vineyard workers, Fourier (1768–1830) and Cahours (1813–91) sons of tailors. Charcot (1825–93) was the son of a wheelwright, Chauveau (1827–1917) son of a blacksmith, Haüy (1743–1822) son of a weaver, Méchain (1744–1804) son of a plasterer, Pasteur (1822–95) son of a tanner and Tisserand (1845–96) son of a cooper, who died when he was very young. Some of these families would be proud tradesmen with considerable skills, as probably was the father of Appell (1855–1930), a master dyer. They would have been socially superior to impoverished artisans like the father of Gambey (1787–1847) and the botanist Turpin (1775–1840), who had been taught by a gardener. As regards actual poverty, we are told that the astronomer Delambre (1749–1822) was at one time so poor that he lived on bread and water. Even Etienne Geoffroy Saint-Hilaire (1772–1844) who, as the son of a lawyer, the procurator of a small town, might seem to have been given a favourable start in life, found himself as the youngest of fourteen children and therefore living in quite modest circumstances.

Indeed there was an image of the *savant* after the Revolution as someone who could exist on a minimum income. The Academician's honorarium of 1500 francs per annum (including *droits de presence* was only sufficient to provide basic subsistence and was not intended to attract the *savant* by high financial rewards. Lalande, the senior astronomer in the First Class, has left us a record of his attitude to money. In his autobiography of 1804 he claimed to live a simple life with few demands:

> I have few servants and no horses; I am sober, my habits are simple. I go on foot. I sleep where I find myself; money is of no use to me. I am not concerned with the pride either of fortune or of rank. The least wealthy men are those whom I most welcome.[13]

This historical evidence would support the norm of disinterestedness,

[13] *Eloge historique de M. de Lalande par Madame la Contesse Constance de S.*, 1810, Appendix, p. 45.

suggested sometimes as one of the characteristics of the scientist, but the evidence has to be placed alongside other evidence of Academicians who used their positions for profit. Gay-Lussac (1778–1850) had started his scientific career with the claim that all he wanted were the basic necessities of life[14] but, by his later accumulation of teaching and consultancy posts, he was to become one of the wealthier members of the Academy.[15]

In the archives of the Academy there is a touching appeal, dated 22 March 1800, from the whole of the Institute to the First Consul, putting on record the inefficiency of the system of payment in the early years of the Institute and the distress this caused.[16] The document states that many members of the Institute depended entirely on their honoraria, which at that time were eleven months in arrears. This delay had caused great distress and had driven some members to desperate expedients. The First Class of the Institute took note of the dire poverty of some of its members in the early years. The botanist Adanson (1727–1806) is often cited as such an example. It is even said that when he was told that he had been elected as a member of the First Class, he had to send his apologies, since he did not own a pair of shoes. On a later occasion when he was ill in 1806 his colleagues made him a grant of 2700f. However, on Adanson's death soon after, his nephew, no doubt trying to save the honour of the family name, denied that his uncle had been in real need. He pointed out that in his final years he had had a special supplementary pension from the Ministry of the Interior and that he owned his own house.[17]

Some humble occupations were actually relevant to the work of the Academy, as we are reminded by the career of the botanist Thouin (1786–1824), successor to his father, head gardener of the Jardin du Roi, which later became the Muséum d'Histoire Naturelle. The botanist Naudin (1815–99) too had been a gardener before he studied at university. Another area where practical skills were of value as a complement to academic training, was in the mechanics section of the Academy. Although 'mechanics' came to be interpreted more as (applied) mathematics, there was still room in that section for the occasional instrument-maker. Thus the aged watch-maker Le Roy (1720–1800), was a member of the mechanics section for the first few years of the new Institute. It is said that the clock-maker and instrument-maker Louis Breguet (1747–1823) might have been a mathematician but the family fell on hard times and he had to learn a trade.[18] The agriculture section might be thought to be an appropriate one for practical experience but many of the members elected to this section were not men of science but landed proprietors, whose association with the land gave them not inferior but superior social status. For some members of the

[14] 'le pur necessaire me suffira toujours', Gay-Lussac to his father, 15 January 1803, M. Crosland, *Gay-Lussac*, p. 267.

[15] *Ibid.*, pp. 228–34.

[16] A.S., *Copie de Lettres, an VI - an XIV (1798–1805)*, 1 *germinal* an 8.

[17] *Observations sur le feu M. Adanson*, B.M. 733.g. 18 (49).

[18] *M.A.I.*, 7 (1827), xviii.

Academy whose fathers had been farmers, like Boussinesq (1842–1929) and Guignard (1852–1928), it is often difficult to estimate the social standing of the family and any attempt to put together in one category those with an agricultural background could be very misleading.

Some Academicians were interestingly the sons of fathers who had risen appreciably in society, making it easier for their sons to have access to higher education. Thus, although the father of Berthollet (1748–1822) was a physician, he had come from a family of ironsmiths. Similarly, the father of Biot (1774–1862) was of peasant stock, but he had managed to obtain a position in the Treasury. Higher education was of course much more widely available after the Revolution but at all times a comfortable income and some appreciation within the family of the value of education would have helped to encourage young people to make the most of educational opportunities available. Middle class parentage is well represented among Academicians and particularly the legal profession. Delafosse (1796–1878) and Gay-Lussac were the sons of lawyers in modest positions. The fathers of Prony (1755–1839) and Poncelet (1788–1867) were more prominent lawyers.

The reintroduction of the class of *Académiciens libres* in 1816 gave a clearly defined representation of the old French nobility, although by the 1840s the interpretation of eligibility for this category was widened to include men of more diverse social origins. Turning to the rest of the Academy, we do find a definite even slightly anachronous representation of the nobility. Thus Henri Ducrotay de Blainville (1777–1850) grew up among the lesser but intensely proud Norman nobility and to the end of his days he showed his distaste for an egalitarian society. The D'Arsonval (1851–1940) family was part of France's ancient nobility, having been landowners in the region of Limoges for centuries. Another naturalist, Lacépède (1756–1825), the only son of a count, became a count in his own right when the new Napoleonic nobility was established. We may also mention that the zoologist Lacaze-Duthiers also had aristocratic links. Yet this did not have the importance in the nineteenth century that it might have had a century earlier.

Having indicated that it was possible to rise in a few cases from even the most socially unpromising position to a career in science and ultimately to attain membership of the Academy, we may now turn to consider another group of Academicians who had the distinct advantage of inside knowledge and support.

5. *Family connections and patronage*

It used to be common in any family for sons to follow the trade or profession of the father. Yet one does not tend to associate this practice with careers in science, since science was hardly established as a profession before the nineteenth century and in science independent intellectual ability, as established for example through examinations and published research, is usually the key to entry to a scientific post. Yet already in the eighteenth century there were one or two families associated with a particular branch of science. If we go back to the foundation of the Royal Academy, we find Colbert enticing Jean-Dominique Cassini (1625–1712) to come from Italy and work in the Academy as an astronomer from 1669. Cassini liked his new situation and in 1694 was able to introduce his son Jacques (1677–1765) into the Academy in the most junior grade of *élève*, later renamed *assistant*. When his father died, Jacques was promoted from *associé* to *pensionnaire* in the astronomy section. Jacques in turn was able to introduce his son Cassini de Thury (1714–84) as a supernumerary Academician in 1735, and he became a *pensionnaire* in due course. This process continued to a fourth generation with Jean-Dominique Cassini (1748–1845), who lived through the Revolutionary period into the nineteenth century and thus enters our story directly. This Cassini (sometimes referred to as Cassini IV) was a Royalist who deplored the Revolution and was reluctant to live under a republic. After several vicissitudes, however, he became a member of the astronomy section of the First Class of the Institute from 1799.[19]

Although the Cassini family was unique in the early history of the Academy, something similar happened with the Jussieu family in the Jardin du Roi, which became the Muséum d'Histoire Naturelle at the Revolution. Antoine du Jussieu (1686–1758) had been appointed professor of botany at the Jardin du Roi in 1710 and managed to get the post of demonstrator there for his younger brother Bernard de Jussieu (1699–1777).[20] Antoine de Jussieu was elected to the Academy in 1712 and his brother in 1725. A third brother, Joseph (1704–79), entered the Academy in 1742, again in the botany section. Antoine Laurent de Jussieu (1748–1836), the nephew of the three listed above, entered the Academy in 1773. By this time it was clearly expected that the botany section should contain at least one member of that famous family. On the foundation of the Institute Antoine-Laurent Jussieu was elected on 9 December 1795 to what was obviously considered his rightful place. The question arises of whether in post-revolutionary France one could still expect to find a member of that family in the Academy. The world had changed in many ways and a right of succession could no longer be achieved by pure nepotism. If the candidate clearly had ability and conformed to the new criteria, such as a good publication record, he stood a good chance. In 1826 Antoine de Jussieu resigned his chair of botany at the Muséum in favour of his son Adrien (1797–1853), thus adding to the latter's qualifications. Finally in 1831, after a few unsuccessful attempts, Adrien was elected to the Academy in the botany section.

Meanwhile Cassini IV entertained scientific ambitions for his own son Henri (1781–1832). Henri Cassini had abandoned the family tradition of astronomy

[19] Cassini was nominated to the astronomy section on 13 December 1795 but was unable to take his place since he had emigrated. He was re-elected on 24 July 1799.

[20] Paul Lemoine, 'Le Muséum National d'Histoire Naturelle, son histoire, son état actuel' *Archives du Muséum d'Histoire Naturelle*, 6e série, 12 (1935), 4–79.

in favour of botany, a subject he had discovered when sheltering in the country from the revolutionary turmoil of Paris.[21] Encouraged by his father, he had carried out botanical studies, written these up and submitted them to the Academy, from whom he received fairly encouraging reports. Under the Restoration he began to stand for election for vacancies in the botany section. When he was unsuccessful in 1820, Cassini père decided to intervene and he addressed a letter of complaint to Antoine-Laurent Jussieu, who was not only the doyen of the botany section but also the current representative of that scientific family and – Cassini hoped – the upholder of ancient traditions. Cassini père hoped that this son could secure entry to the Academy 'with the protection of a famous name'.[22] Not of course that that was his main qualification but he had hoped it would carry weight with another member of a famous family who was in a position to act as his patron. We have also been fortunate in finding another letter which Cassini IV wrote to the Academician Huzard just before a later election. He knew that Huzard, doyen of the agriculture section, was keen to get his own son into the Academy and had even managed to place him first on the list for vacancies in the agriculture section. As one supporter of nepotism to another, therefore, Cassini spoke of the five generations of service to science.[23] He asked Huzard for his vote.

We must report the happy but ironic conclusion to the story. Henri Cassini was finally elected (as *Académicien libre*) in 1827. His principal opponent, whom he beat by a single vote, was comte Daru and in voting for Cassini the Academy was showing its preference not for an ancient family but for someone with clear scientific credentials. Only the previous year he had published the first volume of his *Opuscules phytologiques*. Henri Cassini's death in the cholera outbreak of 1832 effectively brought to an end an almost uninterrupted association of more than a century and a half of that family with the Academy.

Before leaving astronomy altogether, we could mention the case of Lalande (1732–1807), whose life was devoted to astronomical observation. For many years he had trained his nephew Michel Lefrançois de Lalande (1766–1839) in the techniques of patient telescopic observation. A member of the Royal Academy under the *ancien régime*, Lalande was nominated to the First Class in 1795 and it is understandable that in his old age he should want his nephew to join the Academy elite, which he did in 1801. In many of these cases there was no explicit claim for inherited talent. The argument, if it had been

[21] Henri Cassini provides an interesting autobiography in his *Opuscules phytologiques* (3 vols., 1826–34), vol. 1, pp. vii–xxxviii.

[22] A.S., Jean Dominique Cassini dossier, copy of letter from Ramond-Gontaud collection, dated 'Mars 1820'.

[23] 'J'ose donc vous prier de vouloir bien examiner... si le fait unique dans l'histoire littéraire, d'un dévouement aux sciences soutenu par 5 générations successives pendant plus de 170 ans, ne doit pas ajouter quelque poids dans la balance ou vouz pesez les titres scientifiques que présente mon fils et s'il ne lui meriterait pas de votre part quelque préférence sur ses concurrents'. The note is marked 'reçu le 4 mai 1827', i.e. three days before the election. Institut HR5*. Tome 68, no. 10A.

articulated, would have been concerned rather with younger men having been brought up in a scientific atmosphere and having served an apprenticeship. They knew a scientific speciality and the scientific ethos, so to speak from the *inside* and this was never more true than for the staff of the Muséum d'Histoire Naturelle, which provided residential accommodation for its professors. Their children were often born at the Muséum, with 'natural history' in their blood. They would have the support of other Muséum families, who would often be represented at the Academy also.

Clearly dynasties of men of science had a more natural place under the *ancien régime* than in the more democratic world of the nineteenth century. Yet there is one more family which managed to perpetuate itself in the nineteenth century in the Academy, this time in the physics section. The Becquerel family, like the Jussieu family, maintained a power base in the Muséum d'Histoire Naturelle. Pure nepotism was more easily exercised in that institution than in the Academy and three successive generations of Becquerels held the chair of physics there.[24] Yet some explanation seems to be called for why an institution which specialised in the biological sciences should have established a chair of physics. The work of Antoine Becquerel (1788–1878) provides a rationale. His research in physics extended to an examination of electricity in the interior of vegetables (1833). He also had examined the electric eel and made studies on the earth's temperature in different places. This research encouraged the Muséum d'Histoire Naturelle to ask in 1838 for a new chair of physics applied to the natural sciences, and Becquerel was appointed.

A graduate of the Ecole Polytechnique, Antoine Becquerel had had a brief military career under the Napoleonic regime before turning to physics. His work on electricity and the physical study of minerals had won him a place in the physics section of the Academy in 1829. When he moved to take up residence at the Muséum, his son Edmond (1820–91) was eighteen years of age. Although he was originally destined for one of the *grandes écoles*, he turned these down in order to accept his father's offer of a junior post at the Muséum as *aide naturaliste*, whilst at the same time following courses at the Paris Faculty of Science, where he took his doctorate in 1840. He soon set his sights on the Academy but was not elected until 1863, when he was reasonably well advanced in his career.[25] For the next fifteen years there were two members of the same family in

[24] Henri Becquerel, 'Biographies scientifiques. La chaire du physique du Muséum', *Revue scientifique*, **49** (1892). 673–8. Jean Becquerel, 'Les principaux travaux des professeurs de la chaire de physique du Muséum d'Histoire Naturelle', *Archives du Muséum*, 6ᵉ série, **12** (1935). 83–100.

[25] Becquerel's *Notice sur les travaux scientifiques* of 1863 records 52 publications. He was praised by Fizeau, the doyen of the physics section, for the importance, number and extent of his many discoveries and his devotion to research. Fizeau concluded: 'If the name that he bears had been obscure before him, he would not have failed to make it famous, but he has had the no less valuable honour of adding new lustre to a name which has long been dear to the Academy and to all friends of science'. MS *Rapport sur les travaux de M. Edmond Becquerel*, A.S., Edmond Becquerel dossier.

the physics section. But there was another Becquerel on the horizon, Henri Becquerel (1852–1908), who had studied at the Ecole Polytechnique and was to make a reputation for himself for the study of phosphorescence. In 1878 when his grandfather died, his father succeeded to the post of professor of physics at the Muséum and he was appointed *aide-naturaliste*. His father also had the chair of physics at the Conservatoire des Arts et Métiers and he was able to share the duties with his son.

Already in 1876 at the age of twenty-four, Henri had competed for a vacancy in the physics section of the Academy. Although he had only one published memoir to his credit, he could also claim to have presented two other memoirs to the Academy. He had no serious hope of election but he was making clear his future ambitions. Meanwhile he married Lucie Jamin, the daughter of a member of the physics section! In 1884 and 1886 Henri Becquerel was again a candidate but despite the comment of the doyen of the physics section, Fizeau, that he bore a 'famous name', he was not yet successful.[26] Perhaps the pointed comment of Fizeau that his studies of phosphorescence were 'far from being complete' was a hint that his time had not yet come. He was placed second in the list of candidates. Everything seems very fair, although letters have survived which show that his father was not standing idly by. In 1884 his father asked Dumas for support in Henri's candidacy for the Lacaze prize, which he won.[27] As a prize winner he would draw further favourable attention to himself, but he was still too junior. When another vacancy occurred in 1889 he now had forty-three publications to his credit. He was considered to have earned his place and he was elected. His later work on radioactivity was to bring him international fame and helps to vindicate the choice of the Academy.

Powerful dynasties are not always made transparent to the historian by the sharing of a common name. Inter-marriage changed names without weakening alliances. One of the more powerful groups in nineteenth-century French science was that headed by geologist Alexandre Brongniart (1770–1847), who was elected to the mineralogy section in 1815. The son of an architect, he had acted for a time as assistant to his uncle Antoine-Louis Brongniart (1742–1804), who had been professor of applied chemistry at the Muséum d'Histoire Naturelle, which was to be a power house of the family and where also was to be found the chemist Fourcroy, distantly related by marriage. Brongniart also held simultaneously senior posts in the porcelain factors at Sèvres and in the mining administration.[28] He had three children, a son and two daughters. The eldest child was Adolphe Brongniart (1801–76) who in 1822 contributed his first publication in the field of palaeobotany, for which he was later to become

[26] Rapport sur les travaux de M. Henri Becquerel', A.S., Henri Becquerel dossier.

[27] A.S., Edmond Becquerel dossier.

[28] L. de Launay, *Une grande famille de savants, les Brongniart*, 1940. For a family tree, which reveals further involvement of the Brogniart family in the academic web, see A. J. Tudesq, 'Les notabilités intellectuelles', chap. 5 of *Les grands notables en France, 1840–9*, Bordeaux, 1964, p. 459.

famous. His election in 1834 at the early age of thirty-three was not unrelated to merit but it was also helped by his wealthy father, who had already helped him secure the post of professor of botany at the Muséum (1833). The youngest child Mathilde had in 1827 married Jean-Victor Audouin (1797–1841), who no doubt received support from his father-in-law in his career as an entomologist and in particular for his election in 1838 to the Academy (agriculture section).

As early as 1824 Adolphe Brongniart had collaborated with Audouin and a third young man who was later to make a brilliant reputation for himself as a chemist, Jean-Baptiste Dumas (1800–84), to edit a new journal, the *Annales des sciences naturelles*. In 1826 the eldest Brongniart daughter, Hermenie was to marry the ambitious young Dumas. There is abundant documentary evidence of the financial benefits Dumas received from his father-in-law.[29] Apart from a dowry of 40,000 francs and an apartment at the Muséum from the date of their marriage, Dumas was later to receive an adjacent building which he could turn into an important laboratory.[30] Rather less well documented, but of no lesser importance, were many recommendations for positions and honours. An early one was his election to the Academy in 1832 at the youthful age of thirty-two. Dumas was an exceptionally able chemist and administrator but he would hardly have gone as fast or as far without his powerful father-in-law. Dumas' son Charles-Ernest Dumas (1827–90) married the daughter of Henri Milne-Edwards and, when she died, married her sister. It is close family connections such as these which help explain the exceptional power and influence of J. B. Dumas in the Academy. His election as permanent secretary in 1868 merely confirmed his position as the doyen of science of his age. It is interesting that he should later use the occasion of delivering an *éloge* to defend the concept of great scientific families passing on the torch of learning from one generation to the next.[31]

Less spectacular than the dynasties described above but none the less necessary for our understanding of the Academy are several cases of father and son, or two brothers in the Academy. It is worthy of note that it was in 1833, the year after the death of his old enemy Cuvier, that Etienne Geoffroy Saint-Hilaire (1772–1844) was able to sponsor the election of his son Isidore (1805–61) at the tender age of twenty-seven to a vacancy in the zoology section. Since the age of nineteen Isidore had been *aide-naturaliste* for his father at the Muséum, thus serving a useful apprenticeship. Each Academician had greatest influence within his own section and the order of preference was drawn up by the section before submission to the Academy. Some biographers have suggested that Isidore was 'a continuator without originality'[32] and it is certainly true that

[29] L. de Launay, *op. cit.*, pp 144–9.

[30] For a study of the use Dumas made of this laboratory in training a generation of students, see Leo J. Klosterman, 'A research school of chemistry in the nineteenth century: Jean-Baptiste Dumas and his research students', *Annals of science*, **42** (1985) 1–80 (p. 9).

[31] Eloge of Isidore Geoffroy Saint-Hilaire, *M.A.I.*, **38**, clxxvii–clxxviii.

[32] Frank Boudier, 'Isidore Geoffroy Saint-Hilaire', *D.S.B.*, **5**, 358–60.

in this case, as in several others, the father was more distinguished than the son. Another case of mild nepotism in the zoology section was the case of Alphonse Milne-Edwards (1835–1900), who was elected in 1879 while his father Henri Milne-Edwards (1800–85) was a member of that section. Yet Jean-Baptiste Huzard (1755–1838) tried in vain to sponsor his son for a vacancy in the agriculture section. Here the choice of the Academy thwarted transparent nepotism in favour of candidates with strong scientific qualifications like Flourens (1828), and Dutrochet (1831), who had not even been considered by the section. It was here in the agriculture section that Paul Thenard (1819–84) was elected in 1864, having no realistic chance of ever being elected to the highly competitive chemistry section in which his father Jacques Thenard (1777–1857) had been a member. Given the date of election, his father cannot be accused of influencing the actual election, although he had no doubt prepared the way in the last years of his life. The same is probably true of the botanist Claude Richard (1754–1821), whose son, Achille Richard (1794–1852) was only elected to the botany section of the Academy thirteen years after the death of his father.

One of the most flagrant areas of nepotism involving brothers was the election of Frédéric Cuvier (1773–1838) to the zoology section in 1826, when his famous brother 'Georges' (1769–1832) held the key post of permanent secretary. With much more academic justification was Henri Sainte-Claire Deville (1818–81) elected in the mineralogy section in 1861, four years after the election of his elder brother Charles (1814–76). The most suspicious circumstance here is that Henri was really a chemist but found an easier entry to the Academy in the same section as his brother.

Again, however, we need to go beyond those who shared the same surname. We do not know whether Arago was embarrassed when in 1817 he presented a list of candidates for a vacancy in the astronomy section, headed by his brother-in-law, Louis Mathieu (1783–1875).[33] What we do know, however, is that Mathieu was elected. The records of the Academy reveal that in the mathematics section there were two elections in 1856. On 22 April Joseph Bertrand (1822–1900) was elected,[34] followed on 14 July by Charles Hermite (1822–1901). What the minutes do not reveal, however, is that these men were brothers-in-law. This is not to imply that Hermite, one of the great French mathematicians of the nineteenth century, did not deserve election. But it might be unduly naive to suppose that Bertrand was in no way influenced in his vote by family ties. Joseph Bertrand was to become a particularly influential member of the Academy after his election as secretary in 1874. We may note later elections into the mathematics section in 1889 of Emile Picard (1856–1941),

[33] *P.V.I.*, **6**, 188 (19 May 1817).

[34] Bertrand would have been strongly supported by his former teacher, the mathematician J. M. C. Duhamel, a member of the physics section, whose niece he had married.

who happened to be his son-in-law, in 1892 of his nephew by marriage[35] Paul Appell (1855–1930), and finally in 1896 of his son Marcel (1847–1907) into the mineralogy section. Marcel Bertrand was also the son-in-law of Emile Duclaux (1840–1904), a member of the agriculture section.

Thus we see a network of relations within the Academy which helps us to understand certain loyalties which lay beneath the surface. Only exceptionally can one demonstrate unambiguous corruption. The rules of the Academy were much tighter than those of the Muséum and the interests represented were much wider. Yet anyone who wished to consider the different factions in the Academy purely in terms of subject loyalties or adherence to particular theories, would have a very incomplete understanding if he failed to include more personal considerations.

But to consider relationships purely in terms of blood and marriage is to understand only a part of the story of personal favour within the Academy. So much of nineteenth-century French science has to be seen in terms of patronage[36] and an important aspect of this was the relationship between teacher and student. A major study of the research school of J. B. Dumas has identified some thirty students, most of whom not only attended his lectures but also had the special privilege of working for several years in Dumas' research laboratory and are clearly identified with one or more of his research programmes.[37] At least four eminent scientists, who are clearly identified as Dumas' former research students, entered the Academy during the period of the Second Empire: Peligot (1852), Henri Sainte-Claire Deville (1861), Wurtz (1867), Cahours (1868). Also for a whole generation most of the correspondents elected to the chemistry section were Dumas' former students;[38] although not being full members of the Academy, they played only a minor role in the politics of that body.

Dumas protégés were to be found not only in the highly competitive chemistry section but also in agriculture (Peligot) and mineralogy (Henri Sainte-Claire Deville). Other teachers and patrons relate more clearly to one section. Thus although Henri Milne-Edwards (1800–85) was eventually *replaced* in the zoology section by his son Alphonse (1835–1900), it is more relevant that he was *joined* in the zoology section by a succession of former students: Quatrefages (1810–92) in 1852, Blanchard (1819–1900) in 1862 and Lacaze-Duthiers (1812–1901) in 1871. It was by no means uncommon to find former teachers and their students sharing membership of the same section.

[35] Appell had married a niece of Bertrand.

[36] For a preliminary study of patronage in nineteenth-century French science see the author's *Society of Arcueil* (1967), p. 1. 'This is a book about patronage...' See also Robert Fox, 'Scientific enterprise and the patronage of research in France, 1800–70', *Minerva*, 11 (1973), 442–73.

[37] Leo J. Klosterman, *op. cit.*, note 28, esp. Tables, pp. 14–15, 18–19.

[38] Laurent (1845), Malaguti (1855), Gerhardt (1856), Favre (1863), Marignac (1866), Stas (1880). Laurent and Gerhardt (1856), Favre (1863), Marignac their independence and their criticism of the system of patronage.

6. *Political affiliations*

Although religious practice is difficult to analyse, except in broad principle and in some individual cases, the problem with politics is in many ways even greater. The politics of an individual, like his religion, is often very much a private affair. But in addition, Academicians, as members of a state corporation, were expected not to be involved in party politics. Broadly speaking however most Academicians, with a few notable exceptions, seem to have favoured the moderate left and the centre ground rather than the right wing of politics. Yet even if a historian could ever have access to information about how Academicians voted in a particular election,[39] this would not necessarily establish them as life-long adherents to a particular political grouping. As France passed from Republic (1792–9) to Consulate and Empire (1804–14), then successively to absolute monarchy and constitutional monarchy (1830–48), and back to Republic and Empire and yet a third Republic (1871–1947), individuals changed their political allegiances. We can appreciate that there were many times when the contrast in France between monarchists and republicans was very much greater than the difference in Britain between Conservatives and Liberals or in the United States between Republicans and Democrats. Of course monarchists themselves were divided between legitimists and supporters of the Orleans line. Also the term 'republican' is a very loose label, although the memory of the Revolution of 1789 exercised a powerful influence in providing an exemplar of a republic and a programme of reform. In a century when there were several violent changes of government, it was prudent for the Academy to take an official stance above politics. In a later chapter we shall consider the whole question of the position of the Academy in relation to different governments. Here we can only note a few of the more notable cases where members of the Academy publicly took a political stance.

Once it is accepted that the majority of Academicians were avowedly non-political, one can take a sample from successive generations to exemplify political enthusiasms and subsequent tensions. Among the first generation of members of the First Class was the chemist Fourcroy, who had taken an active part in politics from 1793 when he was elected to the Convention and joined the Jacobin club. Although he did sterling work in the reorganisation of scientific and medical education and might well have expected to be rewarded by Napoleon by being appointed grand Master of the Imperial University in 1808, he was passed over, possibly because he was known to be a freethinker or because his former association with Jacobins was held against him. He died the following year, thus avoiding the trauma of sanctions at the Restoration. The mathematician Hachette had also been a fervent revolutionary, although he played a less active part in politics. At the Restoration he was deprived of his post

[39] Since for the most of our period the franchise was linked to property rights some Academicians did not have the vote, at least not before 1830.

at the Ecole Polytechnique and when he was elected to the Academy in 1823 the King refused to confirm the nomination because of his well-known republican sympathies, thus delaying his election till the more liberal regime of the 1830s.

Another case of political interference in the Restoration was that of Monge, who had identified himself too clearly with the Bonapartist cause. Although a senior well-respected mathematician, he was expelled from the Academy in 1816. Yet the slippery Laplace, who had also received many favours under the First Empire and had even dedicated one of his books to Napoleon Bonaparte[40], continued in the Academy and was even made a marquis under the Restoration. Another notorious 'weathercock', who ingratiated himself with successive regimes based on quite irreconcilable political principles, was Cuvier. Having thrived under Napoleon, he had no scruples about welcoming the return of Louis XVIII. Both Laplace and Cuvier were major figures in the French scientific establishment. Any attempt to uproot them would have caused a scandal but the question never really arose. They knew where their best interests lay and were prepared to compromise whatever slight political principles they may have had.

More scrupulous was Cauchy. Many people would say he was too scrupulous when in 1830 he resigned his positions in Paris rather than serve under the Orleanist branch of the Bourbon family. His exile was self-imposed and, if he saw himself as a martyr, others saw him as lacking in worldly wisdom. The senior branch of the Bourbon family hardly merited such devotion and in serving as tutor to the royal prince in exile, Cauchy was making a much less useful contribution to mathematics than if had stayed in Paris. After the repression and censorship of the last months of the reign of Charles X, most Academicians sincerely welcomed the 1830 Revolution, but the (temporary) acceptance of a constitutional monarchy did not prevent the building up of a republican group under the patronage of Arago, who was elected secretary of the Academy in the very year of the Revolution. Indeed the years 1830–48 were ones of great political activity for Arago, and in this period it would be quite wrong to ignore his political role and pretend that he merely continued with his scientific research, like so many of his colleagues in the Academy. It is these years which reveal the greatest contrast, for example, with his colleague Biot, who did not agree that scientists had a public role. We might add that Arago was a particularly important figure in the history of the Academy because he had a number of friends within that body who were strongly influenced by him without necessarily sharing all his political views.

Nevertheless Arago became increasingly the centre of a number of supporters of opposition politics.[41] The astronomer Claude-Louis Mathieu (1783–1875), a member of the Academy since 1817, married Arago's sister and was elected as a radical deputy for Macon (1835–7 and 1838–48), taking a position on the

[40] P. S. Laplace, *Mécanique céleste*, vol. 3, 1802.

[41] I am indebted to John Cawood for much of the information in this paragraph.

extreme left of the Chamber. Then there was a group of younger men who entered the Academy in the 1830s with Arago's powerful support. They included Felix Savary (1797–1841)[42], the son of a former member of the Council of 500, who was elected to the astronomy section in 1832. Also in the republican group was the mathematician Sturm (1803–55), elected to the Academy in 1836. In 1837 the chemist Pelouze joined the opposition group in the Academy. His radical credentials were excellent. He had fought on the barricades in the Revolution of 1830 and was a friend of Beranger, poet and songwriter of the opposition. Although a protégé of Gay-Lussac in scientific matters, in politics he was much closer to Arago and even contributed to a journal run by Louis Blanc. Finally we may mention the election in 1839 of the mathematician Liouville (1809–82). Liouville was active in the reform campaign of 1839–40 and became a member of the Legislative Assembly in 1848.[43]

Reviewing the 1840s it is possible to distinguish at least four distinct political positions.[44] On the left there was Arago and his friends, who might have been described as 'liberals'. Then in the centre was a group of moderates, in which Thenard, Dumas and the Brongniart family were prominent. There were also in the Academy representatives of some of the great Orleanist families who supported Louis Philippe: baron Dupin, comte de Gasparin, Adrien de Jussieu and Mirbel, whose wife was at the centre of one of the Orleanist salons in Paris. These might be described as moderate right, as opposed to the extreme right position of the legitimist Cauchy.

Under the Second Empire it was prudent not to display excessive interest in politics although if, like Louis Pasteur, an Academician became a committed Bonapartist this was all to the good for their careers. Pasteur's father had been a sergeant major in Napoleon's army and it seems that for him the Emperor could do no wrong.[45] Louis Pasteur became almost as fervent in his support of Louis Napoleon.[46] Through Dumas, he developed personal relations with the imperial household and sought the direct patronage of Napoleon III for some of his research. Yet he had no real understanding of politics, as can be seen by a brief and naive candidature for the Senate 'on behalf of science' in 1875. He was seen by the electorate as politically conservative and was defeated overwhelmingly by republican candidates.[47] Claude Bernard was equally nonpolitical, but his mild conservatism blossomed in the final year of the Second

[42] Savary was also a relative of Madame Arago and it was on a visit to the Arago family in their country home near the Pyrenees that he died in 1841.

[43] Liouville's election manifesto is in the archives of the Academy of Sciences (dossier Liouville): 'Aux électeurs du Départment de la Meurthe'. This mentions his friendship with Arago.

[44] A. J. Tudesq, *op. cit.* (note 28), vol. 1, pp. 460–2.

[45] R. Vallery-Radot, *The life of Pasteur*, trans., 1906, p. 4.

[46] *Correspondence de Pasteur*, vol. 1, p. 230. A rather different perspective of the *young* Pasteur is provided by Gerald Geison and James Secord in 'Pasteur and the process of discovery', *Isis*, 79 (1988), 7–36.

[47] G. Geison, 'Pasteur', *D.S.B.*, 10 350–416 (p. 354).

Empire. He was made a senator in May 1869 but found the proceedings of the Senate excessively tedious.[48]

Arriving now at the period of the Third Republic, it is not easy to summarise the complex political scene.[49] The election of 1871 actually produced a conservative majority and it was only at the end of the 1870s that the left became dominant. In 1879, as a symbol of a genuine republican Republic, the Marseillaise, battle hymn of the first republic, was made the national anthem. Republicanism in principle appealed to a wide social spectrum. Jules Ferry had the idea of an 'open Republic' which the centre and even the right might find acceptable but, in practice, many old antagonisms remained. There was considerable diversity among republicans, who found it impossible to agree on what should constitute a republican programme. As late as the 1870s there was also a Bonapartist party but the republicans gradually stole their clothes. In the 1870s we can distinguish between the moderate republicanism of Grévy and the more committed Republican Union of Gambetta, one of whose associates was the Academician Paul Bert. In the 1880s Jules Ferry became a central figure in republicanism but in the continuous struggle between competing local and ideological interests, even his various cabinets never lasted more than one or two years. Divisions tended to increase. Moderate republicans who did not pursue republican ideals too far were sometimes described as opportunists. To the left were the radicals and the most extreme of these were the radical socialists. It was only in the 1890s that socialism emerged as an important movement. By 1898 the infamous affair of the Jewish army officer Captain Dreyfus was to polarise French society, yet many prominent scientists, and most notably Berthelot, refused to join other intellectuals in attacking the government and the military establishment.[50] Thus an episode, of great importance in the history of French politics, was merely peripheral in the history of the Academy.

In a speech of 1877, Gambetta had identified 'clericalism' as the common enemy of republicans.[51] Although a law of 1875 on freedom in higher education had permitted the founding of Catholic universities, a succession of laws passed in the 1880s followed a contrary programme of secularisation. Indeed sometimes these ideas were pursued at the expense of social reform. This laicisation reached its most extreme in 1904 under Combes, when all teaching by religious orders was prohibited, their schools closed and their property confiscated. It was in the 1880s in the cabinet of the sometime radical Goblet

[48] J. M. D. Olmsted, *Claude Bernard and the experimental method in medicine*, New York, 1952, pp. 71, 168–9.

[49] Useful sources are Pierre Barral, ed., *Les fondateurs de la troisième République*, 1968, Claude Nicolet, *L'idée républicaine en France. Essai d'histoire critique*, 1982, Jean-Marie Mayeur and Madeleine Reberioux, *The Third Republic from its origins to the Great War, 1871–1914*, trans., Cambridge, 1984.

[50] Mayeur and Reberioux (*op. cit.*, p. 196) call attention to the silence of Berthelot. Jean Jacques speaks of the discretion of scientists, on an issue which galvanised other intellectuals; *Berthelot, 1827–1907. Autopsie d'un mythe*, 1987, p. 247.

[51] Barral, *op. cit.*, pp. 174–6.

that Berthelot became Minister of Public Instruction. A close friend of the lately deceased Paul Bert, his views on the secularisation of education were well known. We may recall that it had been the Franco-Prussian war which had brought Berthelot into politics. The ensuing Commune and civil war of 1871 left many scars on French society. One could not easily forget on the one hand the brutal killing of many republicans nor on the other hand the murder of the Archbishop of Paris in front of a firing squad. Such events tended to polarise society. They also tended to strengthen the association of politics with religion and this requires a brief retrospective review, which will also serve the purpose of leading into the final two sections of this chapter.

If we go back to the eighteenth century, it is clear that the Roman Catholic Church had been weakened by the attacks of the *philosophes*, whose ideas were particularly influential in the revolutionary period. After the secularisation associated with the Revolution, the first major step which reversed the decline of religious practice was the Concordat, signed by Bonaparte in 1802. Yet Bonaparte was principally concerned with restoring a greater measure of national unity. He cynically regarded religion principally as a means of social control and under the Empire the Catholic Church existed under severe constraints. It was only at the Restoration in 1815 that Louis XVIII allowed the Church a greater role in society. In fact there was a significant religious revival in several European countries in the early nineteenth century. In France after the Restoration there was a large rise in the numbers of members of religious orders and the Jesuit-inspired Congregation was associated with a great increase in Catholic power and influence. The Church, which had been obliged to abandon its traditional educational role at the Revolution, now had its own schools again and, under Charles X, Monseigneur Frayssinous served as Minister of Ecclesiastical Affairs and Public Instruction (1824–8). Thus in the 1820s the pendulum swung briefly to the opposite extreme. This produced a strong anti-clerical reaction and was one of the factors leading to the Revolution of 1830.

Although the Church made clear its teaching on moral issues it was nearly always on questions of education that the greatest open conflict arose between religious and secular parties. The Church claimed territorial rights in education, which were opposed by republicans. Often the Church adopted a conservative position, having received more sympathetic treatment from different monarchies than under successive republics. Individual Catholic intellectuals, however, ranged over the whole political spectrum and while many were on the right, Lamennais provides an example of an influential Catholic political theorist of the left. We might also mention Frédéric Ozanam, who founded the St. Vincent de Paul Society in 1833, the Dominican priest Lacordaire, who sat with the left in the National Constituent Assembly in 1848, and Charles Peguy at the end of the century, who was a socialist.

Under Napoleon III the Catholic church received considerable support and it was under the next government of the Third Republic, as we have indicated above, that the greatest split between Church and state occurred. Liberal elements in society were keen to emancipate themselves from authority of any kind and the Catholic church seemed to represent the most pervasive authority. It was the authoritarian claims of Catholicism, including its moral pronouncements even more perhaps than its traditionalism, which not only placed it out of sympathy with many intellectuals but also made it a principal enemy for republicans. Yet the situation was far from static and relations between Church and state became better after 1892, when Pope Leo XIII finally tried to heal the breach by calling on French Catholics to give their full support to the republican regime. The 1890s marked the final end of any Church support for the royalist cause, although a few Catholic intellectuals like Duhem found it impossible to change their politics.

7. *Religious antagonisms*

Nothing polarised the Academy more than religion. It was not a dispute between different Christian denominations of a kind which bedeviled social relations in nineteenth-century Britain, nor even antagonism between different religions. It was a fundamental dispute between Academicians accepting religious (Catholic) principles and a powerful anti-religious group. In what was nominally a Catholic country, one would have expected any Catholic faction, if it existed, to be in the majority. In advanced and exclusively male intellectual circles like the Academy this was far from the case. It has been only too easy for some scholars to assume that someone like Claude Bernard, baptised as a Catholic, should be so classified in academic politics.[52] In fact Bernard, like so many of his colleagues, adopted a strong rationalist position, antagonistic to religion.[53] The rationalist and secular lobby was much more powerful in France in the late nineteenth century than in most other countries. It was reinforced by a general association with left-wing politics, as opposed to the right-wing politics of some outspoken Catholics.

Any Academician with religious affiliations might seem to the majority of his colleagues to suffer from divided loyalties and this was much more the case with Catholics than with Protestants, since supporters of the Catholic church from time to time expressed criticisms of science, or at least of the more extreme claims of science. Pope Pius IX in the 1860s did not help matters by replying to the growing liberalism and secularisation of the mid-nineteenth century by promulgating his *Syllabus of Errors*, which seemed to condemn all modern movements.[54] In a scientific organisation supported by the state, adherence to

[52] J. Farley and G. L. Geison, 'Science, politics and spontaneous generation in nineteenth-century France: the Pasteur–Pouchet debate', *Bulletin of the History of medicine*, **48** (1974), 161–98 (p. 181).

[53] The hostile references to religion in Claude Bernard's *Introduction à l'étude de la médecine expérimentale* (1865) are pointed out by David Goodman and Robert Olby in Unit 10, *The origins of life: discussions in the later nineteenth century*, p. 12, from Open University, A.381, *Science and belief; from Darwin to Einstein*, Milton Keynes, 1981.

[54] J. McManners, *Church and state in France, 1870–1914*, 1972, p. 19.

the Catholic Church could, therefore, be seen as almost a subversive element and this was particularly the case under the Third Republic. Such a fear tended to bring together in opposition men of different political persuasions or none. There is nothing like a common enemy (real or imagined) to create a common purpose. Only the outbreak of the Great War would unite the French people and reduce the division which had so weakened French society.

In some ways it is easier to identify what the secularists opposed than what they supported. Above all they opposed the Roman Catholic Church and clerical power, although it might be mentioned that some anti-clericalism in the narrowest sense of the term was not unknown even among members of the Church. Secularists argued that education should be completely separated from Church influence, and indeed many held that one of the great achievements of the French Revolution had been to lay the foundations of a state education system. Nor should the clergy be able to insist on marriage in church. Marriage should be a civil ceremony which took place at the local town hall. In practice a compromise was often agreed on, with a secular ceremony being followed by an ecclesiastical one, thus satisfying the parties concerned and their families. Even the Berthelot family accepted the compromise.[55] The celebration of 14 July as a national festival from 1880 became a major secular republican festival, intended to overshadow the 'superstitious' marking of saint's days. Under the First Republic there had been an attempt to found a cult of the 'Supreme Being' as an alternative to Christianity. Under the Third Republic no explicit religion was offered. Instead a fully secular philosophy and education were promulgated, which excluded 'old dogmas'.[56]

Many members of the secular movement abandoned any residual faith in God in favour of a faith in science. Following Condorcet, they believed in the unlimited progress of science and its application to society.[57] Thus in 1849 the anti-clerical philosopher Renan went so far as to claim that 'science alone can solve the eternal problems of man'.[58] Following Comte, the positivists claimed that religion had represented an earlier and more primitive state of knowledge which had been replaced in the nineteenth century by positive knowledge of which science was the great exemplar.[59] Few followed Comte in his later career when he founded a new religion, but there were serious attempts from time to time to create a new morality based on scientific principles. It therefore seemed that there was no need for Christianity and the Academician and deputy Paul Bert in 1873 recommended a reform of the university system in which all Faculties of theology, both Catholic and Protestant, would be abolished.[60]

[55] Léon Velluz, *Vie de Berthelot*, 1964, p. 228.

[56] Mayeur and Reberioux, *op. cit.*, p. 118.

[57] Condorcet, *Esquisse d'un tableau des progrès de l'esprit humain*, 1794.

[58] *L'avenir de la science*, *Oeuvres*, 1947–61, vol. 3, p. 814, quoted by H. Guerlac, *Essays and papers in the history of modern science*, Baltimore, 1977, p. 497.

[59] Auguste Comte, *Cours de philosophie positive*, 6 vols, 1830–42. See also D. G. Charlton, 'The cult of science', chap. 5 of *Secular religions in France, 1815–70*, 1963.

[60] L. Liard, *L'enseignement supérieur en France* (2 vols., 1888, 94), Vol. 2, p. 482.

An influential figure who claimed that one could have morality without religion, was Berthelot. He saw the basis of morality (like science) in experience and knowledge, not in divine revelation. He advocated scientific method as the principal source of moral and material progress. Although Berthelot, like so many nineteenth-century French scientists, was influenced by Comte and positivism, he did make a distinction between *science positive*, which gave certain knowledge, and *science idéale*, which allowed for hope and imagination.[61] Nevertheless, when in 1885 he rashly declared that 'the world today is without mystery',[62] he was understood as claiming that there were no more secrets in nature. Like Claude Bernard, he dismissed as superstition the old idea of a 'vital force', characteristic of living things. His research on organic synthesis showed that chemists could make in the laboratory substances that had previously been thought to be unique products of the natural world.[63] He believed that the future possibilities of organic synthesis were boundless. In principle chemists could make anything.

It was such claims that led in the 1890s to a debate on the 'bankruptcy of science'.[64] Brunetière claimed that physics and chemistry could not eliminate mystery from the world. Only religion could explain man's origin and destiny. Thus to some secularists it seemed that religion provided a constraint on the growth of science, not in the literal sense, for example in the building of laboratories, but rather in the extension of the domain of science from its traditional areas to other areas of intellectual activity. Looking back on the situation a hundred years later we may consider the dispute not so much as one between religion and science as between religion and *scientism*, that is the questionable extension of natural science to other areas of thought.

In 1889, the centenary year of the French Revolution, Berthelot was to become secretary of the Academy and the most influential man of science in France till his death in 1907.[65] His influence extended far beyond the Academy and the scientific community. Having a keen interest in republican politics, he was appointed Minister of Public Instruction in 1886–7 and he made use of his position to promote a bill on the secularisation of primary education. This was at a time when the Third Republic was launching a sustained anti-clerical campaign. Berthelot took a leading part in secularisation, perhaps, as one commentator has remarked, 'without fully realising the complexity of a situation, in which more politically astute colleagues used the laicisation

[61] R. Virtanen, *Marcellin Berthelot. A study of a scientist's role*, University of Nebraska Studies, no. 31, Lincoln, Nebraska, 1965, p. 36.

[62] M. Berthelot, *Les origines de l'alchimie*, 1825, p. v.

[63] *Chimie organique fondée sur la synthèse*, 1860.

[64] H. W. Paul, 'The debate over the bankruptcy of science in 1895', *French Historical Studies*, 3 (1968), 299–327.

[65] Even after his death the influence of Berthelot lived on and not only in France. Thus the centenary of his birth was the occasion for a major international commemoration. See the large commemorative *livre d'or: Centenaire de Marcelin Berthelot, 1827–1927*, 1930.

programme to direct attention [away] from the need for social and economic reform'.[66]

In the late nineteenth century it was customary to organise political and ideological rallies as banquets and at the banquet of St. Mandé in April 1895 Berthelot was the guest of honour and gave a speech on science as a liberating force. It has been claimed that the aim of the banquet was 'to honour science as the basis of republican politics'.[67] Among the 800 guests present were the current Minister of Public Instruction, the zoologist and Academician Edmond Perrier, also a Dr. Blatin 'representing freemasonry' and M. Delhet 'representing the positivist school'.[68] At another banquet where Renan and Berthelot were the joint guests of honour, several Academicians were present, including the chemist Cahours (1813–91), the zoologist Lacaze-Duthiers (1821–1901) and the astronomer Janssen (1824–1907).[69]

Among the factions hostile to traditional Christianity in the Academy, one group which might deserve special study[70] is that connected with the Société de Biologie founded, significantly, in 1848, 'the year of revolutions', which saw the foundation of the short-lived Second Republic in France. Its members may well have had an influence on the life-sciences in the second half of the nineteenth century comparable to that of the Society of Arcueil on the physical sciences in the first half of the century. Its first president was Pierre Rayer (1793–1876), who held the post of professor of comparative medicine in the Paris Faculty of Medicine and was elected to the Academy of Sciences in 1843. According to Berthelot,[71] Rayer had been the victim of religious intolerance during the Restoration and he gathered round him a group of young medical scientists and freethinkers whom he helped to launch on their respective scientific careers. It was through Rayer that the young histologist Charles Robin (1821–85) met Emile Littré and Auguste Comte, who won him over to positivism.

Robin became one of the leading activists in the Société de Biologie, which he used to create a school of positivist biology, and which was to have great influence on the subsequent orientation of physiological, medical and zoological research in France.[72] Although many of the most prominent members of the society were qualified in medicine, Robin and his colleagues wanted to distance themselves from what they saw as the excessively narrow and practical focus of the main Parisian medical school.[73] Robin's idea that life depended simply on a

[66] R. Virtanen, *op. cit.*, p. 23.

[67] J. Jacques, *op. cit.*, p. 213.

[68] M. Berthelot, *Science et morale*, 1897, p. 36.

[69] *Ibid.*, p. 50.

[70] We await a detailed modern study of this group. Meanwhile, apart from references given in the following footnotes, see also: E. Gley, 'La Société de Biologie et L'évolution des sciences biologiques en France de 1849 à 1900' in his *Essais de philosophie et d'histoire de la biologie*, 1900, pp. 168–312 and Harry Paul, *From knowledge to power. The rise of the scientific empire in France, 1860–1939*, Cambridge, 1985, pp. 61ff.

[71] M.A.I., 47 (1894), ccxxix–ccxxx.

[72] M. D. Grmek, art. 'Charles-Phillipe Robin'. *D.S.B.*, 11, 491–2.

[73] John E. Lesch, *Science and medicine in France. The emergence of experimental physiology, 1790–1855*, Cambridge, Mass., 1984, pp. 222–4.

particular molecular state constituted a form of biological materialism, a philosophy shared by many members of the society. Although elected to the Academy in 1866, Robin became less interested in research than in politics and administration and in 1875 he was elected to the Senate, where he was associated with the far left as a 'freethinker', constantly engaged in anti-clerical activity.

Prominent among other early members of the Société de Biologie identified as materialists[74] was Paul Broca, and among those later to become members of the Academy of Sciences were the physiologist Vulpian (1826–87) and the neurologist Charcot (1825–93) who, like Broca, vigorously supported the theory of cerebral localisation in man and was also one of the first to demonstrate a connection between physiology and psychology.[75] Probably the most famous member of the society was the great physiologist Claude Bernard (1813–78), who has been described as opposing a strong current of vitalism with his own brand of determinism.[76] Bernard came under the influence of Comte, though not uncritically. He became a professor at the Collège de France, where he was a colleague of Berthelot for many years. Of the same generation in the society was the physiologist Brown-Séquard (1817–94), who also was elected subsequently to a chair at the Collège de France. The physiologist Paul Bert, a disciple of Claude Bernard, became president of the Society in 1878. Other members of the society elected to the Academy in the 1880s were the anatomist Sappey (1810–96) and the pathologist Verneuil (1823–95). Among the colleagues of these men in the Academy was the surgeon Alfred Richet, whose son Charles (1850–1935) was to carry out extensive physiological research in the respective laboratories of Berthelot, Vulpian and Robin. He also came under the influence of Claude Bernard, Verneuil and Charcot and is said to have been unusually suggestible,[77] thus carrying many of the ideas of the early members of the Société de Biologie into the twentieth century. Yet, breaking away from the extreme physico-chemical reductionism of his teachers, he felt that teleology had a place in biology and he even became involved in the early twentieth-century phase of interest in spiritualism.

8. *Religious affiliations*

Although a secular ethic was prominent in French intellectual life throughout most of the nineteenth century, religious sympathies were not entirely absent from the Academy. Yet they were not the official concern of the Academy and were more often beneath the surface. One must certainly not expect religion to be mentioned in the *Comptes rendus*. Only in the published *éloges* is there

[74] E.g. Jacques Léonard, *La vie quotidienne d'un médecin de province*, 1977, p. 235; also H. Paul, *op. cit.*, p. 64.

[75] Andrée Tétry, art. 'Jean-Martin Charcot', *D.S.B.*, 3, p. 205.

[76] M. D. Grmek, art. 'Claude Bernard', *D.S.B.*, 2, 24–34 (p. 32).

[77] Frederic L. Holmes, art. 'Charles Robert Richet', *D.S.B.*, 11, 425–32.

occasional mention of a religious dimension. For example, in the *éloge* of the geologist Academician Albert Lapparent (1839–1908), professor at the Institut Catholique in Paris, it was said that although he had theologians as colleagues, he consistently kept outside discussions of theological matters, preferring to pronounce on scientific fact than to enter into controversy.[78] Of the physicist Gabriel Lippmann (1845–1921), whose parents were both Jewish, it was said that 'he avoided with great care conversations on matters of religion'.[79] Although sympathetic to a spiritual view of mankind, he saw religious dogma as an obstacle to social harmony. Again the mineralogist and cleric René Haüy was remembered as much for his spirit of religious tolerance as for his piety.[80] There were also a few cases of late conversions. Biot, who as a young man had shared the indifference to religion of most of his contemporaries in the revolutionary era, went out of his way to gain an audience with Pope Leo XII when he went to Rome in 1825. In 1846, at the age of seventy, he made a formal return to the Catholic Church.[81]

In France a study of religious affiliation means in the majority of cases Roman Catholicism, although there were also several Protestants in the Academy, and, later in the century, several Jews. The Abbé Haüy was the only cleric who was a member of the nineteenth-century Academy. The son of a poor weaver, he had become a priest in 1770 and taught in the Collège Cardinal Lemoine. His studies of crystals had led to his election to the new Institute in 1795. During the difficult revolutionary times he had refused to take the oath required by the civil constitution of the clergy, but he was protected by friends through his connection with the former Jardin du Roi. Haüy made major contributions to the science of crystallography, culminating in the publication in 1822 of his *Traité de crystallographie*. Mild in character, Haüy was never outspoken with his scientific colleagues on religious matters.

Although among intellectuals there were undoubtedly some intransigent Catholics who looked backwards rather than to the future, in the scientific community and in the Academy there were many liberal Catholics who supported scientific advance on a broad front and contributed to that advance with the only reservation that there were certain ultimate questions, which science alone could never answer.

It would be possible to imagine a spectrum of Catholicism in the Academy, ranging from the most extreme on the one hand to those on the other who were no more than nominally Catholic. One might begin by classifying the Academicians Cauchy, Binet and Duhem as ultra-Catholic, which has a political (right wing) connotation as well as a religious one. Cauchy was a legitimist with a deep personal commitment to the senior line of the Bourbon dynasty. Even

[78] *M.A.I.*, **57** (1922), lxxi.

[79] *M.A.I.*, **60** (1931), xxviii.

[80] 'aussi tolérant que pieux', *M.A.I.*, **8** (1829), clxxiii.

[81] F. Lefort, 'Un savant Chrétien, J. B. Biot', *Le Correspondant*, Nouvelle Série, **36** (1867), 955–95.

stronger were his religious sentiments, which he carried into social work, for example, in poor relief through the St. Vincent de Paul Society. In 1846 he organised a petition to the Pope asking for material aid for the victims of the Irish famine. He was tireless in his request for signatures from his fellow members of the Institute. According to his biographer Valson[82] he frequently made use of his fellow Academicians, not for political intrigue, but as sources for money to relieve individual cases of hardship.

Unlike Cauchy, most of whose creative work was presented to the Academy, Duhem was not only kept out of the Academy until his final years but even out of Paris which, until 1913, meant exclusion from the Institute. Teaching successively at Lille, Rennes and Bordeaux, he was elected a corresponding member in 1900. His religion seems to have inspired many aspects of his work, including his persistent attacks on the authority and thermochemistry of the militantly secular Berthelot. His discovery of medieval mechanics was pursued in order to show that a form of early science flourished under the patronage of the Catholic Church several centuries before Galileo. His positivist philosophy of science provided a defence from the conviction of many of his contemporaries that science represented absolute truth which could be applied to the spiritual and moral understanding of man.

Less extreme than Cauchy, and perhaps Duhem, was a second group of Academicians, who might be described as 'committed Catholic'. This rather arbitrary label might be attached to men like J. B. Dumas, Le Verrier, Lapparent and Henri Poincaré. The fact that Dumas was a practising Catholic earned him the nickname of 'Jesuit' among his enemies.[83] When a Catholic university (later the Institut Catholique) was founded in Paris in 1875, Lapparent accepted the chair of geology and mineralogy. In 1879 he was forced to choose between a career in the Bureau des Mines where, as his biographer says,[84] he could have looked forward to a brilliant future, and the Institut Catholique, which had very few resources. He chose the latter. Another Catholic scientist associated with the Institut Catholique was the physicist Edouard Branly, inventor of a device for detecting Hertzian waves, who was elected to the Academy in 1911. A more famous contemporary was the brilliant mathematician Henri Poincaré, whose interests extended to the philosophy of science and who pointed out that science had not disproved the possibility of miracles.[85]

But in the course of the nineteenth century there were also several Academicians who had been brought up as Catholics, who accepted a Christian

[82] C. A. Valson, *La vie et les travaux du Baron Cauchy*, Paris, 1888, reprinted Paris, 1970. For a less partisan biography of Cauchy, see Bruno Belhoste, *Cauchy, 1789–1859. Un mathématicien légitimiste au 19e siécle*, 1985.

[83] The term 'Jesuit' was used pejoratively by Liebig and adopted by Pelouze and others, who were jealous of Dumas' rapid rise to fame. Pelouze to Liebig, 9 March 1838, 14 June 1838, Liebigiana, Bayerische Staatsbibliothek, Munich, nos 43, 47.

[84] Franck Bourdier, *D.S.B.*, 1, 471–3.

[85] 'Le libre examen en matière scientifique' (1909), chap. 3 of *Dernières pensées*, 1926, pp. 325–42 (p. 337).

code of ethics and who probably attended church but often without any special enthusiasm for the niceties of religious ritual and doctrine. They had obvious counterparts in Britain among members of the Church of England. It was only when extreme positions of atheism or materialism were advocated that they occasionally spoke out in defence of religion. Such a person was Louis Pasteur, who insisted on an absolute separation between matters of science and matters of faith or sentiment. It was only in exceptional circumstances that he would come out and attack atheism and materialism and the so-called 'free-thinkers'.[86] His own personal philosophy has been described as 'entirely of the heart'[87] but on a few occasions he protested strongly against those who denied the existence of a spiritual realm or the immortal soul. In 1864 he had protested against the appointment of Berthelot to a chair of organic chemistry at the Collège de France, which he had interpreted as a conspiracy and 'one of the manifestations of that school, impatient and dangerous, personified by the names of MM. Renan, Taine, Littré, etc.'[88] Yet, although those named were certainly friends and supporters of Berthelot, it is not clear that this was quite the positivist conspiracy that Pasteur imagined. Pasteur's belief in a spiritual dimension to life was probably one motivation for his campaign against the idea of spontaneous generation, which he viewed as gross materialism.

Among other practising Catholics in the early nineteenth century we may cite Ampère (1755–1836), Fresnel (1788–1827) and Blainville (1777–1850). Of the permanent secretary Elie de Beaumont (1798–1874), it was said that he was piously faithful to the Catholic teaching of his childhood but he was also noted for his religious tolerance.[89] The Academician Tulasne devoted his life to botany and Christianity and, when he retired, left his library to the Institut Catholique in Paris. Astronomers Wolf (1827–1918) and Bigourdan (1851–1932) were both men of deep religious convictions and the chemist Paul Sabatier (1854–1951) was a Catholic who, like Duhem (1861–1916), became a member of the Academy in 1913 when the rules were changed to admit provincial members in a special category. Both these Catholic scientists[90] were out of sympathy with the militantly secular ideals of men like Berthelot. Yet there does not seem to have been an organised Catholic faction within the Academy, religion remaining a purely private matter. Only occasionally, as we shall see in Chapter 6, did it act as an additional bond between a patron and his protégé but such a relationship, of course, applied equally to Protestants and presumably to atheists.

[86] Discussion on fermentation at Academy of Medicine, 9 March 1875, Pasteur, *Oeuvres*, vol. 6, part 1, pp. 56–7.

[87] G. Geison, art, 'Pasteur' in *D.S.B.*, **10**, 350–416 (pp 354–5).

[88] Pasteur, *Correspondance*, vol. 2, pp. 151, 154. For a useful summary of the three major positivists mentioned here, Littré, Renan and Taine see T. Zeldin, *France, 1848–1945*, vol. 2, Oxford, 1979, pp. 600–8.

[89] *M.A.I.*, **39** (1876), xxxiv.

[90] Mary Jo Nye, 'Nonconformity and creativity: A study of Paul Sabatier, chemical theory and the French scientific community', *Isis*, **68** (1977) 375–91.

A few words are necessary about the position of Protestants, since a number of scientists of Protestant origin were also to be found among the members of the Academy. Although accounting altogether for less than two per cent of the total population of France, there were many Lutherans in Alsace, and Calvinism was particularly strong in some parts of the south and of course in Geneva which, although returned to an independent Switzerland in 1815, continued as a centre of French culture. Whereas the Catholic church, with the power it had exercised over higher education as recently as the 1820s and its ongoing claim to be involved generally in education, was seen as a threat by secularists Protestants generally escaped this opposition. They themselves had been a minority grossly discriminated against by the social system of the old regime. They had therefore almost unanimously welcomed the Revolution as providing them with their full civil rights.[91] Under Bonaparte, Protestant churches were established on a parallel basis to Catholic churches, each having salaried pastors or priests.[92]

The most famous and influential Protestant Academician was 'Georges' Cuvier, born in 1769 in Wurttemberg, educated at the Hohen Karlsschule near Stuttgart and originally destined for the Lutheran ministry. Cuvier had become a French citizen in 1793 and sought recognition in the scientific world of Paris. He was elected to the First Class in 1796. Under Napoleon, Cuvier held a position in the Ministry of Religious Affairs and it may have been partly due to the influence of prominent Protestants like himself that Napoleon refused to consider Catholicism as the official religion of France but simply that of the majority of the French people. Religion was one of the bonds between Cuvier and his protégé, the zoologist Dumeril, another Protestant elected to the Academy. Also we should not overlook the naturalist Bosc d'Antic, elected to the agriculture section in 1806, and George Cuvier's younger brother Frédéric, summoned by him to Paris in 1797 to help at the Muséum d'Histoire Naturelle and eventually elected to the Academy in 1826.

Probably a fundamentalist attitude to Christianity and the interpretation of the Bible was one of the factors prompting Georges Cuvier to support so strongly the fixity of species and to oppose Lamarck's theory of transformism.[93] That there was some religious motivation for this opposition may be inferred from the fact that one of the most powerful critics of Charles Darwin in France in the 1860s was Quatrefages, another Protestant. This does no mean, however, that there was not also opposition in Catholic circles and also opposition on grounds quite unconnected with religion.

But Protestant influence was not confined to the biological sciences. The banker baron Delessert (1773–1847), from a prominent Geneva family, was

[91] Strictly speaking, toleration dates from 1787.

[92] W. Doyle, *Oxford history of the French Revolution*, Oxford, 1989, p. 410.

[93] Franck Bourdier writes that 'his respect for biblical chronology prevented him from participating in a new form of thought', art. 'Cuvier', *D.S.B.*, 3 521–8 (p. 527).

elected *Académicien libre* in 1816. He led a crusade for the introduction of a national savings bank into France, which he advocated on moral as well as economic grounds. In his official *éloge* it was said that he seldom spoke directly on religious matters but he made a habit of reading his Bible daily.[94] Another co-religionist of undoubted sincerity was the organic chemist Wurtz, son of a Lutheran minister, who had attended a Protestant school at Strasbourg. In the words of a recent biographer, 'throughout his life Wurtz remained true to his Lutheran heritage'.[95] There is little evidence that any of these scientists, although belonging to a religious minority, suffered from any serious discrimination in the scientific world. Indeed Wurtz, with his co-religionist Friedel, thrived in Paris at the centre of a multi-national school of chemistry which extended its influence through the foundation of the Société chimique (1857).

It might be thought that members of this religious minority would tend to represent certain strict religious principles but it has been suggested that Protestantism in France, rather than necessarily corresponding to a specific theological position could sometimes provide a useful position of independence. In a recent study of Cuvier it has been remarked that

> Cuvier may well ... have tended after the Revolution to see Protestantism as a series of opportunities rather than as a body of belief.[96]

Cuvier's own belief apparently consisted of minimal deism. Later in the century there was a further dilution of belief and many who originated from Protestant families became free thinkers.[97] Thus they were easily assimilated into the secular movement. Liberal Protestants were quite prominent in the anti-clerical period of the Third Republic from 1880–1914. Given that one of the constant problems was the organisation of education, they obviously had little sympathy for Catholic schools and in fact strongly supported the secularisation of education in the late nineteenth century.

Finally, we should consider the Jewish dimension. Although the Jewish population of France in the nineteenth century was tiny, only about 80000 in 1900 out of a total of some forty million, there were several Jewish scientists in the Academy.[98] In a largely meritocratic system of higher education they had been able to enter such elite schools as the Ecole Normale (Lippman and Hadamard) and the Ecole Polytechnique (Michel-Lévy), which served as an important stepping stone to further advancement. Many of the Jews in Paris in the late nineteenth century were recent immigrants with modest trades. But

[94] Flourens, *M.A.I.*, **22** (1850), cxli.

[95] J. H. Brooke, *D.S.B.*, **14**, 529.

[96] D. Outram, *Georges Cuvier. Vocation, science and authority in post-revolutionary France*, 1984, p. 145.

[97] C. R. Day, 'Protestantism' in Patrick H. Hutton (ed.), *Historical dictionary of the Third French Republic*, 2 vols., Westport, Ct., 1986.

[98] The archives of the Academy are often reticent on matters of religion and race. Much of this section is accordingly based on information in the *Encyclopaedia Judaica* (16 vols., Jerusalem, 1971–2).

there was also a moneyed bourgeoisie which had become more integrated into French society. Most Jews felt some attachment to the legacy of the French Revolution, which had granted them the full rights of citizenship. Accordingly they tended to support the secular Third Republic.

Although there was some anti-semitism (often allied with xenophobia) in everyday life, the world of higher education offered some relief, providing a marked contrast with the army, which was the scene of the notorious Dreyfus case. It is worthy of note that a brother-in-law of Captain Dreyfus, unjustly condemned as a spy, was the leading mathematician Jacques Salomon Hadamard (1865–1963), appointed professor of mathematics at the Collège de France in 1897 and professor at the Ecole Polytechnique and Academician in 1912. He had been preceded in the Academy by the Jewish physicist Gabriel Lippman, who had been elected in 1886 and was later to win the Nobel prize. Another Jewish Academician was Auguste Michel-Lévy (1844–1911), director of the French geological survey and one of the most distinguished petrologists of that period. In the field of astronomy we should mention also Maurice Loewy (1822–1907), born in Vienna but later becoming a French citizen. He was an astronomer at the Paris Observatory and a pioneer in the use of photography in astronomy. In 1894 he was elected president of the Academy. A close friend of Loewy was Raphael Bischoffsheim (1823–1906), born in Amsterdam but given French naturalisation in 1880. Bischoffsheim came from a prominent family of international bankers but he had a strong amateur interest in astronomy. After endowing an observatory at Nice, he was made an *Académicien libre* in 1890.

— *Chapter 6* —

ELECTIONS: 'GREEN FEVER'

Election is the only valid method of creating Academicians.
(Arago, Eloge of Monge, *M.A.I.*, **24** (1854), lxxx.)

The influence of an academy depends to a large extent on elections.
(Darboux, *M.A.I.*, **47** (1904), ccclxvii.)

The prospect of a place in the Academy is a stimulus, which more than once has encouraged young men of science at the beginning of their career.
(*Journal des Débats*, 19 September 1832.)

1. *Introduction*

The reference to 'green' in the title of this chapter is to the dark green embroidery on the black costume of members of the Institute. The 'fever' may be more difficult to understand a century later. It refers to the excitement often generated by elections, an excitement which could occasionally reach fever pitch. Of course there were some elections that were hardly contentious, such as those of many correspondents where the names of provincial or foreign candidates were hardly known in the Academy generally and the members had to depend heavily on advice from the relevant section. More passion usually went into elections for full (i.e. resident) membership. Many of the contestants, who had made a practice of presenting their research to the Academy over a period of many years, would be quite well known to members. They would be resident in Paris and be fairly well advanced in their careers.

Several would be the holder of a position at one or more of the great scientific institutions of the capital, such as the Muséum d'Histoire Naturelle, or one of the *grandes écoles*. If so, they would have an obvious constituency among Academicians for their initial support. Among senior scientists election to the Academy was regarded as the single most important step and the highest honour in a scientist's career. Appointment to a chair or the award of the Legion of Honour were trivial occasions in comparison. Although it was the members of the Académie Française who were traditionally known as 'the immortals', most French scientists would have considered election to the Academy of Sciences as constituting a sort of immortality. In the nineteenth century to be an Academician was widely regarded in the scientific community in the way Nobel Laureates came to be regarded in the twentieth century. An Academician had great visibility. He was often regarded as a role model and the choice of such heroic figures was a matter of the greatest importance.

Whereas under the old regime scientists brought into the Academy would fill the junior grades before being considered for further promotion, in the nineteenth century election became an all-or-nothing event with all members in theory being of equal rank. This made the process of election much more crucial after the Revolution and we may note in passing that by its levelling up, the Academy may have come to consider that it had lost a potentially valuable means of management. In the nineteenth century elections, once made, could not be rectified. Inevitably, therefore, the Academy had to carry a proportion of 'passengers' who hardly lived up to earlier expectations.

The elections of the Academy can be approached in a number of different ways. They could, for example, be simply presented at the lowest level as occasions for intrigue. Although no one could deny the lobbying which regularly took place before elections, and the occasional emergence of factions, one should not exaggerate this aspect. Despite a few glaring examples of nepotism, discussed separately, one would like to think that in most cases personal considerations were secondary and that candidates were being judged almost entirely on their scientific merits. But even within science there was room for considerable debate. How should two candidates be compared if one were an experimentalist or field worker and the other a theoretician? How should a candidate, say of fifty-five years of age, who had been contributing regularly but in an unspectacular way to his field for three decades, be compared with a comparative youngster of, say thirty-five, who had made a single spectacular discovery? From the candidates' point of view the very existence of elections constituted an enormous incentive to make major contributions to science. As Pasteur at the age of thirty-four wrote to his mother after a preliminary election skirmish in 1857:

> 'This election ... has given me a great deal of ardour for work.'[1]

Although his candidature was unsuccessful, he nevertheless saw it as a source of inspiration. He would work with renewed energy to enter the Academy. He would recapture the fervour of his very first research.[2] From the point of view of the Academy, it came more and more in the nineteenth century to regard elections principally as a reward system. Indeed election to the Academy was the ultimate reward. As the physiologist Magendie said, when he was elected in 1821 at the age of thirty-eight:

> All my trouble has been repaid, my goal is reached.[3]

[1] Pasteur Vallery-Radot (ed.), *Correspondance de Pasteur*, (4 vols., 1940–51), vol. 1, p. 425.

[2] *Ibid.*, p. 423.

[3] Flourens, 'Eloge de Magendie', *M.A.I.*, 33 (1861), xv. We shall see that in fact much of Magendie's major research belongs to the period *after* his election.

Yet despite undercurrents of potential interest to the sociologist, there is another dimension which needs to be considered, that of administration and bureaucracy. There were specific regulations which had to be followed; in addition, certain procedures slowly evolved until they became standard practice. But although we need to understand the basic rules and regulations they should not blind us to the potential for drama in elections. Very few elections could be regarded as foregone conclusions. Unlike some of the appointments and sinecures of the old regime, they could not be fixed in advance and the very uncertainty of the result added to the drama. It is true that evolving scientific norms usually prevented competition between professionals degenerating into open conflict. But sometimes candidates were identified with particular ideologies and the public became involved. Of course the election of scientists never aroused quite the same intensity of passion as when the candidates stood for a political platform. The Academy elections, to which the newspapers devoted the most attention, tended to be those of the Académie Française rather than of the Academy of Sciences. Nevertheless we should not forget that at least part of any Academy election was conducted in public. The names of candidates were known to the public and their merits sometimes openly debated, complementing the more knowledgeable debates of the Academy held in secret sessions.

Control over elections was probably the main power held by Academicians. How valuable a prerogative they considered this to be can best be judged by the few cases when they were deprived of this opportunity. The establishment of the Institute in 1795 began with the nomination of one third of the new members. Given the abnormal circumstances, there was general acquiescence, particularly as the *savants* nominated were given the immediate opportunity of electing further members. The greatest resentment came with the Restoration of the Bourbons, when in 1816 two new scientists, who had not previously been members of the First Class, were nominated to the renamed Royal Academy of Sciences: Cauchy and Breguet. The fact that both were exceptionally talented in their respective specialities of mathematics and mechanics (and would almost certainly have been elected within a few years if they had not been nominated) was seen as irrelevant. A fundamental principle had been violated and the crime was compounded by the simultaneous expulsion of Gaspard Monge and Lazare Carnot for blatantly political reasons. This made it appear that the two nominations were political appointments and this view was supported by the royalist sympathies of both nominees. In support of the claim of the Academy's knowledge *at the time* of Cauchy's exceptional talents, we may refer to the *grand prix* he was awarded in 1816. (In July 1815 at the age of twenty-five he had presented to the First Class a solution, some 300 pages long, to the subject proposed for the prize competition of 1816, wave propagation on the surface of a liquid.) Most of the disfavour fell on Cauchy but even Breguet, a plain clock-maker less likely to be associated with political intrigue, was resented. It was never forgotten by his colleagues that they had not elected him and even at his funeral, when one expects compassion rather than outspokenness, Dupin on behalf of the Academy remarked that it would have seemed 'purer and nobler' if his colleagues had been permitted to elect him, however much he deserved the honour.[4]

In so far as scientists and historians of science are aware of the elections of the Academy, it is often in relation to one particular individual. Typically the person concerned would not have been elected at his first candidature, despite his genius. Hence elections are regarded (if considered at all) with mistrust. But it is quite misleading to discuss an election without taking into consideration the other candidates and their seniority. Only with hindsight can one say that the genius of *X* should have immediately been recognised. Quite apart from being bad history, to focus entirely on *X* takes no account of the possible genius of *Y* and Z. Often there were several contestants of the first rank. It is only in the case of repeated failure that one has to raise questions of justice or poor judgement. It may be worth emphasising that all elections were in principle genuine elections with a multiplicity of candidates. The original constitution of the Institute stipulated a minimum of five candidates. Thus, unlike some allegedly democratic countries in the twentieth century where the number of (official) candidates used to be exactly equal to the number of vacancies, there was a desire in the Academy for the operation of a genuine choice. An unfortunate corollary, however, was that in elections there would always be several disappointed candidates.

The problem of hindsight can be well illustrated by the election of 1813 to replace Lagrange in the mathematics section. The successful candidate was Poinsot, then aged thirty-six. Poinsot lived on until 1859, proving to be one of the least productive members of the Academy in the 1820s, 30s, 40s and 50s.[5] If only the Academy had chosen Ampère or Cauchy, who were also candidates! In fact Ampère was elected the following year and Cauchy became a member in 1816, so they were not excluded for long. Also it must be remembered that in 1813 Ampère had not yet begun his work on electromagnetism, which retrospectively constitutes his major claim to fame. Cauchy for his part was only twenty-four years old and at the very beginning of a career which turned out to be prolific, although the members of the Academy were hardly to know that at the time of the election, which explains why he only attracted two votes as against the thirty by which Poinsot was elected.

2. *Nationality and residence*

Election was one of the two principal rites of passage administered by the Academy, the others being the official farewell pronounced after the death of a member. Election was certainly much more of a rite than a right for an able scientist. There were a number of important formalities which had to be followed

[4] *Funérailles de A. L. Breguet*, 18 September 1823.

[5] *M.A.I.*, **45** (1899), lxxxvi.

and it behoved the able and ambitious scientist to observe all the niceties and not suppose that ability alone would secure his election.

First, to be considered for election the candidate had to be a French citizen. It is true that corresponding members included many foreigners and the small and select group of *associés étrangers* were, by definition, not French. But these two categories were largely honorary. For full membership one had to be a full Frenchman.[6] Often the scientist had studied in Paris but for what might now be regarded as a technicality, he was not a French citizen. Thus Despretz (1791–1863) and Henri Milne-Edwards (1880–85) were technically of Belgian nationality after 1830 and had to apply for naturalisation before they were eligible to be full members of the Academy. The mathematician Libri (1803–69) had been born in Florence and came to France in 1830 as a refugee. He was naturalised in February 1833, a few weeks before his election. The peripatetic physiologist Brown-Séquard (1817–94) was the son of an American naval officer and a French mother. A British subject by birth, he studied in Paris and divided his life between Mauritius, France, the United States and England. The opportunity of succeeding to Claude Bernard's chair at the Collège de France in 1878 made him apply for French nationality. Deciding now to spend most of his time in Paris, he was eligible for consideration by the Academy, which elected him in 1886. Other Academicians naturalised included the astronomers Burckhardt (1773–1825), Bouvard (1767–1843), Loewy (1833–1907) and Radau (1835–1911), thus making the section of astronomy the most open to foreign (and notably Germanic) talent.

Members of the Academy were expected to be resident in Paris but residence requirements were sometimes rather lax, for example in the Napoleonic period[7] and early Restoration. In 1816 when the chemistry section was trying to persuade Proust (1754–1826), as a senior chemist living in provincial retirement and a corresponding member since 1804, to allow his name to go forward for full membership of the Academy, Berthollet wrote to him in the following terms:

> We have just lost Guyton. The chemistry section wishes to present you first on the list of those who may replace him and there is no doubt that the Class is eager to do you justice; but as the senior member of the section, I need to be sure that you will not refuse the place to which you will be nominated. You will only need quite a short stay in Paris each year to establish your residence and you would be able to spend the greater part of the year in the country. Try to overcome your repugnance to living among us.[8]

[6] Although the regulations did not stipulate that Academicians had to be male, this was the universal practice in the nineteenth century. When Marie Curie stood for election in 1910, the fact that she was a woman was only one of several considerations acting against her. See the discussion in Section 11 of this chapter.

[7] Although Cassini (1748–1845) sought re-election as a resident member in 1799, he interpreted this as acknowledgement of his professional status and he was so seldom in Paris that he could have been mistaken for a correspondent.

[8] A.S., Proust dossier, letter from Berthollet to Proust dated: (illegible) 1816.

Berthollet and Proust had held opposing theoretical views on the theory of definite proportions of chemical compounds but this did not prevent Berthollet and the other chemists from wishing to honour a colleague whose career, partly in Spain, had not been an easy one. A further letter from Chaptal,[9] reassuring him on the laxity of residence requirements, helped to win Proust over and he received a majority of votes in the election of 12 February 1816. Both Proust and his wife were in poor health at the time. It is obvious that the chemistry section was using its powers to confer a belated honour on a colleague and it expected little in return. In 1820 Proust took over the pharmacy in Angers belonging to his brother, who had fallen ill. In fact Proust never took his seat in the Academy and he created something of a record[10] as the register of the Academy showed him as a nominal member for the last ten years of his life without ever attending a single meeting.

As, however, the pressure on the limited number of places in the Academy increased, the condition of residence in Paris came to be taken more seriously. In extreme cases it meant that ambitious young men with salaried posts in the provinces were prepared to give up their very livelihood in order to come to Paris and take their chances in an election. Thus in 1856, when Pasteur was dean and professor at the Faculty of Science at Lille, he wrote to the Minister of Public Instruction, optimistically giving notice of his possible resignation[11], so that he should not be handicapped in his candidature for a vacancy in the mineralogy section of the Academy. Another candidate in another election, the plant physiologist Dutrochet, who lived with his mother in the remote department of Indre et Loire, formally renounced his residence there before a notary in 1828 to assume a Paris address. In a letter of candidature to the current president of the Academy, however, Dutrochet had to admit that he only spent part of the year in Paris.[12] Generally speaking, young scientists could only live in Paris if they had employment there, but to find a job in science commensurate with one's talents was not easy unless of course one was already a member of the Academy, in which case doors would usually open. For many aspirants this provided a 'Catch 22' situation.

The residential qualification produced a situation in which many scientists preferred to accept a fairly junior post in Paris, perhaps as a teacher in a lycée, in preference to a rather more senior post in the provinces, such as in a Faculty of Science. Added to the other attractions of Paris, the residential qualification had the effect of concentrating ambition and a preponderance of talent in the capital to the great detriment of the provinces. The mere possibility of election was a constant spur to middle-ranking scientists and counted for more, in the opinion of many, than a good salary. This led to a situation in which many

[9] A.S., Proust dossier.

[10] A record for an ordinary working member. Several *Académiciens libres*, contemporaries of Proust, never attended.

[11] *Correspondance de Pasteur*, vol. 1, p. 409.

[12] A.S., Dutrochet dossier, letter of 26 May 1831.

hangers-on, like so many satellites, revolved around the great Parisian scientific institutions and especially the Academy. When the chemical manufacturer Chaptal took up residence in Paris in March 1798, the vacancy at the Institute resulting from the recent death of the pharmacist Bayen is said to have been 'probably a major factor in his move'[13] and, when the election took place in May, he was successful. His residence in Paris not only enabled him to make a greater contribution to science but also to government, since during the period of the Consulate of Napoleon Bonaparte he was for fours years a very successful Minister of the Interior.

Although many candidates had some choice between educational posts in Paris and the provinces, this did not apply to members of the armed forces or the civil service. An area of administrative history which would be worthy of systematic exploration is the way in which the army, navy and various branches of the civil service agreed to give postings in Paris to men of outstanding scientific ability and aspirations. Thus Fresnel, as a member of the state corps of engineers, would normally have expected to spend most of his time on construction work in the provinces. It was through the patronage of Arago that he was given several months' leave in 1816 which he was able to spend in Paris where he had better conditions for research. Later a permanent posting in Paris enabled him to stand as a candidate for election to the Academy.

The evidence is clearest for Dupin (1784–1873), who was able to write in the following terms to the President of the Academy on 6 September 1818:

> Encouraged by the kind reception, which the Academy has given to my various works, I dare to announce my candidature for the place now vacant in the mechanics section. Several members feared that I would not be able to reside in Paris, which in view of the regulations would have been an insurmountable obstacle. I have the honour to inform you that his Excellency the Minister of the Navy by a decision of 1 September has just decided to place me on active service in Paris. If at some later date by unseen circumstances I was given duties which would be impossible for me to carry out and incompatible with my normal residence in Paris, I would, whatever the advantages of these duties, unhesitatingly refuse them rather than the honour of sitting in the Academy. I undertake this formal obligation and ask you, M. le President, to make this declaration known to the Academy.[14]

This flattering decision by their corresponding member to place the Academy before employment had the desired effect and on 28 September 1818 he was elected. On the other hand Poncelet, who was elected to the mechanics section in March 1834, still had official duties at Metz, and only twelve months later was his position regularised by his being attached to the fortification committee in Paris. Likewise Admiral Roussin was sent to Constantinople soon after his election in 1830 and was obviously unable to attend meetings in Paris. With

[13] John Graham Smith, *The origins and early development of the heavy chemical industry in France*, Oxford, 1979, p. 47.

[14] A.S., dossier Dupin.

some embarrassment he explained his difficult position in a letter to the Academy.[15] Such a situation encouraged the section of geography and navigation subsequently only to elect senior naval officers when they were securely in retirement.

3. *Electoral procedures*

An election may seem to be a straightforward process but in practice there were almost a dozen stages; some of these were little more than formalities but others were of considerable importance in weighing the scales in favour of certain candidates. One would hope that these were the best candidates, made more obvious by suitable deliberations. The fully articulated and complex mechanism of Academy elections could be regarded by outsiders with some awe, drawing as it did on the experience of more than a century. Alternatively, it could be regarded with considerable impatience by candidates anxious for a quick decision.

Since the process was one of filling dead men's shoes, it always began with the death of an Academician. Respect for the dead meant that a period of several weeks was allowed to elapse before the Academy officially considered the subject of election. Originally this was intended as a procedure to seek confirmation that there were immediately available a number of highly suitable candidates. If, for any reason, there were not, the Academy did not wish to appoint a second-rate candidate simply because he was the best available at the time. Almost invariably the Academy agreed to go ahead with the election but in a few cases there was a postponement for six months. The opportunities thus provided for manoeuvre will be discussed in the following section.

If the Academy was to proceed to an election, all members would be warned that an election was imminent. This announcement would often produce a flurry of activity. Ambitious scientists of middle or junior rank might not only ask the Academy to place their names on the list of candidates but would give considerable thought to their credentials. They would often hurriedly complete some work in progress and present a memoir to the Academy. They might also ask for an official report on the memoir, hoping not only to bring their name to the attention of the electors, but also to receive some words of commendation. Meanwhile the practice developed of having privately printed a *Notice*, setting out the candidate's publications in the subject, together with some words of explanation. The sudden desire to present memoirs to the Academy sometimes meant that candidates in an election had to be given priority over the other scientists.[16] Occasionally it meant that several meetings were required to accommodate all aspirants and the election would therefore be postponed to allow everyone to have a turn. This gave more time for the detailed scrutiny of the candidates by the section. The section was required to examine the credentials of the candidates and then draw up a list in order of merit for the whole Academy to consider. The section took its duties very seriously, delegating

[15] *P.V.I.*, 10, 461 (3 March 1834).

[16] E.g., *P.V.I.*, 10, 674 (9 March 1834).

to one or two of the more senior members the detailed study of the *Notices* of candidates. Originally no doubt the presentation would be based on notes, but it has recently been shown that, at least by the 1860s, a full written report was usually drafted on all serious candidates.[17] Such a report might extend to ten pages or more and would review the candidate's publications as well as his other claims to being an authority in his field.

The list establishing a hierarchy of merit based on the professional judgement of peers was presented to the next meeting of the Academy in secret session and the reports justifying the list were read out and discussed. Elections obviously occupied a large amount of the Academy's time; they involved members in prolonged discussion. An Academician who showed reluctance to immerse himself in long debates was considered exceptional. Such an exception was Admiral Paris, obviously a plain-speaking seafarer:

> His part in the discussion of the qualifications [of candidates] did not take long. 'I am no advocate', he said 'I do not know how to make a speech. M. So-and-so is the strongest. I will vote for him.' And that was the end of it.[18]

Sometimes it was felt that the section had taken too narrow or too partisan a view of suitable candidates and one or two further names might be added to the list. There was a full discussion of the merits of the respective candidates but no election yet. The actual election was postponed until the following week to allow time for mature reflection. A note was sent to the private address of any Academicians not present, warning them of the impending election. Finally, assuming that at least two thirds of the members were present, the election was held by secret ballot. If there were no absolute majority at the first ballot, the candidates at the bottom end of the list would be withdrawn and a further ballot held. The winner would be announced, although it was understood that this election was subject to confirmation by the government. After the 1820s this was no more than a formality and the new member might take his seat at the following meeting. It should be noted that under the old regime the Academy had drawn up a list of three names for each vacancy, with the King making the final choice. Thus the Academy had to be informed each time who their new colleague would be. For the scientists themselves to make the choice was a significant alteration in the balance of power between professional and political authority.

Yet the above is far from an exhaustive account of the stages of an election. It has, for example, omitted any mention of lobbying in the early stages or the 'visit' which was commonly paid the Academicians by candidates in order to solicit their vote. We will postpone a discussion of this until after we have described further aspects of elections within the walls of the Academy.

[17] Maurice Crosland, 'Assessment by peers in nineteenth-century France: The manuscript reports on candidates for election to the Académie des Sciences', *Minerva*, **24** (1986–7), 413–32.

[18] C.R., **117** (1893), 883.

4. *The intermittent character of elections*

One feature of elections which must have been particularly trying for candidates was that, since they depended on following in dead men's shoes, one could rarely predict when an election would take place. Even then the death of an Academician did not always mean an immediate election. A power of the Academy not well understood was the power to postpone an election. The regulation that the Academy could postpone any election for a period of up to six months was originally intended to avoid the possibility of electing someone who was not an outstanding scientist. It was quite possible that from time to time *within a particular speciality* the likely candidates were no more than mediocre. Of course the news of an impending vacancy would influence many scientists within a particular field to remind the Academy of their talents by presenting a paper. If there was a postponement, the number of potential candidates would increase. Thus when the botany section decided in February 1820 to postpone an election[19] following the death of Beauvois, a stream of botanical memoirs immediately followed. When the section finally drew up a list of possible candidates, it could not fail to take into account their work and the list was a long one!

Moreover, when an election was adjourned for a period of six months, this might give enough time for a versatile young man whose main qualifications lay in *another field* to make some contribution in the area where the vacancy had occurred. Of course an astronomer would be unlikely to turn in desperation to botany or vice-versa, but in the case of neighbouring fields it was sometimes not too difficult to make a relevant contribution. Thus Poisson, working in the area between mathematics and physics, could present himself plausibly as a candidate for either, according to where vacancies arose.[20] It might often be in the interests of a patron to urge a postponement. He might even need to arrange a Paris residence for his protégé to make him eligible. The fact that the system was abused is not only supported by a large number of individual cases but is also hinted at in comments of a commission appointed in 1830 to draw up a list of candidates for the prestigious position of *associé étranger*. It pointed out that since the field included men of science in all fields and from the entire world there could never be any question of postponing an election on the grounds of insufficient talent![21] If a postponement was called for, therefore, one might suspect more questionable motives.

Normally one might expect a vacancy in each section every few years and certainly every decade. Although this was true on average, there were also examples of extreme irregularity in vacancies. Thus in the botany section there

[19] *P.V.I.*, 7, 11 (cf. p. 38).

[20] See R. W. Home, 'Poisson's memoirs on electricity: Academic politics and a new style in physics', *B.J.H.S.*, 16 (1983), 239–259.

[21] *P.V.I.*, 9, 407.

was only one vacancy between 1808 and 1830, the successful candidate being Du Petit-Thouars, who was elected in 1820 at the age of sixty-two. Du Petit-Thouars had been quite productive but had become increasingly impatient with the lack of any opportunity to be considered. In 1817 he had written a long letter to the Academy calling attention to the fact that he had presented some forty-five memoirs to that body.[22] Although he is hardly remembered today, it could be argued that he deserved his place on the basis of persistent hard work of a reasonably high standard.

Many other botanists would have been in despair. They were not to know that as many as five out of six members of the botany section were to die in the 1830s, creating almost a surplus of vacancies. Once five new Academicians had been installed in the botany section, there was not a single vacancy in the 1840s. Then inevitably Time's Grim Reaper produced a whole series of vacancies, with four occurring in the fifteen months between October 1852 and January 1854. This clearly shows the system of dead men's shoes at its very worst. It was in such a situation that the young Pasteur, up to then unsuccessful in his candidatures in the sections of mineralogy and chemistry, tried to present himself as having qualifications relevant to botany. In such a time of surplus one might expect entry qualifications to be less stringent than in times of want. Yet Academicians could not so easily be deceived by a young man whose principal qualifications lay elsewhere and, when Pasteur was finally admitted in 1862, it was in the section of mineralogy, corresponding to his most brilliant work up to that date.

The unprecedented succession of deaths in the botany section in the years 1852–4 produced a positive outcome. Several Academicians interested in botany, led by Adolphe Brongniart, who at fifty-three was left as the senior French botanist,[23] decided that botany was in danger of extinction and decided to found an independent Botanical Society, which held its inaugural meeting on 24 May 1854.[24] Brongniart pointed out that geography, geology, meteorology and horticulture all had their own specialist societies and he felt that it was time to bring together persons seriously interested in botany into a formal society. It suggests that Brongniart felt that the future of their branch of science could not necessarily be entrusted to a small elite in the Academy. Botany could only really be safeguarded by the creation of a specialist society with membership throughout France. In fact more than a hundred people had joined by the time of the first meeting.

Although the system of waiting for other men to die is far from ideal, it has been argued that in practice it often worked better at the Academy than in the

[22] *P.V.I.*, 6, 217 (8 September 1817).

[23] Brongniart was elected as the first president of the Société botanique, the vice-presidents, also Academicians, being Decaisne (agricultural section), Moquin-Tandon (botany section), J. Gay (botany section) and F. Delessert (*Académicien libre*).

[24] 'Fondation de la Société Botanique de France', *Bulletin de la Société Botanique de France*, 1 (1854), 4–7.

Muséum d'Histoire Naturelle.[25] Thus many zoologists were elected to the Academy in mid-careers, when they still had much to contribute, and only succeeded *later* to a chair at the Muséum. This is probably because succession at an educational institution was normally to a unique vacancy, whereas in the Academy there were six members in a section and, in most cases, the death of *any one of the six* would produce the required vacancy. Yet the Academy could hardly have been very proud of a system of election depending entirely on the filling of dead men's shoes. The alternative method, now adopted by the Royal Society of London, of electing a small number of new members *each year* is certainly preferable, since it ensures a constant influx of 'new blood'. The adoption of this method, however, requires the acceptance of some flexibility in the total number of members, something the Academy refused to permit.

5. *Adherence to a speciality*

It is an understandable but regrettable aspect of human vanity that when scholars have the opportunity of choosing future colleagues they tend to prefer candidates like themselves. Of course such homogeneity might reduce potential conflict within an organisation but it also eliminates a creative diversity. In the Academy, members tended understandably to favour candidates with an orthodox educational background. If a candidate were a graduate of the elite Ecole Polytechnique, this *in itself* was an important qualification. It would imply a scientific education of a high standard and probably also the acceptance of certain social norms. Academicians who were former *polytechniciens* would tend to favour others from the same school. So, the greater the number of *polytechniciens* within the Academy, the more likely it was to favour future candidates who had studied at the Ecole Polytechnique, a truly self-perpetuating oligarchy. (See Fig. 3, p. 229.) This could be considered as a form of patronage.

But quite apart from any patronage, there was an inbuilt tendency for Academicians to favour their own subject (or branch of a subject) at the expense of other subjects. Thus, although the mineralogy section became increasingly open to geologists as that branch of the subject developed, a mineralogist might tend to favour a candidate whose work lay within the field of mineralogy rather than geology, unless it was accepted that certain places were reserved for geology. In the chemistry section, inorganic chemists viewed with some alarm the burgeoning sub-discipline of organic chemistry, which was adding to the already severe competition within the section. An organic chemist would have to be shown to be outstanding in order to be preferred to someone like themselves. Such problems had to be resolved by an agreement within any section to reserve places so that different aspects of the subject could be represented. In practice this took time and there was a great deal of inbuilt

[25] Toby Appel, 'The Cuvier–Geoffroy debate and the structure of nineteenth-century zoology', Ph.D. thesis, 1975, p. 52.

conservatism. The fact that the geometer Chasles (1793–1880) had to wait until he was nearly sixty to be elected is to be explained by the tendency of early nineteenth-century mathematicians to scorn geometry as old fashioned. Only the presence of a geometer within the section of mathematics would have made it possible to present a strong case for this able mathematician at an earlier stage in his career.

Above all it is necessary to consider the *section* in which a vacancy occurred. The Academy was not simply an assembly of distinguished scientists, it was a confederation of distinguished specialists. At the election of April 1852 the defeated candidates were Claude Bernard and the young Louis Pasteur, names to conjure with, as opposed to the successful Quatrefages, much less well remembered by later generations. But to be fair, we should point out that the section in which the vacancy occurred was that of anatomy and zoology and, purely as a zoologist, Quatrefages deserved to win. The physiologist Bernard followed the precedent of his teacher Magendie in a subsequent election to the section of medicine (1854), and it was as a crystallographer that Pasteur was later to enter the Academy in the section of mineralogy.

People who have not studied the working of the Academy often wonder why a genius was not immediately successful when he became a candidate for membership of the Academy. A case in point is Pasteur. We now have a transcript of the speech made by Fizeau on 1 December 1862 in considering the rival merits of Pasteur and Des Cloiseaux for a vacant place in the mineralogy section.[26] The majority of the section preferred Des Cloiseaux but, out of special consideration for the undoubted merits of Pasteur, they presented both names in the first rank. They considered Des Cloiseaux as the most distinguished mineralogist in France still outside the Academy. The names of the Cambridge crystallographer and mineralogist W. H. Miller and also of Gustav Rose of Berlin were mentioned in support. Fizeau's report continued:

> The qualifications of Pasteur are on the other hand of a very different order. The subjects of the various researches which he has begun (*abordés*) up to now, mostly very happily and with superiority, are clearly for the most part outside the domain of mineralogy and belong rather to that of chemistry. One might even add that those of his works that have been cited as relevant to mineralogy, are precisely those which are considered to-day – and I believe very justly – as the most open to doubt in relation to the very general consequences which were first drawn from them. ... The discovery of isomerism lies within the domain of chemistry. No-one will challenge the fact that M. Pasteur is universally considered as a chemist. To sum up, since it is a question of filling a place in the mineralogy section, I am looking for a mineralogist.

This presentation shows most clearly the difficulties experienced by anyone like Pasteur, whose research related to several fields, even if his superior ability

[26] A.S., dossier Pasteur.

was acknowledged. The other candidate, however, had seemingly perfect qualifications since he had led 'a life dedicated (*consacré*) entirely to the study and teaching of [mineralogy]'. By contrast Pasteur is presented almost as a dilettante, someone who began research in one area and then did something quite different. For the mineralogists he was clearly labelled as a chemist and that was enough to condemn him. The decisive vote, however, was that taken by the whole Academy. For them his general ability was a sufficient recommendation and the challenge to the validity of his speculations on crystalline structure was not taken seriously. In the election of 8 December 1862 Pasteur was elected by thirty-six votes to twenty-one for his mineralogist rival.

6. *Election as replacement*

The common idea that elections should have been occasions to appoint the most brilliant scientist overlooks not only the rigid demarcation of sections within the Academy but also the very concept of an election as an opportunity to *replace* a deceased member. Given that one could never have the same person back again, the Academy often tended to look for a carbon copy. In other words the wonderful opportunity of introducing new blood was constrained by a certain conservatism. Yet perhaps a fairer way of expressing this would be to say that the Academy sought stability. Even the radical Arago felt constrained to urge that, when correspondents were being elected, Academicians should always consider whether the previous holder of the position was a French national or a foreign scientists and they should restrict their choice accordingly.[27] This would guarantee a balance. Similarly within the elections, particularly those representing more than one branch of science, it was relevant to consider whether the specialism represented by the deceased member should not be continued. Sometimes the section made this very clear when they were debating the relative merits of different candidates, by drawing up two lists according to subject. But even if the section represented only one science, consideration had to be given to the representation of its different branches. In the chemistry section we have mentioned the dominance of inorganic chemistry being challenged by the burgeoning new science of organic chemistry. In France physical chemistry developed earlier than in many other countries and so it too claimed representation. Thus when in 1878 Troost aspired to succeed Regnault in the chemistry section, the rapporteur claimed that his research in physical chemistry 'followed the path opened by Lavoiser and Laplace, which had been taken up with so much success by many physicists and mathematicians of our period, and especially by M. Victor Regnault'.[28]

It follows from what has been said that it would be a mistake to consider elections purely in terms of personalities. Often what was being replaced was

[27] A.S., *Comité secret, 1837–44*, p. 96 (25 April 1842).

[28] A.S., dossier Troost, M. Debray, 'Rapport sur les titres de M. Troost'.

more a speciality than a person. Thus in the *éloge* of the botanist, Palisot de Beauvois (1752–1820), it was said that

> His place in the Institute has been given to M. Dupetit Thouars, who was elected to it more or less on account of the similarity of his work and because of his great voyages.[29]

This surely is the conservative philosophy of the carbon copy, producing a situation in which a botanist who had never left his native land would be at a distinct disadvantage, however distinguished his work. A similar case is found in the mechanics section in the election of 1807 to succeed the distinguished watch-maker Berthoud. Given the many possible interpretations of the term 'mechanics', it was surely more than a coincidence that no less than four clock-makers presented themselves as candidates to replace him.[30] In a few cases a peculiar mystique was attached to a particular place in a section, sometimes described as a 'chair' (*fauteuil*). Thus in the early twentieth century, it was a matter of pride that the place of the great mathematician Lagrange (d. 1813) had been occupied solely by graduates of the Ecole Polytechnique: Poinsot (d. 1859), Serret (d. 1885), Laguerre (d. 1886) and Poincaré (d. 1912).

7. *Seniority*

Studies of the operation of the Royal Academy of Sciences in the late eighteenth century have revealed that, in a very hierarchical system, internal promotion to the senior rank of pensioner was often made purely on grounds of seniority, thus avoiding possible unpleasantness in the Academy which might have arisen as a result of an open discussion of the relative merits of the different candidates.[31] After the Revolution of 1789 the hierarchical system of three grades of ordinary membership was abolished, thus practically eliminating internal promotions.[32] Nevertheless in the nineteenth century seniority often provided an objective criterion for distinguishing the claims of good candidates without splitting the Academy into rival camps. Thus in January 1823, when the physics section had the task of drawing up a list of candidates in order of merit, they were faced with two outstanding scientists, Dulong (1785–1838) and Fresnel (1788–1827), both of whom had previously won Academy prizes. On grounds of seniority they decided to place Dulong above Fresnel on the list.[33] Although Dulong was only three years older than Fresnel, he had also started publication of research before his rival and he had won a prize (with Petit) in 1816, whereas Fresnel's prize was in 1819. It was, therefore, Dulong who was elected in January 1823. Yet by a happy chance a second vacancy in the physics section occurred within a

[29] *M.A.I.*, 4 (1824), cccxlvi.

[30] *P.V.I.*, 3, 560 (3 August 1807).

[31] Rhoda Rappaport, 'The liberties of the Paris Academy of Sciences, 1716–1785', in Harry Woolf (ed.), *The Analytical Spirit. Essays in the History of Science in Honor of Henry Guerlac*, Ithaca, N.Y., 1981 pp. 225–53 (p. 241).

[32] The only possible exception being the election of correspondents to full (i.e. resident) membership.

[33] *P.V.I.*, 7, 419.

few months. This time the physics section nominated Fresnel, not only in first place but labelled 'outstanding', as if to compensate for their previous partiality. When the election was held on 12 May Fresnel had the distinction of receiving the unanimous vote of the fifty-two electors present.[34] All is well that ends well!

The manuscript *Rapports* on candidacies confirm what we have suspected about the importance of seniority to the result of an election. No-one was going to be recommended simply because he had published a certain amount of scientific research. All candidates were normally in this position. It was not enough to be good; one had to be *better* than rival candidates, but how was this to be assessed? The consideration of seniority was a method which had the merit of objectivity. One of the better examples of the importance of seniority found in the archives refers to an election in the chemistry section in the 1880s.[35] Debray is explaining why three chemists have only been placed in the third rank:

> Being younger than the candidates who have been placed first or second, they do not yet have to their credit the same total (*somme*) of discoveries, and several of their works have not yet received the precious consecration of time [sic!]. In the circumstances it seemed pointless to the section to proceed to a true ranking order among them, which did not seem easy. When their hour comes they will have added to their qualifications.

The report ends with the admission that despite their junior status (the candidates were about fifty years old), by the quality of their work they were already worthy of belonging to the Academy.

Seniority did not, however, mean simply the age of the candidate. The relevant seniority was *in relation to the Academy*: how long since the candidate had presented his first memoir, how many times he had been an unsuccessful candidate and how had he been ranked previously? Thus Sturm, in presenting the candidacy of Chasles in 1851, emphasised that he was already a corresponding member and had been one for a long time.[36] These were obviously points in favour, but what counted for more was his track record as a candidate in the mathematics section. In the previous election in 1843 he had been recommended as joint first with Binet, the successful candidate. Thus justice would suggest that this time he should be successful.

A similar argument was used by Fremy in 1891 in relation to the candidature of Moissan.[37] In the previous election Moissan had been placed joint first[38] in the list drawn up by the chemistry section. The other candidate, Gautier, had been elected. His colleagues in the chemistry section now placed Fremy as joint first again, this time with Grimaux, but Fremy protested that it was now clearly Moissan's turn. Behind the general concept of seniority we see here some rivalry

[34] *Ibid.*, 499 (5 May), 500 (12 May).

[35] A.S., dossier Gautier, Rapport de M. Debray sur les titres de M. Gautier, Séance du 30 juin 1884.

[36] A.S., dossier Chasles.

[37] A.S., dossier Moissan.

[38] In the 1888 election he had been placed second.

between different branches of chemistry. Fremy, like Moissan, was an inorganic chemist, whereas Gautier and Grimaux represented organic chemistry. The problem of sub-specialities adds a further complications to the task of understanding the procedures of the Academy.

Poincaré in his *Notice* of 1886/7 reminded the Academy of previous candidatures. He had been presented by the mathematics section:

in 1881	5th in order of preference
in 1884	4th in order of preference
in 1885	3rd in order of preference
in March 1886	2nd in order of preference

One did not need to be a mathematician to conclude that the logical continuation of this series in the latest election would bring Poincaré to first place on the list. Camille Jordan in his *Rapport* did not fail to point out to the Academy that although Poincaré was still young (only thirty-two years), this was the fifth time he had presented himself as a candidate. He was, of course, exceptionally lucky that there were so many vacancies in the 1880s in the mathematics section. Poincaré, together with other mathematicians elected in that decade, such as Camille Jordan, was to do much to restore the fame of French mathematics on the international stage.

It must have been considerations of acquiring later seniority that prompted Edmond Becquerel to persuade his son Henri in 1876 to pose his candidature at the early age of twenty-three.[39] At that time he had presented two memoirs to the Academy but he only had one publication to his credit. Henri Becquerel could hardly have expected any votes. With no possibility of success, his candidature can only be explained as laying foundations, on which later candidacies could rest.

A corollary of the qualifications of seniority was perseverance. This is most explicit in a report recommending Payen as a candidate in the section of agriculture. It is stated that,

> Since the last election ... M. Payen has continued his theoretical and practical work on agriculture with constant perseverance.[40]

This perseverance was reflected in the updating of the successive *Notices* prepared by the candidates. One might regard the lure of membership of the Academy as one of the most powerful incentives for many fairly senior scientists to keep up active research.

[39] *Notice sur les travaux scientifiques de M. Henri Becquerel, élève-ingénieur des ponts et chaussées*, 10 April 1876, 3 pp.

[40] A.S., dossier Payen, n.d.

8. *The overthrow of sectional interests*

Although we have considered it necessary to stress sectional interests within the Academy, in many cases the Academy as a whole ignored the particular wishes of the section in order to elect what they saw as the best scientist. Thus the election process contained a number of checks and balances which assured that entry to the Academy could never be guaranteed, not even to a candidate who had the support of an entire section. The candidate also had to stand up to the wider criteria applied by the Academy as a whole. We shall give an example, taken from the agriculture section.

Although the details of successive ballots in elections in the Academy are often a mystery, we know in a few cases what actually transpired, thanks to the surviving archival evidence. Thus by a lucky chance we have unusually full details of a particular election in November 1828 to fill a vacancy in the agriculture section.[41] The evidence depends on the habit of one of the senior members of the section, J. B. Huzard (1755–1838), of keeping all his notes and papers, which are fortunately preserved in the library of the Institute. There were no fewer than nineteen candidates for this election. Many were practical men involved in agriculture or forestry; there were also several botanists and two young physiologists, Dutrochet and Flourens. Also posing his candidature was Hachette, whom one would have normally associated with the mechanics section. Hachette had previously been elected by the Academy[42] but the government had refused to confirm the nomination on political grounds. The agriculture section, no doubt feeling that it was striking a blow for academic freedom, placed Hachette in the first rank of candidates but, with an equally keen eye to the political situation under Charles X, placed as joint first the agricultural proprietor the vicomte d'Harcourt. In second place we find Huzard's son, who had veterinary qualifications, a clear example of nepotism. Also on the list were the names of five men with practical experience in agriculture. Often the recommendation of the section would be accepted by the Academy but in this case the Academy was far from happy with the exclusion of candidates with a strong scientific background. It asked that the names of the two physiologists Flourens and Dutrochet (and two others) be added to the list. At the first ballot Flourens had the most votes (twenty-three) but as this did not constitute an absolute majority a second ballot was held. This time Flourens had thirty votes, Dutrochet coming second with seventeen.[43] The candidates singled out by the section, Huzard *fils* and d'Harcourt, received four and two votes respectively.[44] This demonstrates the power of the Academy as a whole to overturn the recommendations of a section but it is an extreme case that serves to illustrate

[41] Institut, HR5*, vol. LXXII, No. 89.

[42] 10 November 1823.

[43] Although Dutrochet (1776–1847) was the senior of the two, Flourens (1794–1867) had the advantage because of strong support from Cuvier.

[44] Hachette received no votes.

the dichotomy in the Academy between the scientific majority and a practical minority.

We can speculate on the discussion in the secret session of the Academy but unfortunately, in most cases, there is no written record of what was said. This is particularly true of informal lobbying. Exceptionally one finds a note by a candidate referring to a visit to an Academician who has not been at home, and a brief written solicitation results. But when nearly everyone concerned was resident in Paris there was little need to take up one's pen. Nearly everything was done by word of mouth. An election for a correspondent, however, might be managed by a resident member and, if he reported to his protégé, we have the occasion for a letter. This was precisely the case, for example, in 1851 when Decaisne (1807–82), a recently elected member of the agriculture section and a professor at the Muséum, reported in a letter[45] to Alphonse de Candolle (1806–93), a botanist resident in Geneva and the son of the more famous Pyramus de Candolle. He gave a list of twenty-six supporters, headed by the two secretaries, Dumas and Elie de Beaumont. A second list, containing fifteen names, was headed 'uncertain'. Finally ten names were given as being probably against de Candolle: these included Milne-Edwards, Quatrefages and Claude Bernard. Decaisne writes that there was still some hope of winning over the doubtfuls and, when the election was held, de Candolle was successful. One of the interesting features of the lists given in this letter is that they cut completely across the sectional divide. Mathematicians and astronomers seem to be canvassed no less than chemists or zoologists and they are considered as a single constituency, which indeed they were in the final stage of any election. In this case the combination of a famous name with vigorous lobbying was sufficient to win the election, since de Candolle was elected by forty-eight votes out of forty-nine. Rival candidates who were not francophone, like William Jackson Hooker and Asa Gray, had to wait for a later opportunity to be elected, Hooker in 1856 and Gray in 1878.

9. *The notorious 'visit'*

The formal procedures of the Academy were complemented by a number of informal practices which had grown up among candidates to all the Academies. Probably the most notorious of these was the 'visit'. Candidates felt that they could increase their chances of support if, just before the election, they were to call obsequiously at the respective private addresses of the members of the Academy. Even under the old regime the visit had become something of a scandal, particularly in relation to the Académie Française.

Regulations had been introduced in the eighteenth century to forbid solicitation and these regulations were repeated in the reorganisation of the Académie Française in 1816. Article XIV of the official constitution stated

[45] The letter on the notepaper of the Muséum d'Histoire Naturelle is dated simply 'samedi', A.S., dossier A. P. de Candolle.

explicitly that the Academy should stand above cabals and solicitations. It was important that no Academician should promise his vote in advance to one of the candidates. Indeed, Article XV went on to state that any Academician who admitted promising his vote in advance of the official discussion would be disqualified from voting. Unfortunately, the practical problem here was to get Academicians to own up. The habit of solicitation had become so well established that it took more than a regulation to stamp it out. Some Academicians who had felt obliged to debase themselves to get elected actually took pleasure in receiving solicitations which gave them a sense of power. On the other hand it was the candidates who took the initiative, and not all Academicians would have welcomed their visitors.

One critic of the visit, writing in the 1860s,[46] was prepared to admit that in the Académie Française and the Académie des Inscriptions et Belles Lettres, it might serve a useful purpose. It was understandable that Academicians who could have read some work of a candidate might, nevertheless, welcome an opportunity to meet him personally. They would be able to decide if he were not only a scholar but an agreeable person. In other words the ideal member of an Academy was not simply a clever person. He should have certain manners and social graces. Ideally he should be seen as a desirable colleague. A visit could therefore be useful, but it should not be necessary to repeat it if the candidate was unsuccessful and was obliged to stand in later elections.

Although such considerations were relevant also in the Academy of Sciences, they had less force there. In literature it was often extremely difficult to make objective judgements about the relative merits of rival candidates. In such cases personal considerations might play a larger part in deciding between candidates. In science, unlike literature, there was an additional stage in which candidates were studied by a small group of specialists, which provided a more objective comparison of their relative achievement.

By the very nature of the occasion, documentation of visits by candidates is exceedingly rare. It was an opportunity for visual inspection and for oral communication strictly off the record, although occasionally a visiting card might be left by a candidate if the Academician was not at home when he called. Only exceptionally would a candidate describe to a third party the humiliating round which he had undertaken. A terse but graphic account of a late nineteenth-century visit has been left by Henri Poincaré in a letter to his wife.[47] At the age of thirty-two the brilliant young mathematician had already been a candidate for vacancies in the Academy four times when the recently-elected Laguerre died in 1886. Up to now Poincaré had never been placed higher than

[46] Baron Roget de Bollequet, *Petition addressée à l'opinion publique pour la réforme des élections de l'Institut*, Paris, 1862, pp. 6–12.

[47] *Le Livre du centenaire de la naissance de Henri Poincaré, 1854–1954* (Paris, 1955), pp. 284–5 in *Oeuvres*, vol. XI. This source gives the date of the letter as 1886 but internal evidence connects the contents with the election of January 1887.

second on the list of recommendations of the mathematics section. For the election of 1886/7 he was to be placed first.

A crucial figure in the election was the doyen of the mathematics section, Hermite, a member of the Academy for some thirty years and the man who would present the candidates in order of preference. Should Poincaré pay a call on Hermite to solicit his support directly? Darboux, a more junior member of the section, advised against it. He told Poincaré that Hermite had told him that he would vote for him and that he would speak up for him in the secret session in which the merits of the candidate would be discussed but he found it all rather painful and he did not want Poincaré to visit him. A visit to Camille Jordan, another member of the mathematics section who had spoken to Hermite the previous day, confirmed what Darboux had told him. He also visited Picard, not yet himself a member of the Academy but a useful contact, being the son-in-law of Hermite. Picard was also a friend of the botanist Duchartre, who believed in following the recommendation of the section in which the vacancy occurred.

Another mathematician, Halphen, had told Picard that he would vote for a rival candidate, Mannheim, but he expected Hermite to persuade him to vote for Poincaré instead. Picard also said that the physicist Lippman would vote for him, and finally wished him good luck. Bertrand, a mathematician, but now one of the two permanent secretaries of the Academy, had gone to the country when Poincaré called. The chemist Debray was also out, as was the astronomer Wolf, but in any case Wolf was reputed to be a supporter of Mannheim. Poincaré wanted to call on the mathematician Boussinesq but no-one could tell him his address, not even at the Faculty of Science where he was a professor. Poincaré also called on Maurice Levy, a member of the mechanics section, who said that he had not yet made up his mind. Another member of the mechanics section, Reseal, felt that he could not give Poincaré his vote, probably because of his comparative youth. Poincaré, as Ingénieur des Mines, knew the mineralogist Daubrée and, on visiting him, was advised to call also on vice-admiral Jurien de la Gravière, a member of the geography section. Thus we see that although it was natural to focus on the section where the vacancy occurred and to make some contact in related subjects, the *visite* could lead to solicitations among all members of the Academy. When the vote was finally taken on 31 January 1887 Poincaré obtained thirty-one votes to Mannheim's twenty-four and was declared elected.

The science journalist abbé Moigno presented a depressing portrayal of the visit as a sheer waste of time and effort, especially as a persistent candidate might over many years be involved in hundreds of visits.[48] One unfortunate candidate claimed that he had reached his 1500th visit but, if this were to be believed, it would suggest a lifetime devoted to the process; being involved in at the very least more than twenty elections if he had visited most of the members of the Academy at each candidature. Some candidates went further and tried to

[48] 'A propos des candidatures académiques', *Les Mondes*, 4 (1864), 386.

influence Academicians through powerful friends or even occasionally their servants. This not only made demands on candidates but on Academicians too. But there were some members of the Academy who actually revelled in these solicitations and Moigno quoted an Academician who said:

> A candidacy is for me the occasion to meet a very large number of people, which my humble income would never allow me to meet otherwise. It gives satisfaction to my self esteem to receive archbishops and senators and to provide an antechamber for marshalls of France. One day I even had the signal honour of receiving a prince of the blood. My collection of autographs has been enriched by several curious documents, recommendations of the highest kind in favour of candidates.

This sounds much more like the Académie Française than the Academy of Sciences, since the literary elections were always of greater interest in high society. Indeed Moigno did not claim to be speaking only about the Academy of Sciences but of academies in general and, in his desire to abolish the visit, he showed it at its most absurd. It suggests that in an area where great *social* significance was attributed to membership, combined with the absence of objective criteria of deciding merit, for example, in literature, personal pressure might be important in advancing the claims of candidates. In the Academy of Sciences, however, the visit was less important and it rarely extended beyond the members of the Academy. If the visit persisted, it was because it was part of the tradition of the French academic system.[49]

10. *Some external influences on elections*

We have spoken of the influence of factors *internal* to science, notably scientific specialisms, in elections. We may now briefly consider the influence of external factors, especially religion and politics, education and institutional loyalty. As a generalisation we may say that politics alone had comparatively little influence on the choice of Academicians in most cases. All the famous examples of purely political interference in elections relate not to the choice of the Academy but the decision of governments to expel certain members or to refuse to confirm an Academy election. As far as the members of the Academy were concerned, it would only be in cases where candidates were associated with extreme political positions that this aspect might seriously influence their choice. There is the case of the mathematician Binet (1786–1856), who waited as a candidate for thirty years (1813–43) before he was successful. The mathematics section always placed him as their first choice but, according to a standard history, 'Binet was a zealous Catholic and devoted to the Bourbons'[50] and therefore unacceptable to the majority. Binet's friend, the mathematician Cauchy, provides a further example of prejudice, combining as he did extreme attachment to the Bourbons

[49] For a further perspective on the visit, see Charles Richet, *The natural history of a savant* (1923), trans. 1927, pp. 32–3.

[50] *Ecole Polytechnique, Livre du centenaire*, (3 vols., 1895), vol. 1, pp. 104, 411.

with outspoken Catholicism. Cauchy's unpopularity was the greater because he had entered the Academy by royal nomination rather than through the normal process of election. We shall argue that religion was much more a relevant factor in elections than pure politics and we shall follow one particular line of patronage where it seems that strong religious commitment constituted a serious handicap in the strongly secular Academy stakes. Of course, there were probably other quite different cases of prejudice in elections. The examples mentioned below, however, illustrate a common thread.

In 1843 Cauchy acted as the sponsor of another Catholic mathematician, Barré de Saint-Venant (1797–1886), engineer, and follower of the Poisson school of physical mechanics. Although in the recommendation to the Academy the mechanics section agreed to place Saint-Venant in equal first place with Morin, the manuscript record shows that Cauchy revealed his partiality by writing a very favourable report on Saint-Venant, twice the length of the report on his chief rival.[51] Moreover, only two weeks before the election, Cauchy, as rapporteur of a commission, had heaped high praise on the work of his protégé.[52] For many Academicians, however, a candidate who received fervent support from the Catholic royalist Cauchy was to be marked down with great suspicion, a suspicion probably increased by the possession of a name redolent of the *ancien régime*. Saint-Venant, who at the age of forty-six had a considerable record of important publications, received no more than six votes out of a total of fifty-four.[53]

Another fifteen years were to elapse before there was another vacancy in the mechanics section. The section again placed Saint-Venant in equal first place but he received no more than twelve votes from the Academy, compared with forty-three for his rival Clapeyron.[54] Cauchy had died the previous year but alliances were not easily forgotten. When there was a further vacancy in 1864, the mechanics section overlooked completely the candidature of Saint-Venant, who had to write in[55] as if he were some junior unknown mathematician. Only in 1868 did the Academy agree to admit Saint-Venant,[56] who had by then reached the age of seventy-one! Of course one cannot prove absolutely that if the mathematician had not been a devout (but not outspoken) Catholic he would have been elected sooner but the evidence points clearly in that direction. One cannot prove that the support of Cauchy was almost fatally counter-productive, but it seems to have played a significant part in this episode which is not entirely to the credit of the Academy.

In the few years of power granted to him as an Academician, Saint-Venant acted as the patron of another Catholic mathematician of the first rank,

[51] M. Crosland, *op. cit.*, note 17.
[52] C.R., 17, 1234 (27 Nov. 1843).
[53] *Ibid.*, 1327.
[54] *C.R.*, **46** (1858), 545, 564.
[55] *C.R.*, **58**, 469 (7 March 1864). In December 1864, however, Saint-Venant wrote again to the Academy, withdrawing his candidature. The delay in the election raises suspicions of foul play.
[56] *C.R.*, **66**, 791 (20 April 1868).

Boussinesq (1842–1929). Boussinesq was to make important contributions to nearly all branches of mathematical physics and notably to hydrodynamics. He did not have the advantage of belonging to any of the *grandes écoles*. He was largely self-taught, owing much of his early education to his uncle, a priest. His thesis of 1867 on the diffusion of heat won him the favourable attention of Saint-Venant, who henceforth acted as his patron. In 1870 Saint-Venant, acting as rapporteur for a memoir by Boussinesq, praised the author for his 'remarkable spirit of invention, constantly supported by great analytical ability'.[57] Boussinesq made repeated, but unsuccessful, attempts to enter the Academy in 1868, 1871, 1872 (twice), 1873, 1880 and 1883. By this time Saint-Venant had become the doyen of the mechanics section and thus the person with responsibility for comparing the merits of rival candidates. His report for the election of 1886 has survived.[58] It shows an impassioned advocacy of his protégé who, he pointed out, was unique in being the only survivor of these early elections not to have been elected to the Academy. He must have privately compared the fate of his protégé with his own situation earlier. Saint-Venant's advocacy was one of his last acts. He died on 6 January 1886 without knowing if Boussinesq would succeed. When the vote was taken on 18 January 1886 Boussinesq was elected by a narrow majority, receiving twenty-nine votes compared with twenty-six given to his rival, the engineer Deprez.[59]

Although Boussinesq lived a simple existence devoted to mathematics, he was also passionately interested in the relevance of science to religion.[60] In conversation he would go out of his way to raise difficult theological problems[61] and he even extracted philosophical conclusions from his mathematics.[62] In his study of solutions to differential equations he drew attention to the possibility of indeterminism, which he suggested applied uniquely to living systems. Taking this further, he argued in favour of human free will.[63] It was in a spirit of Christian humility that he spoke of:

> the smallness of the ensemble of our unclouded knowledge lost in an ocean of darkness.

If a strong Catholic identity combined with outstanding ability attracted patronage from like-minded Academicians, such ideological ties tended also to produce a reaction, and in the above cases the reaction clearly had an adverse effect on the candidate's prospects of election.

57 *C.R.*, **70** (1870), 367.

58 Rapport, dated 4 January 1886, A.S., Boussinesq dossier.

59 *C.R.*, **102** (1886), 156.

60 Licienne Felix, art. 'Boussinesq', *D.S.B.*, 2, 355–6.

61 Elie Douysset, *Un grand savant, un grand Chrétien*.

62 'Sur la conciliation de la liberté morale avec le déterminisme scientifique', this memoir was presented to the Academy by Saint-Venant, *C.R.*, **84** (1877), 362–4.

63 Mary Jo Nye, 'The moral freedom of man and the determinism of nature: The Catholic synthesis of science and history in the *Revue des Questions Scientifiques*', *B.J.H.S.*, 9 (1976), 274–92 (pp. 280–1).

Thus, just as there had been a strong anti-liberal movement under the Restoration which delayed the admission of several able candidates, so was there later a pervasive anti-Catholic movement, which was particularly strong under the Third Republic. In the field of higher education loyalty to the secular ideology of the republic was paramount.[64] It was his failure to conform to this ideology and indeed his well-known opposition to it, that kept Pierre Duhem (1861–1916), an avowed ultra-Catholic, a permanent exile in the provinces. At Bordeaux he had as a colleague the Jewish mathematician Hadamard (1865–1963), who had no difficulty in obtaining a teaching post in Paris, where he qualified for election to the Academy.[65]

Duhem drew attention to his scientific ability by his frequent contributions to the *Comptes rendus*, which only drew from Berthelot as secretary the rebuke that they were too frequent![66] Yet, for fear of offending Berthelot, Academy prizes were withheld from him.[67] Within a few months of Berthelot's death, however, belated recognition was conferred on the provincial physicist by the award in December 1907 of the Petit d'Ormoy prize of 10 000f. for mathematical sciences, followed two years later by the Binoux prize for history of science. If only Duhem had not been such an independent character, daring to criticise the secretary of the Academy! It is even possible that Duhem actually enjoyed controversy, which was fatal in an academic system that more than any other imposed conformity.

In order to understand better the academic politics involved in elections, it may be worth mentioning the inside testimony of Berthelot. In his *éloge* of the mathematician Bertrand, Berthelot remarked that no-one was more opposed than the deceased to 'those little cabals, which arise all too frequently in Academies, where people combine together to reject or to hinder men of superior talents'.[68] This reveals far more about what went on behind the scenes in the Academy of Sciences than it tells us about the subject of the *éloge*. It might even be argued that the main control that the Academy possessed over the future of French science lay not in its reports or even its prizes but in its choice of future members of that elite body. If there was a filtering process, if in a few cases there was even blatant discrimination which resulted in the failure (or at least the serious delay) of certain categories of able scientist to be elected, this has some relevance to our main theme of the control of science. Any body that controls scientists controls science.

[64] It was customary for the Ministry of Public Instruction to keep detailed records of the professional and private lives of members of the University. For the case of Duhem see Stanley L. Jaki, *Uneasy genius: the life and work of Pierre Duhem*, The Hague, 1984. Despite the hagiographic tone of the book, it contains invaluable documentation and some persuasive arguments.

[65] Jaki, *op cit.*, p. 91.

[66] *Ibid.*, p. 177.

[67] *Ibid.*, p. 150. A comparison may be drawn with an earlier case in the Academy of Medicine, where the antagonism of the bombastic Broussais (1772–1838) intimidated his colleagues so much that it delayed the delivery of an *éloge* of his great enemy Laënnec (1781–1826) until the death of the former eleven years later.

[68] M. Berthelot, '*Science et education*', 1901, p. 131.

Of course there were other loyalties and antagonisms than those connected with religion. Education can provide a powerful bond. The Ecole Polytechnique in particular assumed a growing importance in the affairs of the Academy as its graduates gradually won a significant proportion of the membership particularly among the mathematical sciences. The representation of the Ecole Polytechnique[69] in the Academy was particularly strong in the sections of mechanics, mathematics, physics and astronomy, although there were also several graduates of the school in the mineralogy section and, later in the century, in geography, agriculture and chemistry. Also from the 1840s onwards the school was also quite well represented in the section of *Académiciens libres*.[70] In the mechanics section the Polytechnique was so well represented that, for two periods of approximately eighteen years each (1847–64 and 1868–85), the entire membership of the section was filled with *polytechniciens*. A similar situation occurred in the history of the astronomy section; when Delaunay was elected in 1855, this marked the beginning of a period of seventeen years when the subject was represented exclusively by *polytechniciens*, of whom Le Verrier is probably the most famous. Yet the successive deaths of all these *polytechniciens* in the last decades of the century was not balanced by replacements. Only in the section of *Académiciens libres*, where qualifications were more general and included service to the state, and also in mathematics, where elections in the 1880s brought in four brilliant *polytechniciens*: Camille Jordan, Laguerre, Halphen and Henri Poincaré, did the representation of the Polytechnique remain as high as ever.

The graph (Fig. 3) shows a steady increase in the total numbers of *polytechniciens* in the Academy from the beginning of the century. In the 1830s and 1840s the numerical strength of the Polytechnique grew till it reached its maximum (thirty-one members) around the mid-century. It would be naive to assume that in elections the chances of candidates from the Polytechnique were not improved by the existence of a large number of Academicians who were former graduates of the school. The influence on elections would be felt at two levels. First in a section which was dominated by *polytechniciens*, it is likely that the candidates from that school would be received more sympathetically. The bias of the section would be reinforced by the actual election, where the whole Academy was involved, since here too the Polytechnique constituted one of the larger interest groups. In an obituary notice of the engineer Fourneyron, who was defeated in 1843 for a vacancy in the mechanics section by the *polytechnicien* Morin, the view was taken that 'he would have been elected if he had been a *polytechnicien*'.[71] This favouritism helped to create a self-perpetuating oligarchy,

[69] Albert de Lapparent, 'L'Ecole Polytechnique et l'Académie des Sciences', Ecole Polytechnique, *Livre du centenaire*, vol. 1, pp. 407–455.

[70] *Académiciens libres* had a lesser franchise than other Academicians.

[71] Obituary by Jules Guillemin quoted by E. Kranakis, 'Social determinants of engineering practice'. *Social Studies of Science*, 19 (1989) 5–70 (p. 35).

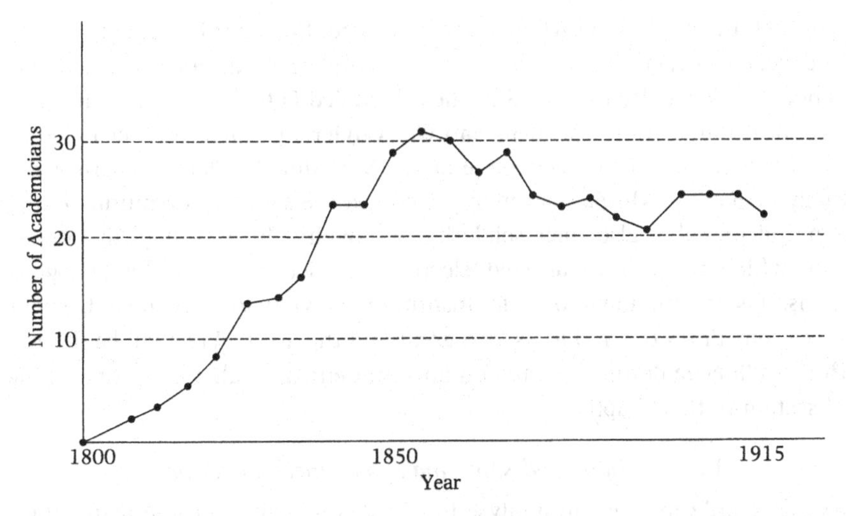

Fig. 3. Graph showing the number of Academicians who were former students at the Ecole Polytechnique.

which resulted in the Polytechnique having perhaps rather more seats in the Academy in the second half of the century than it strictly merited. Despite a decline from the 1870s, the Polytechnique membership of the Academy remained consistently above twenty up to World War I. Altogether ninety-six graduates of the Polytechnique became members of the Academy in the period up to 1914.[72]

The Polytechnique was of course not the only interest group associated with higher education. From the 1870s the Ecole Normale began making a significant contribution to membership, which increased from six in 1877 to thirteen in 1884.[73] A dozen members of a society who share a fierce common loyalty may affect the workings of that society but, because the Ecole Normale covered the whole range of learning, it never equalled the Ecole Polytechnique with its special emphasis on mathematics and physical science.

The only other major academic group was that associated with the Muséum d'Histoire Naturelle. Although it produced no graduates it provided a fair number of academic positions, mainly in the biological sciences. Out of the twelve professors at the Muséum at any one time, as many as eleven could also be members of the Academy between 1810 and 1850.[74] In the majority of cases they had been elected to the Academy before obtaining a chair at the Muséum, but it should be pointed out that there were also a dozen junior positions of *aides*

[72] John H. Wigmore (ed.), *Science and learning in France*, 1917, p. 100.

[73] 1884 was a record year for *normaliens*, no fewer then three being elected (Darboux, Troost and Mascart), *Le centenaire de l'Ecole Normale, 1795–1895*, 1895, pp. 693–4.

[74] The exception was the professor of iconography, who was not eligible, Camille Limoges, 'The development of the Muséum d'Histoire Naturelle of Paris, *c.* 1800–1914' in R. Fox and G. Weisz (eds.), *The organisation of science and technology in France, 1808–1914*, Cambridge, 1980, pp. 211–40 (pp. 216–217).

naturalistes, the majority of whom were both potential chair holders and future Academicians, depending on their seniority and the existence of a suitable vacancy. In the faculty list for 1820 these included Latreille, Valenciennes and Chevreul. Another future Academician was Cuvier's younger brother Frédéric, who was in charge of the menagerie at the Muséum. The fact that professors actually *lived* at the Muséum strongly reinforced the sense of community of the group and probably also their solidarity or 'loyalty' in providing an interest group, which must have affected elections for vacancies in the biological sections. It is not unknown today for institutional loyalty to sway the judgement of those sometimes referred to as 'the great and the good'. We shall be dealing with the influence of the Academy on appointments in higher education in the final section of this chapter.

11. *Elections and some major advances in science*

It would be only too easy to analyse the academic politics on the assumption that once an election had taken place that was the end of the story. But we need to consider the relation between elections and scientific activity. First, let us remember that the very existence of an Academy provided a panel of potential judges to whom much research was particularly addressed. Without an appreciative audience for research and without a reward system such as that provided by the Academy, much research would never have been undertaken in the first place. But we can distinguish the constant encouragement provided by the Academy with special opportunities for the display of expertise, which were provided by the announcement of prize competitions and the announcement of forthcoming elections. Here we shall confine our attention to elections.

Of course great discoveries are not to be commanded, not even by the most extreme career pressures. What could be commanded were effort and activity, which often led to scientific memoirs, even useful memoirs, just before an election. In a few cases the stimulus to impress the members of the Academy and to provide something exceptional seems to have led to outstanding results. One of these occasions was the discovery of a new planet, Neptune, not by direct observation but by inference from observed perturbations of the planet Uranus.[75] This was discovered independently about the same time by Le Verrier in France and by a Fellow of St John's College, Cambridge, John Couch Adams (1811–92) in England. Adams had great difficulty in convincing senior astronomers in England of the importance of his calculations. Le Verrier in Paris was much nearer to the centre of power in France.

Urbain Le Verrier (1811–77), a graduate of the Ecole Polytechnique, had first thought that chemistry offered the best career prospects, but he was known to be skilled in advanced mathematics and, when a junior post in astronomy became vacant in 1837, he was offered the post and accepted it. Two years later

[75] See Morton Grosser, *The discovery of Neptune*, Cambridge, Mass., 1962.

the presentation of his first memoir in theoretical astronomy to the Academy marked his arrival as a serious astronomer. He then worked successively on the theory of Mercury and on comets. In the summer of 1845 Arago, who was both secretary of the Academy and director of the Paris Observatory, suggested that he should investigate the perturbations of Uranus. We know that about this time Adams was working in Cambridge on the subject but nothing of Adams' work was known in France. Indeed, it was hardly known in England.

Le Verrier was an ambitious young man who wanted to make a name for himself in astronomy. Something that must have given him particular encouragement was the prospect of a vacancy in the astronomy section of the Academy. The veteran astronomer Dominique Cassini, then aged ninety-seven, was known to be failing in health and on 18 October he died. For Le Verrier this presented an opportunity which might not recur for another decade. He redoubled his efforts and produced a memoir which he presented to the Academy on 10 November. Not only was the memoir clearly one of exceptional importance but it was also politically acceptable, since Le Verrier took due care to pay tribute to his patron, the doyen of French astronomy, Arago. He also made a point of relating his research to the great *Mécanique céleste* of Laplace. His paper was appropriately entitled 'First memoir on the theory of Uranus'. In it he said that if the cause of the perturbations could not be discovered, the very law of universal gravitation would be in question. No doubt the Academy commission appointed to report on the memoir and headed by Arago, would have emphasised the exceptional importance of this research. At the meeting of 12 January 1846 the astronomy section presented a list of four candidates with Le Verrier in the first place. At the following meeting the election was held and Le Verrier received forty-four votes out of fifty-five.

Of course, Le Verrier had not completed all the necessary calculations in a few weeks. It was not until 1 June that he felt able to claim that the existence of a new planet had been established beyond any doubt,[76] and it was not until 25 September that the planet Neptune was actually observed by Galle in Berlin in the position predicted by Le Verrier. By this time the English work was known and journalists in the two countries began to make rival nationalistic claims. It is the aspect of 'simultaneous discovery' which has understandably received most attention from historians of science. That the Academy remained throughout the centre of debate has been largely taken for granted. Yet, by providing a framework for the presentation and reward of work of great merit, the Academy itself was able to share in the general astronomical euphoria of 1846. On the other hand recrimination on 'the failure of the British scientific community to seize the initiative in the discovery of Neptune' still reverberates over a century later.[77]

[76] *C.R.*, **22** (1846), 907–18.

[77] See Allan Chapman, 'Private Research and Public Duty: George Biddell Airy and the Search for Neptune', *Journal of the History of Astronomy*, **19** (1988), 121–39 and Robert W. Smith, 'The Cambridge

It has been argued that the Academy election, while not crucial, was an additional spur to Le Verrier, who was elected at the early age of thirty-five. It could be pointed out that the work he presented in the months after the election was even more important than his introductory memoir, presented no doubt as a marker. Indeed, it is probably easier to find examples of brilliant scientists elected at an unusually early age producing outstanding work soon after the election rather than just before. It seems that under the old regime the young Laplace had 'framed the main questions that his enormous life work refined, extended and largely answered *during a crucial period of about a year following his election to the Academy* in March 1773'.[78] Probably some young scientists would seek to confirm their place among their new peer group by producing outstanding work. Such a case is provided by the physiologist Magendie.

Magendie was quite well known in the Academy as a young man since he frequently presented memoirs and received favourable reports. In 1821 he had received an honourable mention for his submission for the physiology prize. His greatest strength was the patronage of Laplace, who approved of his programme of reducing physiology to physics and chemistry, but he still had to overcome the problem that there was no section officially concerned with physiology. In July 1821, after an argument in the section of zoology and anatomy, he had been considered as a candidate under the head 'anatomy', but when the election came he did not receive any votes.[79] In November of that year a further opportunity arose in the medical section. The section itself wanted a clinical physician appointed and submitted six appropriate names. At the very bottom of their list, however, perhaps in response to pressure from outside the section, they added the names of Magendie and Orfila 'as authors of works useful to medical science'.[80] The medical section placed on record their regret that they had had to omit from the list of names other distinguished physicians such as Laënnec and Broussais. The election turned out to be an occasion where the preference of the Academy as a whole overcame that of the section. In this case the Academy showed that it preferred the study of general scientific principles that could be applied to medicine, rather than a purely clinical approach. This preference, combined with the strong personal support of Laplace, carried the day and Magendie was elected at the second ballot.

But Magendie was in a rather difficult position. Aged thirty-eight, he was considerably younger than most of his colleagues. He found himself in a section which had clearly stated that it did not really want him as a colleague. Furthermore he represented a branch of science which still had to win

network in action: The discovery of Neptune', *Isis* **80** (1989), 395–422. Even the Royal Society of London gave one of its greatest honours, the Copley medal, to Le Verrier alone.

[78] C. C. Gillispie, art. Laplace, *D.S.B.*, Supplement, **15**, 286 (my italics).

[79] *P.V.I.*, 7, 210–11 (23, 30 July 1821).

[80] Magendie had written a text-book of physiology: '*Précis élémentaire de physiologie*, 2 vols., 1816–17.

acceptance. Although a weak character might have been daunted by such a situation, a stronger character might be inspired to prove his mettle. There is a further dimension to the story. Within a few months of his election the distinguished Academician Hallé died (11 February 1822). Hallé was not only a member of the medical section but also professor of medicine at the Collège de France, a post which combined considerable prestige with a good salary. Sooner or later the Minister of Public Instruction would ask the Academy's advice on filling the chair. The expected letter arrived on 8 July 1822.[81] The medical section was asked to draw up a list of candidates and it gave the great physician Laënnec as its first choice, adding (perhaps under pressure?) the name of Magendie as a second choice. The actual election would take place at the meeting of 22 July. Before this part of the agenda was reached Magendie made an announcement about his research. When the election was held Magendie obtained the majority at the second ballot.

What was the announcement made by Magendie at the crucial meeting of 22 July? It was that as a result of many vivisection experiments he had found that if he cut the posterior roots of the spinal nerves of an animal it affected only sensations, whereas if their anterior roots were cut this affected movement only. The discovery, fundamental to the new science of neurophysiology, is the basis of what came to be known as the 'Bell–Magendie Law', since Charles Bell in England was independently, at about the same time, distinguishing between the motor and sensory roots of the medula. It is probably the most famous of all Magendie's work and we have argued that its announcement in 1822 has to be seen in the context of the procedures of the Academy as much as in the internal history of physiology.

There were, therefore, a number of scientists for whom election to the Academy constituted the opening of a door to important research or at least to great activity. In his *éloge* of Bravais (1811–63), Elie de Beaumont wrote:

> Admitted to a seat in the Academy, M. Bravais testified his gratitude by his assiduity, by the number and importance of his communications ... he applied himself without relaxation. He was stimulated by the desire to reply to the kind welcome of the Academy.[82]

But for every such Academician there were several others who seemed to take life more easily after election. Already in the eighteenth century the cynical Mercier had remarked that as soon as an author was elected to the Académie Française, he lost the desire to write.[83] Some members of the Academy of Sciences would have justified their later careers by saying that although they were less active in field work or in the laboratory, they were making important

[81] *P.V.I.*, 7 pp. 347–9.

[82] 'Memoir of Auguste Bravais', trans., *Annual Report of the Smithsonian Institution*, 1869, 145–68 (p. 167).

[83] Mercier, *Tableau de Paris* (1781–8), quoted by John Lough, *Writer and Public in France from the Middle Ages to the Present Day*, Oxford, 1978, p. 242.

contributions to science by sitting on committees and through their judgement of the work of colleagues. It was not, after all, as if they had recently been presented with splendid new laboratory facilities. Some obviously felt that they had entered the hall of fame and could relax their efforts.

There is abundant evidence that *failure* in an election often stimulated unsuccessful candidates. Some wore themselves out trying to qualify for elections in successive vacancies, only succeeding in their old age. Often any decline in scientific productivity can be related most simply to age. It has been said of the mathematician Halphen (1844–89) that his election to the Academy in 1886 was 'an honour which he enjoyed for only three years before he died of what was called "overwork"'.[84] But there were many others who did not overwork and did not die. They lived on into their seventies and eighties, providing an interesting link with the past but at the same time of course blocking the admission of younger and more active members. As we have remarked previously the process of election was, unfortunately, not complemented by a procedure for retirement.

Finally, we might consider a case which in retrospect seems an opportunity missed by the Academy. It would be all too easy to condemn the Academy outright for not electing Marie Curie (1867–1934) to membership. Of course election was only one of several ways the Academy could help aspiring scientists. The basic facility offered was a platform for communications, of which Madame Curie often took full advantage in her early career. The most useful thing the Academy did was to give her and her husband a generous grant of 20000 francs in 1902 to finance the industrial extraction of radium-bearing barium in large enough quantities for her to purify and extract the radium. Yet one would have thought that as a pioneer of radioactivity the joint discoverer with her husband of two new elements (polonium and radium) and a Nobel prize winner (1903), she was fully worthy of membership. We must therefore examine the situation more closely.

The essential question was candidature.[85] The Institute as a whole had decided that women should not be allowed to stand as candidates but the Academy of Sciences insisted on its rights to consider whom it wished. Accordingly, to the credit of the Academy, it accepted her candidature in 1910 for a vacancy in the physics section. She had first hesitated because of the expected publicity and because of the stress suffered by her late husband, who had not been successful first time and had only been elected by a narrow majority at his second candidature. Nevertheless she knew that membership would give her research laboratory greater prestige. The rival candidate was the

[84] M. Bernkoff, art 'G. H. Halphen' in *D.S.B.*, 6, 75

[85] See *C.R.*, **153** (1911), 1268–9. A good account of the election is given by Robert Reid, *Marie Curie*, 1974, chap. 15 'Academic miscalculation'. See also *D.S.B.*, 3 497–503.

sixty-six year old Edouard Branly, who taught at the Catholic Institute in Paris and, having been a candidate twice before, announced that this would be his final candidature. He was known in France as the 'father of wireless telegraphy' on the basis of work he had done in 1890, developing the 'coherer' which was capable of receiving electromagnetic signals. Some felt that he should have had a share in Marconi's 1909 Nobel prize for physics.

Quite apart from the question of feminism, the Academy was known to favour seniority but it was not permitted simply to weigh scientific credentials. The press took an abnormal interest in the election and in many cases, reduced it to a question of personalities. Although Curie was supported by some eminent Academicians, including Gabriel Lippmann, Henri Poincaré and the secretary Gaston Darboux, and was placed at the top of the list by the physics section, she succeeded in getting no more than twenty-eight votes in a second ballot as against thirty for Branly. In many similar cases the candidate who came second would merely have stood again at the next election, with a greatly increased chance of success, according to all the precedents of the Academy. But Marie Curie was deeply hurt by her rejection as well as by the nastier aspects of the election campaign. Not only did she refuse to stand at a future election; for ten years she refused to allow any of her work to be presented to the Academy or be published in the *Comptes rendus*. Probably, if it had not been for the tragic death of her husband in a traffic accident in 1906, she would have acted differently. The whole case has unpleasant overtones not simply of anti-feminism but also of xenophobia, since Marie Curie was Polish by birth. Although it was not customary for anyone to be elected on their first candidature, it is a pity that an exception could not be made for an exceptionally able woman, whose worth had been clearly recognised by the Nobel prize committee.

The French Academy was possibly shamed by the Russian Academy, which had elected the brilliant woman mathematician Sonya Kovalevsky (1850–91) to membership in 1889.[86] Kovalevsky had managed to obtain a higher education in mathematics in Germany, although as a woman she could not be admitted to university lectures. She taught for several years in Stockholm and submitted her memoir 'On the rotation of a solid body about a fixed point' to the French Academy for the Bordin prize in 1888. The judges considered her work so good that they recommended increasing the value of the prize from the usual 3000 francs to 5000 francs. Thus the Paris *savants* played a part in according international recognition to another remarkable woman. A prize however was one thing and full membership of the Academy was another.

This was not the only occasion on which the Academy had given a prize to a woman. We should recall the case of Sophie Germain (1776–1831), who had

[86] Edna E. Kramer, Kovalevsky, *D.S.B.*, **7**, 477–80. See also: *C.R.*, **107** (1888), 1035–6 and A. Rebière, *Les femmes dans la science*, 2nd edn, 1898, pp. 159ff.

been largely self educated but kept up a correspondence with several leading mathematicians of the time.[87] She submitted a memoir on the mathematical analysis of the models of vibration of elastic surfaces, the subject chosen for an Academy prize competition in 1809. When the time limit for the competition expired in 1811 hers was the only entry for the prize. Although theoretically entries were submitted anonymously, her authorship was known to at least one of the judges, Legendre, who acted as her patron. The Academy decided not to award the prize that year but to extend the period of the competition. This allowed Sophie Germain to submit a further memoir and she was then awarded the prize of 3000 francs. Even in Napoleonic France, therefore, an exceptionally able women was able to gain some recognition.

12. *The Academy and higher education*

Finally, we may consider the use of the machinery of elections to make appointments *outside* the Academy. We need first to provide some background information. Among the many educational plans thrown up by the French Revolution was Condorcet's idea of a National Institute which would be responsible for 'overseeing and directing educational institutions'.[88] But when the National Institute was finally established in 1795 this responsibility was omitted from its formal obligations. This of course did not prevent some close informal contact as well as overlap of personnel between the membership of the Institute and the world of higher education. For the first few years of its existence, therefore, the First Class had minimal formal connections with higher education. There was however at least one incident worthy of mention.

In September 1796 the Minister of the Interior, in apparent ignorance of the constitution of the Institute, asked the First Class to decide between two candidates for the post of professor of natural history at the *école centrale* in the department of Calvados. The First Class replied that the Minister should be reminded that the Institute could not decide such a case, since the law gave it no power to nominate to positions in education.[89] Perhaps if the post had been a prestigious Paris position rather than a humble provincial one, the reply would have been more positive.

In December 1799, however, the new charter of the Ecole Polytechnique established a governing body and asked for three Academicians to serve *ex officio*.[90] This connection continued throughout the nineteenth century. The arrangement by which the official body of science provided nominees for a governing body, was later adopted by the Conservatoire des Arts et Métiers. This indirect influence, however, is less important than the wide-ranging powers of

[87] Louis L. Bucciarelli and Nancy Dworsky, *Sophie German. An essay in the history of the theory of elasticity*, Dordrecht, Holland, 1980.

[88] 'une société nationale des sciences et des arts, instituée pour surveiller et diriger les établissements d'instruction' – *Oeuvres de Condorcet*, ed. A. Condorcet O'Connor & F. Arago, 1847, vol. 7. p. 501.

[89] *P.V.I.*, 1, 105.

[90] Maurice Crosland, *The Society of Arcueil*, 1967, pp. 206–7.

recommending candidates for appointment to chairs in higher education, which began in 1802.

The new education law of 1 May 1802 gave the First Class a very definite role in the nomination of candidates for posts in higher education. Altogether there were nearly 200 chairs throughout France for which the Class would henceforth have the right to recommend candidates. The candidate selected by the Academy would be considered together with the nomination of the appropriate 'inspector of studies' before the final appointment was made. Although more than half the chairs were in the Schools (later Faculties) of Medicine, there were also four chairs in each of the three Pharmacy schools (Paris, Montpellier, Strasbourg), five chairs at each of the Veterinary schools of Alfort (near Paris) Lyons and Turin[91] and most important of all, eight chairs at the Collège de France in Paris and twelve chairs at the Muséum d'Histoire Naturelle. Indeed the new power of the Academy was in conflict with the democratic constitution of the Muséum, which had given the professors themselves the right to recommend any new appointment. The power of the Academy as an electoral board to positions in higher education has not been appreciated by historians until recently. It did not extend to the Faculties of Science, which had not been founded in 1802, but it continued for some time in the Faculties of Medicine which replaced the 'Schools of Medicine' in 1808.

The new power not only gave the First Class more influence but meant that it was constantly reminded of the map of higher education throughout France. The new responsibilities gave rise to regular additional correspondence with the minister responsible for both the Academy and education and it reminded the Academicians of colleagues in the provinces. In the case of provincial posts the Academicians were sometimes glad to have local advice on a suitable candidate, but for positions in or near Paris there was nearly always keen competition. One of the most interesting features of any nomination was how it made use of the machinery evolved by the First Class for ordinary elections. Thus, although the Academicians had originally been involved in the selection of candidates on the basis of their supposed ability to recognise merit within particular scientific specialities, the official body of science had a further claim to competence. From the beginning requests from the Minister for nominations were, as with elections, referred in the first place to the appropriate section. The chemistry section was one of the sections most often called upon, since there were chairs of chemistry not only at the Collège de France and the Muséum but also at the Schools of Medicine; chemistry was also the nearest science to pharmacy. The botany and anatomy sections could also give useful advice in relation to several institutions but the expertise of the sections of mathematics and astronomy were not much used until the Ecole Polytechnique (1830), the Bureau des Longitudes (1852) and finally the Observatory (1878) were added to the list of institutions

[91] This is a reminder that the number of chairs was swollen (temporarily) by recent conquest to include new territory over the Alps and across the Rhine.

to which the Academy had the power of nomination. These institutions, together with the Conservatoire des Arts et Métiers (1880), were probably more appropriate for the general expertise of the Academy than the early link with the various Faculties of Medicine, which did not last long.

It took the First Class one or two years to appreciate fully the implications of its new power.[92] It was worried at first that it might not always be notified of a vacancy, in which case it was powerless to make a recommendation. It resolved that it must never make an immediate recommendation, however pressing the ministerial demand. No election should take place without at least a week's notice being given to members.[93] In 1832 it was accepted that since members not present were warned of an impending election to the Academy by letter to their private addresses, the same practice should be followed for elections as the Academy's nominee to a chair in higher education.[94]

The procedure followed closely the practice for elections. On notification of a vacancy, the question of nomination was referred to the appropriate section or sections. A number of names were suggested and often candidates who had heard about the vacancy would write to the Academy offering themselves. The section then discussed the respective qualifications of candidates. At the beginning this would focus on their training[95] but, as the Academy election system developed to include an examination of publications of candidates,[96] a similar interest was shown later in the published work of candidates for nominations. A list was then drawn up in order of merit, excluding any who were thought unsuitable. After warning members that an election was due the list, normally containing at least three names, would be presented to the full Academy meeting in secret session for a discussion of the relative merits of the candidates before the election finally took place.

One implication of the power of nomination was that it increased the patronage of the First Class, usually through the sections but occasionally through a senior Academician. Already in 1803 Fourcroy anticipated a ministerial approach about a nomination by commenting on a book submitted to the Class by Murat, that the book showed that its author would be worthy of a specific chair in the Paris School of Medicine.[97] Three years later, when A. P.

[92] See the report of commission, 28 *brumaire* year 13 (19 November 1804), *P.V.I.*, 3, 153.

[93] 'la Classe ... ne procédera jamais à *ces sortes d'élections* que huit jours après la notification recue de la part du Ministre', *P.V.I.*, 3, 70, 28 *ventose* year 12 (19 March 1804) (my italics). This rule was not always adhered to. Thus in 1810 the Class yielded to pressure to make an immediate nomination to succeed Fourcroy as professor of Chemistry at the Museum, *P.V.I.*, 4, 297–8 (8 January 1810).

[94] *P.V.I.*, 10, 112.

[95] The procedure suggested in 1804 was that there should be a report on the 'studies' of candidates ('le Rapport de leurs études particulières'). *P.V.I.*, 3, 153.

[96] Maurice Crosland, 'Scientific credentials: Publication record as a criterion for election to the French Academy of Sciences', *Minerva*, 19 (1981), 605–31.

[97] *P.V.I.*, 3, 20, 1 *brumaire* year 12 (24 October 1803).

de Candolle was an unsuccessful candidate in an election in the botany section, one of his biographers suggests that the First Class offered him by way of a consolation prize the position of professor of botany at the Faculty of Medicine at Montpellier.[98] Of course the Academy's nomination could never provide an absolute guarantee of appointment. For political or other reasons it might be less acceptable to the Minister than an alternative nomination.

Although the Academy's authority was usually sufficient, it obviously helped if some agreement could be reached with the second nominating body, for example the professors at the Muséum. Even here in difficult times appointment was not certain. Thus in 1827, under Charles X, both the Academy and the Collège de France nominated Magendie for the chair of medicine at the Collège but, because Magendie's politics were considered too radical, the Minister chose Recamier instead. These were, however, exceptional times and the normal pattern of events is revealed in a letter of Gay-Lussac, written in the more moderate and stable political climate of the 1830s, about a chair at the Ecole Polytechnique:

> M. Despretz was presented about a month ago by the school and yesterday by the Institute as candidate for the physics position at the Ecole Polytechnique. The Minister of War is consequently obliged to nominate him and we can regard his nomination as assured.[99]

In the 1840s the power of the Academy was eroded by its being asked to provide more than one nomination for any position. Finally at the beginning of the Second Empire a decree of 1852 stipulated that in the various nominations for (mainly Parisian) scientific institutions, the Academy should always provide two names which would be considered together with two nominations from within the relevant institution.[100] This made it clear that the Academy was not so much *choosing* the professor as offering expert advice on suitably qualified candidates. The change should be understood less as an attempt to reduce the power of the Academy than as a step to increase the power of the Minister of Public Instruction. It marked a slight change in the balance between autonomous science on the one hand and the government on the other. In this respect the power of the Academy was greater in the first half of the nineteenth century than in the second half but whatever changes took place the title 'candidate of the Academy', which had become standard by the 1830s, was one of which the bearer could be proud and very often carried financial rewards.

Although the Academy had an excellent machinery for comparing the respective merits of candidates for election to the Academy, the extension of this

[98] A. de la Rive, 'Notice sur la vie et les ouvrages de A. P. de Candolle', reprinted from *Bibliothèque universelle de Genève*, Nov–Dec. 1844, p. 36.

[99] A.S., Pelouze dossier, letter from Gay-Lussac to Pelouze, 11 October 1831. The purpose of the letter was to inform Pelouze that the promotion of Despretz would create a vacancy for him as *répetiteur*.

[100] Decree of 9 March 1852, *Moniteur*, 1852, p. 345.

machinery to the recommendation for teaching appointments was complicated by the fact that very often colleagues within the Academy were candidates. The impartial stance of the Academy was then strained. A notable scandal arose in 1843 when the Academy was asked to recommend a candidate to fill the vacancy of professor of mathematics at the prestigious Collège de France. Originally three Academicians were interested in the appointment: Cauchy, Liouville, editor of a leading mathematics journal, and the Italian refugee, Libri. For reasons discussed below the first two withdrew, leaving the field clear for Libri.

Now, although in the case of non-members qualifications were discussed, the practice had developed when dealing with a member of simply presenting his name, thereby avoiding the potential embarrassment of saying anything critical about a colleague. Indeed the Academy was generally so concerned not to express any unfavourable opinion about a colleague that the practice grew up, where there were several candidates from the Academy, of refusing to give any order of merit in the way customary for the presentation of non-members. In one case where a preference was expressed by a section for one Academician over another, the Academy subsequently expressed its regret at this discrimination.[101]

In the 1843 nomination mentioned above it seemed as if, given the known partiality for its own members, Libri would be elected without serious opposition. However Sturm, one of the junior members of the mathematics section responsible for drawing up the list, took the unusual step of writing a letter of protest to Dumas, who was president of the Academy in that year.[102] Sturm, who was supported by his colleague Lamé, pointed out that Libri had a reputation as a bad teacher. Indeed it was to be feared that this reputation as a teacher might affect his standing as a mathematician and ultimately throw discredit on the Academy. The two mathematicians, therefore, formally opposed the candidature of Libri and pointed out that although Libri's name might still go forward to the whole Academy, no recommendation of support was implied.

The circumstances of this particular nomination are particularly complex as Cauchy had withdrawn because he feared that although he had a good chance of nomination by the Academy on his mathematical record it might become as much a political election[103] as a scientific one. Moreover, even if he won that election his nomination would probably be rejected by the Minister of Education, since Cauchy was known as a Bourbon legitimist who had opposed the Revolution of 1830 and the subsequent constitutional monarchy of Louis Philippe. Presumably Liouville had withdrawn also because he did not wish to profit from political prejudice. Libri, who had no such scruples, went forward and was ultimately successful.

[101] The case in question was the nomination for the chair of human anatomy at the Museum in 1832. *P.V.I.*, 10, 114.

[102] A.S., Libri dossier, letter referring to a meeting of 19 June 1843.

[103] 's'il s'agit d'une lutte politique'. A.S., Libri dossier, letter from Cauchy to Dumas, 10 June 1843.

Although there is considerable evidence that nomination to chairs in higher education was one of the fringe benefits of membership of the Academy, which helped to compensate for the very modest basic salary or honorarium, the position is complicated by the fact that, as we have said, in order to be elected to the Academy one had to live in Paris, which for most scientists meant already having employment there. The question of which came first, the employment or the election, could be resolved in the following way. A scientist might be content with almost any position, so long as it was in Paris. Once in Paris he could pose his candidature for the Academy and hope for promotion. His chance of promotion or of further employment, making use of the practice of *cumul*, was greatly increased if he could gain election to the Academy.[104] Even if unsuccessful in this big prize he could hope for the lesser prize of a nomination. Berthelot suggested that 'candidatures have become the continued preoccupation in the life of French scientists'.[105] There are many cases where deserving young men from outside were unsuccessful in gaining the nomination, which would be given instead to a member of the Academy. The case of Balard, recommended in 1850 by his colleagues in the Academy for a chair at the Collège de France in preference to Laurent, is fairly well known. The injustice was pointed out at the time by Biot, who reminded his colleagues that Balard already had two Paris laboratories and had no need of a third.[106] This partiality for its own members did not escape public attention. What the Academy gained in patronage, it lost in public esteem. On the other hand one should not assume that the Academy was always lacking in impartiality. All candidatures were carefully considered and certainly by the early twentieth century written reports would be drawn up on the credentials of the main candidates.[107] These elections, often called 'petites élections', were sometimes taken almost as seriously as elections to membership of the Academy.

[104] Maurice Crosland, 'The French Academy of Sciences in the Nineteenth Century', *Minerva*, **16** (1978), 73–102, esp. pp. 95ff.

[105] M. Berthelot, 'L'Académie des Sciences' (1867), *Science et philosophie*, n.d., pp. 201–14 (p. 210).

[106] E. Grimaux and C. Gerhardt, *Charles Gerhardt*, 1900, pp. 588–91.

[107] A.S., *Comité secret, 1903–12*, p. 448 (15 August 1912, election for chair of geology at the Collège de France). See also *ibid.*, p. 147.

— *Chapter 7* —

REGISTRATION, JUDGEMENT AND REWARD

It is above all by its Reports that the Academy can exercise great influence over the direction of people's minds in scientific research.

(A.S., *Comité secret, 1845–56*, p. 124, 2 April 1849.)

The Academy has always reserved for itself the right to establish the laws of science... The Academy is... the implacable enemy of theories [unsupported by good experimental evidence].

(Liebig in private letters to Gerhardt, 1839, 1840, quoted by Schutzenberger, *Inauguration du buste de Balard*, 1896, pp. 11–12.)

In the old days *savants* received pensions; today the Academy awards them prizes; this is surely the most noble and precious privilege of our organisation. It would be of interest to research how this patronage of science, which the Academy exercises and which is so extensive and so useful, has developed over the years.

(Fremy's speech at *Séance publique*, 27 December 1875, *C.R.*, **81** (1875), 1285.)

1. *A registration bureau*

Two of the most important functions of the Academy were to receive and to authenticate new scientific information. In addition the Academy was able to offer a whole range of rewards for good work. We will consider first the Academy's function as a registration bureau. It is useful to have in every advanced country a central agency for the collection of scientific information. The Academy fulfilled this function well and indeed, if it had not existed, it might have been necessary to invent it. That is to say that if there had not existed one elite body of scientists at the centre of scientific communication, there would have been a strong case to establish a government agency, probably staffed by civil servants. This agency could have registered the titles of memoirs but, if it had not been able to call on the expertise of scientists, it would have been quite incapable of distinguishing between the good, the mediocre and the spurious.

In an age of increasing specialisation there was a need for a *single* agency for the whole of science. In England the Royal Society was in a much weaker position than the French Academy in the early nineteenth century and, with the growth of specialist scientific societies, it might well have collapsed had there not been a need for a general scientific body to cater for men of science of all kinds. Moreover, like the Academy, the Royal Society was able to trade on its history and traditions. A central agency was taken more seriously if it had visibility and antiquity and could not, therefore, be dismissed as a temporary expedient. If in addition the agency had a rigorous mechanism for recording receipt of memoirs and publication of basic information, this added to its value. Not only was the Academy superior to the Royal Society as a central registration bureau for the whole of a country but in the matter of rapid and comprehensive publication it was to lead the world by the establishment of the *Comptes rendus* from 1835.

It is worth emphasising the role of the Academy as a *central* collecting point for new scientific work. Some other countries accepted the idea that new scientific work would be communicated to a wide variety of different *regional* academies or to newly-founded societies representing different specialisms. But in France, with a tradition of centralisation going back at least to Louis XIV and enhanced by revolutionary and Napoleonic reforms, it seemed natural that if government and higher education were centralised so science too should be centralised. For individual careers this often meant moving to Paris but even for those who could not aspire to rub shoulders with the *savants* of the capital one could write to Paris and in science this meant writing to the Academy of Sciences. With the publication of the *Comptes rendus* there was further encouragement to communicate with the Academy. Indeed there was a sense in which laboratory experiments or field work remained a purely private matter until the results were communicated to the Academy and summarised in its increasingly famous journal. Unlike the *Mémoires*, which consisted of a selection of major contributions by Academicians, the *Comptes rendus* was a journal of record, reporting the complete proceedings of meetings, including correspondence. It was, therefore, an important adjunct to the Academy in its role as a registration bureau. But publication is really a separate subject and full discussion is postponed till the next chapter.

The first stage of the process of registration was simply to convey information. Implied in the communication, however, would be the idea that this was original research which would have some place on the map of scientific knowledge. The author might consider that he had achieved some slight degree of recognition by simply being in contact with the Academy. He might be indicating a territory which he intended to explore further himself or he might be making a gift to other members of the scientific community to explore, a gift made in exchange for recognition of his own contribution to that scientific community.[1] Though the majority of such reporting could be regarded as low-grade science, something of great significance would occasionally turn up. Academicians sitting in Paris were in the privileged position of being exposed to a constant stream of scientific 'news', which might inspire them in their own research or which they could pass on to their students. Biot described the Academy as 'a kind of free information centre'.[2]

[1] Warren O. Hagstrom, *The scientific community*, New York, 1965, Chapter 1.

[2] 'une sorte de bureau d'annonces gratuite', J. B. Biot, *Mélanges scientifiques et littéraires* (3 vols., 1859), vol. 2, p. 292.

The Academy was a source of news not only about what was happening in France but also in other countries. Among the publications received in 1822 was an open letter from Charles Babbage to Humphry Davy on the application of calculating machines to the printing of mathematical tables[3], and in the following year Prony reported the contents of a private letter he had received from Pouillet. The latter had recently visited London and had been impressed by seeing Babbage's new invention 'which', he said, 'carries out different kinds of calculations with amazing speed.'[4]

Most scientists preferred to convey information openly to the Academy in the form of memoirs or, occasionally, letters. In considering the Academy as a public registration bureau, therefore, the first step was a statement of a discovery or a new phenomenon. In contrast to the equivocations of the old alchemists or the anagrams sometimes used to disguise a new discovery, this statement would be expressed as clearly and concisely as possible. Accompanying any statement of a discovery was an implicit claim to priority. Statements came to the Academy in the form of '*A* has discovered *X*' or, more commonly, 'I have discovered *X*' rather than the impersonal '*X* has been discovered'. The date of the discovery for purposes of priority was usually taken as the date of the meeting of the Academy at which the discovery was announced.

Scientific research presented to the Academy was probably less likely to be overlooked in the nineteenth century than work presented to any other institution in the world. Stories which are told about the neglect of the work of such diverse figures as Avogadro and Mendel,[5] where obscure publication may have been a major factor in the neglect of important ideas, cannot be told in the same way for the Academy and its *Comptes rendus*. The British physicist Joule, although at first content to have his ideas on conservation of energy reported in a local newspaper, the *Manchester Courier* in 1847, followed this up with a more technical account of his famous paddle-wheel experiment on the mechanical equivalent of heat sent to the Academy in Paris and published in its *Comptes rendus*.[6]

Yet it should not be thought that the Academy agreed with the ideas expressed in a memoir simply because it had been read at a meeting and published in the *Comptes rendus*. In 1845, Milne-Edwards came to the defence of the Academy, which in the case in question (a memoir by Duvernois on comparative anatomy)

[3] *P.V.I.*, 7, 348 (22 July 1822).

[4] *Ibid.*, 524 (18 August 1823) – 'avec une promptitude remarquable'.

[5] Mendel published his key paper in the Proceedings of the Natural Sciences Society of Brno.

[6] *C.R.*, **25**, 309–11 (23 August 1847). Donald S. L. Cardwell, *James Joule. A biography*, Manchester, 1989, pp. 79–82, 95, 300. In June 1847 Joule had presented his new theory of heat to the British Association but its importance was not generally appreciated and a brief and inadequate summary, published by the Association in the following year, occupied no more than 3 cm of text. *Report of the seventeenth meeting of the British Association for the Advancement of Science, Oxford, 1847*, Transactions of the Sections, p. 55.

had merely served as a platform for presentation.[7] Milne-Edwards pointed out that it was only when the Academy appointed a commission and subsequently approved its report that it accepted any responsibility. It could not be expected to pass instant judgement on memoirs presented or to censor them. It simply received them and this in itself was an important function.

Recognition by the Academy can be shown to have had far-reaching implications. If, for example, a historian of science wanted to estimate the size of the nineteenth-century scientific community, the most obvious method available has been to make use of official positions: *professeur de mathématiques*, *répétiteur de physique*, and the like. But a more comprehensive index of who might be considered a scientist in nineteenth-century France would be one based on identification of those who had presented at least one memoir to the Academy and had received a favourable report. Moreover, this index is not the invention of a modern sociologist but rather a yardstick of some authority actually adopted in the nineteenth century. When in 1857 Thenard helped to found a charitable society to support the widows and dependents of former scientists without means, the *Société de secours des amis des sciences*, the society was determined to restrict its charity to the families of those who had really made some contribution to science.[8] It therefore needed some means of excluding applicants who were poor and who were close relatives of individuals who might well have had some interest in science without being practitioners. It accordingly made use of the memoir presented to the Academy as an objective criterion.[9] The first two recipients of the charity were the widows of Gerhardt and Laurent who, many felt, had unjustly been excluded from the honours and positions which only Paris could provide. Since the Academy was a magnet, even for those challenging orthodoxy, it could be used to define a wider scientific community than those who simply represented the establishment.

2. *The system of sealed notes*

The Academy operated as a registration bureau in two ways, privately and publicly. The registration aspect was most explicit in the private facility it offered. Anyone, scientist or non-scientist, Frenchman or foreigner, had the right to submit a sealed note to the Academy which would be dated, signed by the secretary and then kept in strict confidence until the author might ask for its contents to be revealed. This was an excellent method of claiming priority for someone unsure of himself and not wishing to expose himself to later criticism or possible ridicule. Under the old regime it was used on a famous occasion in November 1772 by the ambitious young Lavoisier to record his observations of

[7] *C.R.*, 21 (1845), 1322. See *ibid.*, 1301–16 for the original memoir under discussion.

[8] Jules Marcou, *De la science en France*, 2e fascicule, 1869, p. 117. Société de secours des amis des sciences, *Procès verbal de la séance d'inauguration*, 1857.

[9] Even if a member had not received a report, it would not be difficult for the society to decide if the memoir should be considered as a genuine contribution to science.

an increase in weight on burning of such substances as phosphorus.[10] When, a few months later, Lavoisier had confirmed the significance of this discovery, he called on the secretary to produce his note to establish priority ('pour prendre date'). The procedure was particularly appropriate to record experimental work in progress which might have major theoretical implications.

It is not so well known that this procedure continued into the nineteenth century. Indeed in the archives of the Academy there is a register which shows that although the procedure existed in the Napoleonic period it was only in the late 1820s that the number of *plis cachetés* reached double figures for any single year. In each of the years 1826, 1827 and 1828 we find the radical scientist Raspail leaving at least one sealed note, but the habit was not confined to figures on the fringe of orthodox science since we also find the names of Chevreul (1821, 1823, 1829) and Ampère (1824, 1833). The register of *plis cachetés* from 1796 to 1836 lists a total of 329 items.[11] Later in the century the practice expanded, so that a register covering the years 1837 to 1896 lists no fewer than 5269 items, with a maximum usage around the mid-century. In the 1850s among those using the system were Pasteur and Fizeau and in the 1860s J. B. Dumas and Poiseuille. We might emphasise that this was a procedure which conferred potential benefits at all levels of the scientific community. Even the most obscure would-be scientist in the provinces could relate to the Academy. No credentials were asked for and virtually any claim could be made. Confidentiality assured that there was nothing to lose and perhaps much to gain if later work confirmed the original claim.

Let us consider some specific nineteenth-century examples of the use of the sealed note. At the meeting of 20 December 1824, Fresnel asked to retrieve the sealed note he had deposited the previous week. This was permitted, although it was usual for scientists who had blundered to keep quiet and rely on the confidentiality of the Academy.[12] Members usually used ordinary publications to claim priority, considering the *plis cachetés* only as an ultimate safeguard. Only a minority of notes were ever reclaimed and read out for the proud author to claim priority. On 20 September 1875 Lecoq de Boisbaudran asked for a sealed note which he had deposited three weeks previously to be opened. The secretary read out the contents: 'The day before yesterday, Friday 27 August 1875, between 3 and 4 in the afternoon I found signs of the probable existence of a new element'.[13] It was in this way that the new element *gallium* (named after the country of its discoverer) was announced to the world.

[10] 'The sealed note of November 1, 1772', Henry Guerlac, *Lavoisier – The crucial year*, Ithaca, N.Y., 1961, pp. 227–86. This may conveniently be regarded as the starting point of experimental evidence in favour of the oxygen theory.

[11] Pierre Berthon, 'Les plis cachetés de l'Académie des Sciences'. *Revue d'histoire des sciences*, **39** (1986), 71–8 (pp. 72–3).

[12] Even with the passage of time historians of science are not permitted access to *plis cachetés* of a previous century without official authorisation, which involves the appointment of a special commission.

[13] C.R., **81** (1875), 493–5. M. E. Weeks, *Discovery of the elements*, 5th edn, Easton, Pa., 1945, pp. 400–1.

Foreigners also made some use of the system. Thus the American chemist Charles Jackson, wishing to claim priority for the use of ether as an anaesthetic, first wrote two letters on 13 November and 1 December 1846 to Elie de Beaumont in Paris[14] and then deposited a sealed note with the Academy on 28 December 1846; a note which he marked to be opened in the following month,[15] thus stimulating considerable activity. One rival to the discovery, his student, the Boston dentist William Morton, decided that he had to safeguard his claims on the international stage and prepared a memoir for the Academy which was presented at the meeting of 2 November 1847.[16] In the history of recorded sound the French painter Léon Scott from Martinville merits at least a footnote. In the 1850s he had recorded sound as wavy lines, being more interested in analysing the sound waves than in attempting to play back the original sound. When in 1861 he heard that others were working in the field, he asked for his sealed note of 26 January 1857 to be opened and he made a further statement on the subject.[17]

It is interesting to compare the claims made in a sealed note with those later made in a published paper. Such a comparison may be made between the note, sent to the Academy and registered on 30 November 1825 by the young pharmacist from Montpellier, Antoine Jerome Balard (1802–76) on the discovery of a new element, with the paper published the following year in the *Annales de chimie et de physique*.[18] Understandably, the published paper was longer (nearly fifty pages) but not more than twice as long, which shows that the so-called 'note' was almost on the scale of a full memoir[19] and represented rather more than the first thoughts of its author. The published memoir corrects the earlier statement that bromine does not react with phosphorus. The most significant difference, however, is that the published memoir makes an unqualified claim for bromine as a new element and draws out the analogies of its behaviour with chlorine and iodine. In the sealed note the emphasis had been on the experiments and their interpretation had been hedged about with reservations. After presenting the sealed note Balard had obtained a larger amount of his new substance to extend his experiments and had subsequently written up a formal memoir. On 3 July 1826 J. E. Bérard, a corresponding member from Montpellier, presented Balard's memoir to the Academy and a commission was appointed to examine it, consisting of Vauquelin, Thenard and Gay-Lussac. Six weeks later they were ready with their report, verifying that bromine was indeed a new element.[20] Although the Commission praised the work the Academy at the time could offer no immediate means of publication, which explains why it appeared in the *Annales de chimie et de physique*.

[14] S. B. Nuland, *The origins of anaesthesia*, Birmingham, Alabama, 1983, pp. 88ff.

[15] *C.R.*, **24** (1847), 74–9, 18 January 1847.

[16] *C.R.*, **25** (1847), 626.

[17] *C.R.*, **53** (1861), 108.

[18] *A.c.p.*, **32** (1826), 337–81.

[19] The text of the *pli cacheté* has been transcribed and is available in the Balard dossier in the archives of the Academy.

[20] *P.V.I.*, **8**, 414 (14 August 1826).

For a discovery of obvious commercial or industrial application one would expect the inventor to by-pass the Academy in favour of a patent. The French patent system had come into being in 1791, replacing an earlier system of privileges. Although a patent provided a commercial safeguard for those wishing to make money there was no particular prestige attached to the registration of a patent. For those seeking honour rather than money the Academy provided a national platform. This helps to explain why Chardonnet, inventor of artificial silk, used a sealed note in May 1884 to describe his discovery. In November 1887 he asked the secretary to reveal the contents of the note, deposited three and a half years earlier. The note described a method of dissolving cellulose nitrate in equal parts of alcohol and ether.[21] If the resulting solution was squirted through a very fine hole in a platinum cone into a bath of acidulated water, it would make a fine but strong thread which could then be woven like any naturally occurring textile. Chardonnet had made a major contribution to the new industry of synthetic fibres and he had done so within the framework of the Academy.

3. *The continuing interest in dating scientific 'discoveries'*

When the *Annales de chimie et de physique* was re-established in 1816, it announced that it would make a regular habit of reporting on relevant memoirs read at the Academy. Although it mentioned the advantage that scientists not living in the capital could derive from the service, it gave as its first justification that of 'establishing the true date of discoveries, a subject of so much dispute'.[22] The *Annales*, appearing monthly, was probably a better guide at that time to dating claims than the irregularly appearing *Mémoires* of the Academy. Under the old regime, *savants* like Lavoisier had taken advantage of a common two-year delay in publication of the *Mémoires* to update their contributions, thus obscuring the actual date of discovery.

In the nineteenth century the concern of the Academy with priority, the fact that it saw itself as an international agency impartially recording the date on which scientific work was presented, is clearly implied in a private letter from the secretary Bertrand to the young mathematician Boussinesq (1842–1929), then professor in the science faculty at Lille. Boussinesq apparently had the unfortunate habit of abusing the opportunity of correcting proofs to develop further his ideas. Bertrand's comments are worth reproducing *in extenso*.[23]

> Having the responsibility of overseeing the execution of the decisions of the Academy, I have the duty of editing the collection of *savants étrangers*[24] and inserting in it those memoirs, whose publication has been approved. It

[21] 'Sur une matière textile artificielle ressemblant à la soie', *C.R.*, **105** (1887), 899–900.

[22] *Prospectus*, 1816, p. 4.

[23] A.S.., *Copie de lettres, 1875–1902*, p. 6, letter of 19 April 1875.

[24] At the time of the letter (1875) Boussinesq was not yet a member of the Academy. At best, therefore, his work could only be published in the infrequently appearing *Mémoires des Savants Etrangers* rather than in the *Mémoires*.

> is these memoirs themselves, and not a new version, which should appear in our collections. Of course the Academy concedes to authors the opportunity to correct a phrase or rectify a formula, but you surpass greatly in the 40 sheets printed up to now the most generous tolerance which could be shown to any author. The director of the national printing works has drawn my attention to this point and, to mention a secondary but important issue over which he has shown concern, he writes to tell me that your corrections up to now have accumulated an estimated expenditure of more than 16,000f. But there is another more serious reason which obliges me to forbid you absolutely such alterations in future. If a discussion of priority were to be raised one day, would you be in a position to make a claim, through your memoir, both for the approval of the Academy and the date at which it was presented? What would be the value of such an argument for others and for you if you believed yourself to be authorised to introduce not only phrases but pages, which are entirely new?

It is useful to have this official view of the duty of the Academy as an agency concerned with registering priority, especially as it is compared with the expenditure of a large sum of money, seen as secondary. Of course mathematicians might sympathise with Boussinesq, who was trying to present his work in the most highly developed form, but the Academy's position was that its concern with the advance of mathematics had to be modified by considerations of the morality of presenting later work under a format which implied that it had been done earlier. This conflict of interests was brought to a head by the large sums of money involved in resetting whole pages of mathematical symbols. If non-mathematicians were tempted by delays in publication to bring their work up to date at the proof stage the financial implications were considerably less.

4. *Demonstrations*

Although most of the business of the Academy depended on words no less than a political assembly, it also occasionally included artefacts. It was sometimes possible to bring into the arena specimens of new substances produced in the laboratory, notably new elements,[25] or to carry out demonstrations of newly discovered phenomena. The Academicians were often curious to see something new and were privileged to see it before most other people, but it would be misleading to suppose that a demonstration was carried out mainly to pander to mere curiosity. A second motive was related to the propagation of the new knowledge – to make it known to an elite audience – a method which would be supplemented by later publication. But it is important to appreciate the visual element in a demonstration and this brings us to the most crucial aspect of a demonstration, which was implied *authentication* by the Academy.

[25] Lavoisier in his *Traité* of 1789 had re-organised chemistry around a number of simple substances or elements, which were shown in a table. Lavoisier left it to his successors to add to his list of elements and this task constituted a prominent part of inorganic chemistry in the nineteenth century.

Studies of the early Royal Society have emphasised the importance which its members attached to the use of their senses rather than thought in forming judgements about new knowledge. Yet in reporting sensory evidence, the testimony of individuals was always open to challenge. In a group it was more likely that individual prejudices in observation would cancel out:

> Though it was thought that a single investigator inevitably succumbs to his own prejudices in his observations, the collective body of the [Royal] Society could objectively determine matters of fact.[26]

Yet, as in a court of law, there would be not only the question of the number of witnesses but their reliability. Sometimes under the old regime the testimony of witnesses of superior social status would be invoked. Thus when Lavoisier repeated on a larger scale his experiments on the composition of water in 1785 and sought authoritative witnesses, he invited the duc de Chaulnes, the minister Malesherbes and the intendant Villedeuil, as well as several colleagues involved in science.[27] After the Revolution social position was to count for much less than scientific expertise and this was to be found among the relevant specialists within the Academy.

It was possible for an author to make extravagant claims in a memoir but quite another thing to *show* something or *do* something before the critical gaze of the Academy. To achieve this was almost to demonstrate that a new phenomenon was genuine. To demonstrate a machine or a model which worked was very much more of an achievement than to produce a plan of the machine. Any experiment described in a memoir was *in principle* repeatable, but actually to carry out the experiment during a meeting was much more effective. Of course a visual demonstration might require confirmation subsequently. A shiny metallic substance could hardly be proved to be a new element during the meeting but, if it was put on show, this had an immediate impact which was complemented by a description of experiments carried out on the substance. Academicians became more interested in the invention by Planté of a lead accumulator, a landmark in the history of electricity, when he actually showed it to the meeting rather than relying on a verbal description.[28] In the early years of photography it was possible to pass round among the Academicians daguerrotypes, calotypes and other pioneering artefacts, thus emphasising practical results.[29]

The Academy was thus a stage for demonstrating new phenomena or exhibiting new specimens. Often of course a complex experiment could not be

[26] P. B. Wood, 'Methodology and Apologetics: Thomas Sprat's *History of the Royal Society*', *B.J.H.S.*, 13 (1980), 1–16 (p. 9). A most important discussion of the authority of experimental demonstration is given in Stephen Shapin and Simon Schaffer, *Leviathan and the Air Pump, Boyle and the Experimental Life*, Princeton, N.J., 1985, e.g. pp. 56–7.

[27] *Oeuvres de Lavoisier, Correspondance*, Fascicule IV, Paris, 1986, pp. 67ff.

[28] *C.R.*, 50 (1860), 640–1.

[29] E.g. *C.R.*, 10 (1840), 483, 488; 12 (1841), 492.

shown to the whole assembly but was described; it would then be repeated by a commission nominated by the Academy, which would subsequently report back to the main body. A new substance or specimen could more easily be passed round for examination at a meeting and in the post-Lavoisier era new elements were regularly being discovered. When Gay-Lussac succeeded in preparing several grams of the new element potassium, the first time that it had been obtained in a reasonable quantity, he proudly presented it at a meeting of the First Class on 7 March 1808, although he was not ready to read a memoir on its properties and reactions until a month later.[30] Some twenty years later when Balard sent a memoir to the Paris Academy on a new substance (bromine) which he had extracted from sea water, he also sent specimens of the fuming red liquid to assist the members of the commission appointed by the Academy in examining his work.[31] Writing half a century later, Dumas said that he could still remember the curiosity and excitement with which the sealed tube was passed from hand to hand among the members of the Academy.[32]

Probably one of the most famous and influential demonstrations in the nineteenth-century Academy was that given by Arago of Oersted's discovery of a connection between electricity and magnetism. Although Arago had announced this to the meeting of 4 September 1820, shortly after his return from Geneva, his demonstration of the simple experiment at the following meeting was observed with great attention since, it has been claimed, the Academicians had been sceptical of the report and 'they were convinced only by his actual demonstration of the effect on 11 September'.[33] Ampère was present at the demonstration and immediately tried to repeat the experiment himself and investigate its consequences. After building a circular electric pile, he speculated that magnetism may be nothing but electric currents moving in circles. When he presented this idea at the following meeting of the Academy on 18 September no-one took much notice, but he then tried to demonstrate the effect by winding copper wire into flat spirals. At the Academy meeting of 25 September he was able to show his colleagues that these wires behaved like magnets according to the direction of the electric current. This gave considerable support to his theory.

One of the key developments of the electrical industry of the nineteenth century was the dynamo. The principle of the dynamo had been established and published by the Italian Pacinotti in the 1860s, but it was largely ignored. It was the model of the dynamo made by Gramme (1869) and presented to the Academy meeting of 17 July 1871 by Jamin which obtained the necessary publicity. A description of the new invention with a diagram was included in the account given in the *Comptes rendus*.[34]

[30] *P.V.I.*, **4**, 25, 44.

[31] *P.V.I.*, **8**, 400 (3 July 1826), 414 (14 August).

[32] *M.A.I.*, **41** (1879), lxiii.

[33] L. Pearce Williams, 'André-Marie Ampère', *Scientific American*, **260** No. 1 (January 1989), 72–9 (p. 76). *P.V.I.*, **7**, 90.

[34] *C.R.*, **73** (1871), 175–8. Harry Paul, *From knowledge to power*, Cambridge, 1985, p. 157.

The range of the Academy's interests was sufficiently wide for Geoffroy Saint-Hilaire to pass round at a meeting held in May 1858 a selection of flint axes which had been recently discovered in Normandy by Boucher de Perthes in the same stratum as fossils of mammoth teeth.[35] The Academicians had the opportunity of examining the flints and forming their own conclusions as to whether they could be human artefacts. If so, it would seem to have important consequences for estimates of the antiquity of man.

If, when Peligot announced the isolation of true metallic uranium in 1841, he did not pass round a specimen it was probably because it was still partly in the form of black powder, the sight of which would do less to establish his argument than the experiments he described.[36] On the other hand Henri Sainte-Claire Deville on 20 March 1854 proudly displayed to the Academy a leaf of aluminium which he had obtained by electrolysis.[37] Impure aluminium had been prepared earlier but Sainte-Claire Deville was the first to prepare the pure metal and subsequently develop methods to make it more commonly available. By June of the following year he was able to produce large bars of the shiny metal to show the Academy,[38] which was suitably impressed. The sense of participation in the extraction of the aluminium was enhanced by passing round not only the product but the *reactants*: aluminium chloride and metallic sodium. The new metal was chosen for display at the Paris Exposition of 1855 and caught the popular imagination as 'silver from clay'.[39]

When Pelouze brought samples of the newly prepared butyric acid to the Academy in 1843, the reporter of the *Journal des Débats* commented sourly: 'They have exhibited their curious products under the eyes, or should we say under the noses of the Academy; for the hall of the Institute was filled throughout the session with the odour of butter'.[40] When Moissan succeeded in 1886 in isolating the element fluorine, the apparatus was too cumbersome and the product too dangerous to bring it to the Academy and the commission appointed by the Academy had to visit Moissan's laboratory at the Ecole de Pharmacie.[41] However six years later, when he constructed an electric furnace capable of reaching temperatures of 2000 °C, he made a point of showing it to the Academy.[42] The furnace proved extremely useful since he was not only able to produce what he claimed were tiny artificial diamonds but could also use it in the preparation of many less common metals.

It is important to appreciate that a demonstration at an Academy meeting was a demonstration to *experts*. Some of the members would be able to relate the new knowledge to existing knowledge and form a judgement of originality and

[35] *C.R.*, **46** (1858), 902–3.
[36] *C.R.*, **13** (1841), 417–26 (p. 421).
[37] *C.R.*, **38** (1854), 281.
[38] *C.R.*, **40** (1855), 1296–9.
[39] M. B. Hall, 'The strange case of aluminium', *History of Technology*, 1 (1976), 143–57 (p. 146).
[40] *Journals des Débats*, 13 June 1843, cited by F. L. Holmes, *Claude Bernard*, Cambridge, Mass., 1974, p. 495.
[41] M. E. Weeks, *op. cit.*, note 10, pp. 462–3.
[42] *C.R.*, **115** (1892), 1031–3.

authenticity. This was quite different from a public demonstration. The members of the Academy had their professional reputations to think about and, if there was any possibility of fraud, they would be expected to detect it either at the meeting or in some subsequent examination. When Lecoq de Boisbaudran suspected that he had discovered a new element, gallium, he went to Paris and managed to persuade all the members of the chemistry section of the Academy to assemble in Wurtz's laboratory at the end of September 1875 to witness his experiments. He was not ready to present a sample of the new element at a meeting of the Academy till 6 December, by which time none of the chemist Academicians were in any doubt about his claims.[43] It should be noted that it was not customary to display specimens of naturally occurring minerals or plants. These would be regarded as too trivial, belonging rather to an earlier stage in the history of science.

Partly because of the nature of the subject as a rapidly-expanding *laboratory* science, chemistry seems to have been the subject most often represented in demonstrations, but it was usually *inorganic* chemistry, where there were a finite number of new elements to be discovered and which caught the imagination of Academicians, rather than organic chemistry, which seemed to include an almost infinite number of new compounds, of which many hardly required genius to prepare. (A new organic compound would no more be considered a milestone in science than the discovery of a new variety of a plant.) By contrast, astronomy was grossly under-represented in demonstrations. Academicians were rarely asked to look through telescopes. These were kept separately at the Paris Observatory and in any case French astronomy in the nineteenth century was more distinguished for its theoretical contributions than its observations.

The actual demonstration of new substances, apparatus or experiments certainly contributed a dramatic aspect to meetings. If someone actually produced a metal and it was handed round to the country's leading chemists, they could tell either by inspection or by simple analysis immediately after the meeting how it related to existing knowledge. In this way words were reinforced by works. But not many scientists went as far as the surgeon Gerdy, who actually brought along to an anteroom for inspection by the Academicians a patient on whom he had operated.[44] In 1862 F. P. Le Roux, a professor at the Conservatoire des Arts et Métiers, reported to the Academy that he had set up in the park of Neuilly a large piece of apparatus to determine directly the velocity of sound in a gas.[45] Since it included an iron tube thirty-six metres long, there was no question of transferring it to the Academy. Instead he reported his results and invited members of the Academy to examine the apparatus before it was dismantled. Yet there was a limit to the visits which Academicians could reasonably be expected to make. On the other hand when Daniel Colladon, a professor at the Ecole Centrale des Arts et Manufactures, built a high pressure

[43] *A.c.p.* [5], **10** (1876), 100–41 (p. 105).

[44] *P.V.I.*, **10**, 722 (22 June 1835).

[45] *C.R.*, **55** (1862), 662–3.

steam boiler for a steamboat in 1832, he was able to demonstrate its efficiency by sailing the steamboat up the river Seine to the quai Voltaire, next to the Institute. He invited members of the Academy to inspect it, an invitation accepted by Poncelet, Navier and Dupin among others.[46]

5. *Photography*

Photography, one of the great inventions of the nineteenth century, underwent much of its early development in France. Although there is a long prehistory of photography,[47] the first photograph produced in a camera is usually agreed to be that of a courtyard near Chalon-sur-Saône by Nicéphore Niépce in 1822.[48] He used bitumen of Judea as a light-sensitive substance which was used to form a thin coat on a glass plate. He later used metal plates and called the results 'heliographs'. Later Niépce met Louis Jacques Mandé Daguerre, who showed his artistic interest but who became best known as the inventor of the 'diorama', a variety of panorama displayed with changing illumination in a theatre in Paris. Daguerre had carried out experiments with a classic *camera obscura* and was very interested in the heliographs of Niépce. In 1829 the two men entered into an agreement to develop the invention together. In 1831 Daguerre discovered that iodised silver plates could be used to obtain light images and in 1837 he discovered 'by accident' that the latent image produced could be developed by mercury vapour.[49] These photographs became known as 'daguerrotypes'. Unfortunately Nicéphore Niépce had died in 1833 but the collaboration was continued with his son, Isidore. It is clear from the above account that the Niépce family had taken the lead in the 1820s but by the 1830s Daguerre had become the senior partner and most active inventor. Daguerre knew very little chemistry but seems to have been helped by the chemist J. B. Dumas.[50]

The Academy was to play a major part in publicising and encouraging the development of photography and we shall have occasion to reflect on its role. Above all photography was something *visual*. Its final products could be passed from hand to hand; they were well suited to the procedure of *demonstration*. One might think of photography primarily as an invention of great potential commercial value and indeed Daguerre and Niépce *fils* had, in June 1837, made an appeal to art lovers and business men to buy shares in their invention. The authors explained that they would have to keep their process secret until they

[46] John H. Weiss, *The making of technological man*, Cambridge, Mass., 1982, p. 137. Colladon had previously in 1829 been awarded an honourable mention in the Montyon mechanics prize competition for a suggested improvement in the paddle wheels of steamers (*P.V.I.*, 9, 247–8).

[47] Joseph Maria Eder, *History of photography*, New York, 1945, pp. 1–193.

[48] *Ibid.*, pp. 193ff.

[49] *Ibid.*, pp. 223ff. H. & A. Gernsheim, *The history of photography from the camera obscura to the beginning of the modern era*, 2nd edn, 1969, pp. 66ff.

[50] Eder, *op. cit.*, p. 224.

had received a minimum of a hundred subscriptions of 1000 francs each or alternatively when someone had agreed to buy the process outright for 200 000 francs. Given the lack of standing of the sponsors and their refusal to divulge details of the process, it is not very surprising that the public was unwilling to take a leap in the dark. With the failure of the subscription, Daguerre and Niépce decided to offer their process to the government and they chose as their intermediary François Arago, who was a member of the Chamber of Deputies as well as a prominent figure in the Academy. The secret of the daguerrotype was confided to Arago, who was enthusiastic. Although at this stage the details were confidential, Arago as a scientist felt that the first step was to claim priority with the scientific community for Daguerre's invention. He accordingly used the meeting of the Academy of 7 January 1839 to make the claim but in deliberately vague terms – the details were not to be made public until August.

After the January announcement developments took place on two fronts: the political and scientific. On the political front the government drafted a bill (14 June 1839) granting a pension to the inventors. The bill was examined by a joint commission composed of members of the Chamber of Peers, headed by Gay-Lussac, and by members of the Chamber of Deputies, headed by Arago. These two scientists were to seek the approval of their respective Chambers for the bill. But we have neglected to report developments on the scientific front. In England the gentleman scholar and amateur scientist William Henry Fox Talbot received news in late January of Daguerre's work.[51] Talbot had been working privately for some time with light-sensitised paper, and as early as 1835 had taken photographic images of the family estate of Lacock Abbey, which he called his 'photogenic drawings'. Alarmed by the Academy announcement, Talbot wanted to make his own claim for priority, both with the Academy and with the Royal Society, of which he was a member. Talbot's claim to priority was read out at the Academy meeting of 4 February 1839 but his discovery of his new 'calotype' was not presented to the Royal Society until June 1841. Although the inventions were in fact different and independent this was not immediately clear at the time. The existence of an *English* claim inflamed Arago's strong nationalistic feelings. For him it was not only a question of which scientist came first but which country; that is to say that he saw it as the claims of two Frenchmen versus an Englishman.[52] Even Biot however, who was much less nationalistic, pointed out gently to Talbot that Daguerre clearly had priority since he had been working on his process for more than fourteen years.[53] The existence of a rival claim and the rumour, which Arago repeated in his official report, that Daguerre had been made various offers by foreign powers had the effect of enhancing the importance of the invention. Gay-Lussac presented it to

[51] H. P. J. Arnold, *William Henry Fox Talbot. Pioneer of photography and man of science*, 1977, pp. 97–8.

[52] Even when reporting Talbot's claim to the Academy, Arago felt justified in going into some detail of the prior claims of Niépce and Daguerre, *C.R.*, **8** (1839), 170–2.

[53] Arnold, *op. cit.*, p. 99. *C.R.*, **8** (1839), 172.

the Chamber of Peers as 'a title to glory' and as a testimonial of the protection offered to great inventions by the French government.[54]

But if photography had been simply a great invention it might have had little to do with the Academy. The nineteenth-century Academy was not concerned with approving new inventions in the same way as the eighteenth-century Academy had been. We have already seen that in 1839 photography came to be thought of also as a national achievement: it was a French invention.[55] But partly through the intervention of Arago and partly on its own merits it also became a part of science. Although neither Daguerre nor Niépce were men of science, photography could be seen as arising from the *camera obscura* and requiring a study of optics on the one hand and the study of light-sensitive substances on the other. As an interdisciplinary science, therefore, relating both to physics and to chemistry, it deserved to be discussed in the Academy. The public unveiling of the secret of Daguerre's process took place at a meeting of the Academy of Sciences on 19 August 1839. Moreover, given the aesthetic dimension of photography, members of the sister Academy, the Academy of Fine Arts, were invited to the meeting thus providing a further example of interdisciplinarity. It was reported that the presentation was received with enthusiasm by an enormous crowd that filled the meeting and overflowed into the courtyard and on to the banks of the Seine.[56] Credit must go to Arago[57] for handling the whole affair. It not only produced the desired result for the inventors – a generous pension from the government – but it showed that the Academy was concerned with the registration, appraisal and reward of a major invention. Its immediate effect was to present the Academy as the proper place to which later major developments in photography (at least in France) should be addressed. More generally it helped to remind the public of the *relevance* of the Academy.

Daguerre's pension constituted both an honour and financial compensation for revealing his secret to the public. Talbot in England received no such honour. He chose the alternative path of registering a patent and was involved in many exhausting law suits. Whereas Daguerre could feel that he was a benefactor of humanity, many people in Britain felt that Talbot's successive patents had made photography a personal monopoly which they could not challenge.[58] The feeling of frustration was expressed in a joint letter of 1852 to Talbot from the respective

[54] Eder, *op. cit.*, p. 244.

[55] An interesting counter claim that it was really a *German* invention, the work of a certain Schulze in 1727, had to wait for the research of the late nineteenth-century German scholar, Josef Eder (*op cit.*, pp. 60ff).

[56] H. & A. Gernsheim, *op. cit.*, p. 70.

[57] Although it was apparently the Minister of the Interior who proposed the Academy as the platform for the announcement of the 'secret' of photography, *C.R.*, 9 (1839), 227.

[58] It should be pointed out that Daguerre, despite noble statements expressed, also decided to patent his invention *in England* (Gernsheim *op. cit.*, pp. 130–1.)

presidents of the Royal Society and the Royal Academy.[59] By way of contrast with England, there were many reported developments in France in the 1840s.

In subsequent years we therefore find photography as a prominent subject in the *Comptes rendus*. Among the many contributions on the subject were some important ones, for example, on reducing the time of exposure and improving the quality of the print. Some early attempts in colour photography were made by Niépce de Saint Victor[60] but success had to wait till the 1860s. There was at least one major contribution which, although addressed to the Academy, never reached it. It seems that in 1862 Louis Ducos du Hauron, son of a tax official from Auch, attempted to send his ideas on colour photography to the Academy.[61] A family friend, Dr Lélut, was a member of the Academy of Political and Moral Sciences and Ducos rashly confided to him a memoir entitled 'Solution physique du problème de la réproduction des couleurs par la photographie', asking him to pass it on to an appropriate colleague in the Academy of Sciences. The Academician he contacted is not known but, whoever he was, he considered the matter as too speculative and advised against presenting it to the Academy. Ducos, having been rebuffed on the scientific front, registered his method of combining red, yellow and blue pictures as a patent (1868) and subsequently registered other patents. Independently of Ducos another Frenchman, Charles Cros, had been working on three-colour photography and on 2 December 1867 he deposited a sealed note with the Academy giving an account of his experiments. It was only after Ducos had patented his process that Cros was induced to make his own work public.[62]

The main early development of cinematography also came from France. There is some difference of opinion whether the invention should be credited to the two brothers Auguste and Louis Lumière or to Jules Marey.[63] It is beyond dispute, however, that the Lumière brothers from Lyons coined the term 'cinématograph' for an invention which they patented in 1895. It was a simple apparatus in which a perforated film strip was held and moved by a gripper. The effect of persistence of vision between successive images had to be complemented by a shutter which covered the lens of the projector between frames of the film. It was their father Antoine Lumière, a born showman, who was responsible for

[59] They wrote: 'It is very desirable that we should not be left behind by the nations of the continent in the improvement and development *of a purely British invention* (sic)' H. & A. Gernsheim, *op. cit.*, p. 180.

[60] E.g. *C.R.*, **48** (1859), 740.

[61] Ariane Isler de Jongh 'Inventeur-savant et inventeur-innovateur: Charles Cros et Louis Ducos du Hauron. Les commencements de la photographie en couleurs', *Revue d'Histoire des Sciences*, **35** (1982), 247–73 (pp. 252–3); Eder, *op. cit.*, pp. 642–8.

[62] Eder, *op. cit.*, pp. 648–9.

[63] J. M. Eder, *op. cit.*, p. 520. A recent comprehensive history of technology, after mentioning a number of names, including Muybridge, Marey, Le Prince and Edison, says: 'However, credit for the first successful motion-picture projection system is usually given to the brothers Auguste and Louis Lumière...', Ian McNeil (ed.), *An Encyclopaedia of the history of technology*, 1990, pp. 739–42.

showing films to an increasingly enthusiastic public.[64] The two sons, although not averse to the profits from a patent, were also keen to make contributions to science and from 1893 onwards they regularly sent memoirs on photography and chemistry to the Academy.

The other Frenchman concerned, the physiologist Jules Marey (1830–1904), was much more involved with the Academy, to which he had been elected in the medical section in 1878. Far from any interest in entertainment of the public, Marey was concerned with the analysis of the locomotion of men and animals by photographic means.[65] He presented his work regularly to his colleagues in the Academy. Thus in 1884 he provided an 'analyse cinématique' of a person walking.[66] He developed a photographic gun for taking successive pictures of birds in flight, which permitted twelve exposures in a second. Marey mounted the serial photographs obtained on a stroboscopic disc which allowed the phenomenon of motion to be reproduced.[67] By 1887 he had coined the term 'chronophotography' to represent successive pictures of birds in flight which, contrary to the strict rule of the Academy forbidding illustrations, were reproduced in the *Comptes rendus*[68] and may be regarded as a major contribution to the early history of cinematography.

Finally we may mention the case of the Academician Gabriel Lippmann, who was awarded the Nobel prize for physics in 1908 'for his method of reproducing colours photographically, based on the phenomenon of interference'.

6. *A scale of rewards*

Implicit in the presentation of a memoir to the Academy was the hope of recognition, even of reward. It has been argued that recognition by the scientific community itself is an important form of control.[69] Just to have a paper received by the Academy was an achievement for junior scientists. But we would want to argue that the greatest control exercised by the Academy lay in a system which comprised a whole range of possible rewards.

It may be useful to imagine a ten-point scale of reward, ranging from the most cursory recognition to the highest possible honour (see Table 10). The lowest form of recognition was an oral report on a memoir or book. Since the Academy did not normally report on material which had already been published, this was normally the most that the author might expect for a book. It was also a quick way of dismissing a trivial memoir. Secondly, the Academy might provide a written report on a memoir, even a report bestowing praise. Even better was a written report recommending publication in the *Mémoires des Savants*

[64] David Robinson, *The Times*, 28 December 1985.

[65] J. M. Eder, *op. cit.*, p. 507.

[66] *C.R.*, **98** (1884), 1218.

[67] *C.R.*, **94** (1882), 683, 823.

[68] *C.R.*, **104** (1877), 210. See also *C.R.*, 111 (1890), 626–9, where Marey illustrates the successive movements of a man on a horse.

[69] Warren O. Hagstrom, *The scientific community*, New York, 1965, p. 21.

Table 10. *An interpretation of rewards within the gift of the Academy on a 10-point scale*

Action by Academy		Form of Recognition
1. Oral Report		(minimal)
2. Written Report	PUBLICATION	(*a*) at own expense
3. Written Report recommending publication of memoir in *Mémoires des Savants Etrangers*		
4. Publication by Academy of memoir (from 1835 onwards in *Comptes rendus*)		(*b*) at Academy's expense
5. Honourable Mention for entry to prize competition		(*a*) Honour
6. *Encouragement* (later an explicit grant)	PRIZE Title: 'Lauréat'	(*b*) Money
7. Prize		(*c*) Medal (Honour) *or* Money (plus Honour)
8. Nomination to a senior academic post	TITLE	(*a*) 'Professeur' plus salary
9. Election as a correspondent		(*b*) 'Correspondent'
10. Election as a full member		(*c*) 'Membre'

Etrangers.[70] These reports were of great importance and will be discussed in the next section. We might comment, however, that the recommendation that a memoir should be published in the Academy's subsidiary publication reserved for non-members soon became an empty formula, since the publication was many years in arrears and was not in a position to publish all the recommended memoirs. Fourthly, therefore, we might consider publication of a memoir in full in the *Comptes rendus*. After the introduction of this journal in 1835 any memoir submitted to the Academy would automatically be recorded by the title. Often a summary of the memoir would also be published and this is the most that author would normally expect. In exceptional circumstances the *Comptes rendus*, despite the limitations of space, would publish the memoir of an 'outsider' in full. If necessary it could be continued from week to week. Thus when Lecoq de Boisbaudran wrote a speculative paper on spectroscopy in 1869, the Academy (or rather its secretaries) decided to publish the complete paper in parts in several successive issues.[71]

[70] The *Mémoires des Savants Etrangers* was seldom read. What was important, therefore, was acceptance for publication *in principle*, rather than publication in practice.

[71] *C.R.*, 69 (1869), 445–51, 606–15, 657–64, 694–700; 70 (1870) 144–6, 974–7.

We now come to the prize system, which will be discussed more fully later in the chapter. Here we might simply suggest that an honourable mention in a prize competition could be considered a fifth type of reward. Even better would be a monetary reward, often called an *encouragement*, which later in the century became more clearly identified as a grant. Seventh on the suggested scale would be the actual award of a prize, which might consist of a medal or money but in either case would entitle the winner to describe himself as a Laureate of the Academy.[72]

Tentatively we might place as eighth on the scale the right of the Academy to nominate scientists of proven merit to chairs in institutions of higher education. It would be an honour, but more importantly it would be a source of regular income in exchange for duties which need not take up the whole of the time of the professor appointed. Finally we come to the greatest honour, that of election to the Academy. A worthy *savant* who had regularly sent the Academy acceptable memoirs, could aspire to election as a corresponding member. Full membership would probably only be a possibility for a long-term campaigner, someone who had already benefited by the lesser levels of recognition, such as favourable reports and perhaps the winning of a prize.

We are, therefore, suggesting that there was something of a hierarchy in the reward-system of the nineteenth-century Academy. The revolution of 1789 had largely brought to an end the hierarchical society of the *ancien régime* which had been reflected in the different grades of membership within the Royal Academy. In the Institute there were no explicit grades but this did not mean that distinctions had been abolished. As far as science is concerned, what it meant was that the hierarchy had been transferred. Indeed, with a growing scientific community, the paradoxical situation was reached in which the majority of serious scientists were *outside* the Academy. We have referred to the prize-winners, all initially outside the Academy, some of whom were clearly aspirants of different grades of seniority for any future vacancy in the respective sections. Poisson, as a member of the Council of Public Instruction responsible for university matters, referred to the academic hierarchy in terms of university degrees, which were indeed on the same ladder but at the bottom end. As Cousin, Minister of Public Instruction put it, Poisson always offered encouragement to able young teachers of science and mathematics and, when they had passed the *aggrégation*, that notoriously difficult competitive examination, he pointed out to them the next step: the doctorate. For those with a doctorate he pointed to the Academy.[73] Most, of course, would never become members, but they could send in their research papers and they might even win a prize.

[72] 'Nul n'est autorisé à prendre le titre de Lauréat de l'Académie s'il n'a pas été jugé digne de recevoir un Prix.' A.S., *Comité secret*, 9 April 1877.

[73] *Funérailles de Poisson*, 30 April 1840.

7. The system of reports

A very important duty of the Academy was to pass judgement on memoirs submitted to it by non-members. This task can be seen as one of the principal means by which the Academy exercised control over science. The memoir after being presented would be handed over to a commission of two or three experts, usually providing a range of expertise. In many cases this duty involved the repetition of experiments. Occasionally government ministers would ask for a report on a technical matter but usually it was private individuals wishing to make some scientific contribution and perhaps also trying to attract the favourable attention of France's scientific elite. The English man of science, John Herschel, in 1830 singled out the Academy's reports as a particularly praiseworthy feature and one which helped explain the superior achievement of French *savants*:

> 'What author indeed', wrote Herschel, 'but will write his best when he knows that his work, if it have merit, will immediately be reported on by a committee, who will enter into all its meaning; understand it, however profound; and not content with *merely* understanding it, pursue the trains of thought to which it leads; place its discoveries and principles in new and unexpected lights; and bring the whole of their knowledge of collateral subjects to bear upon it.'[74]

Not only were reports thorough, they were also influential. Thus Biot's report of 1803 on meteorites marked the end of the general opinion among men of science that these were to be dismissed as superstitions.[75]

The commission appointed by the Academy to examine a memoir would appoint one of their number as rapporteur and he would be responsible for the actual writing of the report. Yet the report was expected to represent the unanimous views of the commission. It was very unusual for the members to disagree, or rather for the disagreement to continue beyond private discussion and be reflected in the actual writing of a report. A rare example of public disagreement occurred in 1821 over a historical matter rather than a purely scientific one. Paravey had composed several memoirs in order to try to establish that a large amount of modern astronomical knowledge had originated in ancient Chaldea. The astronomer Delambre was very sceptical about the astronomical knowledge of the ancients and he included much criticism in his long report.[76] One of the other members of the commission, probably Burckhardt, strongly disagreed and when the Academy heard of the division of opinion it decided to issue a bland statement giving vague encouragement to the author

[74] *Encyclopaedia Metropolitana*, London, 1845, vol. 4, art. 'Sound' (1830), p. 810n.

[75] *P.V.I.*, 2, 687 (18 July 1803). The unusual step was taken of distributing printed copies of the report to all members of the Academy. It should not be forgotten, however, that only a generation earlier Lavoisier and several colleagues had dismissed meteorites in an Academy report as superstitions.

[76] *P.V.I.*, 7, 132–40 (5 February 1821).

but suggesting that he should improve the presentation of his rather miscellaneous memoirs.

Yet the commissions mounted by the Academy were often prepared to offer very useful advice. Thus when Poiseuille, later to be famous for his contribution to hydrodynamics, submitted to the Academy in 1840 and 1841 a series of papers on the flow of water through capillary tubes, the research was examined by a commission consisting of Arago, Piobert and Regnault.[77] They approved his work but also persuaded him to make further experiments using other liquids, some of which were found to follow the same law as distilled water; others apparently followed a different law.

One did not need any qualifications to present a paper to the Academy. It is sometimes suggested that several people of great potential were unjustly neglected by the Academy. One of these was Marc Gaudin (1804–80), who had some interesting atomic ideas. But when this unknown and untrained provincial sent a letter in 1826 on the subject of caloric to the Academy, it was taken seriously (as was his later work) and given to Ampère and Fresnel to review.[78] Within a fortnight two of France's leading physical scientists had produced a detailed report on his speculations. They could not accept his hypothesis but the report was polite and positive and it had the effect of encouraging him to improve his knowledge of science and make more useful contributions later on. Gaudin's first step was to come to Paris and attend Ampère's lectures. He was to devote a large part of his professional career to elaborating a version of Ampère's theory of matter.[79]

An Academy report of 1799 on one of the first memoirs by the young chemist Thenard (then working as bottle-washer and assistant to Professor Fourcroy) was full of praise. The Academicians recognised a highly skilled young man 'having all the appropriate qualities to advance the science. They think that he should be encouraged to follow a career, which he had begun with so much advantage.'[80] A report could, therefore, act as an encouragement and a boost to the morale of someone on the threshold of a scientific career. A favourable report could be regarded as validation of work done and was all the more important in cases of aspiring scientists without formal training or qualifications. In the Bibliothèque Nationale there is a letter of 1828 from an obscure would-be mathematician to a member of the Academy, saying that he needed a report from the Academy to persuade a publisher to accept his work, which he was too poor to have printed privately.[81] In 1842 a temporary occupant of the chair of zoology at Toulouse explained to a colleague that he would receive a permanent appointment to the chair if the Minister of Public Instruction were to be

[77] *D.S.B.*, *C.R.*, 11 (1840) 961–7, 1041–8.

[78] *P.V.I.*, 8, 398, 401–2.

[79] See Alan J. Rocke, *Chemical Atomism in the Nineteenth Century*, Columbus, Ohio, 1984, p. 105.

[80] *P.V.I.*, 2, 29.

[81] Bardel to Libri, 1828. Bibliothèque Nationale, Fonds Libri.

convinced of his merit by a favourable report from the Academy on his scientific research.[82] A favourable report was therefore a certificate of merit.

The Academy reports could make or break would-be scientists. When the young German mathematics student Dirichlet submitted a memoir in 1825[83] he received a favourable report within a week, and this memoir is said to have made his reputation.[84] Again young Elie de Beaumont presented his first ideas on tectonics at a meeting of the Academy in 1829. Although the memoir went against current geological theory it was carefully examined by an Academy commission and the report of Brongniart was notable for its enthusiastic support. The report emphasised the importance of the new work and its consequences and helped launch Elie de Beaumont on his later successful career as a geologist.[85]

In a period immediately before an election, favourable reports presented to the Academy were sometimes crucial in the success of the industrious and timely author. If the highest honour possible for a non-member was to have his work published in the *Mémoires des Savants Etrangers* with a normal delay of many years in publication, it was sometimes possible to publicise the talents of a young scientist by publishing a favourable report in the select *Mémoires* of the Academy. Thus the candidature of Louis Pasteur for election to the Academy in 1862 was strengthened by his patron Biot, who arranged for favourable reports on his research to be published in the Academy's *Mémoires*.[86] The whole career of the chemist Balard was based on his discovery of the element bromine, presented to the Academy in 1826. When he published his research[87] it was followed by the text of the favourable Academy report.[88] When the engineer, entrepreneur and would-be scientist Gustave Eiffel published as a book in 1907 an account of experiments on air resistance which he had carried out from his famous tower, he included the favourable Academy report not, as was often done, as an appendix, but as a preface.[89] Although the republication of entire reports might seem to display an exaggerated respect for the judgement of the Academy, it did have the advantage of giving further currency to the Academy's work and of doing this at a higher level than by the selective quotation of laudatory remarks. When in 1806 the Minister of the Interior decided to allow the republication of excerpts from a report the Academy pointed out that such

[82] Joly to Libri, 1842. *Ibid.*

[83] *P.V.I.*, 10, 239–40.

[84] Adolf Harnak, *Geschichte der Königlichen Preussische Akademie der Wissenschaften zu Berlin*, 1900, vol. 1, Part 2, pp. 795–6.

[85] *P.V.I.*, **9**, pp. 338–43. J. Bertrand, 'Eloge d'Elie de Beaumont', *M.A.I.*, **39** (1876), xvi.

[86] *M.A.I.*, **23**, (1853) and **24** (1854). After the early years of the Institute very few reports were published in the *Mémoires*. Nor were reports common in the *Comptes rendus* after the first few years of publication.

[87] *A.c.p.*, **32**, (1826), 337–81.

[88] 'Extrait du procès-verbal du lundi 14 aout 1826', *Ibid.*, 382–4.

[89] G. Eiffel, *Recherches expérimentales sur la résistance de l'air exécutées à la Tour Eiffel*, Paris, Librairie Aéronautique (1907). Précédées du Rapport présenté à l'Académie des Sciences par MM. Maurice Lévy et Sébert.

a practice could seriously mislead the public. This would be particularly unfortunate in the case of selective quotations by inventors who might be tempted to omit any critical observations on their work.[90]

The Academy's reports were not always favourable and when, in 1809, the Austrian Winterl, hoping for vindication of his ideas sent specimens of his supposed new element *andronia* to the Academy, a commission consisting of four leading chemists was quite scathing in its comments. It was easy, they said, to show that this new substance was only impure silica, revealing that its author knew very little about chemical analysis. As regards his general ideas:

> In the nineteenth century one ought not to expect to meet a method of philosophising or reasoning, which is so vague, so elusive and so different from that generally adopted in Europe for the last 30 years.[91]

They concluded with brutal frankness that Winterl's methods were more likely to put science back than to advance it.

When given a poor piece of work to review, Academicians usually preferred to forget about it rather than expose it. For this reason completely negative reports are somewhat rare. One might however mention a report of 1832 on a memoir on hernia. Boyer and Larrey, both surgeons and members of the medical section, began their report by expressing regret that they had to present a report but felt duty bound to do so.[92] They totally opposed the procedure advocated and labelled it as dangerous. Their conclusion was phrased in the most polite way possible to say that the memoir contained nothing deserving the approval of the Academy.

The ideal report was written by specialists, not so much for the public as for other scientists who would need not only to have the work appraised but placed in context. The report might begin with a survey of the state of the field. It would then summarise the contribution at hand and assess it for originality and cogency. Finally it would try to give some encouragement. The prize reports often did this with a financial reward called an '*encouragement*', but the ordinary reports usually gave the commissioners' considered judgement, including aspects which called for further examination. The whole thing could have been highly patronising but fortunately this attitude was carefully avoided. Reports were usually written in straightforward language, only occasionally rising to the heights of eulogy.

Whereas an individual scientist was often concerned with the uniqueness of his discovery, an Academy commission was better able to see it in perspective and place it on the map of knowledge. Thus if we may return again to Balard's discovery of bromine, his overriding concern was to establish that it was a new element. (In the light of the paucity of Balard's later achievements it has unkindly been said that it was not so much Balard who discovered bromine as

[90] *P.V.I.*, 3, 470 (22 December 1806). See also *Ibid.*, 523 (4 May 1807).

[91] *P.V.I.*, 4, 220 (12 June 1809).

[92] *P.V.I.*, 10, 177–8.

bromine which discovered Balard, bringing him eventually into the Academy.) The Academy commission appointed in 1826 certainly wanted to verify that it was indeed a new element and it is worth noting that in addition to examining the specimen provided by the discoverer they were able to obtain within a few weeks an independent specimen from a Mediterranean source.[93] The commission felt that it could improve on Balard's provisional name *murine* and proposed *brome* (bromine) from the Greek word signifying a strong smell, thus choosing a term analogous both to the names and etymologies of *chlore* (chlorine) and *iode* (iodine). The commission went further in emphasising the analogy between the three elements. It commented that the new liquid element both in its physical state and chemical reactivity represented an intermediate element or mean between the very reactive poisonous gas chlorine on the one hand and the less reactive solid iodine on the other. The concept of groups of elements with gradually changing properties was to be of fundamental importance in the development of inorganic chemistry in the nineteenth century, particularly after their properties had been linked with the atomic weights of the respective elements. J. B. Dumas was to make some use of these groups in his textbook[94] but the full realisation of the idea came only with Mendeleeff's Periodic Table of 1869–71.

The duty of Academicians to act as referees took up a considerable amount of their time. Eventually it was to lead to a huge backlog and the breakdown of the system, but at least in the early years the examination of memoirs of non-members and the writing of reports was usually accepted as a serious duty. In the revolutionary and Napoleonic period selected reports were actually published in the prestigious volumes of *Mémoires*. Thus the work of the scientist as critic and analyst was placed on a par with the work of the scientist as researcher. As late as 1828, when Cuvier was presenting to the King a report on work done during that year, he pointed out that he intended not only to summarise original memoirs by Academicians but also to review various reports that other Academicians had written, thus reminding the authorities that this was a major part of their work.

The impression may well have been given that the compilation of reports was one of the more tiresome duties which accompanied the honour of being a member of the Academy of Sciences. For although the principle of passing judgement in an area related to one's expertise seems very reasonable, what was a real burden was the sheer number of reports which most Academicians were assigned. The large number of reports soon made a selective approach inevitable and the burden was therefore lessened. Occasionally an Academician had cause

[93] *P.V.I.*, 8, 414.

[94] It seems that the discovery of new elements and the publication in 1826 by Berzelius of a new table of atomic weights were both important sources of inspiration for Dumas (*Traité de chimie appliquée aux arts*, Paris, 1828, vol. i, pp. lxxiv ff). See also Döbereiner, Poggendorf's *Annalen der Physik*, 15 (1829), 301.

to be grateful for information contained in a memoir he was assessing. Thus when Clément in 1813 presented a memoir to the Academy on a new substance from seaweed, he did not realise that he was dealing with a new element, iodine. This was the contribution of Gay-Lussac, who examined the memoir (and the substance) as an official representative of the Academy, but who could not resist going on in a private capacity to take the research to its logical conclusion.[95] There do not seem to be any clear cases where Academicians abused their position to claim for themselves the work of outsiders, but there were several cases, like that of Gay-Lussac, where Academicians took advantage of information which was not fully in the public domain. Quite often aspirant scientists would feel that their work had not received all the praise it deserved. The mathematician Poncelet provided one such case. He probably felt all the more strongly as his memoir of 1820 had apparently cost him six years of work.[96] However, when one turns to the report, one finds that the commission, consisting of Cauchy, Poisson and Arago, was favourable and even recommended publication.[97] Poncelet did however have to accept a certain amount of criticism for his method and, when he ignored this criticism, his later memoirs were received more coolly.[98]

Finally we should not assume that the Academy always contained among its members experts who were capable of solving all the problems submitted. A well-documented history of failure (or at least very limited success) is that of the Gelatin Commission, first established in 1831.[99] In the late 1820s several hospitals in Paris had begun to supply gelatin bouillon to their patients as an inexpensive source of nutrition. The experiments of Donné, head of the clinic at the Charité, suggested that gelatin had little or no nutritive value. The Academy appointed to consider Donné's work a commission consisting of the chemists Thenard and Chevreul and the physiologist Magendie. Later Dumas and D'Arcet were added to the commission. D'Arcet was in an embarrassing position, since he had advocated the virtues of gelatin and had even patented an inexpensive method of obtaining it from bones. One of the problems of the commission was the keen public interest which resulted in a polarisation, some sending testimonials favouring gelatin and others condemning it outright. Faced with such pressure the Academy hesitated to provide a quick report. Instead it spent several years collecting further evidence and in 1836 Magendie began a number of experiments in which dogs were fed on a diet of gelatin. When the dogs refused to eat the gelatin even when starving, the commission tried to achieve a compromise with the conclusion that gelatin might be a nutrient when added

[95] M. Crosland, *Gay-Lussac*, pp. 81–7. The simultaneous private examination of the same substance by Humphry Davy complicates the story but is not strictly relevant here.

[96] J. Bertrand, 'Eloge de Poncelet', *Eloges académiques*, 1890, p. 110.

[97] *P.V.I.*, 7, 55–60.

[98] *P.V.I.*, 8, 341.

[99] F. L. Holmes, *Claude Bernard and animal chemistry*, Cambridge, Mass., 1974, pp. 8–14.

to other nutrients. By 1842 great concern was expressed even within the Academy that the problem had not been solved.[100] Certainly many members of the public would have cited the Gelatin Commission as evidence of the incompetence of the Academy but this is not entirely fair. We must consider the practical difficulties of arriving at conclusive results in nutrition experiments in the early nineteenth century. We see similar difficulties in the early twentieth century in the history of vitamins. Also there were some strong vested interests in the background. We can accept the conclusion of a scholar who has studied this episode in the history of nutrition:

> The Gelatin Commission was suffering the fate that all too often befalls scientists when science is involved to solve problems dictated by social needs rather than problems predicted by the state of science of the time to be solvable.[101]

Yet it is a pity that the Academy's considerable achievement in issuing hundreds of reports on ordinary scientific matters went unremarked both by government and the public, whereas full publicity was given to its shortcomings in one area of potential public concern. Success in this particular area would have done much to boost its image.

8. *Practical problems in the report system*

We must now consider the changing observance over the course of the nineteenth century of the practice of drawing up of reports. The eulogy of Academy reports by John Herschel, quoted earlier, refers to the situation around 1830, which might be regarded as a high point in the system. The practice of drawing up detailed reports had by then become well established and Academicians on the whole took their duty to produce reports fairly seriously. This was not always so. But if there were Academicians who were lazy or preoccupied with making their own original contributions to knowledge, there was also a major problem on the other side – the sheer size of the task. Even if Academicians had worked full time on examining memoirs submitted to the Academy and abandoned their own work entirely (not a situation ever seriously advocated), they would soon have been swamped by the very large number of contributions. Perhaps the system was too open? It could be argued that there was an increasing need for some sort of filter, so that Academicians were not bothered with cranks and low-level work and they could concentrate their attention on really worth-while memoirs. We shall be looking in due course at developments later in the nineteenth century.

Soon after its foundation the First Class of the Institute found several problems with the system of reports; the minutes for the year 1800 contain several

[100] A.S., *Comité secret, 1837–44*, p. 98 (23 May 1842). Cauchy suggested that the Commission should explain to the public what the difficulties were to save the Academy's reputation.

[101] Holmes, *op. cit.*, p. 13.

references to these. Academicians complained that they were asked to report on work of little merit. At the meeting of 11 April.

> A member says that certain memoirs referred to him are not important enough to deserve a report. After a prolonged discussion the Class decrees that such a description cannot be accepted and insists that reports should always be in writing and with the grounds for their judgements.[102]

But two months later the question of quality of memoirs came up in an even more acute way. The reading of memoirs of little worth was taking up the time of meetings. Yet it was recognised that the authors, however weak or mistaken, were doing their best to contribute to the advancement of science. It was agreed that in such cases the president would be justified in interrupting the reading of a memoir. However, the Class would always appoint a commission to examine the memoir in accordance with its earlier decision.[103]

Yet by the end of 1809 the president for that year, Tenon, was complaining that in that year alone there were fifty-eight reports outstanding, with thirty reports still awaited from the previous year. Furthermore there was a total of 165 reports outstanding from the years 1802–8, suggesting an average of about thirty outstanding for each year.[104] However in these early years it would seem that a prominent reason for Academicians not providing a report was that the memoir in question was simply not worthy of a full academic appraisal. Much more serious were later cases, where good scientific research was simply not reported on because Academicians gave the task a very low order of priority.

This was particularly the case after the foundation of the *Comptes rendus* in 1835 encouraged even more scientists to send in memoirs. One of the people to benefit by the publication of the *Comptes rendus* was the young maverick chemist Auguste Laurent (1807–53). He should have been pleased that his memoirs appeared regularly in the same journal as establishment figures from the Academy such as his former teacher J. B. Dumas, with whom he had quarrelled. Yet we find him embittered, reporting in 1843 that not a single memoir of his had received a report.[105] This may seem at first to be blatantly unjust considering the large number of memoirs Laurent had submitted, some of which are now recognised as useful original contributions to knowledge. At the time, however, he represented an avowedly heterodox position in organic chemistry, even so far as to invent his own nomenclature. Any report in which Dumas was involved would have provided a confrontation. The chemists in the Academy were happy to allow such people as Laurent and Gerhardt the rights of publication. They were tolerant but understandably wished to avoid being drawn into a major controversy. The Academy chemists have been accused of a conspiracy of silence. By refusing to report on the work of Laurent and Gerhardt, they seem

[102] *P.V.I.*, 2, 133.
[103] *Ibid.*, 176–7.
[104] *P.V.I.*, 4, 293.
[105] Berzelius, *Bref*, Part VII, p. 181. Letter from Laurent to Berzelius from Bordeaux, 12 May 1843.

to be consistently ignoring the existence of two original French chemists. At least on the strength of their work the two outsiders were considered for membership of the Academy, Laurent being elected a correspondent in 1845 and his younger friend Gerhardt in 1856.

By the mid-nineteenth century a further reason had emerged for the decline in the reporting system. The 1850s witnessed a great expansion in the number of prizes offered by the Academy: eight new prizes were founded and for the rest of the century there was an average of more than ten *new* prizes introduced every decade, with some acceleration after 1870. It should be remembered that the acceptance of any prize implied a permanent obligation of the Academy to examine memoirs submitted, write a report and award one or more prizes. These prize reports constituted an increasing burden on the Academy and the obligation to produce a prize report by the date of the annual public meeting overshadowed the weaker obligation to report on memoirs in general.

Thus, whereas in the early years it was only the lowest level of competence which was refused a report,[106] by the mid-century it was only some of the most outstanding work which was being favoured with a report. The Academy became increasingly selective. In 1856 it coldly informed the Minister of Public Instruction, to whom it was responsible, that it could not regard it as its duty to produce a report on every memoir presented.[107] Even if it nominated a commission to examine a memoir, the commission was within its rights in judging the memoir as unworthy of its attention ('peu digne d'attention') and, therefore, not requiring a report. In 1866 only six reports were submitted out of 357 memoirs sent to commissions.[108] About this time the Academy register, recording the passage of a memoir round the respective members of a commission, was finally abandoned. With the great increase in the number of prizes awarded by the Academy in the second half of the century requiring more and more time for judgement, Academicians came to regard the judging of prizes as a special obligation which largely excused them from issuing other time-consuming reports. To an increasing extent, therefore, the prize reports superseded the ordinary reports.

Although the virtual abandonment of ordinary reports signalled the collapse of the formal system of refereeing, it did not represent the end of all control. Rather what we find is the replacement of the old system of formal judgement by a commission (after presentation of a paper by the author to the Academy) by advance and informal refereeing by individual Academicians. This proved much less time-consuming and, therefore, in that sense more efficient. In the late nineteenth century an increasing proportion of papers from outsiders were submitted to the meeting by individual Academicians. The Academician was

[106] E.g. squaring the circle.

[107] Flourens to Minister of Public Instruction, 30 August 1856, A.S., *Copie de Lettres, 1841–60*, p. 207.

[108] Institut de France, *Académie des Sciences, Troisième centenaire, 1666–1966* (2 vols., 1967), vol. 1, pp. 115–6.

acting as patron and guarantor to some extent of the validity of the claims in the paper.[109] Since a summary of the paper would be published in the *Comptes rendus* with not only the name of the author but also that of the Academician who had presented it, the reputation of the latter was also at stake, at least to a certain extent. In an extreme case the Academician himself might even have repeated the experiments reported, but in most cases the author would be known to him and the contents of the memoir would have been discussed. If the memoir turned out to be an important contribution to knowledge, the patron might share in the glory. The procedure gave further meaning to the system of academic patronage. Academicians valued their power to act as intermediaries. On the whole they would tend to favour work in harmony with their own views and thus orthodoxy was strengthened. By 1900 all papers by outsiders had to be presented by an Academician if they were to appear in the prestigious *Comptes rendus*. Of course, outsiders could still address letters directly to the secretary who might mention the contents when summarising correspondence, but in that case the author could not expect much more than the title of his work to appear in the official proceedings of the Academy. Amateurs, especially those from the provinces who had no personal contact with Academicians, were discouraged. Ambitious junior scientists normally preferred to have their work introduced by a patron. It was appropriately Henri Becquerel, that pioneer in the history of radioactivity, who introduced one of Marie Curie's first papers on that subject.[110]

9. *The prize system*[111]

The award of prizes had been one of the functions of the former Royal Academy of Sciences. It had not been part of the original organisation of 1666 nor of the revised constitution of 1699, but had depended on the idea of a private benefactor, Rouillé de Meslay, a wealthy lawyer who had left the Academy 125000 livres in his will of 1714 for the foundation of two prizes. The prizes were to be biennial, one on celestial mechanics and the other for naval science. The prize system was developed by the Academy in the eighteenth century and used, for example, for the solution of a variety of technological problems. Thus the Crown considered that to give money to the Academy for a prize would be

[109] Lucien Plantefol, 'L'Académie des Sciences durant les trois premiers siècles de son existence', *Ibid.*, vol. 1, pp. 53–139 (pp. 116, 125, 133).

[110] 'Sur une substance nouvelle radioactive, contenue dans la pechblende', *C.R.*, 127 (1898), 175–8.

[111] The standard sources are Ernest Maindron, *Fondations de prix à l'Académie des Sciences. Les lauréats de l'Académie, 1714–1880*, 1881 and, for the later period, Pierre Gauja, *Les fondations de l'Académie des Sciences, 1881–1915*, Hendaye, 1917. Useful recent analyses of the prize system are: Elizabeth Crawford, 'The prize system of the Academy of Sciences, 1850–1914' in R. Fox and G. Weisz, *The organisation of science and technology in France, 1808–1914*, Cambridge, 1980, pp. 283–307 and Antonio Galvez, 'The reward system of the French Academy of Sciences in the nineteenth century', Ph.D thesis, University of Kent, 1987.

a good way of solving the problem of the artificial production of saltpetre, needed for gunpowder.[112]

The idea of regular prizes was adopted within a few months of the foundation of the Institute. By the law of 15 *germinal* year 4 (4 April 1796) the Institute was to give six annual prizes, two for each class.[113] For the First Class this was transformed in 1803 into a single annual prize of 3000 francs, offered alternately by the divisions of mathematical and natural sciences under the title of *grand prix*.[114] Although it was common for eighteenth-century academies to offer prizes, the Paris Academy seems to have been one of the first in which prizes were offered for the solution of specific questions on science and technology. There was, therefore, a firm tradition of the Academy deciding on subjects for prize questions and this was continued after the Revolution. Another feature of the prizes of the Royal Academy which was particularly important to continue, was that members themselves should be ineligible. It was to the credit of the First Class that it continued this outward-looking policy.[115]

The earliest prize questions posed by the First Class had revealed a great lack of co-ordination in administration and had resulted in a very miscellaneous selection of subjects, ranging from plant chemistry to the drifting of warships. None of the prize questions proposed in the years 1800 and 1801 had received satisfactory answers within two or three years and the First Class felt obliged to extend the deadlines. Although the First Class was recognised in 1803, it took some time to clear the backlog and we can pass to 1807, when the first of a memorable series of prize competitions was held. A prize commission led by Laplace and Lagrange decided to offer the *grand prix* for 1810 for the best memoir on the double refraction of light. They chose this subject fully in the knowledge that a young army officer, Malus (1775–1812), was doing brilliant research in related areas of optics, and it is difficult to resist the conclusion that they wanted him to win the prize. If we note that he did indeed win the prize, it is without any suggestion that he did not deserve to win. It was not a foregone conclusion and indeed, fearing the challenge of serious rivals, Malus was prompted to take his earlier work much further. Also it was during this period that he discovered the polarisation of light, a phenomenon which opened up a whole new field of research. Although the discovery was to some extent 'accidental', it was also largely a by-product of the Academy contest.

We have seen that sometimes, as in the case of Malus, the availability of an outstanding potential candidate seems to have encouraged the selection of a particular field for the subject of a prize question. Here the First Class was clearly

[112] Robert P. Multhauf, 'The French crash program for saltpeter production, 1776–94', *Technology and Culture*, **12** (1971), 163–81.

[113] Article XXVIII.

[114] Article 13, *P.V.I.*, **2**, 621. It should be noted that the sum of money available for the First Class was twice that given to the Second and Third Classes.

[115] James E. McClellan III, *Science reorganised. Scientific societies in the eighteenth century*, New York, 1985, p. 65.

taking advantage of the high level of education provided by the Ecole Polytechnique. (Later the Academy was to consolidate the connection with the school by awarding an annual prize to the best student.) In a way it was pursuing an elitist policy, since the prize question could only be answered by a select few but, on the other hand, by knowing in advance the standard of competence of at least one of the candidates, the First Class could set an advanced mathematical question without the fear that the standard would be too high and that the question would remain unanswered to the obvious embarrassment of the official body of science. The latter was therefore using its power and influence to advance science significantly, as in some later cases, such as the Pasteur–Pouchet debate, when a prize question in 1859 provoked a confrontation between known supporters of rival theories to the obvious benefit of science.

Malus' new contributions to optics, made through the prize competition, soon became widely known and helped to increase the prestige of the Academy prizes. Ambitious young men hoped that the Academy would choose a prize question related to their own research and we know that Joseph Fourier himself suggested a prize on the conduction of heat.[116] The Academy commission certainly knew of Fourier's previous work but, when it chose this subject in December 1809 for the *grand prix* to be awarded in January 1812, it seems that far from wishing to reward him simply for his previous work, which would have been improper, it wanted to encourage a brilliant mathematical physicist to extend and improve his theory of heat.[117] Fourier chose to focus on the conduction of heat in solids, a sub-field which allowed him scope for his mathematical talent. The resultant memoir not only won the prize but has become one of the great classics in the history of physics.[118]

Another famous prize was that offered in March 1817 for the *grand prix* of 1819 on the diffraction of light, which was considered a crucial problem for deciding between the corpuscular and the wave theories of light. The text of the prize question actually mentioned the names of two young men who were working in the area and had recently been rival candidates for a vacancy in the physics section of the Academy: Pouillet (1790–1868) and Fresnel (1788–1827), who supported the corpuscular and wave theories respectively. The story of Fresnel's prize-winning entry has often been told.[119] It is worth repeating that

[116] J. Herivel, *Joseph Fourier. The man and the physicist*, Oxford, 1975, pp. 304–5.

[117] Arago in his 'Eloge historique de M. Fourier' says: 'L'Académie, à laquelle [la théorie de la chaleur de Fourier] avait été soumis, voulant engager l'auteur à l'étendre et à le perfectionner, fit de la question de la propagation de la chaleur, le sujet du grand prix de mathématiques...'. *M.A.I.*, 14 (1838), cxii.

[118] Joseph Fourier, *The Analytical theory of heat*, trans. A. Freeman (1878), Dover reprint, New York, 1955.

[119] See, for example, John Worrall, 'Fresnel, Poisson and the white spot: the role of successful predictions in the acceptance of scientific theories', Chapter 5 of David Gooding *et al.* (eds.), *The uses of experiment. Studies in the natural sciences*, Cambridge, 1989.

the commission appointed to judge the prize contained partisans of both theories and that one of its members, Poisson, even suggested a further experiment to Fresnel which turned out in his favour and in support of the wave theory. Several lessons have been drawn from the story of this competition, including the decline of the Laplacian faction, which supported the corpuscular theory.[120] Two points which deserve special mention are the comparative objectivity of the judging commission and, above all, the ability of prize competitions to advance significantly knowledge of a particular field of research both by theory and experiment.

The prize questions announced by the Academy exercised a maximum of control. Candidates were asked not only to work on a specific problem but they were often asked to follow specific instructions. Thus in the famous prize question selected in 1859 for award in 1862 previously mentioned, which was at the heart of the famous Pasteur–Pouchet debate, candidates were asked to investigate spontaneous generation, concentrating exclusively on one aspect (*infusoria*) of a vast field and to do it in only one way, through rigorous experiments involving heat rather than, for example, by the use of the microscope.[121] Candidates were also obliged to do their work within a strict time limit, normally eighteen months, although occasionally the time was extended if no entries of a reasonably high quality had been submitted. Moreover, the *grand prix* was not just any prize. It was the most prestigious of all the Academy prizes, even when some of the other prizes had a greater cash value. To bestow the *grand prix* on a (comparatively) young scientist was the greatest honour the Academy could confer apart from actual membership.

A development of the prize system, centred on the *grand prix* for broad subjects like the biological sciences, was the establishment of prizes for *specific* sciences. There are two important aspects of this development. First was the decision to reward the best work in a particular area but without first dictating a subject of research. In this new approach, which relinquished prize questions, the Academy was abandoning one means of control and substituting another. Thus by accepting in 1802 Lalande's gift of a capital sum to be invested to provide an annual gold medal (value 571 francs) for the best work in astronomy, the First Class was directing particular attention to the subject of astronomy. One would expect other prizes to be founded for work in other sciences recognised by the Academy (which indeed was the case later on) but perhaps the most interesting feature of this development was the award of prizes for *subject areas not explicitly recognised by the Academy for membership.*

Particular attention may be drawn to two quite different sciences for which an anonymous donor in the early period of the Restoration was prepared to

[120] Robert Fox, 'The rise and fall of Laplacian physics', *Historical studies in the physical sciences*, 4 (1974), (89–136).

[121] Antonio Galvez, 'The role of the French Academy of Sciences in the clarification of the issue of spontaneous generation in the mid-nineteenth century', *Annals of Science*, 45 (1988), 345–65 (pp. 348–9).

sponsor prizes. After his death in 1820 the anonymous benefactor was revealed as baron Montyon, a name of exceptional importance in the financial history of the Academy. It seems to have been Laplace who had persuaded Montyon that the subject of statistics deserved special attention.[122] The Academy accepted the offer, which would provide for an annual prize valued at 530 francs, and the prize was awarded for the first time in 1819.[123] Meanwhile, again through the good offices of Laplace, the Academy received the offer of another subject prize from the anonymous donor, this time for experimental physiology. When the prize was offered for the first time in 1819 it aroused such interest that Montyon felt obliged to make an additional gift to the Academy[124] so that, in the event of the prize being shared between several worthy contestants, each might receive a reasonable reward. Among those physiologists whose work was recognised and rewarded by the Academy in the early years of the prize competition were Dutrochet, Magendie and Flourens.

In the early nineteenth century the decision of the Academy to award a statistics prize meant that the Academy had taken an additional science on board.[125] Moreover recognition helped to define this relatively new subject. The prize was often used to reward people living in the provinces who had devoted many years to the compilation of statistics for their region. The concern of Montyon with philanthropy and utility had the effect of stressing the social aspect of statistics at the expense of mathematical theory.[126] Indeed had the Academy of Moral and Political Sciences been in existence at the time there would have been good reason for it to have taken responsibility for the prize.[127]

There was less fear of ambiguity with physiology. The main issue was to give recognition to those who wanted to go further than study the structure of animals and plants and to study their function. Here the sciences of physics and chemistry were of increasing use and it was probably because of the reductionist implications of this study of living things that Laplace was so keen to support it. Physiology received further recognition through the prize system by the foundation of the Lacaze prize, accepted by the Academy in 1869. This prize covered three subjects, physiology, physics and chemistry, and Claude Bernard argued, albeit unsuccessfully, that since the physics and chemistry sections would be involved in judging their respective prizes, it would be necessary to create a section of physiology to provide competent judges for that part of the prize.[128]

[122] *P.V.I.*, 6, 218–20, 257–60.

[123] E. Maindron, *op. cit.*, p. 83.

[124] *Ibid.*, pp. 89–9.

[125] In his report of 1816 Fourier commented: 'La statistique est donc une science de plus'. He explained that this new subject should not be confused with political economy, *P.V.I.*, 6, 258.

[126] For Fourier's report accepting the foundation of the prize, see *P.V.I.*, 6, 218–20.

[127] We may point out that the Academy also awarded a Montyon prize for industrial medicine (*arts insalubres*), and questions of public health might also arguably be the concern of the other Academy.

[128] A.S., *Comité secret, 1870–81*, pp. 172–3 (10 November 1873).

Thus when we speak of the sciences recognised by the Academy we must sometimes go beyond the titles of the respective sections. The establishment of the physiology prize in 1819 predates by two years the election of the first physiologist (Magendie). One cannot help thinking that the recognition of the subject for an Academy prize helped prepare the ground for the subsequent election of a representative of that science. Again the recognition by the Academy of statistics by its decision to award an annual prize in that subject can be seen as a landmark in the history of statistics and certainly predates considerably the foundation of that key organisation, the Statistical Society of London in 1834. A further example of the Academy helping to establish a new discipline through the award of a prize came through its acceptance of a legacy from Dr Jecker who, in his will of 1851 had bequeathed a substantial sum to fund a prize for the author of the best work in organic chemistry. In the first four years of its existence, 1857–60, the money was used to reward such distinguished organic chemists as Gerhardt, Laurent, Wurtz and Berthelot. France was one of the first countries to give institutional recognition to organic chemistry through the creation of a chair of organic chemistry at the Ecole de Pharmacie (1859) and the Collège de France (1865), both however coming after the recognition of organic chemistry by the Academy.

In the second half of the century, and particularly after the Franco–Prussian war, we witness the foundation of a considerable number of new prizes, many associated with large legacies from private donors. In every case ministerial approval was required for the acceptance of a legacy. To be fair, however, the role of the Minister was usually to insist that the precise wishes of the donor should be respected; thus it limited the free use of such legacies by the Academy. The *grand prix* had declined in importance and private funds were coming increasingly to dominate the prize system. What the Academy did not want, however, was to have its hands tied and to find itself, for example, with a large sum of money to reward work in one branch of science without some balancing funds for neighbouring sciences. The Academy therefore began to negotiate with prospective donors; asking them, when leaving money in their wills, to leave the Academy free to decide on how it was to be used. The Lecomte prize of the large sum of 50000 francs (five to ten times a professorial annual salary) was first awarded in 1889 and the Osiris prize of 100000 francs was first given in 1903. One of the problems of these very large awards was to find a single prize winner who really deserved it. Donors often insisted that the prestige of the prize associated with their name should not be reduced by dividing it into several smaller prizes. In the next section it will be argued that what science really needed in the nineteenth century was a multiplicity of cash awards of modest sums and preferably in advance of the relevant research.

10. *From prizes to grants*[129]

The award of a few prestigious prizes *after work had been completed* was not the only way of encouraging science and it may not have been the best way. The prizes of the First Class had usually been gold medals of a specified value, which implied great honour. But gradually the question arose of awarding subsidiary prizes and even money to recompense men of science, who were often of limited means, for the work they had done. In the 1820s monetary awards emerged partly for the reason given and partly in connection with the large amount of money available in the Montyon fund for medical subjects. Also monetary awards had the advantage of flexibility; a large sum could be divided easily to reward several contestants. The sharing of money was also more 'democratic' than focusing the award on a single person. The next step in the evolution of grants would be to provide money *in advance* on the basis of personal promise. This may seem to represent the other end of the spectrum of control from the early days when most prizes were awarded for answers to prize questions dictated by the Academy but in reality the control was only shifted. The persons receiving early grants were likely to be the protégés of Academicians. Moreover, a grant was given for a specific subject of research as much as to an individual. Therefore, the award of a grant to study a particular subject implied the tacit approval of the Academy for that area of research. It would not, for example, have given money for research on something considered to be pseudo-science.

The growing monetarisation of the prize system was associated with a decline in prize questions. Whereas the division of mathematical sciences tended to favour prize questions, the other division, led by the chemists, favoured the reward of general fields of study and, eventually, the award of money in advance. This may be partly explained by greater experimental costs outside the field of mathematics. But from the 1820s the prize system developed less according to some preconceived plan than in response to pressure from the Montyon legacy, where a growing sum of money accumulated as interest on invested capital, was causing some embarrassment in so far as it led to prizes in the field of medicine greatly exceeding the value of 3000 francs stipulated for the *grand prix*. Some of the experimentalists had a vested interest in diverting some of this money away from medicine so that it could be used for general purposes. But before this happened in the 1850s there were two other developments in the use of the Montyon fund. First there was the growth of small monetary awards, which spread the available money widely among medical scientists. Between

[129] Most of this section is based on Maurice Crosland and Antonio Galvez, 'The emergence of research grants within the prize system of the French Academy of Sciences, 1795–1914', *Social Studies of Science*, 19 (1989), 71–100. See also M. Crosland, 'From prizes to grants in the support of scientific research in the nineteenth century: The Montyon legacy', *Minerva*, 17 (1979–80), 355–80 (p. 374).

1825 and 1842 the Academy spent more than 200000 francs in Montyon medicine subsidiary awards in addition to the actual prizes. The second use of Montyon money was to launch the *Comptes rendus*, a development discussed in Chapter 8. It is clear that control over the use of Montyon funds had been relaxed.

By 1850 a large part of the printing costs of the journal was being met by a regular government allocation, thus freeing the Montyon fund surplus for other purposes. The chemistry section, which had been involved in judging Montyon prizes (for example, for pharmacology[130]), and led by the future secretary J. B. Dumas, insisted on the more equitable distribution of prize money between medical subjects (corresponding essentially to only one section of the Academy) and the interests of the other sections. Understandably the majority of money continued to go to medicine according to the wishes of the donor, but for the first time general grants from this fund were awarded privately in areas quite unconnected with medicine.[131] In 1851 a total of 12 500 francs was awarded as *encouragements*. By 1856 the sum awarded for non-medical subjects exceeded 30000 francs and included, for example, a sum of 3700 francs to Fizeau to pay for apparatus to be used in the determination of the velocity of light. Chemistry and the biological sciences too benefited from these awards, as did most other subjects with the exception of mathematics. Most of the recipients probably belonged to the Academy fringe, hangers-on and junior scientists who were able to benefit from the patronage of the Academicians in the same subject area.

From the 1850s the Academy began to accept an increasing number of new prize funds, and most prizes were to be awarded in the form of money rather than the traditional medals. This encouraged the Academy to make medium sized awards several times over to the same person for good work rather than give a large prize. Over the years 1895–1906, for example, Pierre and Marie Curie received no less than seven Academy 'prizes' totalling 55000 francs, which was of great importance in paying the costs of their experimental researches on radioactivity. Earlier Louis Pasteur had systematically exploited the prize system to pay for future research. In other words, he privately converted a 'prize' (for past work) into a 'grant' (for future work). The most significant official Academy policy statement on this subject came with the general shake up after the Franco-Prussian war. At the public meeting of 1874 the president, the astronomer Faye, spoke of the changing needs of science. He went out of his way to criticise the traditional system of *grand prix* and championed the monetary alternative to honorific prizes:

[130] The most important contribution here was the isolation of quinine by Pelletier and Caventou, for which they received the Montyon medicine prize of 10000 francs in 1827. John E. Lesch, *Science and medicine in France. The emergence of experimental physiology, 1790–1855*, Cambridge, Mass. 1984, pp. 126–8, 126–44.

[131] The following information, which the Academy regarded as confidential at the time, is to be found in A.S., *Commission administrative, 1829–77*, pp. 95ff.

> The Academy prefers to help in advance all those [scientists] with talent, rather than to wait indifferently that they succeed exclusively through their own resources and reward them afterwards[132]

For much of the nineteenth century cash awards were usually only consolation prizes (‘*encouragements*’) or disbursement of surplus funds to protégés, as in the Montyon bonanza of the 1850s. For a fully articulated grant fund in a recognisable modern form we have to wait for that founded by Roland Bonaparte (1858–1924) in 1907 on his election as *Académicien libre*. He was not interested in prizes. His purpose was to encourage further research (‘*provoquer des découvertes*’). He offered 100 000 francs to cover grants for several years and, for the first time, the Academy asked applicants to make *written* requests stating the money desired and its intended purpose. Subsequently grantees were asked to report back to the Academy on the use of the money. The flood of applicants, who received sums of between 2000 and 6000 francs, encouraged Bonaparte to increase his donation. But the grant fund can be judged in terms of quality as well as quantity, since more than one-third of the successful applicants up to 1914 were subsequently elected as members of the Academy.

The Bonaparte fund, being open to all subjects, had a stabilising effect in the Academy which had throughout the nineteenth century been continually influenced in one direction or another by prizes for work in limited areas. But it should not be thought that the Academy was finally abandoning control of science. In the first place, by the very existence of the fund linked with a famous name,[133] the Academy received very favourable publicity at a time when the Academy's monopoly of power in the award of grants was threatened by the emergence of other agencies.[134] The Bonaparte fund boosted the prestige and power of the Academy even more than some of its large cash prizes which rewarded work retrospectively, and there was to be no lack of able applicants. Secondly, applicants had to make a good case for their research and the grants would be public knowledge. This contrasts with the early situation when awards were made privately, even secretly, to avoid jealousy and possible criticism. Finally it was suggested[135] that since the personal support of an Academician was usually necessary to obtain a grant, then the Academician himself should share responsibility for the successful completion of the proposed research. This would have imposed considerable pressure all round on the effective use of the money.

[132] *C.R.*, 79 (1874), 1530.

[133] Roland Bonaparte was a nephew of the former Emperor Louis Napoleon, who was himself a nephew of Napoleon Bonaparte.

[134] Notably the Caisse des Recherches Scientifiques, founded in 1901. See Elizabeth Crawford ‘The prize system of the Academy of Sciences, 1850–1914’, in R. Fox and G. Weisz, *op. cit.*, p. 299.

[135] A.S., *Fonds Bonaparte*, vol. 1, p. 112. The success of the Bonaparte grant fund seems to have inspired the wealthy industrialist Loutreuil to bequeath over 3 million francs to the Academy in 1910. Since, however, grants were not awarded from this fund until 1915, it lies outside the scope of this study.

— *Chapter 8* —

THE PRINTED WORD

Make of your ... [*Mémoires*] a classical and selected compilation of the best of what you have done, leaving yourselves time for revision and correction. This is a means of assuring its future reputation.

(Report of Commission of the First Class in 1809, *P.V.I.*, **4**, 228.)

The communication, reading and publication of a paper presented to the Academy is ... an affair of the inside of a week, and it is a certainty.... [It is not difficult] to show what a powerful engine the Academy of Sciences is in the production and encouragement of work.

(Prof. J. Y. Buchanan, F.R.S., *Nature*, **69** (1903–4) 293.)

The deficits for the *Comptes rendus* ... have been made good by means of loans made with the authorisation of the Minister from the surplus from the Montyon account. But these loans, too often repeated, will end up by becoming a real abuse.

(Letter from Academy to Minister of State, 28 October 1861.
A.S., *Copie de lettres, 1861–74*, p. 33.)

1. *Publication*

In the sixteenth century a major figure like Copernicus could devote a large part of his life to the publication of a single book which expounded his ideas. When in the seventeenth century the first permanent scientific societies were founded, the method of publication through great books gradually came to be supplemented by smaller scale communications. Often written originally in the form of letters, they came to be published as papers or memoirs, possibly describing a series of experiments and suggesting a conclusion. In the eighteenth-century French Royal Academy it became accepted that the unit of completed work was the memoir, which would be read at a meeting of the Academy and subsequently published in the annual volume of *Mémoires*, appropriately named. In England the Royal Society produced its *Philosophical Transactions*, which was of varying quality. Consisting largely of the work of its members, there were, nevertheless, a few members such as the wealthy Henry Cavendish, who considered himself under no obligation to share with his colleagues all the secrets of nature he had uncovered. Joseph Black, as professor of chemistry at Edinburgh from 1766–99, felt that his duty to his students excused him from all responsibility for research and publication.

In France there was much less independence of spirit. The Royal Academy was a much more homogeneous and exclusive group than the Royal Society and the members understood that they were expected to investigate the natural world and report their findings in the *Mémoires* of the Academy. Lavoisier provides an outstanding example of a member who published his principal research in a series of memoirs, spanning more than twenty years and resulting in a completely new interpretation of chemistry. Although Lavoisier was also the author of several books, it is to his memoirs that we must turn if we wish to follow the successive experiments which were to be the basis of the great 'chemical revolution' of the eighteenth century.

As science became professionalised in France in the early nineteenth century, its practitioners came to be judged increasingly by the quality and quantity of their publications. Publication is therefore pushed to the forefront of the agenda. It is no longer an optional extra as it had been for Black and Cavendish. Publications came to be accepted as the principal credentials of the nineteenth-century scientist. A process which in the eighteenth century might have been interpreted as no more than satisfying an author's vanity became in the nineteenth century an essential part of a scientist's career.

The idea that publication was not only the right but the *duty* of a scientist was made explicit by Arago, who poured scorn on anyone who,

> in love with his discoveries as the miser is with his treasure, buries them in the ground, takes care even lest his discoveries be suspected, for fear that some other experimenter develops them or applies them. The public owes nothing to someone who has rendered no service to it.[1]

He went on to ridicule those who refrained from publication on the grounds that they were engaged on a large work which required time for completion. Arago favoured openness. He felt that an area of research belonged, at least for a limited time, to the person who had done the basic work in the field and had published his results. The scientific community would show its disapproval of anyone who tried to gather in the harvest in a field which he had not himself sown. In Arago's opinion stealing other people's ideas was worse than stealing their money. Any suggestion about property rights could be avoided by publication attached to a particular date, as in the *Comptes rendus*. Retrospective claims, made by appeals to the testimony of a few friends, could never have the same authenticity as work which had been placed in the *public* domain.

We shall see shortly the problems caused in the Academy by journalists taking advantage of freedom of the press after the 1830 Revolution to criticise unduly the work of the Academy. There is an interesting antithesis between this later freedom and the severe press restrictions under Napoleon, which nevertheless offered facilities for scientific publication. One aspect of the French state which

[1] *C.R.* 17 (1843), 776n., the text taken from Arago's 'Notice sur la vie et les travaux de William Herschel', *Annuaire du Bureau des Longitudes*, 1842, 462–3. The context is a dispute with Libri.

causes difficulties for scholars from other countries is that the First Class was the official body of science, a concept alien to British and American traditions. In Napoleonic France there was also an official press, the *Moniteur* being a government newspaper. In 1800 these two institutions came together in an interesting way. The secretary of state, Maret, wrote to the secretary of the First Class inviting him to send suitable material to the newspaper for publication.[2] Such contributions could be seen as adding to 'national glory'. The Class appointed a commission to examine the implications and reported with a recommendation that they should take advantage of the offer.[3] A newspaper had the obvious advantage of rapid publication and members were given reassurances about accuracy: they themselves would be able to exercise control over the printing of the text. They should give preference to the *Moniteur* over every other newspaper. There would be no charge to the author or the Institute and each author would be sent a copy of the page containing his article.

The *Moniteur* thus became the first newspaper to publish regularly scientific material at the frontiers of research and written not by journalists but by scientists themselves. The agreement only applied to members of the Institute. Thus in Napoleonic France a new meaning was given to the concept of 'official science'. It gave rise to particular resentment in such cases as the rivalry of Gay-Lussac and Davy, who worked simultaneously on iodine in Paris in 1813. Gay-Lussac, as a member of the First Class, was able to publish his research immediately in the *Moniteur*,[4] thereby winning some rights of priority over Davy, a non-member[5] and indeed a national of an enemy power.

2. *From* Mémoires *to* Comptes rendus

For the first forty years of its existence the main publications of the Academy were its *Mémoires* for the work of members, and a parallel though infrequently appearing series for the work of non-members. Although the First Class was given the duty of publication, in the first constitution of the Institute it was treated exactly in the same way as the Second and Third Classes and could not, therefore, claim that it had a special need for rapid publication which was different from that of literary works. The principle of equality between the three Classes was interpreted to mean that it could not publish more volumes or even more pages than political or literary colleagues.[6] However, the reorganisation of 1803 gave the First Class more independence and now it began to publish at irregular intervals a separate series for non-members: *Mémoires présentés à l'Institut des Sciences, Lettres et Arts par divers savants, et lus dans ses assemblées: Sciences mathématiques et physiques*, corresponding to the *Mémoires des Savants*

[2] *P.V.I.*, 2, 100, 16 *pluviose* year 8 (= 5 February 1800).

[3] *Ibid.*, 113–14, 6 *ventose* year 8 (= 25 February 1880).

[4] Sunday 12 December 1813.

[5] A further piquancy is added to the incident by the fact that on the *following day*, Monday 13 December, Davy, at the height of his reputation, was elected as a corresponding member of the First Class.

[6] *Mémoires ... par divers savants*, vol. 1 (1805), preface.

étrangers of the previous Academy and most conveniently known by this shorter title. There were also delays in publishing the main series of *Mémoires*, but in 1806 the First Class decided to follow the practice of the Royal Society of London and publish its annual volume in two parts at six-monthly intervals.

An institutional factor which encouraged annual publication was that the two secretaries were called upon to make annual reports of the work of their respective halves of the First Class and these reports were to be published in the *Mémoires*. Up to 1830 the obligation to compose and publish such a report was a strong, though not always sufficient, incentive to publish annual volumes. The collapse of Napoleon and political trouble in the years 1813–15 had a serious effect on the publication of the *Mémoires*. A single volume for the three years 1813, 1814, and 1815 did not appear until 1818, by which time it was possible to publish the first of the new series under the Bourbon restoration under the title: *Mémoires de l'Académie Royale des Sciences de l'Institut de France*. These *Mémoires* continued the practice of the *ancien régime* of publishing a summary of the work of the Academy (written by the secretaries) under the heading *Histoire*, followed by a collection of memoirs by Academicians. They were also used as the vehicle for the publication of *éloges* of deceased members.

Throughout the eighteenth century the *Mémoires* had been the main publication of the Royal Academy. It had appeared annually (but usually two or three years in arrears) and contained invaluable scientific material. Ambitious *savants* like Lavoisier took advantage of this delay to make corrections, although this practice was obviously more to the benefit of their personal reputation than the interests of the scientific community. When the Academy was restored after the Revolution under the name 'First Class of the Institute' it started another series of the *Mémoires*.

Yet already in 1809 there was fear that the *Mémoires* of the First Class were being bypassed by specialist scientific journals like the *Annales de chimie*, which appeared at monthly intervals and were able to report the most recent research, whereas the Academy sometimes had a time lag of several years. The recommendation to their colleagues of the special commission set up by the Academy to discuss the problem was couched in the following terms:

> Do not exclude any longer from your *Mémoires*, papers which have already been published elsewhere. On the contrary make of your collection a classical and selected compilation (*un recueil classique et choisi*) of the best of what you have done, leaving yourselves time for revision and correction. This is a means of assuring its future reputation when periodical works, which compete with it to-day, will have long been forgotten.[7]

Certainly by the 1820s the *Mémoires* had become *un recueil classique*. What this meant can be seen by comparing, for example, the succession of research papers published by Thenard on hydrogen peroxide in the *Annales de chimie et de*

[7] *P.V.I.*, 4, p. 228.

physique with the final monograph published in the Academy *Mémoires*.[8] The former reflect the excitement (and mistakes) of fresh research at the boundaries of knowledge, whereas the latter was composed when everything had been discussed and checked and the existence, composition and properties of this new compound had all been well established.

A similar situation existed for Ampère and the long title of his Academy memoir summarises the whole situation:

> Mémoire sur la théorie mathématique des phénomènes électrodynamiques uniquement déduite de l'expérience, dans lequel se trouvent réunis les Mémoires que M. Ampère a communiqués à l'Académie des Sciences dans les séances des 4 et 26 décembre 1820, 10 juin 1822, 22 décembre 1823, 12 septembre et 21 novembre 1825.[9]

In fact such memoirs of more than 200 pages might well have been published as a book. By using the *Mémoires* Academicians may have felt that they were at the same time supporting the Academy and also creating impressive memorials to their own reputations.

In the very form of the Academy's publications we see a changing concept of knowledge which was an important part of the development of science. This involved a change from the view of scientific knowledge as something found at leisure in leather-bound tomes on the library shelf to something more rapid and immediate which needed frequent reporting and possible revision. From 1835 the unit of communication was to be a weekly journal, the *Comptes rendus*, an impressive publication in the frequency and regularity of its appearance as well as for the information contained in it. The enemies of the new style could dismiss it as akin to vulgar journalism but it showed a genuine concern for rapid communication, which is part of modern science. Scientists need to know the most recent research on a particular subject in a way that is largely irrelevant to academic colleagues concerned with literary, historical or moral problems. It was inevitable that the new form of scientific publication would be more prone to error but it gave scientific research an immediacy and excitement which provided a stimulus and it had a good effect on the Academy, whose meetings assumed a new importance.

With the coming of the *Comptes rendus* the Academy became responsible for two contrasting methods of publication. For those who did not favour the new brief but rapid method of publication, the *Mémoires* continued throughout the nineteenth century to provide a ponderous alternative. It is significant that Biot, arch-enemy of the *Comptes rendus*, should have continued to make use of the *Mémoires*; publishing as late as 1849 one of his 200 page monographs on the polarisation of light. J. B. Dumas' 'Recherches de chimie organique', read to the Academy in 1833, was not published in the Academy's *Mémoires* until 1838,

[8] M. Crosland, article 'Thenard', *D.S.B.*, 13, 309–14.

[9] *Mém. Acad.*, 6 (1823) [1827], 175–387.

but this did not matter very much since it had appeared earlier in other journals. Sometimes the *Mémoires* provided a means of publication of massive research monographs which would hardly have been commercially viable.[10] The physicist Regnault was also able to make use of entire volumes of the *Mémoires* to publish data on steam engines and the specific heats of gases.[11] In the 1860s and 1870s the *Mémoires* became almost the private preserve of a small number of senior Academicians, notably Chevreul and Antoine Becquerel, but by then they had long outlived their earlier importance.

To go from the *Mémoires* to the *Comptes rendus* is like going from one world to another. Indeed it is difficult to believe that they were published by the same institution. It is clearly a matter of some importance to understand how the Academy, an obvious target for criticism as authoritarian and old-fashioned, could have launched a new enterprise which is a landmark in the history not only of French science but of scientific literature on an international plane.

As early as 1809 a committee of the Academy had seen the desirability of a journal where the members could 'contribute their daily observations'.[12] What finally emerged in 1835, the *Comptes rendus hebdomadaires des séances de l'Académie des sciences*,[13] can be seen as the result of a number of factors. In the first place the system of *Mémoires*, inherited from the science of the *ancien régime*, was not working efficiently. The work of Academicians was too great to be encompassed in the annual volume and there were long delays in publication. In theory the Academy could show its appreciation of a memoir by an outsider by recommending that his work be published in the *Mémoires des Savants étrangers*, but this series had a very low priority in the leisurely administration of the Academy and, as only two volumes of this series had appeared between 1815 and 1830, many scientists whose work was recommended for such publication never in fact achieved it.

3. *Factors leading to change*

We shall trace three distinct factors leading to a virtual revolution in the Academy's method of publication. First in chronological order we shall consider the effect on the Academy of newspaper reporting of meetings, a practice which developed in the 1820s and came to a head in the early 1830s. Secondly, we shall consider differences of opinion within the Academy on matters of publicity, an issue brought into prominence by the election of Arago in 1830 as one of the two permanent secretaries. Finally, we shall return to problems already experienced by the Academy in the traditional form of publication. In particular we shall call attention to the new situation created in 1832 by the sudden death of the other permanent secretary, Cuvier, who had previously exercised a restraining hold on innovatory ideas.

[10] E.g. Serres, Vol. 25 (1860), Dumeril Vol. 27 (1856).

[11] Vol. 21. (1847), Vol. 26 (1862).

[12] *P.V.I.*, **4**, 228.

[13] I.e. *Weekly Proceedings of the Meetings of the Academy of Sciences*.

Apart from the inefficiency of the Academy's own system of publication, one can see a more positive force at work in the 1820s through the interest taken by other scientists and the educated public in the proceedings of the Academy. In 1816 the *Annales de chimie et de physique* began to publish regular summaries of the proceedings of the Academy. In the 1820s reports about the affairs of the Academy began to appear in newspapers and from 1825 *le Globe* published a regular account of its proceedings, contributed by Alexandre Bertrand.[14] The *Journal de commerce* began in 1827 to comment on the meetings of the Academy and was later joined by the *Journal des Débats* (1832) and, most importantly, by *Le Temps*, whose Academy correspondent, Dr Roulin, was later to work on the *Comptes rendus*. The lifting of restrictions on the press after the Revolution of 1830 led to a great expansion in the number of newspapers and several showed an interest in the affairs of the Academy. In 1833 a new journal was founded, which not merely included reports on the meetings of the Academy but made such reports its principal concern. In an attempt to take advantage of the prestige of the official body this entirely unofficial journal called itself *L'Institut*. In its Prospectus it claimed that it would provide:

1st An exact and regular account of all the meetings of the Académie des Sciences.
2nd The text or a very detailed analysis of all memoirs composed and read by members.
3rd The text or a very detailed analysis of all reports on memoirs sent by non-members.
4th A fairly detailed analysis of the memoirs addressed to the Academy, which seem worthy of interest and attention.[15]

But there were not only external pressures on the Academy. There were also pressures from within, and in this connection we must mention the key role of D. F. J. Arago in the matter of reporting the activities of the Academy, a role which he began to play very shortly after his election on 7 June 1830 as secretary for the mathematical sciences. The year 1830 was important both in French national politics and in French science. The great German poet Goethe, who visited Paris in that year, considered the debate in March 1830 in the Academy between Cuvier and Geoffroy Saint-Hilaire on the fixity of species to be of greater importance than the fall of Charles X in July 1830, and he chose to write what proved to be the final article of his career on that debate.[16] But controversy within the Academy in 1830 was not limited to differences of opinion on biological theory. That famous controversy extended to a dispute on the reporting of meetings, a subject on which the new secretary Arago had strong feelings. The dispute arose at the beginning of the meeting of 19 July on

[14] Darboux, éloge of J. L. F. Bertrand, *M.A.I.*, **47** (1904), cccxxv.

[15] *L'Institut*, No. 136 (16 December 1835).

[16] Toby Appell, *The Cuvier-Geoffroy debate and the structure of nineteenth-century French zoology*, Ph.D. thesis, Princeton, 1975, p. 302.

the approval of minutes of the meeting of the previous week, in which the recently extinct bird, the dodo, had been discussed and there had been a further clash between Cuvier and Geoffroy Saint-Hilaire.[17] Only a hint is given in the official minutes of the 19 July meeting:

> Several observations are raised on the question of deciding whether extracts of memoirs read to the Academy should be included in the minutes or whether the secretaries should limit themselves to reporting titles, as had been done up to the present.[18]

There was, therefore, a clash of ideologies between the two secretaries. Arago led the movement for greater information at all levels about the work of the Academy, while Cuvier stood for tradition no less in safeguarding the privacy of the Academy than in matters of biological theory.

The stalemate produced by the co-existence of two secretaries with contrasting views on reporting the work of the Academy was broken by the death of Cuvier in the cholera epidemic of May 1832. Although he was eventually succeeded by his disciple Flourens, his successor had neither the great power of Cuvier nor the will to oppose change. In 1833 we find Arago helping to establish a commission to consider current publication policy and even the possibility of new publications, a commission of which he was a prominent member.[19] Arago was able to take further steps towards a policy of openness. He went out of his way to help newspapers to report Academy meetings.[20] Not only could journalists attend the ordinary meetings but on the following day a room was set aside where the relevant memoirs and letters were displayed on a table so that reporters could copy accurately the relevant documents or make a leisurely summary with every facility for checking details.

Yet Arago found that his openness was abused. Some journalists took the opportunity to emphasise differences of opinion and even to indulge in minor character assassination. By 23 March 1835 Arago was complaining that anyone reading a memoir was exposing himself to potential insults from the press. He went so far as to claim that journalists, by indulging in polemic, had driven away from the weekly meetings a good number of distinguished scientists. If necessary, Arago said, he could cite several learned physical

[17] *P.V.I.*, 9, 473 (12 July 1830).

[18] *Ibid.*, 474. The minute ends: 'This will be discussed later in secret session'. Unfortunately the relevant register is missing from the archives of the Academy.

[19] 'L'Académie procède au scrutin pour la nomination d'une Commission de 5 membres, qui examinera sous leurs diverses faces les questions relatives aux collections de toute nature que l'Académie *pourrait* vouloir former' (my italics). *P.V.I.*, 10, 409 (2 December 1833).

[20] Arago writes to Raspail: 'Les rédacteurs des journaux sont admis à notre secretariat non d'après une délibération de l'Académie, car jamais la question lui a été soumise, mais par un acte de ma volonté'. Raspail, *Petit coup d'état à l'Académie des Sciences*, Institut, HR47 No. 30, extrait du *Réformateur* du mercredi 25 mars 1835, p. 8.

scientists and naturalists who now preferred to keep the best of their laboratory research in their files rather than descend into the arena where there lay in wait enemies prepared to trample underfoot the most basic proprieties.[21] Arago first tried to defend the Academy by substituting a system of personal authorisation of journalists for general press passes, a system to which the extreme Republican Raspail violently objected, claiming that it was an infringement of the liberty Arago claimed to represent.[22]

Previously in Napoleonic France it may have been in keeping with the authoritarian nature of the regime that 'official science' should be confined to the bound volumes of *Mémoires*, the official reports, and the government newspaper the *Moniteur*. But with the restoration of the freedom of the press, and in particular in the more democratic age after 1830, it was more appropriate that the science of the Academy should appear in a new form, preferably controlled by the Academy. In the 1830s the main problem was not political but more related to science itself. It was necessary to overcome a tradition that scientific publications consisted of substantial memoirs, carefully drafted and revised and published many months, even years, after the research had been completed. Even in the eighteenth century, however, the Academy had published as a complement to the *Mémoires* the annual *Histoire*, in which the secretary attempted to summarise and explain the work of the Academy.

Although the *Histoire* belongs primarily to the eighteenth century, this form of publication continued in a desultory way into the nineteenth century. The final attempt by the two secretaries to summarise a year's work of their colleagues was published in 1832,[23] although it related to the year 1828; revealing something of a crisis in publication, an ideal situation for a 'revolution'.[24] The *Histoire* could be justified as providing a summary of research for which there was no room in the *Mémoires*, and in this way it might be seen as preparing the ground for the *Comptes rendus*. Yet it was much more subjective, since it depended on the interest and competence of the secretary.

Passing judgement on one's colleagues gave the secretary considerable power but Arago was happy to forgo this, preferring to allow Academicians to present their work directly to the scientific community and to any members of the public who might be interested.

Yet there was still a burden of editorial work to prepare copy for the printer and when Arago presented his proposal to the Academy in March 1835, he announced that he and his fellow-secretary Flourens would be happy to undertake this additional work. Without such unpaid additional work the foundation of the *Comptes rendus* would have been clearly impossible. Although

[21] *Ibid.*, pp. 6–7.

[22] *Ibid.*, pp. 10ff.

[23] 'Analyse des travaux de l'Académie Royale des Sciences pendant 1828.' Partie mathématique (by Fourier), 118 pp., Partie physique (by Cuvier), 112 pp., *M.A.I.*, 11.

[24] Concepts developed by T. S. Kuhn in *Structure of Scientific Revolutions*, Chicago, 1970.

the two secretaries were soon able to bring in an assistant editor, it required their full authority to overcome the whims and prejudices of their fellow-Academicians. If Arago was interested in power over his colleagues, it was not on the basis of favour and patronage but rather on the basis of firm rules, agreed by the whole Academy, which would dissuade individual Academicians from acting like prima donnas. If his plan was to succeed, publication would have to combine accuracy with speed. Thus, although Arago was keen to include discussion in the proceedings, the only discussion reported would be that which had been put down in writing by Academicians and handed to one of the secretaries not later than at the end of the meeting. In the new accelerated communication of modern science there would be no desire to produce literary masterpieces, only reports of experiments or observations complemented by brief conclusions.

4. *The founding of the* Comptes rendus

It was at the Academy meeting of 23 March 1835 that Arago formally raised the question of the Academy taking responsibility for publishing an account of its own proceedings. One advantage of this would be that if the Academy took control it could ensure accuracy of reporting. This was an important point in convincing Academicians of the desirability of such a new enterprise but some members, including Biot, felt that whatever advantages were to be gained they were outweighed by the disadvantages. Nevertheless, by 13 July 1835 not only the principles but the practical details of the new publication had been worked out. The journal was to appear weekly and consist of forty pages.[25] In a year this might amount to some 2000 pages and it was decided from the outset to provide for two volumes a year, a wise decision which governed its format for the next hundred years. However, this expansion beyond the traditional annual volume of *Mémoires* required from the beginning severe and explicit restrictions on the length of individual contributions, if the *Comptes rendus* was to be an account of the proceedings of the whole Academy rather than a selection of the work of favoured members. Arago spoke of providing extracts and summaries of memoirs presented.

As a few members of the Academy had tended to dominate the *Mémoires*, the *Comptes rendus* decided that it must specify a maximum length for the work of any one Academician in any issue (eight pages) and a maximum for the year (fifty pages). Any discussion at the meeting would not be published unless an Academician who had spoken particularly desired this, and he must then provide the secretaries with a summary of his remarks before leaving the meeting. Non-members could have their work included in the *Comptes rendus*, either through a report by members of the Academy or directly and immediately if they were sponsored by a member. The Academician in question had to provide a summary by the end of the meeting. This requirement was soon

[25] *P.V.I.*, 10, 756.

modified to allow non-members to contribute directly. Although the two secretaries carried responsibility for editing the *Comptes rendus*, the working editor was François Roulin, a medical doctor with scientific interests who had served his apprenticeship by writing weekly reports on the meetings of the Academy for the newspaper *Le Temps*.

The publication of the new journal gave new vitality to the Academy. Indeed its publication became one of its principal activities and even its enemies had a grudging admiration for a journal which appeared regularly after each of the weekly meetings. The *Comptes rendus* had the effect of making the Academy and its procedures even more widely known, since it consisted essentially of an edited form of minutes of the weekly meetings. Whereas the *Mémoires* contained long memoirs describing details of scientific research, the scientific content of *Comptes rendus* is presented within the framework of the business of a meeting. It would therefore be a mistake to condemn the *Mémoires* outright and ignore the bureaucratic implications of the *Comptes rendus*. There is continuity and similarity between the (private) minutes of meetings of the Academy[26] in the early 1830s and the *Comptes rendus* in the late 1830s. Only gradually did the secretaries learn not to take their title too literally (i.e. proceedings of meetings) and to go less for form and more for scientific content. Thus the correspondence, summarised by the secretaries at the beginning of each meeting, was soon transferred in the *Comptes rendus* to the end, thus giving greater prominence to the memoirs and reports presented at the meeting. Only by bitter experience did the secretaries learn that it was inadvisable to publish full details of elections as recorded in the minutes, including the voting figures for unsuccessful candidates.[27] Thus the journal gradually evolved from being the edited minutes of meetings to becoming a modern and impersonal scientific journal. French scientists who felt that it did not provide them with enough space to describe the experimental details of their work, would not bypass it on that account. They used it to obtain publicity and claim priority, leaving a fuller description to a memoir in a specialised scientific journal. Thus the publication role of the Academy in relation to other scientific journals was reversed. The Academy came first not last.

We have discussed at some length the *form* taken by the *Comptes rendus*, but we have said little about the finance which made the whole enterprise possible. Considering that the secretaries of the Academy did not normally dare suggest even a minor change in the Academy which would have occasioned a modest increase in the annual budget, how was it possible to launch a major publication, costing many thousands of francs per annum? Although the

[26] *P.V.I.* [1795–1835], 10 vols., Hendaye, 1910–22.

[27] The practice continued till 1872, when it was pointed out that unsuccessful candidates were embarrassed to find their names in the *Comptes rendus* and it was decided to publish only the names of those successful. (A.S., *Comité secret, 1870–81*, p. 77).

Academy was eventually able to persuade the Ministry of Public Instruction to meet the basic costs, this was only after the *Comptes rendus* had been proved to be a great success, an essential adjunct even to the traditional functioning of the Academy. The answer is that the Academy was able temporarily to draw on its own funds.[28]

In the previous chapter the Montyon legacy with its large sums of accumulated interest was mentioned, a fund which allowed the original prize system to be supplemented by generous *encouragements* and, eventually, outright grants. From the late 1820s the Montyon fund had also been used to pay for the printing of prize programmes, quite a legitimate use of the money. Then the fund had been used for engraving and printing plates, first on medical subjects and later on other areas of science. Finally, by an extension of these practices, a sum of 10000 francs was taken in 1833 to pay for the printing of an outstanding volume of the *Mémoires des savants étrangers*. Here then was a precedent to be used by Arago. Montyon funds could be used for the publication of memoirs presented to the Academy. The accounts for the crucial year 1835 are obscure, perhaps deliberately, but by 1837 it is clear that the Academy was drawing annual sums of about 30000 francs from the Montyon fund to pay for the new publication.[29] But while asking the Minister for his formal approval of this expenditure, Arago did not neglect to point out the desirability of direct government financial support, which began in 1840. Even then the funds provided did not cover the entire printing costs and further demands had to be made for many years on the versatile Montyon legacy.

5. *Rapid publication*

It was a remarkable achievement to have scientific research described at one Monday meeting of the Academy and accurately printed and published before the meeting of the following Monday. We are fortunate to have discovered a report drawn up by the publishing firm[30] Gauthier-Villars in 1875 describing the dedicated and disciplined routine followed by the firm in order to convert the proceedings of the Monday meetings into a perfect printed copy of a journal, which was available to Academicians by the Saturday of the same week:

> *Monday evening*: Copy sent by the secretaries arrives at the printing works by 6 or 7 p.m. It is all hand written and in different hands. Badly written manuscripts are copied out in a fair hand and the manuscripts are assigned to different printers in such a way that the most experienced printer receives the most difficult copy. The publishers make a practice of giving regularly the

[28] See Maurice Crosland, 'From prize to grants in the support of scientific research in France in the nineteenth century: the Montyon legacy', *Minerva*, 17 (1979–80), 355–80 (pp. 367–8).

[29] See e.g. A.S., *Commission administrative, 1829–77*, p. 22 (14 September 1838). It is unfortunate that a minute book covering the vital period is missing from the Academy archives. Fortunately, much of the financial data is duplicated in the minutes of the *Commission administrative*.

[30] A.S., dossier: 'Gauthier-Villars'.

> manuscripts of a particular Academician [for example, the prolific Berthelot] to the same printer each time.
> *Tuesday*: The printers devote most of the day to composing the type of their respective articles so that proof readers can make the necessary corrections that evening.
> *Wednesday*: Typographical corrections of the first proofs. Composition of any new copy that has arrived. Proofs are despatched to the homes of the respective authors and collected.
> *Thursday*: Compositors make further corrections before setting type in pages. Messengers collect any author's proofs not yet returned. Page proofs are sent that evening to the secretary of the Academy.
> *Friday*: Printers collect proofs from the secretary and from those members of the Academy, who have insisted on seeing a second page proof. At 5 p.m., whatever other works are being printed, they are removed from the press to allow printing of the journal.
> *Saturday*: Pages are spread out, dried, cut and collated and bound with a cover. Because of the shortage of time and the large numbers of copies [1 400 by the 1870s] fifty women are brought in for this labour. By 1 p.m. the first copies are ready to deliver to the members of the Academy. By 6 p.m. all the other copies are taken to the post office or the other distributing agents.

This remarkable story ends with the claim that in forty years the printers have never once been late! Apart from the assumption of night work, one may note that the schedule depended on the members of the Academy being resident in Paris and thus available to correct proofs immediately. Yet despite the tight schedule, the technical nature of the material and the tendency of the scientists to make further alterations in their original texts, the system of two proofs and efficient proof-reading ensured the highest typographical standards. Today the journal continues to be produced with a high standard of accuracy but printing takes longer.

6. *Continuing criticism of the* Comptes rendus

It is only too easy to assume that after the publication of the first volume of *Comptes rendus* in 1835 Arago had won the battle and the proceedings would henceforth be published almost automatically at weekly intervals. Such a view overlooks the great financial problems of a major permanent commitment, not allowed for in the government budget, and it overlooks continued hostility to the innovation within the Academy itself.

The main critic was Biot. Although he was involved in various scientific disagreements and also personal quarrels with Arago, his views on scientific publications represent a deep conviction held over a long period. As early as 1817 he was pointing to the merits of the traditional form of publication:

> In these large academic collections, where the slow but continuous progress of the human mind is deposited, and which is destined to last as long as civilisation on earth, the priority of a date is of little interest... This is why

> academic collections have nothing to fear from the rivalry of scientific journals, which normally publish scientific discoveries first.[31]

Such news, thought Biot, was usually based on superficial reporting and was only of transitory interest and he contrasted such journals with the *Mémoires* of the Academy, which would be of permanent interest to scientists. In 1837 Biot used the old established *Journal des savants* to attack the new publication of the Academy.[32] He admitted that the *Comptes rendus* acted as a powerful encouragement to scientists to present their work to the Academy, knowing that it would receive immediate publication, yet publication was necessarily in an abbreviated form. The very possibility of immediate publication would tempt scientists to cut corners. Not only might data not be double checked, but ideas which were only half thought out would be presented and published. Biot therefore supported the alternative method of publication in the *Mémoires* of the Academy. Admittedly they were slower, but the *Mémoires* were more in keeping with the dignity of science:

> These volumes... represent the influence and the lasting glory of the Academy of Sciences.

Biot felt that science was not for everyone. It could only be understood by those who had made a detailed study of it. Science therefore had to be judged by scientists, not by the public. Yet the effect of the *Comptes rendus* was to popularise science. The trouble about this immediate publication was that scientists might occasionally make mistakes and any such errors would be disseminated and perpetuated in the *Comptes rendus*. Biot used the *Journal des savants* again in 1842 to cast doubt on the wisdom of dependence on the *Comptes rendus*.[33] By now he was almost resigned to the continuing existence of this journal. He did, however, remind his readers of the limitations: that long articles had to be drastically abbreviated, thus omitting experimental details.

But Biot was not alone in criticising the *Comptes rendus*. Etienne Geoffroy Saint-Hilaire[34] in 1839 looked back with nostalgia to the period before the revolution of 1830, mentioning some of the great names of the early nineteenth century: Lagrange, Laplace, Berthollet, Cuvier and the rest. The Academy had won an international reputation without anything like the *Comptes rendus*. In that journal petty restrictions were being introduced, for example limiting the length of a paper by an Academician to a maximum of six pages (previously

[31] *Journal des savants*, 1816–17, p. 144.

[32] 'Remarques sur l'institution récente des *Comptes rendus hebdomadaires de l'Académie des Sciences*, et sur la publicité donnée à ses séances', *Mélanges scientifiques et littéraires*, vol. 2, Paris, 1858, pp. 257–64.

[33] *Ibid.*, pp. 265–92.

[34] A.S., dossier Etienne Geoffroy Saint-Hilaire, contains three scribbled drafts of a long speech, which the zoologist intended to make about the *Comptes rendus*. Although notes for formal occasions like elections are sometimes found in the archives, it is most unusual (but fortunate for the historian) to have the full text of a speech made in a secret session of the Academy.

eight pages) – as if genius could be confined to an arbitrary number of pages! This outburst of indignation by the zoologist was prompted by a commission of the Academy, set up in 1839 to consider the running of the new Academy journal.[35] The trouble was that it had become *too* successful. After the first two volumes of some 600–700 pages each the size had soon grown to around 1000 pages, and it continued to grow.

Unwilling to compromise with the principle that publication was open to all, the Academy had to limit the length of papers and in December 1839 the proposal that we have mentioned was circulated that even Academicians, who naturally expected special privileges, should be limited to six pages in any one issue. It was this which had prompted Geoffroy Saint-Hilaire to indignation, saying that Academicians were being regulated like school children. Discussion continued for several months until finally the new regulations were generally agreed.

7. *The final establishment of the* Comptes rendus

The *Comptes rendus* fulfilled two complementary functions. First it reported on what had transpired at the regular Monday meetings of the Academy. This had long been a source of widespread interest and many scientific journals, particularly those appearing at frequent intervals, recognised this and made a point of giving the latest news of the Academy. In some journals, such as the *Bulletin de la Société Philomatique* and, more importantly, from 1816, the *Annales de chimie et de physique*, news of the Academy had been brief, objective and informative. It was only when the general newspapers took over that a conflict arose. We have tended to emphasise the conflict but it would only be fair to point out that some newspapers continued the most amicable relations with the Academy. In 1832 members of the Academy were even offered free offprints from the paper *Le Temps* so that they could read the reports of their meetings.[36] Yet all reports were necessarily brief and selective. They tended to emphasise branches of science of particular interest to the writer or, more excusably, to the readers of a specialist journal concerned at most with only a few of the sciences represented within the Academy.

A second function of the *Comptes rendus*, given the increasing professionalisation of science, was as a journal of record which did more than simply list titles of memoirs, yet avoided the other extreme of chatty journalism. Researchers could derive immediate benefit from the *text* of a memoir, even in an abbreviated form. Misrepresentation was almost eliminated by making authors responsible for their respective texts. And even the humblest contributor could be sure that at least the title of his memoir would be recorded. Often much more was given. The courtesy with which non-members were treated is illustrated by the fact that Azaïs, so easily dismissed as a crank, was given no less than seven full pages of the *Comptes rendus* for a memoir in 1840, rather more

[35] A.S., *Comité secret, 1837–44*, e.g. p. 30, 2 December 1839.

[36] *P.V.I.*, 10, 165 (10 December 1832).

than the usual allowance for members.[37] The chemist Laurent, who complained that his original ideas were not appreciated by the Paris establishment, was a regular contributor to the *Comptes rendus*. In volumes 1–5 we find many examples of his work, appearing both as short notes and as memoirs of up to three pages in length. Looking through later volumes, we find contributions from him in every volume between 9 and 31 (1850), although in the latter few years his status had changed, having been elected a correspondent in 1845. The openness of the *Comptes rendus* may be contrasted with the more exclusive policy of traditional journals. Thus if we consider the British pioneers of the kinetic theory, we find that in 1820 John Herapath had his paper on the subject rejected by the Royal Society for publication in the *Philosophical Transactions* and in 1845 the same fate befell John Waterson, his paper being considered too speculative. But this is hardly a fair comparison. The French analogue of the *Philosophical Transactions* is the *Mémoires* of the Academy. The *Comptes rendus* represented an extremely new concept of scientific publication.

It would be easy to interpret the weekly publication of the *Comptes rendus* in a purely negative way as merely compensating for the slow appearance of the *Mémoires*. Rather it should be seen as a positive move, capturing the initiative from journals such as the *Annales de chimie et de physique* and even from newspapers. By combining rapidity of publication with guaranteed accuracy it was setting a new standard in scientific publishing. The *Comptes rendus* had a few imitations, but to find a national and international journal of comparable importance, one would have to wait for *Nature* (Macmillan, 1869–), whose weekly publication continues more than a century later to be an accepted feature of the scientific calendar.

In the first few years (the late 1830s), the production of two volumes of *Comptes rendus* per annum cost the Academy an average of about 25 000 francs a year, although the amount was always tending to increase as the number of pages increased. Thus the first two volumes contained less than 700 pages each. By 1840 this had risen to over 1000 pages for each volume, reaching an unmanageable 1900 and 1500 pages in 1845. By this time however there was an appreciable income from subscriptions. Whereas in 1836 they had hardly amounted to more than 2000 francs and in 1837 4000 francs, by 1845 this annual income had risen to nearly 12 000 francs, where it stayed for at least a decade.

Circulation had been small at first, not more than a few hundred, and the Academy felt bound to donate copies without charge to leading scientific societies at home and abroad. The minutes record many pathetic pleas from minor provincial scientific societies to be added to the list to receive regular free copies. By the mid-century about 1000 copies were being printed and this had risen to 1700 by 1874. By 1901 the number printed had risen to 2050, of

[37] 'Mémoires sur l'affinité ou puissance de combinaison', *C.R.*, 11 (1840), 226–33.

which 1246 were paid for by subscribers, 644 were given away, leaving 160 spare copies.

The cost of production rose steadily, although in 1840 the government for the first time agreed to an annual subsidy of 15 000 francs, thus relieving the burden on Montyon funds. In 1846 the government subsidy was increased to 30 000 francs, which covered a large part of the printing costs. Yet by the 1870s the cost of production had nearly doubled. This was partly due to increase in size, taxes, and higher printing costs but the printers confirmed that much of the increase was due to corrections (and additions) at the proof stage. By allowing less latitude here costs were reduced from 58 000 francs in 1872 to 40 000 francs in 1901. The chief regulation however concerned the maximum allowance of each contributor. As mentioned previously, members of the Academy could not exceed six pages in any one memoir and fifty pages in any one year.[38] For correspondents this allowance was reduced to four pages, with an annual maximum of thirty-two pages. It was also said that there should be no exception to the rule, although in the early years there were several exceptions. For non-members the maximum length was three pages, not ungenerous for a non-refereed paper. Illustrations were not normally allowed because of financial stringency and this also helped rapid publication. The *Comptes rendus* was therefore ideal for providing a summary of research. Full experimental details would be given later in a specialist journal.

For many Academicians the *Comptes rendus* became their principal journal for publication, although loyalties might be divided between that journal and a specialist journal or the house journal of one of the *grandes écoles*.[39] The number of contributions an Academician made to the *Comptes rendus* would depend in the first place on his productivity. Given that he was reasonably productive, contributions might depend on his personal attachment to the Academy and the appropriateness of alternative journals in his field.

A good example of an Academician overwhelmingly loyal to the *Comptes rendus* was the mathematician Cauchy. Of the 407 memoirs he published from 1836 to his death in 1857, no less than 396 (i.e. rather more than 97%) were destined for the *Comptes rendus*. There is no doubt that he found the Academy journal a convenient outlet, its very existence even a source of inspiration. This did not prevent some of his colleagues considering him too productive, since he tended to show them up. We find that Arago published few original memoirs in the journal he had founded. Having reached his fifties by 1836, he probably felt that he could make the greatest contribution to science in an editorial capacity. His younger colleague Dumas, on the other hand, was already a regular contributor to the *Annales de chimie* by the time the *Comptes rendus* was founded. Although he made good use of the Academy weekly journal, he tended to divide

[38] Règlements adoptées dans les séances des 23 juin 1862 et 24 mai 1875.

[39] Thus after the foundation of a house journal for the Ecole Normale in 1864, Pasteur felt obliged to support that journal.

his chemical publications between it and his first love, the *Annales*. Of the astronomer Le Verrier it has been said that 'most of his scientific work was presented in the form of notes (totalling 230) in the *Comptes rendus*... between 1835 and 1878'.[40] Sometimes an ambitious provincial scientist like Pouchet would send a good proportion of his research to the Academy for publication in the *Comptes rendus*, although provincial loyalties in general were likely to be more diffuse.

In conclusion we may claim that the publication of the *Comptes rendus* was one of the most valuable contributions of the Academy to science in the nineteenth century. It undoubtedly increased the central importance of the Academy, not only in France but internationally. It went against the nineteenth-century fragmentation of science into separate disciplines, since it was concerned with *all* branches of science. It was institution-oriented rather than discipline-oriented and therefore reinforced in the minds of the scientific community the importance of the Academy. The *Comptes rendus* became one of the most important scientific journals in the whole world, but nothing could appear in it unless it had been read, or at least sent, to the Academy. The Academy, so often accused of elitism, pursued an open-door policy on contributions to its *Comptes rendus*. Outsiders were welcome to share the journal with distinguished Academicians, thus providing an important source of encouragement for outside talent. It represented the highest standards of a scientific journal but in an extraordinary way it combined this by means of its weekly publication with the topicality of a newspaper. The *Comptes rendus*, despite all the misgivings of the critics, gave a new lease of life to the Academy.

8. *Publication of the collected works of outstanding Academicians*

By making the *Comptes rendus* rather than the *Mémoires* the focus of the Academy publications, it may seem that the Academy had abandoned its concern with posterity in favour of topicality. On reflection this accusation would be seen to be unjust, in the first place because the *Comptes rendus*, duly bound and placed on library shelves, constituted a sort of memorial to the great activity and diversity of the Academy's work. Admittedly the bulk of material is so vast that one has to employ some principle of selectivity in order not to be confused and overwhelmed. This brings us to the second point. The Academy was very much concerned with its reputation and much of this depended on its history. It therefore needed to call attention to the work of its most illustrious members. It was consistent with the subsidiary role of the Academy as a hall of fame that it should concern itself with the publication of the work of outstanding Academicians.

Since mathematics was one of the fields in which France led the world at the beginning of the nineteenth century, it may not be surprising that the first proposal for a national edition of the work of great French scientists should have

[40] *D.S.B.*, 8, 279.

chosen Laplace,[41] sometimes considered the Newton of his age. The reputation of Laplace was kept alive by the use of his works as an annual prize for students at the Ecole Polytechnique and also by his son, who distinguished himself in a military career, ending with the rank of general. All this, together with the keen interest of the Academy, would have lent support to laws passed by the two legislative chambers in 1842 and 1843, authorising a national edition of Laplace's works. Such high level decisions were necessary in order to authorise a special budget of 40 000 francs for publication. The Academy could only give moral support but, when a second edition was required and the Laplace family offered to meet the costs, the Academy was adamant that it should appear 'under the auspices of the Academy of Sciences'.[42]

The year 1843, in which Laplace's works were first republished, happened to be the year that Dumas was elected as president of the Academy. He took advantage of this position to write to the Minister of Public Instruction, urging that similar treatment should be given to Lavoisier:

> Whilst Laplace prepared the basis of that admirable system of the world, which the country has just adopted under your aegis, his contemporary, his collaborator and his friend, Lavoisier, for his part, laid sound foundations for the principles of the science, to which all molecular phenomena are connected [namely, chemistry].[43]

Dumas went on enthusiastically to develop the comparison between Laplace and Lavoisier. But there was a further moral reason for the posthumous tribute to the founder of modern chemistry. It would be 'an expiation', which France owed to the largely forgotten genius who had suffered such a tragic death.

The Minister asked the Academy to advise him on this project and in particular to state how many volumes were required and what the expense would be. The Academy appointed a commission consisting of the sections of chemistry and physics, together with Arago. The commission in turn asked Dumas to do the necessary research on the Lavoisier material, including manuscripts, a task not completed until 1846. The next few years witnessed great political upheaval and it was not until 1861 that the government agreed to meet the cost of a national edition of Lavoisier's works. Although the impetus had come from the Academy and from Dumas in particular, when the first of the six volumes appeared in 1864, ministerial vanity required that the title page should state that the work was published under the auspices of the Minister of Public Instruction.

In the same year, 1861, in which the Minister Rouland authorised the Lavoisier edition, he also agreed to the publication of the complete works of

41 A parallel proposal to collect the works of the seventeenth-century mathematician Fermat was not proceeded with until 1891.

42 *Oeuvres de Laplace*, Imprimerie royale, 7 vols., 1843–7, 2nd ed, Gauthier-Villars, 14 vols., 1878–1912.

43 A.N., F^{17} 2278, dossier Lavoisier, undated letter received by Minister 29 April 1843.

another prominent Academician, Fresnel, and in 1867 the first volume of Lagrange's collected works appeared.[44] But we must pass on to the Third Republic before we find any further initiatives. It would have surprised the former royalist Cauchy that it should be an ideologically hostile republican government in 1882 which would provide funds to immortalise his work.[45] But the work in question was mathematics not politics, and France after the defeat in the Franco–Prussian war was more in need of national heroes than ever.

The initiative came in a letter dated 6 January 1882 to the Minister from the two secretaries of the Academy.[46] They began by listing some foreign precedents. The Göttingen Academy had published the complete works of Gauss, the Berlin Academy was beginning to publish the works of Jacobi and only a few weeks previously the Norwegian government had proudly distributed the work of their compatriot, the mathematician Abel. It was natural for French mathematicians to look for someone comparable and since justice had already been done to Laplace and Lagrange, Cauchy was the outstanding French mathematician of the nineteenth century on whom their choice fell. What they wanted from the Minister was a large subscription which he could use to distribute to educational establishments throughout France and at the same time would produce a reasonable income. The Minister obliged by taking out a subscription for 300 copies of the twenty-five volumes at twenty-five francs per volume, thus making the project economically viable.

In the twentieth century the Academy has continued to sponsor the publication of the complete works of some of its more outstanding former members. One instance which falls within the period of the book is the publication in four volumes of the work of the mathematician Charles Hermite.[47] It is fitting that the Academy should not only follow respectfully its distinguished members to their graves and deliver suitable eulogies but also construct appropriate memorials. Statues are often useful reminders of past glories but for specialists the republication of the work of a few masters, otherwise inaccessible, has a major contribution to make in establishing a tradition, based not on vague sentimentality but on concrete achievement. Academicians were reminded of the honour of their positions and others were encouraged to aspire to such honour.

There were also other publications of the Academy. Since 1871 the archivist Maindron was urging the Academy to use Montyon surplus funds to publish the minutes of the Academy meetings from 1795 to 1835, the year of publication

[44] *Oeuvres de Lagrange*, Gauthier-Villars, 14 vols., 1867–92. For Academy correspondence with the Minister about the republication of the works of Lagrange, see A.N., F^{17} 3247.

[45] *Oeuvres complètes d'Augustin Cauchy, publiés sous la direction scientifique de l'Académie des sciences et sous les auspices de M. le ministre de l'Instruction publique*, Gauthier-Villars, 1st series, 12 vols., 2nd series, 10 vols., 1882–1903.

[46] A.N., F^{17} 3244.

[47] *Oeuvres de Charles Hermite, publiés sous les auspices de l'Académie des Sciences*, par Emile Picard, 4 vols., 1905–17.

of the first volume of the *Comptes rendus*.[48] In fact it was not until 1910 that the Academy took on this task and then it was interrupted by the first world war.[49] Meanwhile Maindron himself had taken on the responsibility of describing the Academy's prize system, providing tables of prize winners.[50] This task was continued by a later secretary, Pierre Gauja.[51] But these works, invaluable though they may be for historians of science as works of reference, did not make much impact on the science of the time.

For practical contributions to science we might cite at the beginning of the century, *Base du système métrique decimal* (3 vols., 1806–10) or, towards the end of the century, *Recueil de mémoires, rapports et documents relatifs à l'observation du passage de Venus sur le soleil* (3 vols., 1877–85), both of which provided full details of the respective expeditions, laboratory work and calculations. We might also cite the numerous reports of Academy commissions, but it was only in exceptional circumstances that such reports would be published by the Academy itself.

Finally the Academy published the funeral speeches and the longer and more considered *éloges*, to which in principle all former Academicians were entitled. These booklets would help to put a seal on an Academician's reputation. The eulogies would be partly biographical but they would also have a moral dimension. They would tend to present the deceased as a kind of hero, thus conforming to the ideal to which the Academy aspired. One can hardly grudge the Academy a little mild propaganda. Yet the real apotheosis of the nineteenth century Academician lay less in the *éloge* than in the *Oeuvres*. To have one's works published in a national edition was to attain true immortality.

[48] A.S., Carton 38, dossier, 'Notes sur les publications de l'Académie', letter of 9 October 1871 from Maindron to Dumas.

[49] *Procès-verbaux des séances de l'Académie des Sciences, tenues depuis la fondation de l'Institut jusqu'au mois d'aout 1835, publiés conformément à une décision de l'Académie par MM. les Secretaires perpetuels*, 10 vols., Hendaye, 1910–22.

[50] E. Maindron, *Les fondations de prix à l'Académie des Sciences: 1714–1880*, 1881.

[51] P. Gauja, *Les fondations de l'Académie des Sciences, 1881–1915*, Hendaye, 1917.

— Chapter 9 —

AN ACADEMY UNDER GOVERNMENT CONTROL

The Royal Academy of Sciences will continue to remain under the protection of the King *and will receive his orders* through the particular Secretary of State, to whom his Majesty assigns the task.

(Regulations of 1699 from Fontenelle, *Histoire du renouvellement de l'Académie des Sciences*, my italics.)

It is by science that we have been vanquished [in the Franco-Prussian war]. The reason for this lies in the regime which has oppressed us for 80 years, *a regime which subordinates men of science to politicians and administrators*.

(H. Sainte-Claire Deville, *C.R.*, **72** (1871), 238, my italics.)

Since the first days of August [1914], our Academy has only had one thought: to help the government in the defence of the motherland and of liberty.

(Paul Appell, speaking as President of the Academy, *C.R.*, **159** (1914), 824.)

1. *Government control of the Academy?*

The relationship between the Academy and the government was always a rather delicate one. Although Condorcet had used the expression 'fonctionnaires publiques' to describe members of the Institute which he planned,[1] when the National Institute came into being in 1795 its members were not civil servants. Yet in so far as the National Institute was a government-sponsored body, its members obviously had a certain connection with the state, both in fact and in the public mind. They represented 'official', that is to say, government-sponsored science in a way the Royal Society never did, being 'Royal' in name only. The seal was set on the official status of members of the Institute by their receipt of a small salary, even if it could more properly be described as an honorarium.

'Official science' has a disturbing connotation for many people in the Anglo-Saxon world. It suggests not only that the science is paid for by the government but that it is government approved. Many would feel that this interferes with the essential freedom of science. For example, if science is to flourish it must be open to change. Governments too may change but for totally different reasons. Too close an association between science and government might result in a situation in which a government had an interest in maintaining stability in all areas under its control at a time when new experimental work might demand some

[1] Caritat de Condorcet, *Oeuvres* (12 vols., 1847–9), vol. 7, p. 519.

important change in scientific theory or even in scientific organisation. Equally, at a time when governments were changing it was often the scientists (or a large number of them) who wanted to be left undisturbed to continue their previous studies.

The association of science and government took place all the more easily in France because of the early centralisation of the state. The revolutionary and Napoleonic regimes further increased this centralisation based on Paris. In early nineteenth-century England it is inconceivable that a government minister in London should have the slightest interest or influence over the speculations of an Oxford natural philosopher, a Cambridge mathematician or an Edinburgh geologist. But when the leading intellectuals and representatives of science were concentrated in the city which was the seat of government and their positions depended on government approval, at least implicitly, this implied a greater degree of contact and even control.

It could be argued that mere membership of the Academy imposed certain constraints and therefore reduced the individual freedom of a scientist. So much was made of the honour and the privilege of membership that one tends to forget the other side of the coin, the obligations that membership imposed. If there was control by government, one would tend to suppose that this would be purely at the administrative level through the appropriate ministry. There is, however, evidence of intermittent attempts at political control as well, when governments of different complexions expected compliancy from the members of the Academy.

Although the rules of an institution are written down, it is not often that the basic assumptions of any society are laid out clearly and explicitly in any generally circulated work. We have been exceptionally fortunate in discovering a letter, undated but probably written in 1804 or 1805, composed by the surgeon Pelletan (elected to the First Class in 1795), which he had privately printed and addressed to all the members of the Institute.[2] In this he accepted that men of letters in 'forming a corporation under the authority of the government, lose their political liberty', but he argued that in exchange they had gained valuable protection. Pelletan felt that by accepting their honoraria, members of the Institute had surrendered their political independence and he concluded logically but dramatically that 'the man of science or the man of letters who wishes to enjoy ideal and unlimited liberty should renounce Academies and [government] places'. For his part he was content to concentrate on 'true science' which, he claimed, did not disturb the social order.

We have made a distinction between political control and administrative control. Although governments of diverse political complexions came to power in the nineteenth century, administrative control was always in the hands of a

[2] M. Pelletan (1747–1827) membre de l'Institut National des Sciences et des Arts, à ses Honorables Collègues, B.M.733.g.18 (62). A footnote explains: 'Cette lettre, distribuée seulement aux membres de l'Institut, n'a été imprimé que pour éviter l'embarras des transcriptions'. The copy has the personal signature of Pelletan.

specific ministry. At the foundation of the Institute it was responsible to the Ministry of the Interior, but later the relevant ministry was that of Public Instruction.[3]

In a sense all the activities of the Academy required the approval of the Ministry, but most of this was soon reduced to routine bureaucratic procedures and purely formal approval. Thus government control did not have an implication so much for the freedom of science as for the possibility of new initiatives. Once well established a particular pattern had to be followed and it is remarkable that the nineteenth-century Academy was not more static and moribund. It required considerable individual and collective enterprise and sometimes a certain deviousness to introduce improvements into the system.

A reminder that the Institute was ultimately under government control came in 1803, when Bonaparte decided on a general reorganisation. His principal concern was not the First Class but the Second, which included political scientists who were potential critics of his government.[4] Some heads of a totalitarian state would have been content simply to expel their enemies from the Institute, if not from the country. Certainly for Bonaparte the whole idea of social science was potentially subversive and he was particularly hostile to a group of philosophers and political theorists within the Institute, described as the *idéalogues*. He therefore abolished the Second Class completely and, after a few expulsions, redistributed its remaining members, together with those of the Third Class, into a new Second Class concerned with literature and a new Third Class dealing with history and ancient literature. A Fourth Class was to be devoted to the Fine Arts. The First Class was almost unaffected except that it received from the former Second Class three geographers whose interests could be interpreted as 'scientific'. As previously mentioned they were to form a new (half) section entitled 'Geography and Navigation'. When in the 1860s it was thought desirable to increase the membership of the section from three to six, at least the appropriate minister had the tact to consult the Academy before reaching a final decision.

Although it might appear that the relationship between the government and the Academy was a one-way affair, a closer analysis might suggest that it was based on an exchange of services. On the one hand the Academy was available to the various ministers to advise on practical problems and to pass judgement on the authority of claims relating to scientific matters. Occasionally it could be used as a platform for some special government announcement. Thus it was at the request of the Ministry of the Interior that the Academy was chosen as the platform for the announcement of Daguerre's photographic process.[5]

But the Academy also expected to receive certain things in exchange. Apart from the payment of an honorarium, there were other less obvious benefits for members. The Academy had the right of nomination to many senior positions

[3] For a fuller discussion of different ministries, see Section 9 of this chapter.

[4] F. Picavet, *Les idéologues*, 1891.

[5] *C.R.*, 9 (1839), 227.

in higher education and administration, notably at the Muséum d'Histoire Naturelle, the Ecole Polytechnique, the Collège de France, the Ecole de Pharmacie, the Bureau des Longitudes and certain chairs in the Faculties of Science and Pharmacy. The Academy needed the permission of the Minister of Public Instruction to spend money in certain accumulating funds, notably the Montyon fund, although this was to become little more than a formality.[6] In the second half of the nineteenth century, when increasingly large sums were bequeathed to the Academy as prize money, the authorisation of the Minister was needed before such bequests could be accepted by the Academy. As the official body of science, the Academy could expect to be consulted on important matters affecting science, although it was not always consulted on educational matters. A link with the government was sometimes useful to the Academy in planning international research, since the French government could represent its interests to foreign governments. Finally, when there were expeditions organised by the French government the Academy could sometimes send its own representatives. On the other hand the Academy came to be expected to draw up *Instructions* for all major expeditions and this may have been sometimes an onerous duty rather than a privilege. On the whole, however, the Academy had much to expect from government, and if the scientists sometimes seem timid in their approach to government it is probably partly because they modestly felt that they were receiving more than they were giving.

2. *Interference in elections*

The most direct involvement of the government in the affairs of the Institute came in the initial nomination of members in 1795. But as the members nominated constituted only one third of the membership and most of these had previously been elected as members of the respective Academies under the old regime, few objections were raised. Moreover, these nominated members were then completely free to nominate the remaining members of the Institute. Members expected to remain such for life, but interference occurred in a few cases where the government suddenly terminated membership for political reasons. The other main interference came when the government refused to accept the result of Academy elections.

The first political victim was Lazare Carnot (1753–1823), who had taken command of a large undisciplined French army during the period of Jacobin power but later fell foul of new political forces. Carnot's repressive military activities had made him many enemies. After the coup of *fructidor* (1797) he fled the country and the Directory declared that he had forfeited his place in the mechanics section of the First Class. On 17 October 1797 the First Class obediently declared his place vacant, and it is ironic that in the ensuing election in December it should have been General Bonaparte, a protégé of Carnot, who

[6] Maurice Crosland, 'From prizes to grants... ; the Montyon legacy', *Minerva*, 17 (1979–80), 355–70.

should be elected. Carnot returned to France after Bonaparte's seizure of power but had to flee again at the Restoration. His second expulsion from the Institute in 1816 was because the Royalist government could not overlook that he had been one of those who, some twenty years previously, had voted for the execution of Louis XVI. The other regicide expelled by Louis XVIII in 1816 was Monge. Again predominantly a man of science, a mathematician and administrator, Monge had made the mistake of getting involved in revolutionary politics. He, like Carnot, was to die in exile.

The period of the Restoration was a time when the independence of the Academy was most threatened. We see this first in the refusal of the government to confirm Academy elections, where the person elected was someone with strong Bonapartist or Republican associations. The crime of the mathematician Fourier (1768–1830) was to have served as prefect under Napoleon, first at Grenoble and then at Lyons during the Hundred Days. Under the Restoration Fourier hoped to live in Paris and qualify for membership of the Academy on the strength of a powerful mathematical reputation, which included winning a prize competition in 1811. There was no immediate vacancy in the mathematics section but the Academy elected him in May 1816 as one of the new category of *Académiciens libres*. The King refused to confirm the election and the leading officials of the Academy took the unusual step of addressing a protest to the Minister of the Interior.[7] However it could be argued that the section of *Académiciens libres* had been created to reintroduce the old nobility into the Academy and Fourier was being excluded not as a Bonapartist but as a bourgeois. The test came in the following year. The death of Rochon in April 1817 created a vacancy in the physics section, a reasonably appropriate opening for Fourier, considering that his most famous work had been a mathematical study of the conduction of heat. At the election in May, Fourier obtained forty-seven out of fifty votes and this time the King approved the nomination. Only five years later he was to be elected as one of the two secretaries of the Academy.

A second case did not resolve itself so quickly. In the election of 10 November 1823, the mathematician Hachette (1769–1834) was elected to a vacancy in the mechanics section. A month later the Academy received a letter from the Minister of the Interior stating that 'the King had not considered it appropriate to confirm the election of M. Hachette'.[8] The trouble was that although Hachette was admirably qualified on academic grounds, he was considered to be disqualified politically, being well-known since the time of the Revolution as an ardent republican. There was nothing the Academy could do until after the Revolution of 1830. When a further vacancy arose in the mechanics section in

[7] For the text of the letter and a fuller discussion, see John Herivel, *Joseph Fourier, the man and the physicist*, Oxford, 1975, pp. 122–4.

[8] *P.V.I.*, 7, 604 (15 December 1823).

1831, Hachette received thirty-seven out of forty-two votes cast and was declared elected. This time there was no problem about confirming the election.

Although these are the outstanding cases of political interference during the Restoration, they are not the only ones. One can see a strong element of political partiality in the documents relating to the election of other Academicians in this period. The Minister of the Interior in his report to the King did not always limit himself to reporting the Academy's vote. Sometimes he included a brief biographical statement. Thus when Beudant was elected in 1824 soon after the accession of Charles X, the King was told of the positions held by the candidate and also of his very successful book. The report concludes: 'This consideration seems to me such as to justify the choice of the Academy...'[9] Similarly, when Antoine Becquerel was elected in 1829, the Minister felt it necessary to discuss the successive ballots held in the Academy and to say something about Becquerel's scientific eminence before he could bring himself to recommend the Academy's choice.[10] The Ministerial correspondence did not normally comment on the politics of candidates, but occasionally one finds such considerations in correspondence to the Minister. Thus in a letter in support of the candidature of the botanist Auguste de Saint-Hilaire written in March 1830 (just four months before the Revolution which overthrew Charles X), it was said that if he entered the Academy he would bring with him 'religious and monarchial principles in which he has been brought up and which are those of his family'.[11] However the fact that Saint-Hilaire was successful had more to do with the fact that the botany section of the Academy had placed him first on their list rather than with extra-scientific considerations.

In case anyone should think that the Academy was singled out for special disfavour during the Restoration, one might point out that the treatment of the Academy was mild compared with the Paris Faculty of Medicine. This was closed down in 1822 after student disturbances. Several liberal and Bonapartist professors were expelled and their positions given to professors of known Royalist sympathies.[12] The Academy escaped lightly.

Later in the century under the Second Empire there was one Minister of Public Instruction, Fortoul, who made himself particularly unpopular with all the Academicians by regular interference in their affairs. In 1854 he asked the Academy of Sciences in future not only to forward to him the name of the successful candidate in an election, but also an account of his qualifications and publications.[13] This could be interpreted as no more than a slight tightening up of the bureaucracy. Alternatively it could be seen as a refusal to leave the

[9] A.N., F[17] 3578, letter to King, 20 November 1824.

[10] *Ibid.*, letter to King, 23 April 1829.

[11] A.N., F[17] 1543, letter from Bellenave, the candidate's cousin to the Minister, 8 March 1830.

[12] L. S. Jacyna, 'Medical science and moral science', *History of Science*, **25** (1987), 111–46 (pp. 134–5).

[13] Comité secret, 6 February 1854, quoted in *Règlements intérieurs de l'Académie des Sciences*, p. 40. See also A.S., *Copie de Lettres 1841–60*, p. 207 (30 August 1856).

judgement of merit purely in the hands of the professionals. Something which worried the Academy more, however, was the Minister's attempt to reimpose the system of annual reports to the government. The Academy considered this a retrograde step and argued that the reports presented in the early part of the century had been superseded by the *Comptes rendus*. Apart from technical reports required from time to time by the government departments, it insisted that the Academy could not be given special tasks – each member was free to choose his own research.[14]

In 1855 a new section of ten members was added to the Academy of Moral and Political Sciences and these members were nominated by the Minister, violating the principle of peer selection by election. One possible reason for this interference is that the government had not been pleased by a series of recent elections in the Académie Française which had voted in a number of figures critical of the new Napoleonic regime.[15] Yet Napoleon III distanced himself from the initiatives of his minister, who was perhaps trying too hard to please his master and advance his own career. This episode is discussed more fully in Chapter 12.

3. *Government control over individual scientists*

Some governments were concerned not only about the activities of the Academy but also about those of individual members. An extreme example is the concern shown by Napoleon in 1805 that the astronomer Lalande (1732–1807), a member of the Institute, had allowed his name to be linked with an atheist publication.[16] Lalande had in fact contributed to a supplement to the *Dictionnaire des athées*. But atheism was contrary to state policy since the signing of the Concordat, by which Bonaparte had brought about a reconciliation with the Roman Catholic Church in order to increase his own political power. Napoleon asked the Minister of the Interior to summon the presidents and secretaries of the respective Classes of the Institute to tell them of his concern about this supposed threat to public order. A special meeting of the whole Institute was convened for 26 December, at which the aged Lalande agreed to conform in future with the Emperor's wishes. In a later letter Napoleon wrote that he expected the Institute to do everything the government asked.[17] Although the context was literature, the letter reveals an extreme view taken by Napoleon of the duties of all members of the Institute. In practice literature suffered the greatest restriction during

[14] 'L'Académie...n'a pas de travaux qui lui soient spécialement confiés...elle n'en a d'autres que ceux auxquels se livre spontanément chacun de ses membres...' Comité secret, 13 October 1856, quoted in *Règlements intérieurs*, p. 85. Note that the Ministerial request of 14 April 1855 did not receive this reply until October 1856.

[15] Emile Mireaux, *Le coup d'état académique du 14 avril 1855*, 1963, A.S., Carton 32, dossier A.S.M.P.

[16] Napoleon, *Correspondance*, vol. 11, No. 9562. A full account of this episode is given in A. Aulard, 'Napoléon et l'athée Lalande', Part 10 of his *Etudes et leçons sur la révolution française*, Série 4, 1904.

[17] *Ibid.*, vol. 15, no. 12415.

Napoleon's regime, whereas science was much freer. The rationale for this was that laboratory science at best could produce useful results or even results which would add to the glory of the regime, but at worst could do no harm.

A different example of interference relates to the Restoration. In October 1824 the mathematician Legendre, then aged seventy-two, suddenly found himself deprived of a government pension of 3000f which supplemented his modest honorarium as a member of the Institute. Although the reason of a change of administration was given, Legendre alleged that the real reason was that he had failed in a recent Academy election to support the ministerial candidate, Binet.[18]

It was probably through finance that the government could exercise most influence on Academicians. Yet they had only a modest honorarium which was the same for all[19] except the secretaries. One could conclude that there was therefore a minimum of pressure. Alternatively, one can interpret the acceptance of money – even a token sum – as implying an obligation. As the sum involved was too small to allow an Academician to live comfortably, it was natural that he should look for supplementation. Since many of the posts he might fill would be under the control of different ministers, the Academician might be subject to government pressure on several fronts. In practice this usually meant uncritical acceptance of the government of the day. All civil servants – and members of the Academy – had to take an oath of loyalty. Even with a drastic change of government, most scientists did not find such an oath difficult. Only the scruples of Cauchy, after the Revolution of 1830, sent him into a self-imposed exile.

The case of Raspail suggests some political control. In 1833 his friend Etienne Geoffroy Saint-Hilaire, then president of the Academy, suggested that his microscopical researches and his treatise on organic chemistry (partly written in prison) might deserve a Montyon prize of 10000 francs. Raspail, however, was known for his extreme and outspoken republicanism as much as for his contributions to medicine and science and the Minister of Public Instruction, Guizot, is said to have put pressure on the Academy not to favour Raspail in this way.[20] According to the most extreme report of the interview of Guizot with Geoffroy Saint-Hilaire, the Minister said 'I forbid you to line the rioters' pockets'. The official version of the interview published in the *Journal des Débats* was more restrained. It admitted that the question of a possible award to Raspail was raised: 'The Minister emphatically replied that the Academy alone was competent to decide on a purely scientific matter and that he had no intention either of opposing it or of intervening in any way. Since that day the Minister has heard no more about this matter'. In fact Raspail was such a persistent critic both of the government and of the scientific establishment that he felt that the

[18] Institut, HR5*, tome 61, nos. 25, 26.

[19] A small fraction of the honorarium was retained and paid according to attendance.

[20] Dora B. Weiner, *Raspail, Scientist and Reformer*, New York, 1968, pp. 126–7. See also a collection of writings of Raspail, ed. David Ligou, *François-Vincent Raspail ou le Bon Usage de la Prison*, etc., Paris, 1968, pp. 454–61, 'Advertissement historique'.

prize was intended to buy him off. He said that he would accept the prize if awarded, but without any conditions.

Only a few weeks before the Academy's offer of a prize he had been asked to abandon his hostile political activities and give all his attention to science.[21] In the circumstances a friendly letter from the Academy, offering a substantial sum of money, could be interpreted as a bribe. Probably the combination of Raspail's proud independence and government pressure were together sufficient to deter the Academy, since in the end the prize was not awarded to Raspail. Nevertheless he subsequently made the most of the incident, pointing out, for example, that he had not asked for a prize; it had been the Academy which had come to him. Indeed, the whole incident is as much about academic politics as it is about ordinary political independence. It is significant for example that the offer was made in the year after the death of Raspail's old enemy, Cuvier. A fuller account of Raspail's career in relation to the Academy is given in Chapter 10.

4. *Independence from government*

Although Academicians were usually proud of the government recognition of their institution, they also sought some independence. Sometimes this might be no more than a gesture. Thus the Academy was in principle obliged to meet every Monday throughout the year. It accepted this obligation but, on a few special occasions such as the death of a prominent Academician like Laplace in 1827, it immediately adjourned its meeting. Later in the century the Academy automatically adjourned whenever one of its members had died within the last day or two. Not only was the Academy showing respect for the dead, it was also demonstrating its independence. In the final analysis the spirit of solidarity among its own members may have meant more to it than the letter of a government regulation. Yet the main claim of the Academy to independence lay in the specialised nature of its studies. It alone had the authority to pronounce on matters scientific. It alone could decide who among its members were the relevant experts to deal with a particular problem. Thus when in 1796 the Minister asked it to examine a problem and suggested the names of suitable experts the Academy immediately rebuffed him, pointing out that although it accepted the obligation to study the problem only it could nominate the members of the Commission.[22]

Yet pronouncement was one thing and action another. The freedom of action of the Academy suffered considerably, particularly during the first generation, by a small and rigidly controlled budget. Only in the 1820s, with the Montyon legacy, did new possibilities of freedom of action begin to seem feasible. In the early years there are a number of instances when the Minister went out of his way to limit the scope of independent action of the Academy.

[21] F. V. Raspail, *Nouveau système de chimie organique*, 2nd edn, Paris, 1838, vol. i, pp. xix–xx.

[22] *P.V.I.*, 1, 35.

Some of the correspondence between Montalivet, Minister of the Interior from 1809–15, and the Institute illustrates the problem of demarcating the role of the government and the role of the official body of science as a patron of science. Montalivet criticised the First and Second Classes of the Institute for having spent 7500f. on subsidies for publications by different authors. He continued:

> The laws organising the institution charges it with assembling those discoveries which tend to perfect the sciences and the arts and to contribute to the progress of enlightenment; but they do not confer on it the responsibility of distributing favours other than the prizes founded by its regulations.'[23]

If, he continued, the Institute felt that certain authors deserved financial help, it should inform the Minister, who had a special allocation in his budget for such matters. The Minister obviously resented the mild independence shown by the Institute in making its own decisions within its modest budget. In its reply the Institute tried to justify the expenditure. The First Class, for example, had been in the habit of meeting the cost of engraving plates for certain scientific works. The strictest financial control was always exercised and the term 'favour' used by the Minister obviously rankled and was rejected:

> Such use cannot in any way be regarded as 'favours' in the sense the Minister seems to give the term.[24]

In its formal reply, however, the Institute had to be content to remind the Minister that its constitution used the formula 'the advancement ("*perfectionner*") of the sciences and the arts', a phrase obviously capable of a variety of interpretations. Certainly the Institute wanted to take a direct part in the patronage of learning.

As a further example of relations between the Minister and the Academy, let us take a case where the Academy seemed to be in the ascendant. Although one could be misled by the use of literary conventions in letters to suppose that the Academy was servile in its relations with the government, there are occasionally refreshingly direct letters sent to the Minister of Public Instruction. Here is the main text of a letter sent by Elie de Beaumont in his capacity as secretary, to the Minister in 1855.[25] It asks for the necessary authorisation to use the surplus of the Montyon funds in the way the Academy wished:

> Dear Minister,
> The Academy of Sciences has decided (sic):
> 1st at its meeting of 29 October 1855 on the proposal of the combined sections of mineralogy and zoology that a sum of 2000f. should be granted

[23] Letter of 14 September 1813, 4th dossier in A.S., Boite: Commission [centrale] Administrative, 1795–1831.

[24] 'Note pour servir à la résponse à faire au Ministre', *ibid.*

[25] A.S., *Copie de Lettres, 1841–60*, p. 179, letter of 27 November 1855.

to M. Marie Boule (?) to help and encourage him in his studies of geology and palaeontology...

2nd... that a sum of 4000f. should be awarded to [M. George Ville], namely 2000f, to reimburse him for the costs of the experiment which he repeated at the Museum of Natural History in the presence of the Academy commission, and 2000f. to help him in the continuation of his research [on the assimilation by plants of nitrogen from the air]...

3rd... that a sum of 8000f. should be devoted to the acquisition either of chemicals or of natural history specimens [from the universal exhibition]. These funds would be taken from the disposable surplus of the Montyon fund. These decisions of the Academy cannot be put into effect until Your Excellency kindly gives his authorisation. But the Academy has always counted on the interest which you have in the sciences and it hopes again for the support which you have already given in similar circumstances.

Obviously precedent was a useful weapon in persuading any Minister that his signature was only a formality. Under the Third Republic the Academy was writing to the Minister for authorisation of the use of surplus (Montyon) funds in the way the Academy thought suitable 'in conformity with your decision of 13 April 1874.'[26] Effectively all the Academy had to do was to fill in a standard application form.

5. *Reports to the government*

Articles 40–1 of the law of 15 *germinal* year 4 (4 April 1796) stipulated that each autumn in the republican month of *fructidor* the secretaries of each Class would draw up a report on the work of the respective Classes to present to the Legislative Body. On 1 *fructidor* year 4 (18 August 1796) the First Class agreed on the form to be taken by the report. It would contain a review ('une notice') of all the memoirs presented by members and corresponding members ('associés') as well as a reminder of the various reports which the government had asked for and finally, they said, a report of 'all deliberations contributing towards the progress of the sciences'.[27] The same form was adopted for the report prepared to submit to the Legislative Assembly exactly one year later, on 1 *fructidor* year 5, together with a draft budget for the following year.[28] It was actually on the morning of the first day of the next republican year, 1 *vendémiaire* year 6 (22 September 1797) that the respective Classes of the Institute went in turn to the two elected chambers, the Council of Ancients and the Council of 500, to present their report.[29] A printed version of the report was later distributed to members of the First Class.[30]

But the work of the respective Classes of the Institute was of potential interest not only to the government but also to the public, thus providing another dimension to public accountability. The 'notice des mémoires', given respectively by Prony and Lacépède to cover the mathematical and natural

[26] A.S., *Copie de Lettres, 1875–1902*, p. 12, letter of 26 June 1875.

[27] *P.V.I.*, I, 85 (cf. p. 71).

[28] *Ibid.*, 260.

[29] *Ibid.*, 277.

[30] *Ibid.*, 308.

sciences in July 1796,[31] soon became transformed into a 'report' on the memoirs and finally a report on the work of the First Class. After the coup d'état of Bonaparte of 1799 the system of reporting annually to the government continued. Of special note is Napoleon's demand of 1807, when he asked for a report on the progress of science since 1789.

Some principle of accountability was obviously appropriate for the Institute, first in the literal sense that since it spent government money it was required to produce annual accounts. But a more important meaning of 'accountability' is to account for one's actions. What had the First Class, for example, achieved? One might interpret the demand for a detailed report as typical of an authoritarian government, but it could also be argued that a request for a report showed genuine interest in what was being done. In the Napoleonic regime we have a combination of great centralised power with a claim by the head of state to be concerned with the progress of science in France. In the Consular decree of 1802 there was also the opportunity given to the Institute to make representations to the government:

> Article 3. At the same time [as the National Institute presents its report] it will inform the government of its views concerning the discoveries, whose applications it considers may be useful to the public services, the help and encouragement which the sciences, arts and letters need, and the improvement of methods used in the different branches of public education.[32]

Considering the bureaucratic inertia from which the Institute tended to suffer an opportunity to propose new initiatives was of considerable potential value. In practice, however, little came of it.

The two famous reports drawn up in 1808 by the two permanent secretaries of the Academy on the progress of science in France since 1789 contain much useful information. But they were also to some extent political documents, intended to ingratiate the First Class with the head of state. It was true that a great deal of Napoleonic science was pure science and Cuvier seized on this to flatter Napoleon. He suggested that an ordinary prince might have insisted on pursuing utilitarian aspects of science whereas the Emperor, who had had a mathematical education, could appreciate a more theoretical approach which might only be useful later.[33] Cuvier presented the members of the First Class as hardworking men, occupied with the enlightenment of their colleagues and raising the dignity of mankind.[34] At the same time his speech conformed to the standard panegyric of Napoleon by describing him as 'the Hero, who has carried military and political glory beyond all bounds...'[35] The flattery offered by the other secretary Delambre, was more restrained. It consisted in calling special

[31] Institut, HR7, Tome 3, no 6, Séance publique du 15 *messidor* an 4.

[32] Arrêté du 13 *ventose* an 10, Aucoc, *L'Institut de France*, pp. 91–2.

[33] *Rapport historique sur les progrès des sciences naturelles depuis 1789*, 1810, p. 15.

[34] *Ibid.*, p. 16.

[35] *Ibid.*, p. 3.

attention to the translation from Italian to French of a work of geometry which Bonaparte as a young general had brought to the attention of the mathematicians of the Institute.[36] Delambre, however, did not neglect also to flatter Laplace, a favourite of the Emperor.[37] Yet in all this it would seem that the government had much less effect on the actual content of the science done than on its presentation. In other words, provided fine speeches were made on the appropriate occasions, the men of science were largely free to pursue their own research interests.

6. *Public rhetoric on a political tightrope*

On public occasions speakers on behalf of the Academy or the Institute could be relied upon to use the right language and to select historical precedents consonant with the current regime. Thus it is not surprising that Cuvier began an address to Louis XVIII in the first few months of the Restoration with an appropriate reference to Louis XIV, the original patron of the Academy.[38] Similarly nothing went down better under the Second Empire than a flattering reference to the First Empire and Napoleon I. Thus the *éloge* of the mathematician Puissant delivered in 1869 made much of the year of his birth, 1769, which had witnessed the birth of many other brilliant men, including Napoleon. In the same way the Second Republic brought about a flood of references favourable to the First Republic. At the public meeting of 25 October 1848 the Institute was presented as the *republic* of letters.[39] There are several references to the Convention, which had been instrumental in founding the Institute. It was said to possess a *fraternal* spirit and to practise *equality*.

Yet while being shamelessly selective in the choice of antecedents, the Academy had to be careful not to bring itself fully into the political arena. Its claim was therefore that it was above politics and political change. With every major change of regime – and one thinks principally of 1815, 1830, 1848 and 1871 – there was a very real danger that the new government would identify the members of the Institute with the previous government, since it was officially an agent of the state and always carried an establishment aura. The way out of the dilemma was for the Institute to declare in the most outspoken way its loyalty to the new government (of whatever complexion) but to claim at the same time that really the Institute never descended into the arena of practical politics.

While the Academy often seems to have gone too far in ingratiating itself with new governments, it should not be overlooked that it could occasionally play a small part in toppling an old government. The part played by the Institute and

[36] *Rapport historique sur les progrès des sciences mathématiques depuis 1789*, 1810, pp. 5, 47.

[37] E.g. *Ibid.*, pp. 169–70.

[38] *Première séance annuelle des quatre Académies*, le 24 avril 1816, Cuvier, *Eloges*, vol. 3, p. 251.

[39] *Séance publique des cinq Académies*, le 25 octobre 1848, Discours d'ouverture.

the Academy in preparing the way for the Revolution of July 1830 should not pass unmentioned. In April 1830 the engineer Girard, as president of the Institute, took advantage of the public meeting of the four Academies to speak out strongly in favour of free expression of thought. Although much of the speech was delivered in a heavy patriotic vein, Girard bravely spoke out for 'a free press, the first of our public liberties'. He also told his fellow Academicians that, as they had liberty and independence of choice within the Institute, they could express their beliefs by voting appropriately in elections to the Academies. The implication was an unusual politicisation of Academy elections.

The Academy was very soon to be confronted with such an election. The death of Fourier in May 1830 produced a vacancy for the key position of secretary for the division of mathematical sciences. The radical Arago was persuaded to stand for election, the other two candidates, Puissant and Becquerel being political nonentities.[40] In the election of 7 June Arago received an overwhelming vote of thirty-nine out of a total of forty-four votes. One of Arago's first public duties was to deliver an *éloge* of his old friend Fresnel, who had recently died. The *éloge* was to be delivered at the annual public meeting of the Academy, planned some time in advance for 26 July. Only the day before, four Ordinances had been issued to meet the state of increasing political turmoil. These dissolved the political assemblies which had just been elected and further restricted freedom of the press. Nevertheless, despite the pleas of Cuvier, Arago refused to delete some explicitly political references in his *éloge*.[41] He spoke of Fresnel's independent political stance during the brief return of Napoleon in 1815. He spoke of a royalist minister questioning Fresnel on his politics before refusing him the position of examiner. Such references to politics in the very delicate political atmosphere in Paris in late July 1830 were received by the public with great attention and applause. Two days later the Revolution broke out. In the new Chamber of Deputies Arago was to be elected as a republican, a political position for which he was well known.

Arago's political indiscretion of 1830 was provoked by an extreme situation which threatened basic freedoms of speech and of the press and could obviously not become the norm for the Institute. The only prudent position was a neutral one and in this connection we may observe the stance adopted by the spokesman of the Institute after the next revolution, that of 1848. At the annual public meeting on 25 October 1848 Burnouf, from the Academy of Inscriptions, spoke on behalf of the whole Institute. He spoke of the recent revolution but suggested that real *savants* were 'beyond favours of government and the applause of the crowd'. He continued:

> Placed far from all action, in the calm and elevated sphere of theory, the Institute has not descended from the heights where the genius of the

[40] *P.V.I.*, 9, 452, 455.

[41] Arago, *Oeuvres*, 2nd edn, 1865, *Notices biographiques*, vol. i, pp. 103–85.

> Revolution placed it. It has remained what it had been ordered to be, the scientific and literary centre of France. The face of things has changed more than once in our country; more than once governments have been able to touch an institution [i.e. the Institute] very close to them; but the changes which modified the original organisation have never touched its basic principles; the unity of the Institute is as intact as the independence of the Academies, of which it is composed.[42]

The speaker went on to claim that since the foundation of the Institute,

> neither independence nor zeal have been lacking. May it proceed therefore with constancy along the path laid down by its founders... It may also hope in the future with the same security that it enjoyed in the past. It will have accomplished its duty, for it will have remained faithful to the scientific genius of France.

7. *Political commitment or neutrality?*

In order to understand the relation between the Academy and the government, we may consider in turn two opposite extreme possibilities. In the first scenario the sympathies of the Academy would change to conform with any change of government; in the second case the Academy would stand completely aloof from political change.

One of the problems of the first scenario is that there were too many political changes in nineteenth-century France and many of them were quite drastic changes. There was a great deal of difference, for example, between the Republican ideology of the Directory (1795–9) and Napoleon's rule, especially under the Empire (1804–14). The Restoration of the Bourbons in 1815 produced considerable strain in the loyalties of individuals. Further problems of political re-thinking were provided by the proclamation of the Second Republic in 1848, the Second Empire in 1852 and the Third Republic in 1870, representing successively moves to the left, to the right and again to the left. Probably the best example of political interference in the membership of the Academy was under the Bourbon Restoration, when Monge was expelled as a Bonapartist and royalists like Cauchy were nominated as members. The greatest change in the composition of the Academy came in 1816 through the integration of a whole new section of *Académiciens libres* largely drawn from the nobility. Under Louis XVIII and Charles X, therefore, the Academy could be seen as a significantly royalist institution and not simply in name. Royalist candidates for election might think that they now had a much better chance of being elected, although they would have to overcome a hard core of resistance from those with republican and Bonapartist sympathies within the Academy.

For scientists in a world of bewildering political changes, a stance of political neutrality could well have been a safe refuge. If required, most scientists were quite willing to sign a pledge of loyalty to whatever government happened to be

[42] *Séance publique des cinq Académies*, le 25 octobre 1848, p. 6.

in power, the scruples of Cauchy on the right and Arago on the left being exceptional. In 1852 Napoleon III had sufficient wisdom to excuse both. More than a signature was expected of the office-holders in the Academy, although the loyal addresses of the secretary Cuvier, first to Napoleon and later to the Bourbons, may seem to us unnecessarily obsequious. Under the Restoration the unwillingness of Louis XVIII to confirm the election of the avowed republican Hachette was passed on to the Academy by the Minister of the Interior.[43] Yet less than six months later the same minister announced that the state would be happy to pay for a bust in marble of the recently deceased Berthollet.[44] Berthollet, although a personal friend of Napoleon and a member of the Napoleonic Senate, had always been considered as non-political.

There is an obvious distinction to be drawn between official acts of the Academy and the actions of its individual members. Yet if its members brought politics into their discussions, they would inevitably have involved the Academy. Any public political statement by an Academician had implications for the institution to which he belonged. This may seem unduly restrictive to us, but it was a part of the price paid by scientists when they accepted membership of the Academy. The majority of Academicians seem to have been happy to get on with their work, whatever the political regime in power. When there was a change of government, their main concern was that this should not affect the traditions and prerogatives of the Academy. Many Academicians were apolitical or at least were so moderate and discrete in their political beliefs that anyone beyond their closest friends would be forgiven for thinking that they had no political opinions. This could give support to the view that the Academy was above politics. Such was the opinion that Jamin wished to convey in the public meeting of 1838 when he described the Academy as

> 'alone immobile in the midst of political storms...'[45]

And Brewster after the Revolution of 1848 could describe the Academy as:

> Unshaken and active amid all the revolutions... an institution of order[46]

There are several different interpretations one could give to the claim of being 'unshaken' or 'immobile'. It could have meant that the Academy opposed innovation. A corollary of this conservative thesis might be that it tended to support the ideology of previous rather than future political regimes. But the most plausible intention of the speakers was that the Academy represented a centre of scientific culture, an ivory tower of learning remote from the conflict of political change.

Yet this image too must have been at variance with the actual situation in

43 *P.V.I.*, 7, 604 (15 December 1823).
44 *P.V.I.*, 8, 88 (24 May 1824).
45 *Séance publique des cinq Académies*, 2 May 1838.
46 Brewster, *Report of the 20th meeting of the British Association*, Edinburgh, 1850, p. xlii.

many instances. Given the active involvement of a significant number of Academicians in politics, the Academy cannot be said to have lived in a different world. If it often stood to one side it nevertheless provided a reflection of the political state of the nation. Yet, although the Academy might have found it expedient to welcome almost any change of government in order to ensure its own continuity as a protected organ of the state, it could not afford to be too clearly identified with a particular brand of politics, since it was the experience of Frenchmen that no type of government was likely to last longer than a decade or, at most, a generation. Eschewing political ideology therefore the Academy went out of its way to appeal to a higher ideal and to assert its love of country. Patriotism and nationalism were able to override narrow political considerations. Particularly after the Franco–Prussian war when a new wave of patriotism swept over France, the Academy found that it could gain public support by expressing its not insincere patriotic feelings. Some of its members could then, under the Third Republic, come out as avowed republicans. Certainly the Academy could not afford to appear as anti-republican. Yet most of all the Academy tried to transcend politics by stressing patriotism.

8. *Personal inter-relationships*

The simplest model of the relations between the Academy and government is one which treats these two institutions as two quite separate entities. The reality is rather more complex. The involvement of scientists in politics is an old tradition in France, dating back to the Revolution of 1789. Both Guyton and Fourcroy were members of the Convention, and Fourcroy played a leading part in founding new medical schools and a new university administration, including Faculties of Science. It would therefore be quite wrong to think of the scientific community as simply receiving instructions passively from above.

There were several cases where the scientists and the government ministers were actually the same persons wearing a different hat. Thus when the Minister of the Interior asked the Academy in the winter of 1804 for a report on the dangers to public health of factories producing unpleasant smalls, Chaptal and Guyton were appointed to examine the questions and Chaptal drew up the report.[47] Yet until July of that year Chaptal had been the Minister of the Interior!

When Louis-Philippe rose to power in 1830, the agronomist Gasparin (1783–1862) was elected to the Chamber of Deputies as a government supporter. He was soon made a prefect for the Rhône department, where he took a tough line in a workers' uprising in Lyons. He was rewarded by elevation to the Chamber of Peers in 1834 and in 1836 he became Minister of the Interior in a cabinet lasting seven months. In a later cabinet crisis in 1839 he held for six weeks the post of Minister of the Interior, Public Works, Agriculture and Commerce. Already a correspondent of the Academy in the agriculture section,

[47] *P.V.I.*, 3, 152, 165.

he was elected a full member in 1840. He had ample credentials for election as the author of several important books in the field, but some of his future colleagues may well have thought that it would not do the Academy any harm to have in its midst someone of cabinet rank, who was favoured by the political powers of the time. He would not have approved either of the Revolution of 1848 or of the coming to power of Louis Napoleon.

Under the Second Republic Arago was successively Minister of the Navy (February 1848) and Minister of War (March–May 1848) in the Provisional Government. Under the Third Republic science and scientists were even more in favour and in Gambetta's Ministry of 1881–2 his friend the physiologist Paul Bert, became Minister of Education, a post occupied for six months in 1886–7 by Marcellin Berthelot. It was reassuring for the Academy to know that its members could occasionally reach ministerial rank but for one of its members to hold the key post of Minister of Education was something of a triumph.

A second and more important factor in the relationship between government and science was that a fair number of French administrators and politicians had received a scientific education; they were early technocrats. There is an obvious contrast with England, where for most of the nineteenth century a classical education was considered the only education appropriate for future careers in administration and government.[48] In Paris the Ecole Polytechnique provided a common education for future scientists, engineers and administrators. During the nineteenth century it produced two presidents of the republic, Cavaignac and Sadi Carnot (1837–94), not to be confused with his relative of the same name who had been a pioneer in thermodynamics. The Ecole Normale was later to challenge the Polytechnique in producing a cultural elite. One could take the argument further in terms of a 'ruling class', as has been done in a study of the Ecole Polytechnique.[49] Certainly the sons of workers and peasants were under-represented in the Academy, as in the Ecole Polytechnique, but if the middle classes contributed disproportionately to science and learning, the political world, which demanded less by way of formal qualification, was a more open one and there would be less reason for overlaps. Nevertheless, in many bourgeois families there would be both political and scientific contacts. Many such families would encourage their sons to enter the civil service, and some branches of the higher civil service came only second to higher education as a recruiting ground for Academicians. One thinks particularly of the Corps des Mines and the Ponts et Chaussées.

Not only had leading politicians and Academicians often attended the same school, whether a top Paris lycée or a grande école, but they were sometimes

[48] An exception is the mathematical education provided by the University of Cambridge. Yet paradoxically several ex-Cambridge mathematicians later became bishops in the Church of England yet hardly any became engineers. It is, therefore, difficult to see Cambridge in this sense as the British counterpart of the Ecole Polytechnique.

[49] Terry Shinn, *Savoir scientifique et pouvoir sociale. L'Ecole polytechnique, 1794–1914*, 1980.

related, either through birth or marriage. This was particularly the case in the late nineteenth century. The mathematician Henri Poincaré (1854–1912) was the cousin of Raymond Poincaré, President of the Republic. A letter asking the Academician to use his influence with his cousin to obtain advancement for a friend has survived in the archives of the Institute. Henri Poincaré in his reply said that his very closeness to his cousin was a reason for refusing![50] Adolphe Carnot (1839–1920), elected as *Académicien libre* in 1895 to succeed Lesseps, was the brother of the President of the Republic, recently assassinated. Adolphe Carnot was elected largely because of his contributions to geology and agriculture, but one of the effects of his brother's death was to make him take up politics. He presided over the *Alliance démocratique* party, which had considerable influence between 1901 and 1914.[51]

The greatest influence on government however came through having friends at court. Under the Bourbon Restoration the Academy was probably most remote from ministers, often chosen from the old nobility. Under the two Empires, however, personal channels seemed to have worked. It was quite exceptional that Napoleon Bonaparte had been elected as an ordinary member of the Academy before he came to power. Yet after the coup d'état of November 1799 the Academy could at least feel that it had a patron. If Bonaparte no longer attended ordinary meetings after the early years of the Consulate it was well known that some Academicians, notably Monge, Berthollet and Laplace, were favourites of the Emperor. Yet most administration went through the proper channels via the Minister of the Interior. Probably it was under Napoleon III (1852–70) that a friend at court was most useful. Marshal Vaillant, aide de camp to the Emperor, was elected *Académicien libre* in 1853 (being elected at his first candidature virtually unopposed) and served as a most useful intermediary between the head of state and the senior members of the scientific community in the Academy.

We have mentioned the common education of many successful scientists, administrators and politicians, and the possibility of a common bond through 'the old school tie'. But if their paths had not crossed in their youth they might well cross in their maturity as members of the Institute. Although very few ministers were members of the Academy of Sciences, several distinguished ministers had been elected to the sister Academy, the Academy of Moral and Political Sciences and could, therefore, be addressed as a fellow-member of the Institute and a colleague. After Fortoul, Minister of Public Instruction, had been elected, we find him addressed by the Academy of Sciences in official communications as 'cher collègue'.[52] We later find Pasteur addressing Thiers in November 1873 as 'Monsieur et illustre confrère'.[53] Thiers had actually been

[50] Institut, MS.2720, dossier 8–9, letter of 13 May 1894.

[51] *Biographie française*, vol. 7, p. 1179, art. Carnot, Marie-Adolphe.

[52] A.S., e.g. *Comité secret, 1845–56*, p. 398 (13 August 1855).

[53] Pasteur, *Correspondance*, vol. 2, p. 573 (20 November 1873).

head of government from February 1871 to May 1873. Although out of office at the time Pasteur wrote, he was still a potentially powerful ally for a man wanting a state pension, and Pasteur asked him to approach the new president MacMahon. In 1872 the Academy seems to have used its 'colleague' Barthélemy St. Hilaire, secretary to the President of the Republic, as a means of gaining an audience with the President.[54] On 4 May 1899 the secretary of the Academy addressed the War Minister Freycinet as 'cher Confrére',[55] but to little avail as, in the context of constantly changing cabinets, he lost his post two days later! We must not however expect full documentation for informal contacts. There would be very good reasons for not recording for posterity some of the conversations which took place in certain Parisian social circles.

9. *Relations with different ministries*

Although for simplicity we have sometimes referred to the Academy's relations with the government or with 'the Ministry', there were of course a number of different ministries which had dealings with the Academy. At the foundation of the National Institute in 1795–6 it was placed under the giant Ministry of the Interior. This ministry was responsible for a wide range of services including education, religion, public works, customs and excise, state functions, theatres, palaces, museums, hospitals and prisons. In the Napoleonic period and the Restoration there were a number of modifications in the responsibilities of the different ministries. Under Louis Philippe, responsibility for academies was given to the Ministry of Public Instruction and this was the relevant ministry for most of the remainder of the nineteenth century.[56] When one of the other government departments wanted some information or service from the Academy they occasionally approached it through the ministry directly responsible, but usually they wrote directly.

Probably if the Academy had been concerned exclusively with pure science it would have been ignored by all ministries except that formally responsible for it. Since however it had expertise in agriculture, medicine, engineering, and several other practical areas the different ministries turned to the Academy for advice. The Academy was often approached by the Ministry of Commerce, later the Ministry of Agriculture and Commerce, or Agriculture, Commerce and Public Works. It corresponded with the Academy on the subject of alcohometry for, since the work of Gay-Lussac,[57] accurate density measurements had become the established basis for taxation of alcoholic liquors. The Ministry of Finance wanted the Academy's advice on how to stop special legal papers which had

[54] A.S., *Copie de Lettres, 1861–74*, pp. 249–50.

[55] A.S., *Copie de Lettres, 1875–1902*, p. 216.

[56] In 1824 Public Instruction had been linked with 'ecclesiastical affairs' under Bishop Frayssinous. The decree of 5 December 1860 placed the Institute under a newly appointed Minister of State (*C.R.*, 51 (1860), 955). In 1863 responsibility for the Academy reverted to the Ministry of Public Instruction.

[57] Maurice Crosland, *Gay-Lussac*, pp. 190–3.

been stamped and taxed from being bleached and used again fraudulently.[58] The War Ministry wanted advice on horses[59] and on coating iron military utensils with zinc.[60] When in 1873 the Minister of War asked the Academy to nominate one of its members to the special gunpowder and explosives committee, the Academy replied that it would prefer to nominate two experts, one from the mechanics section and one from among the chemists.[61] This political dispute took several years to resolve.[62]

Another example of ministerial involvement with the Academy was in relations with foreign governments. Thus when the British government in 1854 wished to have the benefit of French experience in the field of public health, the British ambassador in Paris asked the French Ministry of Agriculture, Commerce and Public Works, which consulted the Academy.[63] The Academy sent a suitable reply to the Ministry for transmission to the British government. This does not mean of course that scientists were incapable of corresponding directly with each other, but rather that in areas of public concern with a scientific dimension there existed formal government channels which could be used. This government/ministerial aspect gives a further dimension to our understanding of the Academy as representing 'official science'.

Yet it must not be thought that contact with various ministries was always a matter of giving rather than receiving. From time to time throughout the nineteenth century the Ministry of Foreign Affairs passed on information about science in other countries. This ministry could be particularly useful to the Academy in time of war. The Academy used the good offices of the ministry to send its *Mémoires* to the Royal Society in exchange for the *Philosophical Transactions*,[64] and it communicated with Jenner, of vaccination fame, through the same ministry.[65] In the winter of 1799–1800, the First Class wanted to take up experiments by the Prussian chemist Achard on the extraction of sugar from beet, experiments which he had described in a letter published in the *Annales de chimie*.[66] He claimed that he had developed an economical way of obtaining sugar identical with cane sugar, which could no longer be imported from the West Indies. The First Class wrote to the Minister in the following terms:

> The National Institute is concerned at this time to follow the experiments of the chemist Achard of Berlin on the means of obtaining sugar from several indigenous plants. The commission entrusted with this task will need for its

[58] *C.R.*, **4** (1837), 180; **29** (1849), 236.

[59] *C.R.*, **10** (1840), 73, 618.

[60] *C.R.*, **23** (1846), 314.

[61] *C.R.*, **77** (1873), 1288; **78** (1874), 338.

[62] *C.R.*, **84** (1877), 81.

[63] A.S., *Copie de Lettres, 1841–60*, pp. 130, 139.

[64] Jussieu to Ministre des relations extérieurs, 21 *nivose* an 7 (10 January 1799), A.S., *Copie de Lettres, an VI–an XIV*. Copies of the *Mémoires* were also sent to the presidents of the Academies of Berlin, Turin and Madrid.

[65] Cuvier to Ministre des relations extérieurs, 16 *thermidor* an 9 (4 August 1801), *ibid.*

[66] *Annales de chimie*, **32** (1799), 163–8. For the Prussian experiments see Karl Hufbauer, *The formation of the German chemical community (1720–95)*, Berkeley, Cal., 1982, pp. 206–7. For a summary of the French contribution see Maurice Crosland, *The Society of Arcueil*, pp. 32–6.

> research the same variety of beet, which this chemist has used for this purpose. The Institute has recourse to your intervention and we beg you in its name kindly to write to the agent of the republic at the Berlin court, asking him to send us as soon as possible about thirty kilograms of this root crop, taking all possible precautions for it to arrive in good condition and sheltered from frost. It would be helpful to place them in a bed of very dry sand.[67]

Here was a matter not simply of scientific interest but of public utility and the First Class did not neglect to point out to the Minister 'the importance of this work for national prosperity'. The beet arrived in due course and the sugar was extracted with hot water. The sugar beet industry as developed by scientists in Napoleonic France was to lay the foundations of the modern industry world-wide. But it is a story which depended not only on the expertise of chemists and agriculturalists but also on communications through the official body of science and the French government.

The Ministry of Foreign Affairs throughout the nineteenth century sent information to the Academy on volcanic eruptions and other dramatic natural phenomena in different parts of the world. It also helped Lacaze-Duthiers when he wanted to establish marine biology stations in the Mediterranean.[68] The Ministry of the Interior with its network of prefects in each department throughout the country was in an ideal position to provide the scientists with national data on extreme meteorological phenomena or even earthquakes.[69] The Ministry of the Navy was in regular contact with the Academy, which provided instructions for scientific observations on naval expeditions. The final result of this exchange was that scientists were provided with valuable data from different parts of the globe. Sometimes, at the conclusion of an expedition, naval personnel were asked to stay some time in Paris to collaborate fully with members of the Academy.[70] It was this ministry which provided a free passage for D'Abbadie (1810–97) to visit Brazil on a scientific expedition.[71]

10. *The Academy's expert advice on safety*

Safety was a common factor in many of the approaches of government to the Academy. The government had a responsibility for the life and limb of French citizens and, in so far as this was threatened either by forces of nature such as lightning, or new technology, notably steam engines, this was a matter of concern for the appropriate Ministry and one in which the official body of science might give invaluable advice. Government concern with public health was a tradition going back to before the Revolution and here again the Academy was consulted, as in the terrible cholera epidemic of 1832. Here both the concern of

[67] Lacépède and Cuvier to Ministre des relations extérieurs, 6 *frimaire* an 8 (27 November 1799), A.S., *Copie de Lettres, an VI–an XIV*.

[68] *C.R.*, 77 (1873), 528.

[69] *P.V.I.*, 7 (1 July 1822), 345.

[70] E.g. *C.R.*, 11 (1840), 607, 706.

[71] *C.R.*, 3 (1836), 200.

central government and the immediate appeal to the established scientific body for advice provides something of a contrast with Britain. The Academy therefore served occasionally as an agent for government in France. One might go further and view it as an agent for social stability; potentially disruptive forces must be studied and controlled.

In the eighteenth century the traditional sources of wind and water power had begun to be supplemented by steam power. The most common engine was the Newcomen engine, grossly inefficient but safe, since it depended on atmospheric pressure. One of the improvements of Watt was to use steam pressure to help to drive the piston, and in the early nineteenth century steam engines, using increasingly high pressures, were being widely tried. Such a development was necessary to develop the locomotive, but it is mostly the stationary steam engine with which we are concerned here. Britain was leading the world in steam power at the end of the eighteenth century and one consequence of the Revolutionary and Napoleonic wars was to put France even further behind. It was, therefore, a problem for the governments of the Bourbon Restoration to try to catch up with the British lead in power technology. This explains why the Minister of the Interior should have approached the Academy to ask for a report on the dangers arising from the use of high pressure steam engines. There had been several cases where boilers had exploded causing injury and, in one case, loss of life. In its best tradition the Academy appointed a high level commission consisting of Laplace, Ampère, Prony, Dupin and Girard, thus representing mathematics, physics and civil and mechanical engineering expertise.

The commission finally reported in April 1823, the report being drawn up by Dupin.[72] It explained first that high pressure steam engines were necessary because they were more efficient and also they took up less space in factories or mines. They reported on the achievement of high pressure engines in Britain and America and on their own trials. To ensure safety, boilers should have safety valves and the manufacturers should specify the pressures at which the boilers were to be used. In case of an accident publicity should be given to the unfortunate manufacturers. The thickness and distance of the factory wall should also be specified.

This was not the only occasion on which the Academy was asked for advice on steam engines. In 1825 Dulong was involved in a study of boilers at high pressures and complained privately of the dangers involved in the work.[73] In 1836 the Ministry of Commerce asked for the Academy's opinion on fusible plugs as a safety measure in boilers, and it obediently appointed a commission to investigate.[74] In 1847 the Academy published in the *Mémoires* the results of

[72] The conclusions of the report are given verbatim in *P.V.I.*, 7, 470–9.

[73] Letter of Dulong to Berzelius, Paris, 10 November 1825, Berzelius, *Bref.*, Part IV, p. 63.

[74] *C.R.*, 3 (1836), 620–1.

a long and painstaking piece of research undertaken by Regnault: *Relation des expériences entreprises par ordre de M. le ministre des travaux publics et sur la proposition de la commission centrale des machines à vapeur pour déterminer les principales lois et les données numériques qui entrent dans le calcul des machines à vapeur.*[75] Publication in this form was probably intended as a reminder to a wider audience outside the Academy of its value at a utilitarian level.

If the safety of high-pressure steam engines was one example where the Academy could give advice on the possible dangers of technology to society, the widespread introduction of coal gas was another area. The large scale distillation of coal in sealed retorts in urban areas raised the spectre of major explosions, as the resulting coal gas formed an explosive mixture with air. The gas holders themselves seemed to constitute a permanent potential danger to the neighbourhood. The Academy commission actually recommended large gas holders as preferable to a battery of smaller ones on the grounds that they were easier to control.[76] Householders were recommended to ventilate rooms where gas lighting was used. The report of 1824 urged the government to encourage the extension of gas lighting, which was already common in England, and it played down the theoretical dangers of manufacturing and storing large volumes of potentially explosive material. The combined expertise of Gay-Lussac, Prony, Dulong, and D'Arcet represented chemistry, physics and engineering. Whereas it was usual for such reports to be approved and forwarded to the relevant ministry, in this case many other members of the Academy wished to add their own experience or advice and there was a long discussion at the following meeting.[77] The Academy agreed, however, with the reassurance of the commission that, with appropriate safeguards, the manufacture of coal gas in urban areas should be permitted. It was neither a threat to life nor to public health. The advice of the Academy was incorporated into new legislation that same year.[78]

In 1817 a special prize for statistics was established through the interest of Laplace and the generosity of baron Montyon. Looking through the *Comptes rendus* one finds frequent papers on applied statistics, which must have been of great interest to the relevant ministry, for example, on mineral resources in France,[79] on crimes committed in Corsica in the 1830s,[80] on a comparison of the crime rate in Britain and France,[81] on pauperism,[82] on the main rivers of France[83] and so on. Occasionally a critic might even use the Academy as a base to comment on the accuracy of government statistics.[84] Cholera statistics,[85] and many agricultural and industrial statistics were provided for the whole country.

[75] *M.A.I.*, **21** (1847), 3–748.
[76] *P.V.I.*, **8**, 14–24 (2 February 1824).
[77] *Ibid.*, 25.
[78] Ordonnance du 20 aout 1824.
[79] *C.R.*, **3** (1836), 22.
[80] *C.R.*, **5** (1837), 67.
[81] *C.R.*, **6** (1838), 160.
[82] *C.R.*, **6** (1838), 822.
[83] *C.R.*, **8** (1839), 677.
[84] *C.R.*, **2** (1836), 34.
[85] *C.R.*, **34** (1852), 65.

Studies of the population of the whole of France[86] were supplemented by studies of individual towns and departments.

Among the forces of nature studied in the Academy earthquakes were a comparatively minor phenomenon in France and mainly confined to the far south.[87] Lightning on the other hand could apparently strike anywhere, even buildings which were equipped with lightning conductors. In 1822 several major churches were struck by lightning and the Minister of the Interior considered that the Academy would be able to give appropriate advice to prevent a recurrence. The Academy felt that the problem lay entirely within the province of physics and took the unusual step of appointing the entire physics sections, headed by Gay-Lussac, to examine the question. The report[88] did not raise any new problem of physics, since the principle of lightning conductors had been established seventy years earlier by Benjamin Franklin. It did, however, provide practical details in plain language. It pointed out that the earthing of the conductor in moist ground was as important as seeing that the pointed terminal rod was several metres above the highest point of the building. When a powder magazine at Bayonne was struck by lightning in 1828 (despite the fact that it had been fitted with a lightning conductor), the Minister of War called on the Academy for an explanation. In 1854 the Academy was again called on for advice on lightning conductors and Pouillet produced a supplementary report. The original report of 1823 by Gay-Lussac, together with the supplement, were translated into German and into English (as late as 1881) and were influential for the remainder of the nineteenth century. Thus an investigation which originally had been called for in a report to a Ministry on a specific problem became the standard authority on the subject throughout much of Europe in the nineteenth century.

11. *The utilitarian role of the Academy*

Some of the tasks undertaken by the Academy at the request of various ministries were challenging problems on the frontiers of knowledge, but others were much more routine and we may wonder why the great intelligences in the Academy did not rebel and why the Academy seemed content to investigate almost any rational problem put to it. The answer in the first place is that it was a part of the duty of the Academy. But a second more interesting reason is that the Academy wished to demonstrate its utility. The scientific enterprise is always difficult for the public to understand. One way that the Academy could justify its existence in the public eye and to the state, its paymaster, was to undertake

[86] E.g. *C.R.*, **50** (1860), 683.

[87] Despite a steady stream of reports of earthquakes it was only towards the end of the century that the interest increased to the point of establishing an Academy Commission, *C.R.*, **100** (1885), 438; **104** (1887), 612.

[88] 'Instruction sur les paratonnèrres adopté par l'Académie des Sciences le 23 juin 1823 et publié par ordre du ministre de l'intérieur', *A.c.p.*, **26** (1824), 258–98.

tasks of demonstrable utility. In the very first year of its existence the First Class of the Institute in drawing up a report on its activities, included not only a list of memoirs but also 'a summary of all the reports made in the First Class, *particularly those which have been asked for by the Legislative Assembly or the Government.*'[89] One early enquiry carried out by the Academy was on vaccinations and it has been claimed that 'it was in the Academy of Sciences that the scientific merit of vaccination was decided.'[90] When the Academy was consulted on the subject of chemical pollution, the advice given was used in preparing suitable legislation.[91]

One of the most tedious tasks which successive governments gave to the Academy was in connection with water supply. Probably the justification for the Academy being given a task which might well have been delegated to a less august body was the precedent established in the eighteenth century, when there was a great burgeoning of interest in mineral waters and men of science and Academicians, particularly chemists, had been asked to analyse different samples. In 1835 the Minister of Public Instruction submitted samples of water from different places around Bordeaux for analysis, so that the city could decide which was preferable for a general water supply.[92] In the early 1840s the Minister of Commerce passed on to the Academy an enquiry from the municipality of Grenoble about piping hot water from a thermal spring over a distance of several kilometres without serious loss of heat.[93] By 1847 the Ministry of Public Instruction saw nothing improper in passing on to the Academy a request from the municipal council of Castelnaudray, a town near Toulouse, for general advice on its domestic water supply.[94] A serious study of the problem would have involved the Academy in sending a group of scientists, including a geologist and a chemist, to the other end of France. But the problem hardly justified the time of the elite of French science. In such cases the Academy's best defence was delay and in this particular case it was probably saved by the Revolution of 1848!

Yet the Second Republic and the Second Empire were to bring new demands. A small example is the excellent and seemingly very modern idea of a certain Leclaire that white paint should no longer be permitted to be made from lead (carbonate), but that non-poisonous zinc oxide should be substituted.[95] When the Academy had not replied within three weeks, the Minister of Public Instruction, on behalf of the Minister of Public Works, asked the Academy to

[89] *P.V.I.*, **1**, 85, 1 *fructidor* an 4 = 18 August 1796 (my italics).

[90] Ann F. Le Berge, 'The early nineteenth-century French public health movement', *Bulletin of the History of Medicine*, **58** (1984), 363–79 (p. 366).

[91] *P.V.I.*, **4**, 256, 268–73.

[92] *C.R.*, 1 (1835), 124.

[93] *C.R.*, **16** (1843), 1171. Arago rashly undertook to carry out large-scale experiments.

[94] *C.R.*, **24** (1847), 874.

[95] *C.R.*, **28** (1849), 151. Modern advances in chemistry have recently led to the substitution of non-poisonous titanium oxide for lead carbonate.

hurry, pointing out that it was a matter of considerable importance to the health of working men.[96]

One of the many ways in which France differed from Britain economically in the nineteenth century was that it had an important silk industry. In 1837 the Minister of Commerce asked the Academy to study the problem of disease among imported silkworms.[97] By the late 1840s much of the south of France was threatened by a mysterious disease which was destroying whole nurseries. Alternative sources of silk worms in the East were investigated. In 1854 several memoirs were presented to the Academy on acclimatisation of oriental silkworms in western Europe. At the end of a memoir presented to the Academy on 19 October 1857 by Isidore Geoffroy Saint-Hilaire on silkworms, Marshall Vaillant, who had experience of Algeria, proposed that a Commission should be established to draw up instructions on the breeding of silkworms, and Duméril, Milne-Edwards and Quatrefages were elected.[98] They planned to go to the midi the following spring and they were able to use some of the surplus prize money of the Academy to pay their expenses.[99] The Academy had volunteered to investigate the problem, so that when representations from the Hérault and Gard departments were made in 1858 the Minister of Commerce forwarded them to the Academy.[100]

By 1865 the silkworm blight was even worse and Pasteur, recently elected to the Academy, was asked by the Minister of Agriculture to investigate the disease.[101] Encouraged by his patron Dumas, Pasteur undertook a thorough study of the breeding cycle of silkworms. After two years' investigation of the very complex problems involved, he was able to show that two diseases were involved,[102] which depended on the temperature and humidity of the silkworm nurseries for their progress. Pasteur recommended that the silkworms and eggs should be continuously monitored by microscopic examination and gradually the situation began to improve. His work was not only a model of scientific method in a complex area involving both heredity and environment but it served as a well-publicised example of a scientist helping the national economy.

Another agricultural problem of great economic importance in which the Academy was involved was the disease of vines called phylloxera that became a serious threat to the wine industry in the 1860s. The problem was soon brought to the attention of the Academy, but it was not until 1871 that it treated the question as one of priority. A special commission was established to

[96] *C.R.*, **28** (1849), 268; **30** (1850), 49.

[97] *C.R.*, **4** (1837), 1002, **6** (1838), 18, **7** (1838) 702.

[98] *C.R.*, **45** (1857), 560.

[99] The Minister of Public Instruction authorised the Academy to spend 3000 francs of its own (surplus) funds. *C.R.* **46** (1858), 845 (3 May 1858).

[100] *C.R.*, **47** (1858), 593, 707.

[101] Letter of Pasteur to Raulin, 19 May 1865, *Correspondance de Pasteur*, vol. 2, p. 194. The letter makes clear the confidential nature of the mission. The Minister was using the expertise of Academicians without approaching the Academy officially.

[102] Gerald L. Geison, art. 'Pasteur' *D.S.B.*, **10**, 350–416 (pp. 372–6).

examine the problem.[103] In 1872 it asked for and received 10000 francs from the Ministry of Agriculture to meet the immediate costs of research. Various junior scientists were sent to Bordeaux and other wine-growing areas. News that the Academy had funds for research on phylloxera brought many requests from scientists outside the Academy for financial help. The Academy made a public statement in 1874 emphasising its limited funding,[104] but at the same time it asked the Ministry of Agriculture for further funds and 20000f. was granted. The publicity attending the work of the phylloxera commission also brought funds from outside sources including a railway company, the Chemin de fer du Midi. By 1875 two research laboratories set up by the Academy at Cognac and Montpellier, were asking for 10000 francs each to cover their expenses.[105] In 1876 the Academy Commission published an *Instruction*, a booklet explaining to wine growers how they should treat their vines with carbon disulphide. The work of the Commission continued with increasingly bold requests for money, so that by 1877 it was asking for a regular budget of 20000 francs.[106] Part of this was to pay for publication of the *Mémoires des Savants Etrangers*, much of which was being devoted in the 1870s to publications by non-members on phylloxera.[107] There is no doubt that the Academy did good work in helping to eliminate this blight on vines.

Having given some examples of utilitarian tasks undertaken by members of the Academy at the request of the government, it is important that we should correct the possible impression that membership of the Academy involved Academicians in constant drudgery. There were a few Academicians, notably Pasteur, who had an absolute genius for applying science at the frontiers of knowledge to solve practical problems, something which had special appeal for Napoleon III. But the majority of Academicians spent only a very small fraction of their time on utilitarian tasks. In principle they were obliged to investigate a problem whenever asked, but the truth is that they were not asked very often, particularly in the second half of the century.

An unpublished study of consultancy work carried out by the Academy for different ministries over successive periods of fifteen years starting in 1835, shows the highest number of tasks undertaken from 1835–50, and then there were only sixteen requests from the Ministry of Public Instruction (an average of one per year) or forty-nine requests from all Ministries combined, giving an average of approximately three per year.[108] By the period 1896–1910 the total number of requests had dropped to five, of which three were from the Ministry of War. It is clear that other agencies were being used for consultancy work. Science having proved its utility, it made sense for ministries to appoint their

[103] A.S., Registre: *Commission du Phylloxera, 1871–74*.

[104] *C.R.*, 78 (1874), 555 (meeting of 23 February 1874).

[105] A.S., *Copie de Lettres, 1861–74*, pp. 310–11.

[106] A.S., *Copie de Lettres, 1875–1902*, pp. 40–3.

[107] E.g. five out of seven memoirs in volume 25 (1877) were devoted to phylloxera.

[108] A.S., Carton 34, dossier 'Académie, Conseille du Gouvernement'.

own experts. At the end of the nineteenth century, therefore, the Academy was no longer justifying itself in utilitarian terms. It was more important that it should justify itself ideologically, a task which it found comparatively easy under the Third Republic. But we have still to look at the possible military applications of science.

12. *Response to war*

One of the first wars which systematically made use of scientific expertise was the French Revolutionary war and particularly the early period of 1793–4, when France was being invaded. In Chapter 1 we have seen how scientists gave advice and helped in the production of gunpowder, saltpetre, cannon, etc.[109] When copper for cannon was in short supply they could use the new chemistry to obtain bronze for cannon from bell metal from church bells. A hydrogen balloon was used for observation at the battle of Fleurus, the hydrogen having been obtained by the action of iron on steam, the method previously demonstrated by Lavoisier. All this ingenuity had the effect of increasing the prestige of scientists. Yet the main activity by scientists came before the foundation of the National Institute and would, therefore, seem to lie outside our immediate concern.

Yet the memory of the Revolution lived on and the role of scientists in national defence was a theme exploited more than once by Arago, as secretary of the Academy and a committed republican. In his *éloge* of Lazare Carnot in 1837 Arago naturally called attention to his military exploits, although here it was less a case of science being applied as a man of science playing a second role in military affairs.[110] In his *éloge* of Monge in 1846 however, Arago was able to describe at some length the services rendered by Monge in 1793–4 to the manufacture of cannon and munitions. Here was a reminder in peacetime of the *potential* utility of science.[111] Although France was involved in a few military incursions, notably in North Africa and the Italian peninsula, there was no war on native soil between 1815 and 1870.

War between Prussia and France was declared on 19 July 1870. The well-trained Prussian troops quickly gained the advantage and on 1 September Napoleon III, at the head of an army of 80000 men, surrendered at Sedan. This marked the end of the Second Empire, but it was only the beginning of French resistance. The Prussian army marched on Paris and by 19 September it had completely surrounded the capital. Now there began a siege which was to continue into the bitter winter of 1870–1. The question of interest to us is what part science, and particularly the Academy, was to play in the siege. Since French science was so centralised there was in the capital a unique concentration of talent. There was also a memory of the revolutionary crisis of 1793–4, which acted as a source of inspiration.

The Academy continued to meet weekly through the siege and the meetings

[109] See Chapter 1, Section 4.

[110] *M.A.I.*, 22 (1850), xlvii–xlviii.

[111] *M.A.I.*, 24 (1854), xxxvii ff, lx.

were generally well attended.[112] Hardly a meeting went by without discussion of problems relating to the war. The Academy established a commission on 'military arts' and another on 'culinary arts' to deal with the growing food crisis. Yet the Academy's organisation was such that it was not able to deal efficiently with the flood of ideas which came from outside.[113] Much more effective work was carried out, for example, by a special committee set up by the Ministry of Public Instruction, the Commission scientifique pour la défense de Paris, of which Berthelot was a prominent member. Berthelot, not yet an Academician, also mobilised the Société chimique de Paris and did important work on explosives. The nineteenth-century Academy was concerned primarily but not exclusively with pure science and its procedures, like a court of law, were slow but thorough. The siege required the immediate solution of practical problems drawing on all available talent, thus cutting across the distinction proudly maintained by the Academy between members and non-members who were never allowed to discuss problems together. If the Academy failed to respond to the crisis it was not for lack of good will. The members moved in the right direction even if with little practical effect. In so far as the majority of members had not fled from the capital they could continue to hold their heads high. In April 1871 a book was even written to justify the position of the Academy during the siege.

A further opportunity for the Academy to apply scientific expertise to military matters came at the outbreak of the first world war. Germany declared war on France on 3 August 1914, which happened to be a Monday. After the ordinary meeting of the Academy that day it met in secret session and proposed setting up a number of commissions to study special areas such as telegraphy, aviation, explosives and hygiene, and names were put forward for membership of these commissions.[114] The following week in a further secret session, having had a little time to think about the proposals, a revised scheme was put forward to establish six commissions under the following titles:

1. Mechanics (including aviation)
2. Wireless telegraphy
3. Radiography
4. Chemistry (including explosives)
5. Medicine, surgery and hygiene
6. Food

Although the titles of some of these commissions clearly reflect a new practical dimension, in the titles for commissions 1, 4 and 5, we can see the perpetuation

[112] M. Crosland, 'Science and the Franco-Prussian War', *Social Studies of Science*, **6** (1976), 185–214.

[113] Many of these were naive or at least impractical, but it is interesting that no less than 74 communications were received relating to balloons, G. Grimaux de Caux, *L'Académie des Sciences pendant le siège de Paris*, 1871, p. 11.

[114] A.S., *Comité secret, 1912–18*, pp. 141ff. A brief summary of these activities was reported in *Nature*, **94** (1914–15), 572–3.

of the traditional Academy division of science. The Academy also decided to collaborate with the Red Cross in the installation of a military hospital and one of the Academy prizes was to be given to the hospital.

At the meeting of 7 September the Academy declared that its expertise was completely at the disposal of the military governor of Paris, General Gallieni, who happened to be a corresponding member of the geography section of the Academy. In October there were allegations of atrocities by the Germans against civilians, including children, but it was pointed out that the truth of such stories would need to be double checked. Thus despite their patriotic fervour, some of the scientists maintained a certain detached scepticism. In relation to propaganda, the secretary Darboux suggested that it was important that the Academy should supply information to neutral countries to counter any misinformation supplied by the other side.

— Chapter 10 —

'OUTSIDERS': THE SCIENTIFIC FRINGE AND THE PUBLIC

Out[side] of the pale of the Institute and those who aspire to its honours, there is infinitely little.

(R. Chevenix, *Edinburgh Review*, **34** (1820), 411.)

I demand as an amendment to Article 1 of the regulations [of the Institute] that the meetings of every Class should be held in public. The public will derive great advantages for their education. Also, if members of the Institute are placed under the public gaze, they will not fall asleep in their academic chairs.

(Dupuis, *Conseil de 500*, 15 March 1795.)

For two years an idea had been absorbing all my research and a hope had been consuming me. I had finally captured the idea and I thought that fortune was smiling. In those days young men knew only one dream, a king of glory, whose temple was the Academy of Sciences, guardian of truth...I remember how I trembled when I dared to speak to one of [the Academicians] in the courtyard of the Institute.

(Raspail, quoted by Dora B. Weiner, *Raspail, scientist and reformer*, New York, 1968, p. 84).

1. *On the fringe of official science*

All groups make distinctions between members of that group and non-members. Such distinctions apply to nationalities, to races, to schools and to learned societies. If the learned society in question was one of ancient lineage with government sponsorship and the very highest standards of membership, one should not be surprised if the distinction between members and non-members was almost an absolute one. This seems to have been the case with the Academy and we have seen that membership was the ultimate ambition of very many able and enterprising French scientists. It is true that some French men of science living in the provinces were reasonably satisfied to be corresponding members, and prize-winners could always content themselves with the title of *lauréat de l'Académie*. There was, therefore, a recognised penumbra but outside this, many would have felt, all was darkness.

It would be nearer the truth to say that beyond these inner circles, there was a circle of aspirants, including the occasional critic and, even further from the centre, there was the general public. As regards aspirants, we have already looked at many who were successful and it might be useful to complement this perspective by looking at a few who were not, even to some who wanted to communicate with the Academy yet who clearly did not understand what modern science was about. In any historical situation where there are powerful institutions one will always find evidence of murmurings against the power of the institution, and the Academy was no exception. But we intend to mention one or two cases where would-be *savants*, having been rejected by the Academy, decided to appeal over its head to the Minister or, more interestingly, to the public. The Academy had always been sensitive to public criticism. Thus it was reluctant to admit the general public to ordinary meetings, although we have seen that at one time a compromise was reached for the admission of certain categories, as it were, a 'deserving public'. In the last part of the chapter we pass from the reluctant admission of the public to ordinary meetings to the institution of annual public meetings, which were not only accepted but even welcomed as useful opportunities to influence public opinion.

First, however, we must consider the issue of 'fringe science'. If one of the basic functions of the Academy was to act as the 'gatekeeper' for science then it is important to consider not only what science was accepted but also what was rejected. In trying to understand the Academy, there is something to be said for consideration of a view from the outside as well as a view from the inside. We may also understand better the senior scientific community which made up the Academy if we can contrast it with the situation of outsiders, who had little or no chance of aspiring to membership.

In any account of the work of a scientific institution it may be tempting to focus exclusively on major scientific achievements, that is those which, with the passage of time, have proved to be important or even epoch-making. It is certainly understandable that modern scientists should be tempted to pay special attention to contributions which have later proved to be 'correct'. Yet a better perspective is given if mention is also made of work which, with hindsight, is clearly pseudo-science, such as ideas for perpetual motion machines or phrenology. We do not want to adopt an explicit label of 'pseudo-science' since to do this would be to impose modern categories on the nineteenth-century material. To do justice to the total situation one needs to consider a reasonably representative sample, ranging from the brilliant to the pitiful. While bestowing due praise in its place on, say, Ampère's work on electromagnetism or Pasteur's work on spontaneous generation, we must not overlook a host of other contributions presented to the Academy, such as Raspail's miscellaneous microscopic observations, the naive theory of colours of Opoix or numerous memoirs on the quadrature of the circle.

This section therefore includes quite a broad spectrum from the weak or trivial to the lunatic fringe, omitting most major scientific advances only because these are discussed elsewhere in the book. Of course the sources for investigating fringe science are more restricted, since the Academy's archives and publications focus exclusively on work accepted by that body. Yet in so far as the minutes record indiscriminately the receipt of memoirs of all kinds, and there are even in some cases written reports on what were apparently the least deserving contributions, it is possible to say something about this area, often ignored in previous, more heroic, histories of nineteenth-century science.

The Academy is sometimes thought of as an exclusive club for top-level scientists. Although this may be part of the truth, such a view would conceal the essential openness of its proceedings. After several years of hesitation about the admission of the public, its meetings were essentially open to all, not only as listeners but as potential contributors. Obscure figures from the provinces had the right at any time to address to the Academy a contribution which would be read or at least summarised in the part of the meeting devoted to correspondence. Of course it helped to have a member of the Academy as a patron and more notice would be taken of a memoir if it were submitted through the good offices of an Academician. Yet one could get a fair hearing even without a patron. Thus at an Academy meeting in January 1835 the outsider Auguste Comte began to read a memoir on the nebular hypothesis. Since his presentation came late in the meeting and the respective merits of candidates in a forthcoming election were due to be discussed that day in a secret session, Comte was stopped at 4.30 p.m., but he was invited to continue the reading of his memoir at the following meeting.[1] Only at the time of another election, when a large number of candidates discovered a pressing need to read memoirs, was a specific time limit temporarily imposed on non-members and then they were given a generous fifteen minutes.[2] When in the 1860s Robert Houdin (later known as Houdini, the magician) wanted to communicate to the Academy his studies of physiological optics and some new optical instruments, should the Academy have turned him away? In any case his later memoirs were submitted on his behalf by the Academician Cloquet.[3]

2. *Practical men*

Since the Academy was predominantly concerned with theory there was always a gulf between its members and purely practical men, whether their background was agricultural, commercial or industrial. For practical individuals it was often enough if something seemed to work; there was no point in wasting time by considering general principles. Generally speaking, purely practical men only rarely came near the nineteenth-century Academy. Although under the *ancien régime* the approval of the official body of science had been often sought by many artisans, the establishment of a patent system and of new institutions for

[1] *P.V.I.*, **10**, 650 (19 January 1835), 653 (2 February 1835). The Academy also took Comte seriously enough to appoint a commission to report on his memoir. For a fuller discussion of Comte in relation to the Academy, see Section 6 of this chapter.

[2] *Ibid.*, 674 (9 March 1835).

[3] See e.g. *C.R.*, **62** (1866), 617–19; **63** (1866), 865–8; **66** (1868), 630–3. Also André Keime Robert-Houdin, *Robert-Houdin, le magicien de la science*, Geneva, 1986, pp. 66, 69, 72.

technology after the Revolution meant that the First Class was spared the constant problem of passing judgement on new inventions. Nevertheless there were occasional problems, such as embarrassment involved in confronting artisans, who came from time to time claiming to have discovered perpetual motion. Many were undoubtedly sincere, and the nineteenth-century Academy had to find a way of dealing with this issue that, on the one hand, would not waste the valuable time of its members and, on the other, would not cause unnecessary offence. Practical men, for their part, expecting Academicians to be fully in touch with current technology were sometimes disappointed. Thus when Daniel Colladon in 1832 invited members of the Academy to inspect his steamboat, the invitation was accepted by Poncelet, Navier and Dupin. He was surprised at the ignorance of these distinguished scientists, all members of the mechanics section, about marine steam boilers. He remarked:

> I was astonished to see how little they know about the construction of machines. I was obliged to demonstrate each piece to them.[4]

But let us return to the perspective of the Academy as the judge. As early as 1775 the Royal Academy of Sciences had resolved not to accept any further communications on the following subjects:

> The duplication of the cube, the trisection of the angle, the quadrature of the circle, or any machine as showing perpetual motion.[5]

The resolution went on to speak of mechanics who had wasted their means, their time and their genius on this will-o'-the-wisp when it was well known that friction alone in the real world would reduce any initial force. No power, therefore, could produce an effect equal to its cause. The Academy also rejected the argument that any of these questions, although misconceived in themselves, might give rise to some interesting problems in their solution.

Unfortunately in the revolutionary turmoil, this regulation, like many other things, was forgotten. And even if it had been remembered, the new political climate of equality and fraternity made it unacceptable, at least in the forthright language of the old regime. Also the new Institute wanted to be seen to be keeping its distance from the old and supposedly corrupt Royal Academies. Such regulations could be seen as arbitrary and authoritarian. In some ways, therefore, we shall see that the nineteenth-century Academy had to re-live the experience of the eighteenth century. Also, with increasing industrialisation in

[4] John H. Weiss, *The making of technological man*, Cambridge, Mass., 1982, p. 137.

[5] *Histoire de l'Académie Royale des Sciences*, 1775, 61–6, quoted by Y. Elkana, *The discovery of the conservation of energy*, Cambridge, Mass., 1974, pp. 19–30. A standard source on the subject of perpetual motion is: Henry Dircks, *Perpetuum mobile; or search for self-motive power during the seventeenth, eighteenth and nineteenth centuries*, 1861. For a list of references relating to the Academy of Sciences in the period 1838–58, see pp. 103–5.

the nineteenth century, there were further pressures in society to look for new sources of power.

In 1822 we find a certain M. Ferrand, a landowner from Tournon who had previously written to the Academy on the subject of balloons and boats, submitting the plan of a wheel which would serve as a source of power for machines.[6] Instead of rejecting his memoir as no more than a perpetual motion fantasy, the Academy appointed a commission consisting of Girard and Dupin, who persuaded the inventor to produce a model and they were soon ready with a report.[7] The inventor had devised a series of moving weights and levers with which he hoped to keep the wheel in perpetual motion. The inventor admitted that his model did not actually work and Dupin tried to convince him that this was because of a fault inherent in the system and not because the model was too small or badly made. It was only too easy for optimistic inventors to extrapolate to an ideal world, where their practical problems would be solved by building on a large scale or by eliminating friction.

There were other areas where it was possible for conflict to arise between the practical and the theoretical approach. One of these was the manufacture of iron and steel. Even in the twentieth century it has proved possible to be a good foreman in an iron works without formal instruction in chemistry and physics. If we put to one side the question of the analysis of the iron ore, where chemistry has unquestionably a major contribution to make; in the manufacture of iron and steel it is often practical experience of the feel and appearance of the material, which is crucial in making a good product.[8] Yet the industry (unlike the chemical industry) was not carried out anywhere near Paris and ironmasters would not normally have had contact with the Academy. In 1852, however, a certain C. E. Jullien, who described himself as an 'engineer', began to send communications to the Academy on the subject of his trade. Jullien had wide practical experience in the manufacture of steel, including some time at the great steel works at Le Creusot in Lorraine. In the late 1840s he had also worked for one of the new railway companies. In his early days, however, he had been a student at the Ecole centrale des Arts et Manufactures and therefore had a basic scientific education and perhaps some academic ambitions.

On 5 April 1852 we find Jullien presenting to the Academy a note on the hardening of steel.[9] The Academy immediately appointed a commission to examine the memoir, a commission consisting of Poncelet, Combes and Séguier. A few weeks later Jullien, hoping to gain the approval of an eminent chemist, submitted a further memoir on the subject to J. B. Dumas.[10] But it was not Dumas but Combes who seems to have been the Academician most interested

[6] *P.V.I.*, 7, 70, 76, 311.

[7] *Ibid.*, 337–8 (17 June 1822).

[8] D. S. L. Cardwell, 'Science, technology and industry' in G. S. Rousseau and Roy Porter (eds.), *The Ferment of Knowledge. Studies in the historiography of eighteenth-century science*, Cambridge, 1980, pp. 449–83 (464–5).

[9] *C.R.*, **34** (1852), 530.

[10] *Ibid.*, 655.

in his work. Combes (1801–72) had been a member of the mechanics section since 1847 and was a professor at the Ecole des Mines. On 7 June we find Combes, acting as the patron of Jullien, presenting a revised version of his original memoir and asking that one or two chemists be added to the commission previously appointed to examine his work.[11] The names of Dumas and Berthier were accordingly added. Encouraged by what he interpreted as the interest of the Academy, Jullien presented a further memoir the following week.[12]

He had now advanced into the field of chemistry and wanted to argue that the so-called carbides of iron and steel are no more than solutions of carbon of variable composition. Jullien, not knowing that reports on ordinary work had now become the exception in the Academy rather than the rule, waited for a report on his work. He suffered many years in silence. In 1861 he finally made a formal protest to the Academy,[13] but to no avail. The most he could do to attract attention to his ideas was to address further memoirs from time to time to the Academy.[14] He also studied and tried to engage with the work of members of the Academy, notably Fremy, Pelouze and Chevreul.

It seems to have been the failure of the chemists to take him seriously which annoyed Jullien most and this explains why in his major attack on the scientific establishment, a book published in 1870, he singled out that science.[15] He accused the leading chemists of poor understanding of the difference between a compound and a solution, an accusation which, with hindsight, we might say had some justification in the case of metal alloys. Yet he insisted on going beyond metallurgy and turned his book into an explicit attack on the 'new chemistry' of Lavoisier and the related nomenclature. If such a book had appeared in the 1790s or even early 1800s it would have been understandable, but the fact that it appeared as late as 1870 suggests how far removed Jullien was from the scientific community. He concluded with a general attack on the honours, titles and decorations which, he suggested, were all symbols of corruption. He had failed to storm the citadel of the Academy and in despair he offered an original, speculative and highly personal science which was, of course, generally ignored. Yet however much of a crank he may have appeared, the Academy scrupulously recorded his repeated interventions.

3. *Out-of-date science*

We have mentioned the experience of the former Royal Academy. In the eighteenth century the Royal Academy of Sciences had often been approached by people claiming to have solved the age-old problem of the quadrature of the

[11] *Ibid.*, 879.
[12] *Ibid.*, 913.
[13] *C.R.*, **52** (1861), 640–1 (1 March 1861).
[14] E.g., *C.R.*, **59** (1864), 1083.
[15] *Chimie nouvelle, ou le crassier de la nomenclature chimique de Lavoisier*, Paris, 1870. It may be noted that '*crassier*' (slag heap) was his favourite term of abuse (see pp. x, xi).

circle, that is finding a square whose area is the same as that of a given circle. Much of the enthusiasm with which this problem was pursued was related to a persistent rumour that the Academy was prepared to reward handsomely any solution to the problem. There was no foundation for the rumour but it persisted in popular thought well into the nineteenth century.[16] It was not long after the foundation of the National Institute that the First Class was approached with a supposed solution and it accordingly took the opportunity in 1797 of saying that it would not examine such memoirs. Yet the 'solutions' continued to be sent and the Academy was not consistent in its refusal to examine them. Thus in 1820 Arago was asked to give a quick oral response to a memoir on the subject, such a response counting as the lowest level of report.[17] Indeed, there were other cases during the Restoration of the Academy not refusing such memoirs outright.[18] In 1823 such a memoir was forwarded to the Academy by the Minister of Public Instruction and Cauchy was consequently obliged to employ the politest terms in dismissing it.[19] When a further collection of geometrical problems was presented in 1830 a quick perusal by Poisson during the meeting confirmed that they involved the quadrature of the circle and the Academy declined to accept the offering.[20] Supposed solutions to the problem continued to be sent after the publication of the *Comptes rendus* in 1835, but an editorial policy was later adopted of excluding any mention of this problem[21] in the hope that it would simply die away.

Sometimes the science presented to the Academy represented not simply ignorance but ideas of a bygone age or a discredited theory. Such was the case when, in 1824, the elderly pharmacist Opoix (b. 1745) presented a memoir on the theory of colour and of inflammable bodies.[22] Opoix still accepted some aspects of the theory of phlogiston of the previous century. Phlogiston had been used by some chemists to explain both combustibility and colour. Opoix belonged to a tradition which believed that colour was to be explained in purely qualitative terms as an essential principle in bodies, rather than as a physical effect of the surface. But the mathematisation of light was a Newtonian achievement going back to the seventeenth century, and there is some irony in the fact that the person chosen by the Academy to report on the work of Opoix was Fresnel, whose advanced mathematical analysis of light had opened yet another chapter in the history of mathematical optics.[23] Fresnel and Opoix represent two extremes. They could have been separated by at least a century, yet they happened to face each other in Paris in the year 1824. Fresnel's demolition of Opoix was polite but it was no less definite. The sophistication of

[16] *C.R.*, 67 (1868), 792–3.

[17] *P.V.I.*, 7, 71.

[18] See, for example, *P.V.I.*, 6, 309 (20 April 1818).

[19] *P.V.I.*, 7, 546.

[20] *P.V.I.*, 9, 474 (9 July 1830).

[21] *C.R.*, 67 (1868), 793.

[22] *P.V.I.*, 8, 38, 134. The text of Fresnel's report (wrongly dated 1826) is given in his *Oeuvres* (3 vols., 1866–70), vol. 2, pp. 724–8.

[23] Geoffrey Cantor, *Optics after Newton. Theories of light in Britain and Ireland, 1704–1840*, 1984, pp. 150ff.

mathematics had succeeded the instinctive common-sense approach to nature typical of the earliest science. With the dawn of the nineteenth century and the development of specialised disciplines, one would not really expect a useful contribution on optics from a pharmacist. Yet the Academy was obliged to consider each case without prejudice.

It may seem surprising that in the nineteenth century some people were still debating issues which are normally considered to have been settled in earlier centuries. What is really interesting, however, is not the continuance of popular beliefs or the obvious failure of the world of science to communicate with many literate Frenchmen. Rather it is the juxtaposition of the two totally different worlds: the first almost timeless, the second representing the greatest refinement of modern professional science. A particularly bizarre communication was addressed to the Academy in 1827. It was not simply that M. Hurault, who also had the title marquis de St Denis, should have something of his own to communicate on the calculation of the distance between the sun and the earth; rather it was his demand that a public subscription should be opened for his benefit, after which he would agree to communicate further details of his 'discovery'.[24]

Again, although we might expect attacks on Newton in the early eighteenth century, it comes as more of a surprise to find this in France a hundred years later. In May 1824 Jean-Baptiste Souton addressed a manuscript to the Academy, the first part of a 'refutation of Newton'.[25] Souton felt himself inspired by the religious sentiment of the period to accuse Newton of atheism. Ampère and Cauchy were delegated to deal with the memoir and might well have conveniently forgotten it, had Souton not presented a further memoir on the subject in June. By July he was demanding some response from the Academy. The commission had to indicate as politely as possible that the memoir contained a variety of naive scientific claims. The author, unwilling to accept such criticism, tried to present to the Academy a critique of the report but the official body of science carefully refrained from engaging in debates with members of the public.

4. *Exercises in speculative cosmology*

Although the Academy encouraged the submission of scientific research at the highest level, it was also presented from time to time with research at the other end of the spectrum of excellence. Some of these memoirs were ambitious works which purported to reduce the whole of the natural world to a few general principles, thought out imaginatively by the author but without any reference to the scientific knowledge of the time. Thus in 1818 M. Déan sent the Academy a copy of his book entitled *Système de l'univers*. Déan, however, had no monopoly on such ambitious titles and was disgusted to find that another private individual called Paris had a book with exactly the same title published in 1819. By 1820, having waited in vain for two years for some comment on his book,

[24] *P.V.I.*, 8, 551.

[25] *P.V.I.*, 8, 78, 93, 101, 112, 136–8, 149.

Déan reminded the Academy of the delay, a complaint which was formally passed on to Laplace and Poisson[26] as the members of the Commission originally appointed to report on the book, a duty which they had conveniently overlooked. Yet another *Système du monde*, this time by Desrivières, was submitted to the Academy in 1834. Ampère and Savary discharged their duty of reporting on the book by simply quoting from it. The sun was said to lie at the centre of the universe and its rays were said to extend to the edge of the universe, whence they were reflected to the centre. Given such wild speculations, the memoir should not take up any more of the Academy's time.[27] In 1828 and 1829 a certain Nicholas de Boeuf addressed memoirs to the Academy on the relative movements of the earth and sun, apparently to deny the annual motion of the earth. Biot and Arago were formally appointed to examine the work but could not take it too seriously. The author appeared at the Institute, demanding to be heard, but was turned away.[28] All of these authors were obviously ignorant of science but not all the statements they made were necessarily wrong. Also in a subject like cosmology we may reflect that the actual knowledge then possessed by orthodox science was minimal.

It was in relation to a booklet by Demonville entitled *Vrai système du monde*, sent to the Academy in October 1831, that the minutes give most details about the problems of the Academy in dealing with low-level independent scientific 'contributions'.[29] The Academy first tried to ignore the contribution, but Demonville persisted. He even had a face-to-face discussion with the astronomer Bouvard, who tried in vain to persuade him that his ideas were totally erroneous. Demonville then decided to involve the Ministry of Public Instruction, which forwarded his memoir to the Academy. Forced to take the memoir more seriously, Bouvard, on behalf of the Academy, decided to take the easy way out and he devoted the greater part of his report to quotations from those parts of the book which would most readily be seen to be absurd, for example, speculations about an ether with 'all the properties of glass'. Although the earth, moon and sun were accepted as real, the planets were dismissed as optical illusions! For good measure Demonville even included a demonstration of the quadrature of the circle. The report ends with the unusually harsh conclusion that the Academy should refuse its approval. A more common formula for such contributions was to proceed as if the work had not been sent.

[26] *P.V.I.*, 7, 9.

[27] *P.V.I.*, 10, 432.

[28] The relevant meetings mentioned in *P.V.I.* were on 22 December 1828 and 2 February, 13 April, 10 August, 31 August and finally 30 November 1829: M. Le Boeuf demande à être admis, séance tenante pour défendre son Mémoire contre le mouvement de la terre. Cette demande soumise à l'Académie n'a pas de suite. (*P.V.I.*, 9, 362.)

[29] *P.V.I.*, 10, 167–9. The published minutes, 1795–1835, contain a fair number of pseudo-scientific submissions, usually mentioned only briefly. Such contributions tend to drop out of sight after the founding of the *Comptes rendus*, although it was the rule that titles of all books and memoirs received should be recorded.

No study of the fringe of the Academy would be complete without some mention of Hyacynthe Azaïs (1766–1845). Azaïs, philosopher and writer, had been an obscure tutor in the South of France under the old regime. Responding to the revolutionary turmoil, he felt that he had discovered a principle of compensation between good and evil according to a providential justice. Under the Empire he came to Paris. Azaïs first appeared before the Academy on 15 September 1806 to read a memoir on caloric. After a quarter of an hour the Academy had heard enough but, undaunted, he resolved to have his work printed privately.[30] Having failed to obtain any encouragement from the official body of science, he was fortunate enough to have a private conversation with Laplace, who told him frankly that his work was valueless;[31] yet he insisted on interpreting this hostile interview as patronage and asked Laplace to present his work to Napoleon. He also wrote to the First Class claiming the grand prize founded by Napoleon for a major discovery on galvanism.[32] While making regular incursions into science, Azaïs also thought of himself as a philosopher and wrote on everything from politics to education.

Under the Restoration he gave courses of philosophy at the Athénée des Arts where his colleagues included Pouillet, professor of physics, Francoeur, professor of astronomy, Blainville, professor of natural history and Magendie, professor of physiology, all to become sooner or later members of the Academy of Sciences. Azaïs published his lectures in 1824 in eight volumes under the title *Cours de philosophie générale, ou Explication simple et graduelle de tous les faits de l'ordre physique, de l'ordre physiologique, de l'ordre intellectuel, moral et politique*. He saw all parts of philosophy, natural and moral, as linked together. He said that he welcomed digressions in his lectures, since he wanted to range widely over human experience. Disciple of Plotinus and Rousseau, he found it difficult to come to terms with nineteenth-century views on the demarcation of knowledge. He offered a system of universal explanation[33] in place of the limited explanations of the scientists. Thus he claimed to have discovered that the world represented a position of equilibrium between the two opposing forces of expansion and compression.

Enthusiastic partisan of phrenology, Azaïs gave lectures in his own house and in May 1831 he formally requested the Academy of Sciences to appoint a commission to attend a new course which he was giving in his garden.[34] Rebuffed by the Academy, he still sent it various memoirs which were also ignored. He was not too bitter, although he did complain in a letter published in the *Moniteur* that he had not been taken seriously.[35] If the Academy were to accept his 'universal system' it would have to change its theories, its language and its books, which was not likely. He saw the Academy as representing a

[30] *Mémoire sur le mouvement moléculaire et sur la chaleur*, 1806. This incident is mentioned by Robert Fox, *The caloric theory of gases from Lavoisier to Regnault*, Oxford, 1971, p. 119n.

[31] Azaïs, *op. cit.*, p. 93.

[32] *P.V.I.*, 3, 439 (27 October 1806).

[33] *Cours d'explication universelle*, 1834.

[34] *P.V.I.*, 9, 635.

[35] *Moniteur*, 6 February 1839, p. 229.

federation of special interests: 'how many enlightened men in the Academy of Sciences would not declare themselves in favour of the universal system if only they were not Academicians!'[36] If the Academicians would abandon their conspiracy of silence ('coalition silencieuse') and join him in the search for Truth, the human mind would rise to new heights and understand clearly the constitution of the universe.[37] In 1839 he even read extracts from a manuscript with this title to the Academy and Dumas afterwards spoke to him politely about it. Easily encouraged, he published it as a book, dedicated to the Academy.[38] At this late stage he claimed to be indebted to various figures of the scientific establishment[39] but at the age of seventy-four it was perhaps too late for him to start taking lessons in orthodox science.

5. *The Academy's judgement of fringe science*

Some mention has been made of the subject of phrenology. The Academy experienced an early contact with one of the founding fathers of phrenology, Franz Joseph Gall (1758–1828) who, after suffering a ban on his controversial ideas in Vienna, arrived in Paris in 1807. He gave a series of popular lectures which were well attended, but he was particularly concerned to obtain the approval of the French scientific community. Accordingly within a year he and Spurzheim presented a memoir to the Academy entitled 'Research on the nervous system in general and on the brain in particular'. The memoir was submitted on 14 March 1808 but, due to pressure of other business, it was carried over to the meeting of 21 March when it was read and a commission appointed consisting of Tenon, Portal, Sabatier, Pinel and Cuvier. It was an indication of the efficiency of the Academy that by 25 April the five commissioners, with Cuvier as rapporteur, had drawn up a full and detailed report, occupying some fifteen large quarto pages in the version of the minutes of the First Class, which was later published.[40]

In so far as the memoir put forward the doctrine of cerebral localisation, relating directly different parts of the brain with the moral and intellectual qualities of individuals, the Commission began the report with a disavowal. No Academy of Sciences could be expected to pronounce authoritatively on such questions. In other words, to the extent that the memoir was a scientific study of the brain they were happy to examine it, but in so far as it raised philosophical issues, such as materialism (the aspect which had brought Gall into conflict with the authorities in Vienna), it was hardly the duty of the Academy of Sciences to make a pronouncement. The Commission focused on the anatomical as opposed to the physiological aspects of the memoir. By refusing in principle to consider the relationship between structure and function, the Commission was implicitly undermining the basis of Gall's doctrine yet, as a modern historian has pointed

[36] *De la phrénologie*, 1839, vol. i, p. 44.

[37] *Ibid.*, pp. 45–6.

[38] *Constitution de l'univers*, 1840.

[39] *Ibid.*, pp. 303ff, 418.

[40] *P.V.I.*, 4, 48–63.

out, neither he nor Spurzheim had presented any compelling evidence of that relationship.[41] In its conclusion the Commission spoke favourably about some parts of the memoir but repeated its refusal to comment on the issue of cerebral localisation. It should be noted that it did not condemn the theory.

It is understandable that the First Class of the Institute should have made the most of its great scientific expertise and should have declined to become involved in any ideological discussion. Yet the claim was made at the time, and has been repeated uncritically by modern historians, that Cuvier was rapidly won over by Gall's philosophy but that 'subsequently under evident pressure from Napoleon, a report was drawn up by the Institute in which Cuvier...made an about face and declared against Gall and Spurzheim's work'.[42] Looking at the official report today many would consider it to have erred on the side of caution, but it could also be said that the Institute and the Napoleonic regime allowed Gall a fairer hearing and a more constructive appraisal in Paris in 1808 than he would have received in any other European capital at that period.

It would be a gross error (but perhaps a pardonable one) to suppose that the task of the Academy was simply to distinguish between good and bad science. 'Good science' is of many kinds and so is 'bad science'. 'Good science' presumably means scientific work carried out according to certain rules and giving results with some contemporary validity, even if subsequently disproved. This would include the simple repetition of work which had previously been carried out by others without this fact necessarily being known to the author. Such work might have educational value but would be of little interest to the Academy, which was interested in *original* work.

Similarly it is dangerous and possibly misleading to speak of 'bad science'. At its most extreme it could be fraudulent, but there were few cases of blatant dishonesty in the nineteenth century. Delusions were usually self-delusions due to excessive enthusiasm and the absence of critical self-appraisal. Another major problem was simple ignorance. Many people today would like to believe that science is democratic – that any reasonably intelligent person could, if they so wished, make some contribution, even if only very minor – to the study of the natural world. Yet already by the early nineteenth century the number of areas in which this was possible had been greatly reduced. Not only was some basic training necessary in scientific procedure, but it was also necessary to know what problems had already been tackled and with what degree of success. Too often well-meaning figures emerged with some naive statement which might

[41] Robert M. Young, 'F. J. Gall', *D.S.B.*, 5, 250–6 (p. 252).

[42] Michael Shortland, 'Courting the cerebellum: early organological and phrenological views of sexuality', *B.J.H.S.*, 20 (1987), 173–99 (p. 176). A fairer perspective is provided by Robert M. Young, *Mind, brain and its adaptation in the nineteenth century*, 1970, p. 25. Disappointingly, however, Young also repeats the anti-Napoleonic story (p. 56). For another myth about authoritarian science in Napoleonic France see Maurice Crosland, 'Humphry Davy – an alleged case of suppressed publication', *B.J.H.S.*, 6 (1973), 304–10.

well have been made in ancient Greece for all the knowledge it showed of early modern science.

The Academy was primarily concerned with work at the frontiers of knowledge and here contributions could be of several kinds. They could be a simple extension of known experiments along a scale of, say, temperature or pressure. They could be the discovery of further instances or species which fitted in well with existing knowledge. Alternatively, they could be the presentation of an entirely new method, for example in mathematics, or the discovery of a new experimental phenomenon, unlike anything known previously. It is hardly surprising that it was in the examination of completely new fields that the Academy was at its weakest. Often the authors were unknown young men who *at the time* were not clearly distinguished from cranks. A mathematical memoir, as in the case of Galois, might introduce a whole series of new concepts which might not necessarily be clearly explained. (Mathematical geniuses are not known for their patient exposition of intermediate stages in calculation.) To examine such work requires the investment of a considerable amount of time. If the referee is himself a distinguished mathematician, it is understandable that he might give priority to the development of his own ideas. To examine patiently an obscure and very original mathematical memoir requires an act of faith which only hindsight makes reasonable. For one genius there were dozens of very ordinary and sometimes muddled mathematicians.

We may consider two cases of mathematicians who presented their work to the Academy hoping in vain for recognition. There are interesting parallels although the first, Wronski, is hardly known and could easily be dismissed as a crank while the second, Galois, is now honoured as a brilliant mathematician and, if his talent was not recognised by his contemporaries, this might seem to be entirely their fault.

The first character, Hoëne-Wronski (1776–1853), had some mathematical training, although he was also something of a philosopher. He was later to write works on Messianism, the philosophy of history and the union of philosophy and religion. Originally a Polish artillery officer, he acquired French nationality in 1800. In 1810 he moved to Paris and submitted his first memoir on the foundations of mathematics to the Institute.[43] The First Class appointed a commission consisting of Lagrange and Lacroix, who had considerable difficulty in understanding the complex memoir. Nevertheless, after commenting on the problem of intelligibility, they produced a mildly favourable report.[44] Although the commissioners accused the author of some obscurity and vagueness, they thought that the First Class should beware of repudiating new ideas. They therefore urged Wronski to develop his ideas and apply them to specific cases which might show their value. The following year Wronski was back with a 'refutation of the theory of functions' of Lagrange, and again a commission was

[43] *P.V.I.*, 4, 373 (13 August 1810).

[44] *Ibid.*, 385–7 (15 October 1810).

appointed.[45] Arago was less encouraging in his report.[46] First he accused Wronski of trying to introduce a new language, which is of course a common feature of the work of ambitious pioneers. Much of the memoir, he said, was unintelligible. (It may be that in mathematics there is greater justification than in most other subjects for the introduction of new symbolism, provided that it is related to the accepted symbolism and is clearly explained.) Arago also complained that Wronski made an annoying habit of introducing formulae as a kind of enigma, which he invited other mathematicians to solve. On the main theme of the paper Arago defended Lagrange from the criticisms offered.[47] It may be tempting for outsiders to seek to make a place for themselves by attacking established figures but it is usually counterproductive.

The simplest way for the modern writer to deal with Wronski would be to ignore him, yet this has not been the policy of the *Dictionary of Scientific Biography*.[48] A fellow Pole, Jerzy Dobrzycki, argues that Wronski made definite contributions both to mathematics and to the philosophy of mathematics. Yet he admits that he had 'a marked psychopathic tendency' and was given to 'grandiose exaggeration of the importance of his research, [and] violent reaction to the slightest criticism'. It would seem that Wronski was a gifted mathematician whose contribution to the subject was unfortunately obscured by the author's insistence on using a private language and by his increasingly dogmatic and all-embracing philosophical ideas, which were not appreciated by the professional mathematicians of the First Class. In desperation Wronski went to England to seek recognition, but was even less successful there.[49]

Turning now to Galois (1811–32), the best known story about him is that he died in a duel, having devoted his final hours to writing down a summary of his mathematical ideas.[50] This romantic story, although basically true, needs to be put in context. That Galois was an intellectual genius is agreed, but such was his originality that he failed the entrance exam to the Ecole Polytechnique because he refused to follow the methods proposed by the examiners. In 1829, while still a student, he presented a memoir to the Academy but his rapporteur, Cauchy, did not present a report. Instead he advised the young man to re-submit his work, taking into account the parallel work of Abel. When Galois did submit a new memoir in February 1830, hoping to win the *grand prix* in mathematics, it passed into the hands of Fourier, who died shortly after and the memoir was lost.

Galois' bad luck was to continue and he now began to believe that he was

[45] *Ibid.*, 522 (9 September 1811).

[46] *Ibid.*, 553–4 (11 November 1811).

[47] Wronski in iconoclastic mood later published an attack on the work of the other outstanding mathematician, Laplace: *Critique de la théorie des fonctions génératrices de M. Laplace*, 1819.

[48] *D.S.B.*, **15**, 225–6.

[49] Hoëne Wronski, *Sur l'imposture publique des savants à privilèges*, London, 1822. He even quoted one sentence from an Academy report out of context in order to support his claims (p. 12).

[50] This account of the life and work of Evariste Galois is heavily dependent on the article by René Taton in *D.S.B.*, **5**, 259–65.

being persecuted by the authorities. Already of a liberal persuasion like his father, who had tragically committed suicide the previous year, Galois became extremely politicised. Making friends with the republican leaders Blanqui and Raspail, he divided his time between political activities and mathematics. A further mathematical memoir submitted to the Academy in January 1831 was given a report by Poisson, who coldly reported that part was unintelligible and that some of the remainder was similar to work previously published by Abel.

Poisson wrote:

> We have made great efforts to understand the demonstration of M. Galois. The reasoning is not clear enough, nor well enough developed, for us to be able to judge its value and we are not even in a position to give a clear idea of it in this report. The author explains that the proposition, which is the subject of his memoir, is a part of a general theory capable of many other applications. It often happens that the different parts of a theory throw light on each other and are easier to understand together than in isolation.[51]

Poisson went on to say optimistically that it would be necessary to 'wait until the author has published his work in its entirety' for them to give a definite opinion. Without a clear understanding of the memoir, the commissioners could not ask the Academy to give it its approval.

It was unfortunate that Cauchy, who would probably have understood the importance of Galois' work, had left France after the Revolution of July 1830. If the Academy did less than justice to Galois it was partly because he was too original and quite uncompromising. It was not until at least fourteen years after his death that his work became available to mathematicians and the development of mathematical research in the intervening years created a climate favourable to its reception. The posthumous reputation of Galois is, therefore, totally at variance with the slight recognition he received during his tragically short life.

In the mid-century Chevreul became involved in several cases of apparent pseudo-science. It seems to have been a letter by a certain M. Pons, read to the Academy in May 1853[52] on the alleged phenomenon of table turning, which prompted Chevreul to study the question. By July he had added the divining rod to his investigations[53] and by the following year he was reading memoirs to the Academy on the subject.[54] Yet these memoirs were always reported in the *Comptes rendus* by no more than the title and we may suppose that the secretary had ruled that this was not a proper subject for the official journal of the Academy. Meanwhile Chevreul had collected his studies into a book, and when he presented this to the Academy in August 1854 the secretary relented to the extent of providing a brief summary of the book in the *Comptes rendus*.[55]

[51] Institut, Fonds Joseph Bertrand, MS.2031, f. 71 (4 July 1831).
[52] *C.R.*, **36** (1853), 921.
[53] *C.R.*, **37** (1853), 139.
[54] *C.R.*, **38** (1854), 295; **39** (1854), 169.
[55] 'De la bagette divinatoire, du pendule explorateur et des tables tournantes au point de vue de l'histoire, de la critique et de la méthode expérimentale', *C.R.*, **39** (1854), 321.

Chevreul explained these phenomena as due to involuntary muscular movements of the subject and denied that they had a supernatural cause.

The early nineteenth century has been chosen as the most fruitful period in which to look for examples of the Academy's reactions to claims on the fringe of orthodox science because this is the period when reports were most conscientiously undertaken. In the second half of the century there are very few reports relevant to 'pseudo-science'. Yet we may take one episode which comes at the end of our period which will serve to complement the earlier examples. Fortunately the case of Gustave LeBon's 'black light' has been the subject of a meticulous modern study which carefully places this curious episode in the full context of *fin de siècle* intellectual history.[56] We may therefore be content with a brief account.

Gustave LeBon (1841–1931) was a psychologist and popular writer, best known for his book on the psychology of crowds which characterised the mentality of crowds as non-rational. He also dabbled in other sciences and became interested in X-rays. It is surprising that an amateur physicist like LeBon should ever have been taken seriously by the Academy at the turn of the century, when physics had become so specialised and mathematical. But the discovery in the late nineteenth century of cathode rays, X-rays and radioactivity had broken down many of the old barriers. Moreover, LeBon had become friendly with several members of the Academy, and among members of the dining club he organised were the mathematicians Henri Poincaré and Emile Picard, as well as the influential politician Gabriel Hanotaux, who considered LeBon to be a great physicist who had 'destroyed matter'. This latter comment is a hint of the wave of anti-materialism at the turn of the century which welcomed certain philosophical implications of the new physics.

It is against this background that we turn to the first communication to the Academy by LeBon in January 1896 on 'black light'.[57] He claimed to have discovered a new invisible radiation which affected a photographic plate in a closed box. LeBon's paper was presented to the Academy by the physiologist D'Arsonval and Henri Poincaré and confirmation was soon sent to the Academy from different parts of France.[58] Doubt, however, was cast on the validity of the phenomenon by Auguste and Louis Lumière, since they were unable to reproduce it.[59] In 1897 LeBon extended his experiments using a sheet of ebonite, but his report to the Academy[60] was almost immediately criticised by Becquerel, who carried out a demonstration before LeBon's supporters.[61] Although Poincaré, one of the most sympathetic of LeBon's supporters, invited him to the Academy to settle the matter, LeBon seems to have kept his distance. When he

[56] Mary Jo Nye, 'Gustave Lebon's black light; A study of physics and philosophy in France at the turn of the century, *Historical Studies in the Physical Sciences*, **4** (1974), 163–95.

[57] *C.R.*, **122** (1896), 188–90.

[58] *Ibid.*, 500–1.

[59] *Ibid.*, 463–5.

[60] *C.R.*, **124** (1897), 755–8.

[61] *Ibid.*, 984–8.

reappeared a few years later with further ideas on radiation, he prudently refrained from presenting his work to the Academy; instead it appeared in the *Revue scientifique*. Few reputable scientists had any dealings with LeBon after about 1900. He antagonised even his friend the Academician and chemist Armand Gautier, by making fantastic claims for priority in connection with radioactivity but this did not prevent him standing several times as a candidate in the physics section of the Academy in the 1920s. He was not successful. It might be appropriate to end this story by quoting from a letter to LeBon from Gautier in 1903. Gautier said that he was not so committed to the idea of the stability of matter as to refuse to look at evidence to the contrary. But should one not look for an explanation *within* the existing framework before overturning it? He continued:

> Moreover, does not *official science*, which is obliged to take some responsibilities to separate ideas from generalities, always show a little fear? But this very fear keeps it healthy by keeping it on its guard. Those who question it do so at their own risk and peril.

Fortunately the Academy did not make the mistake of officially recognising 'black light', any more than it accepted 'N-rays' when these were reported by Blondlot, one of its correspondents, in 1903.[62] In both cases other observers claimed to have confirmed the new phenomenon, but this tells us more about human psychology than about physics. Both 'discoverers' received support in the Academy from Poincaré and D'Arsonval rather than from the Academy as a whole.[63] The two alleged phenomena were reported at a time when everyone was excited by the new forms of radiation which were being discovered, and ambitious men on the fringes of the Academy inner circle wanted to make a name for themselves. Unfortunately their ambition got the better of their judgement.

6. *Some critics*

We have discussed at some length the barriers imposed by the Academy on unorthodox science. But we must not think of criticism and rejection as an entirely one-way process. Indeed we have already had a few indications that not all the would-be *savants* who were turned away accepted their rebuff meekly and without comment. It was a natural reaction of someone excluded to say that there was something wrong with the selection process or perhaps even with the very concept of an Academy. We therefore now come to consider a few cases where individuals who originally approached the Academy as supplicants gradually emerged as major critics. It is interesting to observe how what began

[62] 'Sur de nouvelles sources de radiation', *C.R.*, **136** (1903), 1227–9. Mary Jo Nye 'N-rays: An episode in the history and psychology of science', *Historical Studies in the Physical Sciences*, **11** (1980), 125–56. For further discussion see pp. 432–4.

[63] Although Blondlot received the Le Conte Academy prize in 1904, the Academy report, drawn up by Poincaré, states that this was to reward him 'pour l'ensemble de ses travaux' and the work on N-rays was played down.

as a dispute about science was transformed into a discussion about the organisation of science. In other words, there was a transference from the cognitive to the social plane.

We may understand the Academy better if we take into account not only its members and those who aspired to membership, but also its critics. One of its most famous critics, and a critic of the nineteenth-century scientific establishment in general, was François-Vincent Raspail (1794–1878). Raspail, who came from a poor family, originally trained for the priesthood. He refused to take his religious vows, but his seminary training provided him with a good classical education. In the early Restoration he went to Paris, where he became an outspoken republican. As he tried to support himself by tutoring and writing, he began to teach himself science and, although he never quite reached a rigorous enough standard for entry to the Academy, he carved out a definite field for himself in organic chemistry and biology, particularly in microscopy and the study of cellular structure, which was to be an important area of advance in his generation. He achieved enough in science to merit a few pages in the multi-volume *Dictionary of Scientific Biography*.[64] Raspail managed to combine a political and scientific career. Following the Revolution of February 1848, he was one of the left-wing candidates in the Presidential election of December of that year but only received 36 920 votes, compared with over 5 million for Louis Napoleon. At least he received twice as many votes as the writer Lamartine. Although he was one of the most prominent republicans in the reign of Louis Philippe, he did not give up scientific research entirely and indeed also had a career as a writer. He was the author of text-books, scientific papers and political tracts. Between the world of politics and the world of science there was academic politics, and in one of his text-books of organic chemistry he inserted a fifty-page long preface criticising the scientific establishment, including the Academy.[65]

Raspail later reflected on the optimism with which, at the age of twenty-nine, he had presented his first completed piece of research (on the physiology of certain grasses) to the Academy in 1824:

> For two years an idea had been absorbing all my research and a hope had been consuming me. I had finally captured the idea and I thought that fortune was smiling. In those days, young men knew only one dream, a king of glory, whose temple was the Academy of Sciences, guardian of truth...I remember how I trembled when I dared to speak to one of [the Academicians] in the courtyard of the Institute.[66]

After he had presented his paper, Geoffroy Saint-Hilaire had gone out of his way

[64] Marc Klein, 'F. V. Raspail', *D.S.B.*, 11, 300–2. See also Dora B. Weiner, *Raspail, scientist and reformer*, New York, 1968.

[65] 'Avertissement historique', *Nouveau système de chimie organique*, 2nd edn, 1838, vol. 1, pp. xvii–lxiv. An extract from this preface relating to the Academy is reprinted in a modern collection of Raspail's works: D. Ligou (ed.), *François-Vincent-Raspail ou le bon usage de la prison*, 1968, pp. 456–61.

[66] Quoted by Dora B. Weiner, *op. cit.*, p. 84.

to say something encouraging, but Raspail was disappointed that the scientific world went on much as before. So much for the idealism of the young and the ambitious!

In 1833, the year after the death of his former protagonist Cuvier, Geoffroy Saint-Hilaire was elected president of the Academy of Sciences. Although this office was purely honorary and hardly compared in power with the position of permanent secretary, the new president felt that he could take advantage of his position to offer Raspail patronage and he proposed offering him a monetary reward to be taken from the generous Montyon legacy. He had already discussed the possibility of using Academy money in this way with several of his colleagues and had obtained their agreement. Raspail was quite well known to the Academy since he had often presented papers there and indeed Geoffroy Saint-Hilaire (unlike Cuvier) had shown himself well disposed to the outsider. Raspail, who had very little money, would have been glad to accept a sum which might have been as much as 10000 francs, provided that there were no conditions attached. He would in no way alter his outspoken criticisms of the government.

The award might well have taken place, given the exceptional freedom the Academy enjoyed in the disposal of the Montyon funds,[67] if news of the plan had not reached the ears of Guizot, then Minister of Public Instruction. He considered it unacceptable that the official body of science should reward generously an outspoken critic of the government. The Academy delayed taking any action. Meanwhile Raspail was arrested for taking a leading part in an alleged anti-government conspiracy. In fact the fatal meeting, in which he had played a prominent part, was concerned with press freedom and was attended by many members of the Chamber of Deputies. Raspail was therefore soon released but with increasing bitterness since he now saw how vulnerable the Academy could be to government pressure. He saw all academies as 'depending on the government, functioning according to its orders or with the permission of the government and in the interest of the views of the government.'[68] Although academies might contain men who were privately out of sympathy with the government of the day they were 'forced to stifle their emotions and their sympathy'.

When Raspail went on to claim that the Academy of Sciences made whatever sounds were required by the government[69] he was going too far. In purely scientific matters the Academy yielded to no-one. Yet when he spoke[70] of a group focused on the Academy and including among its members, candidates for election and prize-winners, patrons and their protégés, with their access to funds, there is more than a grain of truth in the implication of corruption. In so far as

[67] Maurice Crosland, 'From prizes to grants...the Montyon legacy', *Minerva*, 17 (1979–80), 355–80 (p. 366).

[68] Raspail, *Nouveau système*, p. xxvi.

[69] *Ibid.*, p. xxvii.

[70] *Ibid.*, p. xxxii.

the Academy passed judgement on the merit of scientific work, its members should keep scrupulously apart from the authors of these memoirs.[71] Just as in courts of law judges were forbidden to eat or drink with those whose cases they were considering, or receive presents from them, so it should be in the Academy. Nor should science journalists be in a position to be rewarded by the Academy.

What Raspail valued above everything was his *independence*, and his attitude, although extreme and often unreasonable, does serve to suggest that members of the Academy were, in a sense, 'scientists under control'. They were bought men. There was no question of course of Academicians automatically becoming spokesmen for the government of the day, but there might be issues on which, as part of the establishment, they might be expected to keep quiet.[72] One could put it more politely by saying that they had a greater sense of responsibility. They accepted a certain ethos, a certain discipline, which scientists outside the Academy did not have to accept. To enter the Academy was to accept a system of patronage and, therefore, to lose a certain amount of independence. But for Raspail this was doubly unacceptable because a part of the patronage related to the government. Membership or close association with the Academy therefore placed restrictions on the level of one's political activity and Raspail often felt obliged to shout his republicanism from the roof tops. Others such as Arago managed quite well to reconcile no less sincere but more moderate republicanism with wholehearted participation in the affairs of the Academy. Arago tangled with Raspail more than once. In 1835, in a report on the work of the Academy, he accused him of mischief making.[73] For Raspail criticism was almost a sacred duty.

Raspail clearly represented an extreme. He had had the opportunity to follow university courses in Paris but had never persisted to the level of taking a degree.[74] Unqualified, therefore, and without institutional affiliations or friends in high places, he was only too ready to see himself as a lone fighter for liberty and truth. He too easily allowed emotion and political sentiment to rule his judgement. His independence extended to founding his own journal, the *Annales des sciences d'observation*, which fought a losing battle to compete with other scientific journals, some of which were supported by state subsidies like the journals of the Academy. Much of his life he tried to foster a kind of science parallel to, but independent of, 'official science'. The fact that he was trying to combine chemistry, botany, physiology and microscopy in a new way did not make it easy for orthodox scientists to appreciate him. Only in the light of modern knowledge could later historians claim him as the founder of histochemistry. Not only did he attack much of the science of his time, he poured even more scorn on the medical profession as ignorant, venal and vain[75] and,

[71] *Ibid.*, p. liii.

[72] The case of Lalande's atheism is discussed in Chapter 9, Section 3.

[73] Muséum d'Histoire Naturelle, MS.2388, No. 2, Arago to Raspail, 21 mars 1835.

[74] Weiner, *op. cit.*, p. 77.

[75] *Ibid.*, p. 119.

when tried in court for the illegal practice of medicine, claimed to represent a new and superior system of medicine. He antagonised both scientific and medical establishments by appealing over the heads of these professions to the people. It was Raspail's politics rather than his science which made his opposition to the Academy inevitable.

After Raspail, one of the best known names among the critics of the Academy was that of the philosopher and mathematician Auguste Comte (1798–1857), probably best remembered as a founder of positivism and one of the major figures in the early history of the social sciences. He had entered the Ecole Polytechnique in 1814 but was expelled with several other students for political reasons in 1816. He was always something of an outsider whose economic life was a continual struggle. Employed for a few years as secretary to the utopian comte de Saint Simon, he later earned money by lecturing, examining and writing, with periodic subsidies from a circle of admirers which included J. S. Mill. Members of the Academy sympathetic to his ideas included Fourier (d. 1830), Blainville and possibly Poinsot.[76]

Comte was a competent mathematician and, like many others on the fringe of science, followed the work of the Academy with interest but with no direct involvement until 1831, when his name appears in the minutes. Already an examiner at the Ecole Polytechnique, he applied to be considered as a candidate for a vacant chair in analysis and mechanics in that institution.[77] It was the prerogative of the mathematics section of the Academy to make recommendations for the chair, but when they considered possible candidates they deliberately omitted Comte's name and he made a vigorous protest.[78] He was known to a few members of the Academy as the author of the (now famous) *Cours de philosophie positive*, the first volume of which he had presented to the official body of science in the previous year. However, this could hardly have been considered a qualification for a chair in mathematics. When in 1840 there was another vacancy for a professor of mathematics at the Ecole Polytechnique, Comte, now in his forties, thought he had a good chance of being nominated by the Academy since he had had the experience of having given mathematics lectures at the school on a temporary basis, but his hopes were again dashed. This time he not only sent a letter of protest to the Academy,[79] but took advantage of the publication of the final volume of his monumental *Cours de philosophie positive* to include in the text a public attack on the Academy.

Comte thought the Academy had too much power. He considered it unreasonable that it should decide not only on its own membership but also the membership of institutions of higher education. In the first place there was a

[76] Comte describes some half-hearted support from Poinsot in his *Cours de philosophie positive*, vol. 6, 1842, p. 471n.

[77] *P.V.I.*, **9**, 583 (21 Feb. 1831).

[78] *Ibid.*, 587 (7 March 1831).

[79] A letter dated 3 August 1840 is cited by Comte in his *Cours*, vol. 6, p. 469n. Understandably the secretaries refrained from reading out this critical letter at the relevant meeting (*ibid.*, p. 470n).

strong possibility of corruption in a system in which members of the Academy could recommend their colleagues for appointment in preference to non-members, a situation which the Academy avoided in its prize system by disqualifying its members from competing. He spoke of Academicians holding a monopoly in higher scientific education. These teaching posts were in danger of becoming mere sinecures, not much better than the pensions awarded under the *ancien régime*.[80] Secondly, Comte made a very persuasive distinction between teaching and research ability. He was the author of several text-books and treatises and he considered them more relevant qualifications for a teaching position than any number of research memoirs.[81]

After making further criticisms of the Academy, Comte pointed out that there was no representation of philosophical ideas.[82] A modest request that philosophy of science might somehow be represented within the Academy could have been regarded as reasonable, but Comte's ideas extended to megalomania, since he claimed that the new section would be so superior to the others that it should always provide from its members both a president and a permanent secretary of the Academy![83]

We have seen how Comte was driven by successive rebuffs to criticise the Academy. Many of his criticisms were probably justified, but he was so impractical a man and he took his ideas to such extremes that they gained little support. The final years of Comte's life were devoted to 'the cult of humanity' and the foundation of a positivist religion.

After dealing with two well-known names on the fringes of French science, we may mention a third critic who belongs to a later generation and in fact went further than his predecessors in writing a detailed and constructive critique of the Academy in 1869, the penultimate year of the Second Empire.[84] Inevitably the critique indirectly involved the government as well as of the organisation of science. The author, Jules Marcou (b. 1824) was a geologist and field worker, and for the previous fifteen years he had regularly been submitting papers to the Academy. He had even competed for the Cuvier prize. Although the Academy had dutifully recorded his various 'opuscules', it had never offered him much encouragement. Marcou's main argument was that the Academy was tied to its organisation of 1803 and that science had changed considerably since that time. Criticising the sections in turn, the mechanics section should, he said, include engineers and even inventors. Similarly the astronomy section should sacrifice some of its mathematical expertise to include observational astronomers. He considered that some sections should be enlarged, for example the chemistry section, which he judged to be one of the strongest sections of the whole Academy. The mineralogy section, to which he himself aspired, was

[80] *Ibid.*, pp. xxv–xxvi, 471.
[81] *Ibid.*, pp. 469n., 472.
[82] He mentions one possible exception, probably his supporter Blainville.
[83] *Ibid.*, pp. 475–6n.
[84] *De la science en France*, 1869, Deuxième fascicule, *L'Académie des Sciences de l'Institut de France*.

another area he wanted to see expanded to include a designated number of places for geologists and palaeontologists. He considered the zoology section weak; he wanted to see more expeditions and the inclusion of anthropology. Since the foundation of the Academy of Medicine in 1820, there was no longer justification for a section of medicine and surgery. Instead he would have liked to see physiology explicitly recognised.

Marcou then went on to criticise other parts of the organisation. He pointed out that the value of elections was sometimes reduced by nepotism and occasionally mediocrities with strong patronage would enter the Academy at the expense of more able candidates. When he suggested finally that membership of the Academy should carry a salary of 4000 francs, we can see again that he was in a position to appreciate another inherited weakness of the official body of science. Despite its special pleading, as a book of some 200 pages Marcou's work was one of the most sustained and perceptive criticisms of the Academy. Although he had originally been invited to prepare a private report for the Minister Duruy,[85] in the end Marcou preferred to address the public directly, thus distancing himself from a government which, in order to implement the recommendations, would have needed to make a considerable increase in its budget. War in the very next year was to mark a major crisis in French history and Marcou's critique was forgotten. Yet its existence serves to remind us that the Academy had its critics in every generation.

7. *Admission of the public*

Having discussed the vulnerability of the Academy to criticism, it is appropriate to consider the Academy as being generally on the defensive in its relations with the general public. There was always a tension in the Academy between those who felt that science was too precious and delicate to be exposed to the gaze of the uninitiated and those who felt that the Academy should be more open. In the second quarter of the nineteenth century these respective positions were held by Biot and Arago. The problem, however, goes back to the very foundation of the Institute.

Originally there were to be public meetings, four times a year, of the whole Institute coming together as a single body. However, an objection was made in the Council of 500 by Dupuis, who wanted *all* the meetings of each Class to be public:

> The public will derive great advantages for their education. Also if members of the Institute are placed under the public gaze, they will not fall asleep in their academic chairs.[86]

This comment was first a reflection on the original idea of the Institute as an educational institution and second on the latent hostility to any privileged body.

[85] *Ibid.*, fasc. 1, pp. 22–3.

[86] Conseil de Cinq Cents, 25 *ventose* an 3 (15 March 1795).

Most members of the Institute, however, felt that quarterly public meetings were sufficient and the remainder should be private. When a non-member wished to submit a memoir, he was encouraged to find an Academician willing to present it to the meeting. If he attended personally, he was normally admitted only to submit the paper and was then expected to leave. The First Class published its work and the presence of the public was generally considered to have more disadvantages than advantages. It was said that ordinary working meetings could not be expected to be polished performances but would involve preparatory work, even periods of silence.[87] Despite this determination to exclude the general public, some exceptions were made for members of certain societies such as the Société Philomatique, which in some ways represented a junior Academy of Sciences. Members sometimes invited guests but by 1800 this privilege was thought to have been abused and members were limited to one guest each, who would be required to sign in.[88] Yet this regulation was gradually forgotten and, when an enquiry was held in 1809 to find out why members were reluctant to read their memoirs at meetings, one of the reasons given was the presence of too many strangers who, some thought, might steal their ideas or at least criticise their work outside the Academy before it had been published.[89]

Once again the Academy decided that it should specify the restricted group of outsiders that would be allowed to attend its ordinary meetings. This privilege was restricted to correspondents, representatives of certain scientific societies and those who had merited the favourable attention of the Academy either by presenting two memoirs or who had won a prize. This group of some fifty names might be considered to have been aspirants to membership of the Academy.

Despite such regulations, toleration of members of the public being present at ordinary meetings was gradually extended. In 1825 the Academy, meeting in secret session, appointed a commission to examine 'the means of admitting to meetings only persons who wish to attend for love of science and with the intention of educating themselves and in such a way as not to cause any disturbance or interfere with free discussion'.[90] In other words the Academy was not so much concerned to restrict the admission to a sub-elite as to have reassurance on the *bona fide* motives and good conduct of the members of the public. The Academy was beginning to be worried about inaccurate or unfavourable reporting by journalists and the worry expressed in 1809 that Academicians' work might be stolen was still present.

The next stage in the admission of the public was to allow journalists from certain newspapers to attend meetings. When Arago became secretary in 1830 he decided that the press should have every facility for reporting accurately what

[87] F. J. L. Meyer, *Fragments sur Paris*, 1798, vol. 2, pp. 31–2.

[88] *P.V.I.*, **2**, 284, 6 *nivose* an 9 (27 December 1800).

[89] *P.V.I.*, **4**, 227.

[90] *P.V.I.*, **8**, 206 (4 April 1825). No report of this commission has been found and it seems that it abandoned its difficult task.

had been said and done at the meetings, and he had a room set aside where all correspondence and memoirs relating to the meeting were freely available for consultation by journalists. By 1842 'the ever-increasing number of people' attending the meetings of the Academy of Sciences (but not apparently the other Academies) created a further problem, that of adequate ventilation. There were numerous complaints about the air in the Academy's meeting place and some members even threatened not to attend during the summer months for fear of fainting.[91]

Yet not every Academician was happy with the new freedom of information and in particular with the freedom of members of the public to attend any meeting. In 1848 Thenard, then one of the most senior members of the Academy, suggested a return towards the concept of a privileged audience. He suggested that priority of admission should be given to those who had previously submitted at least two memoirs to the Academy which had received a favourable report, or one memoir if it had received the great honour of a recommendation for publication in the *Mémoires des Savants Etrangers*. His idea was that a list of such people (similar to the list of 1809) should be pinned up, and places should be reserved for them until a quarter of an hour before the session.[92] The commission appointed to examine this proposal was generally sympathetic to Thenard's ideas and agreed in particular that it should be regarded as an honour to be allowed to attend meetings. Unfortunately the whole report system, which had been one of the features of the Academy in the first few decades of the century, was no longer fully in operation and it could hardly, therefore, be used to regulate admission. But if the Academy accepted a public audience and it permitted non-members to submit memoirs, it insisted on complete segregation of these categories. It forbade any discussion between members and non-members or the public and in 1857 it reaffirmed this rule.[93] Thus in 1866 during one of the debates involving Pasteur the journalist Victor Meunier, sitting in the public benches, asked him whether he accepted his work and Pasteur kept silent.[94]

8. *Public meetings and some of the problems of presenting science*

According to its first constitution of 15 *frimaire* year 4 (6 December 1795) the entire Institute would hold a public meeting four times a year. The dates later chosen were the fifteenth day of the republican months of *vendemiaire, nivose, germinal* and *messidor*. Accordingly, on 15 *germinal* year 4 (4 April 1796) the Institute held its first public meeting in the Louvre.[95] Among the speeches

[91] A.S., *Commission administrative, 1829–77*, p. 38, 30 May 1842.

[92] A.S., *Comité secret, 1845–56*, p. 117, 11 December 1848; pp. 120–4.

[93] A.S., *Comité secret, 1857–69*, 14 December 1857.

[94] 22 January 1866. Plantefol in: *Académie des Sciences, Tricentenaire de l'Académie*, vol. 1, p. 104.

[95] E. Maindron, 'La fondation de l'Institut National', *Revue scientifique* (3), 1 (1881), 106–10. This version of Maindron's works provides more information than the corresponding chapter of his book.

Daunou, on behalf of the Institute, spoke of a desire for peace and harmony within the Republic. Even if the country was still at war with other nations, he was concerned more with the influence of the Institute on the general public in France, which he saw as the victory of Enlightenment over ignorance and prejudice. The secretaries of the three Classes spoke in turn about what had been achieved by their respective groups. A poem was read about the reunion of a great family, and *éloges* were delivered of recently deceased members of the Institute. The main contributions representing science were a memoir by Fourcroy on the detonation of potassium chlorate and another by Cuvier on different species of elephants. Potassium chlorate had been tried as a substitute for potassium nitrate in gunpowder, with fatal results, but it could still be presented to the general public as an example of the power of science and in particular the new science of chemistry. The choice of explosions and elephants suggest no great subtlety in appealing to the public's supposed delight in the spectacular. After an ambitious programme representing different aspects of the work of the Institute and lasting for four hours, the session ended with practical demonstrations of the detonations. In its first public presentation science seems to have been reduced to little more than pyrotechnics, but perhaps this was an example of the importance attached to direct impact on the senses, deriving from the philosophy of the *idéalogues*.

Yet the public interest in the work of the Institute was put under a strain by a programme lasting as long as four hours and for the next public meeting in July considerable effort was devoted to shortening the individual contributions so that none lasted more than a quarter of an hour and several only lasted five minutes.[96] The total session occupied a very reasonable two and a half hours. This time science came first on the programme with a very brief summary by Prony and Lacépède of memoirs presented in the two divisions of the First Class. Later in the programme Prony read an *éloge* of the recently deceased astronomer Pingré, and Delambre explained the work being carried out by the First Class on measuring the arc of the meridian. Scientific contributions at later public meetings included a memoir by Fourcroy on the problems of painting on porcelain,[97] a memoir on the new metric system by Van Swinden and experiments with diamonds by Guyton,[98] a memoir on the bleaching of cotton by Chaptal[99] and a contribution on public health by Guyton.[100] Subjects like bleaching and the new metric system came up more than once, suggesting not only that they were matters of special interest to the First Class at that period, but that these subjects were considered particularly suitable for public presentation.

From 1796–1802 the public meetings of the First Class were always held as

[96] Institut, HR7, vol. 3, no. 6, 'Séance publique du 15 *messidor* an 4'. This copy of the programme is annotated with the time occupied by each of the contributors.

[97] Séance publique, 15 *nivose* an 6.

[98] 15 *messidor* an 7, B.M. 733.g.13 (49).

[99] 15 *nivose* an 8, B.M. 733.v.13 (50).

[100] 15 *nivose* an 9, B.M. 733.g.13 (51).

part of the public meeting of the entire Institute. Thus the science contribution was always limited by the understandable but conflicting demands of the other Classes. From 1803, however, according to article 10 of the new constitution,[101] each Class of the Institute was to hold a separate public meeting, although, as a concession to unity, members of the other Classes were encouraged to attend as spectators. A separate meeting meant that previous difficulties of compression of material no longer applied. The First Class of the Institute could concentrate on science in its own right without the obligation of relating it to the interests of the other Classes. The announcement of the names of prize-winners and of the choice of future prize questions could be made in an unhurried way and it is interesting that although a *grand prix* had been founded in 1796, it was only under this new constitution that the prize system realised its inherent potential. *Eloges* of deceased members could now be longer and more elaborate. A more distinct impression could be conveyed to the public of the nobility, and even occasional heroism, of the scientific enterprise. The fact that *éloges* were now entrusted to two *permanent* secretaries chosen partly for their literary ability, rather than to temporary secretaries elected on a rotating basis, also meant an improvement in the standard of *éloges*.

Yet a major part of the annual public meetings of the individual Classes after 1803 was given over to reports to the government, thus providing an interesting variation on the theme of public accountability. Yet it was important in such a context not simply to report on work done but to explain scientific achievements and to suggest their future potential.

It was particularly after the Restoration that a serious effort was made to interest the public in science. In 1819 the Academy, mindful of the growing interest in Egyptology, asked Latreille to speak about the insects depicted on ancient Egyptian monuments. There were also some attempts to keep the public informed of the latest scientific developments. Thus on 2 April 1821 Ampère was already reporting briefly to the public on his recent experiments on electromagnetism. At the public meeting of the following year, he was able to give a fuller account of his research. In 1823 the physiologist Magendie spoke about the function of the nervous system with particular reference to what came to be known as the 'Bell–Magendie law', enunciated independently the previous year by the two named medical scientists.

We gain an impression of the atmosphere at the public meeting of the Academy of 1834 from the report given by the *Moniteur*:

> Yesterday the Academy of Sciences, under the presidency of Gay-Lussac, held its annual meeting devoted to the distribution of funds. This solemn occasion attracted a large and brilliant audience, composed in large part of notabilities of all kinds and a group of very elegant ladies. Lord Brougham [an *associé étranger* of the newly-established Academy of Moral and Political Sciences] attended the meeting, sitting next to M. Dupin senior. M. Arago read a biography of James Watt, which was heard with great attention.[102]

[101] *P.V.I.*, 2, 619–25 (p. 620).

[102] *Moniteur*, 10 December 1834, p. 2201.

This account suggests the occasion was a fashionable one but also one which was taken seriously. The 'public' may have largely excluded Parisian working men who might have been interested in the life of Watt, but what was said reached more easily the section of society with the greatest political power.

9. *The public meetings of the combined Academies*

The annual meeting of the Academy was only one of two occasions on which the scientists presented themselves to the public. The second occasion was the joint meeting at all the Academies, which perpetuated the revolutionary idea of bringing together all the classes of the Institute. We shall see that the Academy of Sciences was called upon to present itself rather differently. The technical dimension was reduced and the ideological component was correspondingly increased. It was in 1816 that Louis XVIII decreed that there should be an annual joint meeting of all the Academies to be held on 24 April to mark the anniversary of his return to France,[103] a date which was later to be changed according to the wishes of successive political regimes. It fell to Cuvier, with only a few weeks' notice, to make a speech on behalf of science. He chose as his title: 'Reflections on the current state of the sciences and their relationship with society'.[104] After a tactful reference to Louis XIV as founder of the first Paris Academy of Sciences, he launched into an extended and imaginative history of mankind from the most primitive times to the present. Man's greater understanding and control of the natural world was his constant theme. He argued that war was caused by poverty, inequality and jealousy, whereas science acted as an equalising factor, overcoming climate and natural obstacles for the greater benefit of mankind and bringing an end to previous exploitation. In the modern world, he suggested, people were usually rewarded suitably for hard work, thus increasing the stability of society. In the difficult days of the early Restoration after the many political changes and the defeat of the French army at Waterloo the previous year, Cuvier's audience was ready to think about peace and greater social stability. If science could be shown to have a part to play so much the better.

When the Academy of Sciences contributed to a joint public meeting with the other Academies, it was not only addressing the general public but its colleagues within the Institute. There was, therefore, a conscious decision from time to time to stress the *cultural* aspects of science. Thus Dupin in 1819 spoke of the influence of the sciences on the 'humanity' of European nations and Jomard spoke in 1827 on the different 'degrees of civilisation' of the African peoples, as reported by geographical expeditions. The technical side of science was not totally neglected however particularly if aspects could be found of potential general interest, such as the determination of the shape of the earth (Biot, 1818) or progress in natural history in the absence of war (Cuvier, 1824). Most of the

[103] *P.V.I.*, 6, 41–3. (p. 42).

[104] Cuvier, *Recueil des éloges historiques*, nouvelle édition, 1861, vol. 3, pp. 251–70.

technical contributions tended to stress progress. Finally there were a number of practical subjects which helped to justify science in the eyes of their colleagues: navigation (Biot, 1817), machines and industry in Britain and France (Dupin, 1821), the recent applications of chemistry to society (Cuvier, 1826) and even savings banks (Navier, 1830). In the years 1816–30 we find an approximately equal division between the choice of cultural, technical and practical subjects for presentation in the annual joint meeting. A constant theme in these joint meetings was the unity of the Institute, reflecting eighteenth-century encyclopaedic ideals. As one president remarked, such meetings helped to 'tighten the bonds of the great academic family and to consecrate the unity of the Institute'.[105] Yet there was also a tension between the rhetoric of unity and the display of the *different* perspectives provided by the respective Academies.

After the July Revolution of 1830 the date for the joint public meeting of the four (later five) Academies was changed to the feast of St Philip, to mark the accession of the new King, Louis Philippe. In 1830 and 1831, as after most of the other great political changes of the early nineteenth century, the representatives of the Academy of Sciences felt it appropriate to express their appreciation of the new government, the greater freedom it expected and, by implication, the greater patronage for which it hoped! Again Dupin appeared as a prominent representative of the Academy of Sciences, for example on the economic resources of France (1831). It was important for the Academy to convince the public that science was a national resource. Dupin compared the industrial potential of France with its agrarian economy with the advanced industrial economy of Britain.[106] He advocated the use of statistics in order to give a true measure of French national resources, from which the government could produce a progressive economic plan. Another prominent speaker was Antoine Becquerel who spoke, for example, on the relationship between physics with chemistry and the biological sciences (1835). During the Second Republic the public meeting was held in October and then, under the Second Empire, on the feast of St. Napoleon. Babinet now became the representative of science on many of these public occasions, speaking of different natural phenomena of interest to the public: comets (1853), earthquakes (1855) and droughts (1858). In 1860 Claude Bernard spoke about the adaptation of natural poisons like curare to medicine, and in 1866 Louis Pasteur spoke about the diseases of vines and their cure, subjects on which they were respectively experts of international standing. Under the Third Republic the meeting was transferred back to October. Admiral Mouchez spoke in 1875 about the transit of Venus expedition of the previous year and Lesseps, builder of the Suez canal, about the progress of geography and navigation in 1878.

Every public meeting of the Academy provided an opportunity to represent the

[105] *Séance publique des Cinq Académies*, 2 mai 1833, B.M. 733.g.13 (12).

[106] Dupin, 'Mesure de la richesse française', *Séance publique annuelle des Quatre Académies*, 30 avril 1831, pp. 29–46.

broader meaning of science to the public. A common theme was the continuation of the Enlightenment idea of science as the exemplification of progress. There may even have been some smugness among members of the Academy of Sciences that such an exercise was relatively easy and objective compared with the problems faced by their colleagues in the other Academies who may have wished to claim corresponding progress for literature or art. Although several of the speeches delivered by Cuvier in public meetings in the Restoration made claims for progress through the advancement of knowledge, perhaps it was in the late nineteenth century that the most sustained and convincing case was made for progress affecting the lives of ordinary people. The presidential address of Jamin in 1883 claimed that the public was receiving new benefits from science, as exemplified by electricity and bacteriology.[107] This at least is what he wanted the public to believe. It would be several more years before many French people would receive these benefits.

10. *Eulogies of former Academicians*

An important part of the annual public meeting of the Academy of Sciences was devoted to an *éloge* or eulogy of former members of the Academy by one of the permanent secretaries.[108] *Eloges* covered the life and work of an Academician. They are valuable biographical sources, often carefully researched, yet they are not simple biographies. They are heroic presentations of the life of science with a firm moral dimension, a pattern which was already well established by the early eighteenth-century *éloges* of Fontenelle for former members of the Royal Academy. We have had occasion to comment several times on political speeches made by Cuvier as secretary. It seems appropriate to take a few other examples, and Arago, who held the post of permanent secretary in the period 1830–53, provides an influential counter-example of more consistent political commitment.

It is worth commenting first on the choice of subjects for *éloges*. In the time available at the annual public meeting one could hardly do justice to the memory of more than one Academician, or at most two. Yet the mean number of deaths in the Academy considerably exceeded this allowance. Thus, as a backlog grew, the secretaries had a considerable choice of subjects. Ideally they would speak about those who had died recently, but it is notable that in the years 1837–46, Arago chose on four separate occasions to speak about Academicians who had played a prominent part in the Revolution of 1789 and the First Republic. In the middle years of the bourgeois monarchy of Louis-

[107] *C.R.*, **96** (1883), 875–9.

[108] There have recently been a number of useful studies of *éloges* of the Academy of Sciences and other academies. See e.g. Dorinda Outram, 'The language of natural power; the funeral *éloges* of George Cuvier', *History of Science*, **16** (1978), 153–78; Charles B. Paul, *Science and Immortality. The Eloges of the Paris Academy of Sciences (1699–1791)*, Berkeley, Cal., 1980; George Weisz, 'The self-made mandarin: The *éloges* of the French Academy of Medicine, 1824–47', *History of science*, **26** (1988), 13–40.

Philippe there was a danger that men like Lazare Carnot, Condorcet, Bailly and Monge would be forgotten and Arago was determined to take advantage of his position to rehabilitate them. He presented them all as giants, privileged not only to live through one of the great periods of French history but actually to participate in those stirring events and arguably to help change the course of history.

Certainly there could be great interest in the eventful life of Lazare Carnot (1753–1823), applied mathematician and 'organiser of the victory' of the republican armies.[109] For his biography of Condorcet (1743–94), mathematician and Girondin, Arago took infinite pains, going through the family papers to arrive at a proper appreciation of his subject's life and work.[110] Condorcet had died before the Institute had been founded but in his enthusiasm for a fellow republican, Arago brushed this aside as a technicality. The same could be said of the astronomer Bailly (1736–93), who had been mayor of Paris and president of the National Assembly before going to the guillotine as a victim of a political movement further to the left.[111] Monge (1746–1818) had contributed to the war effort as well as being the virtual founder of the Ecole Polytechnique.[112] His death in the early period of the Restoration had prevented the Academy from paying him a suitable tribute earlier, but Arago was determined to honour a great mathematician who had been a republican before he had fallen under the spell of Bonaparte.

Whereas other secretaries might have been cautious about Academicians who had had political careers, for Arago a political career not only provided greater interest but also more fulfilment in the life of a man of science. For him the Revolution of 1789 was a 'glorious revolution'[113] and he could even find a certain glory in the guillotine.[114] He spoke of the Revolution with an enthusiasm which others reserved for Napoleon's victories. Arago's model of a man of science was someone who did outstanding work but not in an ivory tower. He had to involve himself in the life of his fellow citizens.

But while often idealising the French Revolution, Arago could also come down to earth. In his *éloge* of Ampère, after praising him for his intellectual brilliance, he had no compunction about revealing how poor a teacher he had been.[115] He mentioned his religious leanings and rather mischievously laid greatest emphasis on Ampère's credulity.[116] Ampère had little understanding of business affairs or of politics. He obviously had a completely different character from that of his biographer, sharing with him only an outstanding ability in

[109] 'Carnot' (1837), Arago, *Oeuvres*, 2nd edn, 1865, *Notices biographiques*, vol. 1, pp. 511–633.

[110] 'Caritat de Condorcet' (1841), *ibid.*, vol. 2, pp. 117–246 (119).

[111] 'Bailly' (1844), *ibid.*, vol. 2, pp. 247–426.

[112] 'Gaspard Monge' (1846), *ibid.*, vol. 2, pp. 427–592.

[113] 'notre glorieuse révolution', *ibid.*, vol. 2, p. 206.

[114] He spoke approvingly of the resignation of many of the victims, *ibid.*

[115] 'Ampère' (1839), *ibid.*, vol. 2, p. 107.

[116] *Ibid.*, p. 95.

physics. As secretary to the Academy Arago did not feel himself obliged to adopt a neutral position. He spoke as a moderate republican and he had a considerable following. He helped create a precedent for the more extreme ideology of Berthelot as secretary from 1889. Arago having died at the beginning of the Second Empire, the Academy would have been embarrassed to pronounce an *éloge* immediately. It was appropriately under the Third Republic that this omission was made good.[117]

But secretaries did not all come in the same mould. While for Arago politics was an important consideration, for the Catholic Dumas religion provided an extra dimension to some of his *éloges*. For example, in the case of two foreign associates belonging to the Protestant tradition, he did not neglect to comment on their religious practice, despite the normal taboo within the Academy on such subjects. Of Faraday (1791–1867) he wrote that:

> Fidelity to a religious faith and constant observation of the moral law were the guiding principles of his life'.[118]

Faraday's religious faith may not have directly inspired his great discoveries, but it gave him the courage to overcome his humble origins and not to be diverted from this path when prosperity came his way. Of the Calvinist *associé étranger*, Arthur Auguste de la Rive (1801–73) from Geneva (regarded by some Catholics with the suspicion that some Protestants reserved for Rome), Dumas emphasised how far removed his life had been from the time of the wars of religion. Normally gentle and tolerant, only occasionally would de la Rive speak out as a Christian when, for example, it was a question of combating materialism.[119]

Secretaries were very conscious that in presenting a life of a deceased Academician they had to show not only the Academician in a favourable light[120] but also science itself. Thus when in 1861 Elie de Beaumont chose to present an *éloge* of the mathematician Legendre (1752–1833), he used him to illustrate the thesis that many men of science devote themselves to science, even to the most difficult science, purely for intellectual satisfaction.[121] It was true that some scientists had been known to follow the more lucrative applications of science but they were in a minority. Very much on the defensive, Elie de Beaumont would not even accept that one of the benefits of science was to increase man's material well-being. On the other hand, some secretaries felt that this aspect of science was a valuable justification of the work of the Academy for ordinary French people. It must have been this consideration which prompted Bertrand as secretary to choose the obscure Eugène Belgrand (1810–78) as the subject of

[117] *Eloge* by Jamin, *M.A.I.*, **44** (1888), LXXIX–CXXII.

[118] J. B. Dumas, *Discours et éloges académiques* (2 vols., 1885), vol. 1, p. 56.

[119] *Ibid.*, vol. 1, p. 299.

[120] The notable exception being Cuvier's *éloge* of Lamarck, which so offended his relatives that they tried to prevent it being published. G. Cuvier, *Recueil des éloges historiques*, (3 vols., 1861), pp. 177–210 (180, 209); *P.V.I.*, **10**, 171–2.

[121] *M.A.I.*, **32** (1864), xxxvii–xxxviii.

an *éloge*.[122] It was not so much that he was a great scientist as that he had contributed to the material welfare of Parisians by his work on water supply and sewage disposal.

When Berthelot was elected secretary in 1889 he showed a constant concern to emphasise republican values. He was particularly anxious to emphasise that science was democratic in so far as scientific achievement depended on effort rather than birth. Thus in the *éloge* of Brown-Sequard (1817–94), it was said that he had been raised by his mother 'amidst privations and misery' and to complete his medical studies he had to work in a garret, sometimes living on no more than bread and water and without a fire in the depths of winter.[123] In the case of the botanist Decaisne (1807–82), his original poverty was emphasised so that he could provide a model of

> those scientists who are both sincere and without pretension, who have risen by hard work alone from the lowest ranks of society to the supreme honour of the Institute.[124]

Both Decaisne and Naudin (1815–99) had begun as gardeners at the Muséum d'Histoire Naturelle before being promoted to *aide-naturaliste* or studying for a doctorate. Scientists were not men of superior birth but men who had made themselves superior by their dedication to the pursuit of truth.

Secretaries, therefore, always had an eye open to recruitment of future scientists. The poorest young man of talent could rise to the heights of the Institute if only he set his mind to it. And there was no doubt that the pursuit of 'scientific truth' was the noblest of careers. In one *éloge* of 1900 Berthelot went beyond this, reflecting on some of the power he himself had wielded:

> Scientists of our period and our state of civilisation hold an important place in human society. The services which they render not only to the theoretical culture of the human spirit but to the practice of industry and agriculture are daily better appreciated and some of them are called to occupy the highest positions in Education, Administration and Politics.[125]

Dumas had urged his audience to 'glorify genius' and to 'honour our great men'.[126] Berthelot went much further in the public meeting of 1893 in speaking of heroes:

> 'Science ... has its heroes, such as Newton and Lavoisier which their genius has elevated to the rank of demi-gods'.[127]

[122] *M.A.I.*, **42** (1883), ix–xxxvi.

[123] 'Notice sur Brown-Sequard', Berthelot, *Science et éducation*, 1901, pp. 247–83 (250, 253–4).

[124] *M.A.I.*, **47** (1904), lxiii–lxiv.

[125] Curiously this passage occurs in Berthelot's *éloge* of Naudin, to whom it was least applicable, 'Notice sur Charles Naudin', Berthelot, *op. cit.*, p. 285.

[126] J. B. Dumas, *Discours et éloges académiques* (2 vols., 1885), vol. 2, p. 326.

[127] *M.A.I.*, **47** (1904), lxiii.

11. *Rhetoric*

It was particularly under the Third Republic that we find the public meetings of the Academy used not simply as occasions to explain scientific research but as opportunities to urge the importance of science as a cultural value. Previously if the *éloges* presented were not particularly inspired, then the public had been left with little more than a list of prize-winners and the announcement of future prize subjects. Something more was wanted to give a sparkle to the occasion. It was the Franco–Prussian war and the subsequent French reaction to defeat which put new life into the organisation of French science. One aspect of this was a reform of the public meetings of the Academy. No public meetings were held in 1870 or 1871, but from 1872 onwards all public meetings of the Academy began with a speech on general matters by the president for that year or, exceptionally, the vice-president.

Faye, speaking at the meeting of 25 November 1872,[128] could hardly ignore the tragedy which was uppermost in people's minds at the time. He pointed out that French science and scientists had played a part in the war and particularly during the siege of Paris, when the Academy had continued to meet regularly every week. Although some of his colleagues were to claim that the war had been lost because of the superiority of 'German science', Faye would not accept that French science had been eclipsed. At the outbreak of the war an international conference on the metric system had been held in Paris. During the war French astronomers had taken part in international observations of a solar eclipse and after the war the French were determined to play a major part in the observation of the transit of Venus. France had no reason to be ashamed of her scientists. On the contrary, Frenchmen should regard them as a major national asset. It is time we began to consider the public meetings as opportunities for the moulding of public opinion. We should remember that the meetings were widely reported in the press.

The new format of the public meetings gave successive annual presidents greater freedom to advance their own interests and those of the Academy. Presidents could reflect on the events of the past year without any obligation to give a report. They could express sadness at the recent deaths of senior colleagues without involving themselves in the research necessary for an *éloge*. They could invoke history and compare the Academy of the late nineteenth century with some earlier period. This was a device used, for example, by Faye, when he was called upon for a second time in 1874 to give an introductory address. He was able to explain an important change in the Academy's prize policy from retrospective prizes to grants. He said that the Academy now preferred to help able scientists in advance with their research rather than reward them afterwards.[129] This was a splendid opportunity to appeal for more private benefactors to add to the Academy's power and influence through

[128] *C.R.*, 75 (1872), 1293–7.

[129] *C.R.*, 79 (1874), 1530.

grants. The subject of prizes was often mentioned in the general reviews by the annual president. Not only did this provide a convenient link with a later part of the proceedings but it furnished an easily accessible and readily intelligible example of recent progress in the Academy, with at least a dozen new prizes being founded every decade, some of them involving very substantial sums of money.

In 1883 and 1884 the French contribution to the international efforts to observe the transits of Venus were very topical. In the 1880s there were frequent accounts of the contributions of Pasteur to medicine, for which the Academy wished to claim some credit. Other subjects chosen ranged from polar expeditions (1880) to the development of electric lighting (1885). The public would have been interested to hear in 1896 of research on cathode rays and the X-rays of Röntgen, a reminder that the Academy did not focus exclusively on science produced in France. In 1894 Loewy had made a powerful claim for the pursuit of *pure* science, pointing out that the significance of any discovery, and certainly its possible applications, were usually not at all obvious until much later. Although most of the public addresses were in an optimistic vein, Bouchard in 1909 chose to speak about the declining population of France. This was because the subject had recently been raised in the Chamber of Deputies and had caused considerable public alarm. The speaker concluded that little could be done to increase the birth rate but he held out the hope that science and medicine had further important contributions to make in decreasing the death rate. The subject of this speech was well chosen. It was a matter of great public interest and it was made at a time when, due to the successive work of such people as Pasteur, Lister and Koch, science and medicine were at last beginning to make a tangible contribution to minimising infection and eradicating disease.

Only one year after the Franco-Prussian war France began to resume her role as the host country in the international organisation of the standardisation of the metre. In 1872 the representatives of most European states, together with the U.S.A. and several South American countries, had met in Paris. The president interpreted this as 'homage rendered spontaneously...to French science' but he also described it as 'an ecumenical council of science'. Although Fremy in the public meeting of 1875 was to say that French astronomers on expeditions to observe the transit of Venus were 'like true soldiers of science',[130] metaphors of war and conquest, often used earlier in the century, were relatively uncommon in the Academic rhetoric of the late nineteenth century. Much more common were the themes of science and patriotism, science and civilisation and science and peace.

The rhetoric of the Third Republic often stressed the role of France among civilised countries. Faye in 1872 was concerned with 'civilised man'[131] and the place of France among 'nations where science is honoured'. International co-operation in the metric system was to be seen as the concern of 'the civilised

[130] *C.R.*, 80 (1875), 1458.
[131] *Ibid.*, p. 1295.

world'.[132] For Faye, the French astronomers who had observed the transit of Venus had

> worthily upheld the honour of French science in this great co-operation of civilised nations.[133]

When it fell to Janssen, who played a leading part in the transit expeditions, to speak on behalf of the Academy, he claimed that the French 'influence in the world' was a 'mission of civilisation'[134] in which of course science played a large part. For Jamin, in Comptian vein, modern science was the final stage in modern civilisation 'crowning the progress made by humanity since its beginnings'.[135] It took another war with Germany to force the successive spokesmen of the Academy to develop the theme of French science and civilisation. In December 1914 the mathematician Appell suggested that 'modern civilisation has its roots in scientific research'.[136] He went on to claim that the pursuit of scientific truth had a kind of nobility and moral beauty. He contrasted this concept of science in France as part of a 'harmonious culture' with the 'learned barbarism' of specialism which, he alleged, was dominant in Germany. The Academy stood for the reign of justice as opposed to the reign of brute force. This interpretation can only be understood in the context of several decades of rhetoric on French science and patriotism.

At the end of the century we find the scientists of the Academy increasingly concerned with peace and the contributions which science might make to bringing mankind together. Jamin pointed out in 1883 that electricity through the telegraph, and more recently the telephone, could bring together men from neighbouring states which might be at war;[137] in this respect science was more powerful than politics. In 1894 Loewy concluded his address to the public meeting of the Academy by claiming that it exercised the most beneficial influence throughout the civilised world. It awarded prizes to scientists of all countries, thus propagating ideas of 'peace and concord which should precede the unification of all human families'.[138] A spirit of impartiality, blind to political events, helped increase the international prestige of the Academy. But it was the 1900 meeting of representatives from academies from all over Europe to found an international Association of Academies which prompted the following year's president to speak of recent movement towards 'peace and the brotherhood of peoples'.[139] He pointed out that scientists, sharing the same love of truth, were particularly well disposed to collaborate on an international scale. Thus the Academy, which served as a focus for the most blatant nationalism in scientific matters, paradoxically was also genuinely concerned with international co-operation. The French combined nationalism with a brand of internationalism by advocating greater international communication with Paris as the centre.

[132] *Ibid.*, p. 1296.
[133] *C.R.*, **79** (1874), 1525–6.
[134] *C.R.*, **107** (1888), 1038.
[135] *C.R.*, **96** (1883), 875.
[136] *C.R.*, **159** (1914), 821.
[137] *C.R.*, **96** (1883), 875.
[138] *C.R.*, **119** (1894), 1049.
[139] *C.R.*, **133** (1901), 1055.

12. *The audience*

Although it is useful to look at the language used in public meetings, we need finally to complement this perspective by considering the audience addressed. We might conjecture that the public meetings of the Academy were aimed principally at the general French public, with some thought for the government and perhaps a consideration of the wider scientific community. This may not be too far from the truth but it would be helpful if we could find supporting evidence. The astronomer Janssen, addressing the public meeting of 26 December 1887 as President of the Academy, hoped that he would be heard by three distinct audiences:

> First by those who have seats in the councils of the nation and who have responsibility for the future and for the glory of France; then by citizens with large and generous dispositions like our benefactors, who desire the good of their country; and finally by our admirable young people, who are looking for a career compatible with their activity and talents.[140]

Something might be said about each of these three audiences in turn. Members of the Chamber of Deputies might have an important influence on the government. The days of direct reports to the government had passed but these annual meetings did provide a public occasion which would be reported in the newspapers and journals, and could be used to address a message indirectly to the government. As the meetings were widely reported in the press, any praise or blame would be taken up by those with political influence, who would gladly use it for their own purposes.

We come secondly to the most miscellaneous category, the 'general public'. In the late nineteenth century the Academy realised that its audience on these occasions included wealthy citizens with some interest in science. It therefore tried more and more to use the occasion to ask for donations. In the second half of the nineteenth century an increasing proportion of the income of the Academy came from private sources, mainly for prizes. On the negative side this was a result of the Academy's problems in finding more money from government sources. But on the positive side it provided wonderful opportunities for the Academy to express its needs, and for donors to gain immortality for their family names by association with a prize fund.

Finally, we can interpret the public meetings as being concerned with the recruitment of future scientists. The third category was described above by Janssen as 'young people' but, at the age of sixty-nine, he could well have meant anyone under forty. For the most able the Academy itself was presented as the ultimate goal. But a concern which was even more basic was the representation of what it meant to be a scientist. It was, of course, a career or profession and had been at least since the Revolution. Also, it was an honourable profession.

[140] *C.R.*, **105** (1887), 1306.

It might not be very lucrative, but the career of the *savant* in France always commanded respect. Under the Third Republic, Academicians like Berthelot repeatedly held out the moral values of science as being concerned with truth.

At a later public meeting in 1903, when the palaeontologist Gaudry was president, he made an appeal for more manpower:

> Gentlemen, ask your friends to come to our aid[141]

He presented his branch of science as a combination of pleasure, philosophy and poetry and he made special reference to a recent trip he had made to the USA to study the bones of dinosaurs. Again at the public meeting of 1907, the physiologist Chauveau was particularly concerned with recruitment:

> Young people are sometimes frightened by the abundance of harvests yielded each day by the culture of the terrain of science. Indeed it is easier for them to assess the riches, which are constantly taken out of it, then to appreciate the value of what remains. Let them be reassured! This terrain is inexhaustible! The ardent and passionate curiosity of researchers is not about to run out of raw material. In the world of scientific thought there will always be unknown and mysterious regions to transform into fine luminous spaces, richly coloured by the splendour of the rays given off by a new-born truth![142]

[141] *C.R.*, **137** (1903), 1093.

[142] *C.R.*, **145** (1907), 980.

— Chapter 11 —

THE INTERNATIONAL DIMENSION

It is an accepted opinion [among some people] that the Academy constitutes the central point, to which all discoveries and inventions from the entire world converge as soon as they are made. These discoveries will take the trouble to come to us to verify their authenticity and it is only after being supplied with a passport, signed by us, that they can decently circulate in the republic of science.

(Victor Meunier, *Scènes et types du monde savant*, 1889, p. 214.)

The title of *associé étranger* is reserved for those scientists who, from all countries except France and in all the sciences which the Academy studies, have reached the first rank. By this happy association the Academy is in a way universal, and the history of all the great discoveries, with which the sciences have been enriched since its foundation, belongs to its own history.

(Elie de Beaumont, *éloge* of Plana, 1872, *M.A.I.*, **38**, cvii.)

I bequeathe to the Academy of Sciences the sum of 20,000f., the interest from which will be given every two years as the Delalande-Guérineau prize to the French explorer or scientist, who has rendered the greatest service to France or to science.

(Will of 1872, from Pierre Gauja, *Les fondations de l'Académie des Sciences (1881–1915)*, Hendaye, 1917, p. 270.)

1. *Internationalism and nationalism*

A certain universalism emerged in the eighteenth century from the ideas of the Enlightenment. The Enlightenment helped to spread the idea of man learning from nature. The natural world was thought of as providing models for law and politics, as for other human concerns. In France men of science turned to nature to provide the basis for a new system of measurement to replace the chaotic feudal system, which accepted so many different and purely arbitrary measurements of length, weight and volume. If a standard of length could be found, derived from nature rather than the decree of a particular government, it would provide the basis of a system of measurement capable of universal adoption. The standard of length finally chosen was related to the size of the Earth, the common home of mankind.[1] Measurements of large distances of the

[1] A valuable source for all aspects of the metric system is G. Bigourdan, *Le système métrique des poids et mesures*, 1901. See also: Maurice Crosland, '"Nature" and measurement in eighteenth-century France', *Studies on Voltaire and the eighteenth century*, **87** (1972), 277–309.

Earth's surface by the French, notably in Peru and in Lapland, go back to the 1730s but it took a political revolution to introduce anything so radical as a new system of measurement.

In the year 1789, the Royal Academy of Sciences appointed a commission to consider the problem of uniform weights and measures. When news reached France in 1790 that this was also being discussed in the British House of Commons, Talleyrand even suggested co-operation between the two countries. Unfortunately the political climate between Britain and France was soon to deteriorate and the French decided to proceed unilaterally. An Academy Commission reporting on 19 March 1791 proposed as the new unit of length one ten-millionth part of the quarter of the great circle of the earth or meridian passing through Paris, that is the line of longitude stretching from the equator to the north pole. The Commission enumerated a number of basic tasks which would be necessary, including the accurate measurement of base lines and the measurement of the distance between Dunkirk and Barcelona by surveying, using the method of triangulation. By September 1792 the Academy had made a start and had established a provisional size for the new unit of length, the metre. This, together with the concept of a decimal scale, was accepted by the Convention in August 1793.

Yet it was just at this time that the Academy was suppressed and, despite the nomination of some of the members to a temporary commission to continue the work, the active measurement of the meridian ceased, only to be taken up again in the calmer political atmosphere of 1795.[2] Much work was left, including the determination of the unit of mass, the gram. Field work on the measurement of the southern part of the Dunkirk–Barcelona meridian was not completed until November 1798, when Méchain and Delambre arrived back in Paris. Thus basic work on the metric system went on intermittently right through the revolutionary period and included work done by members of the old Royal Academy, by some of the same scientists during the eighteen months when no Academy existed, and finally by the same scientists who found themselves elected to the First Class of the National Institute. Whatever view one may take of the difference between the First Class and the old Academy, there is a certain inescapable continuity in the work on the metric system.

The French Revolution, despite its moods of intense nationalism, was a revolution with universal implications, since it developed into an attack not only on one king, but on the monarchy in general; a feature of great concern to other European powers. Also the slogan of 'liberty, equality, fraternity' was intended to apply outside France as well as within. The revolutionary armies representing the new nation state claimed to be bringing 'liberty' to their neighbours. Probably it was under Pitt's government in Britain that the greatest reaction took place. The British had their own ideas on liberty and did not want to take

[2] 'Reprise des travaux de la méridienne. Loi du 18 *germinal* an 3 (7 Avril 1795)', *ibid.*, Chap. VI.

lessons from the French. But it was particularly the excesses which took place under Jacobin rule, usually described as 'the Terror', which strengthened the British reaction. Edmund Burke had even helped to form British opinion with his prophetic *Reflections*, in which the revolutionaries were portrayed as monsters. The war declared in 1793 and later the rise to power of Bonaparte and the threatened invasion of Britain all helped to accentuate nationalistic feelings. The French for their part were encouraged to be equally hostile to the British. Members of the Institute at one time were even asked to subscribe to the war against Britain.[3]

It has, however, been claimed that the 'sciences were never at war' since science transcended national and political disputes even in the revolutionary and Napoleonic wars.[4] Certainly claims were made that men of science should not be held prisoner since they were above politics, being concerned with knowledge of Nature, an activity valued by both sides. Thus in 1807 the minutes of the First Class read:

> M. Delambre reads a letter from M. Banks [President of the Royal Society], in which he thanks the Institute for the trouble that it has taken to secure the liberty of several English men of science, who had been detained in France.[5]

Yet it was at the height of the Napoleonic wars, when patriotic fervour on both sides was at its highest pitch, that we find an intense rivalry between two exact contemporaries, Gay-Lussac (1778–1850) and Humphry Davy (1778–1829). Gay-Lussac was supported by the Arcueil group under the patronage of Berthollet and Laplace, by his former school, the Ecole Polytechnique and by the Academy, which he joined in 1806. Davy was supported by the Royal Institution and the Royal Society. The two chemists followed closely each other's work on electrolysis, the isolation and study of potassium, sodium and boron. Even when Davy visited Paris in November 1813, they both managed to become involved in elucidating the nature of iodine. In this case personal rivalry was accentuated by intense nationalism under wartime conditions.

But strong feelings of nationalism were to continue in peacetime. One of those loudest in his 'patriotic' sentiments in the first half of the nineteenth century was Arago. As patron of Daguerre, he presented the daguerreotype as a French invention, 'the discovery of our fellow-countryman'. He continued:

> This discovery has been adopted by France; from the beginning she has shown herself proud to be able to give it freely to the whole world.[6]

A second occasion in which Arago was involved, was connected with the discovery of Neptune by Le Verrier in 1846. Arago, who had first proposed to Le

[3] A.I., Registre de la Commission des Fonds, 1D1, p. 7 (2 *nivose* an 6).

[4] Sir Gavin de Beer, *The sciences were never at war*, 1960.

[5] *P.V.I.*, 3, 557 (3 August 1807).

[6] *C.R.*, 9 (1839), 267.

Verrier that he should study the perturbations of the planet Uranus, was delighted when his protégé discovered a new planet, which he first named *Neptune*. By October that year Le Verrier aspired to have the planet named after himself and he asked for Arago's support. This Arago gave, saying that it followed 'the inspiration of a legitimate patriotism'.[7] Unfortunately the French claim was complicated, as we have seen in Chapter 6, by the simultaneous prediction of the planet in England by Adams. It was to be Arago who took on himself the duty of defending Le Verrier's priority, which he did at a meeting of the Academy on 19 October 1846.[8] Arago emphasised that Le Verrier had indisputably published his discovery first. The fact that the Englishman had not published, led him to refer to Adam's work as 'clandestine'. He continued emotionally:

> What! M. Le Verrier has made his research available to the entire scientific world; following the formulae of *our learned compatriot*, everyone has been able to see the new planet ... and to-day we are called upon to share this glory, so loyally and legitimately acquired, with a young man who has communicated nothing to the public ...
>
> In the eyes of every impartial man, this discovery will remain one of the most magnificent triumphs of astronomical theory, one of the glories of the Academy and *one of our country's noblest titles to the gratitude and admiration of posterity*.[9]

The newspapers found inspiration in this speech. Unable to understand the finer points of Newton's law of universal gravitation which had led to perturbations of Uranus being interpreted as being due to a remote planet, hitherto unobserved, they needed no-one to instruct them in patriotism. On 21 October *L'Univers* carried an article attacking England, which it accused of 'an odious national jealousy, which appears determined not to let us have the last word'.[10] On the same day *Le National* began a campaign which lasted for several weeks. It accused the English of 'treating France as a stupid nation, M. Arago as a humbug', and of inexcusable nationalism in proclaiming 'Adams and England forever'. By 7 November, when the journal *L'Illustration* published offensive caricatures of Adams, Arago and Le Verrier had publicly disassociated themselves from the press campaign.

Although Arago, as one of the two secretaries, was seen as the spokesman of the Academy, a second Academician vied with him in proclaiming the superiority of the French. As Dupin was principally concerned with applied science, he could not forget that it had been Britain rather than France which had been the scene of the great industrial revolution and this circumstance had

[7] *C.R.*, **23** (1846), 662.

[8] 'Examen des remarques critiques et des questions de priorité que la découverte de M. Le Verrier a soulevées' par M. Arago, *ibid.*, 741–54.

[9] *Ibid.*, 754 (my italics).

[10] Quoted by Morton Grosser, *The Discovery of Neptune*, Cambridge, Mass., 1962, pp. 133–4.

left France at a great economic disadvantage compared with its neighbour. At the funeral of A. L. Breguet on 18 September 1823, he claimed that Breguet's success in making watches and clocks had placed French industry ahead of other nations, and he compared industrial success to (the recent Napoleonic) military success. Speaking of many great scientists recently dead, Dupin asserted that France was 'inexhaustible in her heroes'. Perhaps he was implying that although the time for military victories had passed there were battles to be fought on the intellectual and industrial fronts. The parallel was drawn explicitly by the comte de Salvandy, Minister of Public Instruction in 1846, when he wrote to the Academy to inform it that he was promoting Le Verrier to the rank of officer in the Legion of Honour, although strictly speaking he did not have the proper seniority. This was because 'in science...as in war' there were brilliant actions which cried out for reward: 'The work of M. Le Verrier...honours our century and France'.[11] This was said not in a private letter, but in an official communication read to the Academy at its meeting of 5 October 1846 and published in its *Comptes rendus*. The Minister said that he expressly wished Le Verrier to be honoured as a member of the Academy before his peers.

But to return to Dupin, who used the public meeting of 1835 to argue that science made a major contribution to 'the glory of France'. Perhaps he reached the height of his eloquence when he was describing the mathematical precision of French maps:

> Here is the perfection which it is the duty of the Academy to claim as a national glory and which should be looked upon with religious respect.[12]

At the public meeting of 22 March 1852 Dupin spoke of the French contribution to the Great Exhibition of the previous year, for example the work of Balard on the extraction of salts from sea water. France could not rival Britain in her merchant navy but she could make better use of natural resources. Of a total of 1760 French exhibitors, a thousand had received at least an honourable mention. A table of French inventions would be 'a monument of national glory'.[13]

The Frenchmen of the early to mid-nineteenth century, who still saw Britain as their traditional rival and old enemy, were soon to be succeeded by others who looked rather to Germany, first with curiosity, then envy and finally hatred, which inspired the strongest nationalistic sentiments. Hatred of Germany, formed around the nucleus of Prussia, reached a peak after the defeat in the Franco-Prussian war of 1870–1.[14]

[11] *C.R.*, **23** (1846), 676.

[12] *C.R.*, **1** (1835), 564–74 (p. 571).

[13] 'Notice sur quelques tributs des Français à l'Exposition Universelle', *C.R.*, **34** (1852), 441–8 (p. 448).

[14] Harry W. Paul, *The sorcerer's apprentice. The French scientist's image of German science, 1840–1919*, Gainesville, Florida, 1972.

2. *Paris as an international centre*

We have mentioned one of the first international exhibitions. In the second half of the nineteenth century Paris became increasingly important as the venue for international exhibitions and congresses. After the Great Exhibition of 1851 in London it was Paris which was repeatedly chosen for later major international exhibitions, notably the 'Expositions Universelles' of 1855, 1867, 1889 (to celebrate the centenary of the French Revolution) and 1900.[15] Whereas the 1851 Exhibition had attracted 6 million visitors, all records were broken with the last two Paris exhibitions mentioned, which were attended by 32 million and 48 million visitors respectively. Of course Paris was, by the 1860s, at the centre of a formidable national rail network, with connections to most other European countries. Unlike London, it did not suffer the disadvantage of imposing a sea crossing on foreign visitors in addition to rail travel.

In the second half of the nineteenth century a growing number of international congresses were organised and many of these were held in Paris.[16] Thus international sanitary conferences were held in Paris in 1851 and 1859. An international statistical congress was held in 1855 and an international congress on telegraphy in 1865. We should not forget France's colonial interests, which helps to explain the choice of Paris for an international congress of geographical sciences there in 1875. After the recent war with Prussia, however, this was not a very favourable period for international conferences to be held in Paris. Indeed France went through a period of isolation. It was really only the memory of the revolution of 1789 which again raised French efforts to a high pitch, so that in 1889 we find a whole series of congresses in Paris, ranging from aeronautics to zoology and including chemistry and photography; nor should we omit to mention the welcome given in the same year in Paris to electricians and meteorologists. In fact, of a total of ninety-seven international congresses on every conceivable subject held in the world in that year, eighty-seven were held in Paris. The international exhibition held in Paris in 1889 acted as a powerful magnet and a similar effect was noticed in connection with the Paris international exhibition of 1900. It should be pointed out that whereas Switzerland and Belgium were also the scene of many international congresses these were shared out between major cities, whereas nearly always a meeting in France meant Paris.

An important function of international meetings was to agree on a common set of technical terms within a particular science. No longer could a Linnaeus dictate the language of botany nor a Lavoisier and his colleagues the language

[15] Eugene S. Ferguson, 'Expectations of technology, 1851–1900' in M. Kranzberg and C. W. Pursell (eds.), *Technology in western civilisation*, New York, 1967, vol. 1, pp. 706–26.

[16] Union des Associations Internationales, *Les congrès internationaux de 1681 à 1899, Liste complète*, Brussels, 1960. See also F. S. L. Lyons, *Internationalism in Europe, 1815–1914*, Leyden, 1963.

of chemistry. In geology the Americans in particular felt that it was necessary to reach an internationally agreed system of geological classification and nomenclature. A committee established in 1876 decided to take advantage of the fact that Paris was to be the scene of the International Exhibition in 1878 to hold the first international geological congress there. Invitations were widely distributed and some 300 participants from twenty-two countries came. Curiously among the smaller delegations were the German and the British (three each). This can be explained largely in terms of linguistic and cultural affinities, since French was the only official language of the meeting and Germany had not forgotten the war of 1870–1. Two thirds of the participants were French and members of the French Geological Society. The president of the congress was the Academician Hébert (1812–90) but it was a notable feature of the meeting that it was French geologists outside the Academy who predominated in the discussions. Here then on the international stage was an occasion when the traditional differentiation between Academicians and other scientists broke down. In such an international context the longstanding pre-eminence of the Academy was reduced. Thus it was not only the advancement of science, but developments in scientific *organisation* which reduced the power of the Academy in the late nineteenth century. Given the necessity for consultation on the international front, science was now much less under the control of the Academy than formerly.

3. *International collaboration on metrication and terrestrial magnetism*

What was arguably the first international scientific conference was held in Paris in 1798–9.[17] By early 1798 the labours of most of the groups previously appointed to establish the basis of the metric system were nearly complete. Yet what had originally been conceived as a potentially international system of measurement had been constructed entirely by Frenchmen. At the meeting of the First Class on 20 January 1798, it was suggested that in order to impose a stamp of international approval on the new system the French government should be asked to invite scientific representatives of foreign countries.[18] In a war situation the French government obviously was not going to ask enemy powers, but it was happy to extend invitations to neutral or allied states. Representatives came from Holland, Denmark and Spain, and the various states making up the Italian peninsula were prominent. Much of western Europe was represented with the notable exception of Britain. The Danish representative was the astronomer Thomas Bugge, who wrote a detailed account of his visit[19] which provides a commentary on Parisian scientific institutions of this period.

[17] Maurice Crosland, 'The congress on definitive metric standards, 1798–9: the first international scientific conference?', *Isis*, 60 (1969), 226–31; see also Bigourdan, *op. cit.*

[18] *P.V.I.*, 1, 335.

[19] Maurice Crosland (ed.), *Science in France in the revolutionary era described by Thomas Bugge*, Cambridge, Mass., 1969.

The foreign delegates were invited not merely as observers but as collaborators in the final stages of the establishment of definitive metric standards. They were therefore in a position to make criticisms of some of the details of the final determinations. Several committees were appointed with a membership intended to strike a balance between French and foreign representation. After considerable discussion and some delay the scientists presented their work to the First Class and then to a meeting of the whole Institute on 17 June 1799. On 22 June they presented a report on their work, together with a standard metre and kilogram, to the two legislative councils. Foreign delegates were asked to promulgate the new system in their own countries and several later attempted to do so.

In France the metric system was taught in schools but was not adopted in commerce until prescribed by law in 1837. In Britain the system was sometimes referred to as the French system and was largely ignored. It was the Great Exhibition of 1851, held in London and bringing together objects described by a whole range of different measures, which made people think about the desirability of an international system of measurement. At the exhibition the French provided specimens of metric weights and measures but it was not until the 1867 Paris exhibition that there was a movement to consider these as internationally acceptable measurements. This raised again the question of precise definition of standards, which was considered by various bodies including the French Academy of Sciences and the Academy of Sciences of St Petersburg. There was then a distinct possibility that France might be by-passed in any international agreement and the French Academy of Sciences, led by its secretary Dumas, decided it was necessary to urge the French government to take the initiative and invite representatives of foreign governments to come to Paris to agree on metric standards.[20]

Accordingly the government of Napoleon III invited representatives to come to Paris for a meeting on 8 August 1870. But already on 19 July war had been declared between France and Prussia, and it was understandable that although some foreign delegates attended the first meeting, the Germans were absent. A full meeting was not possible until September 1872, when representatives agreed to take the metre and kilogram from the French Archives as their starting point. Copies of these standards would have to be made using a suitable metal not subject to corrosion, such as an alloy of platinum and iridium. It would also be necessary to specify the temperature at which the length was to be measured. Among further problems was the multiplication of standards for international use. A diplomatic conference on the metre was called for 1875 in Paris, which agreed on the establishment of a permanent international bureau of weights and measures at Sèvres, just outside Paris. It was now well out of the hands of the Academy, but the Academy could justly claim that it had played a vital part in the early history of the international system of weights and measures.

[20] 'Rapport sur les prototypes du système métrique: le mètre et le kilogramme', *C.R.*, **69** (1869), 514–18.

Another area of potential international collaboration was terrestrial magnetism.[21] French scientific expeditions in the eighteenth century had included measurements of magnetic inclination and declination of the earth's field at different points of the earth's surface. After the Revolution interest in the earth's magnetism seemed to fall within the responsibility of the newly-founded Bureau des Longitudes, but the most influential person was Alexander von Humboldt, who spent the years 1798–1804 in Central and South America on a major geographical and natural history expedition, in which he also made regular magnetic observations. Although Prussian by birth, Humboldt felt that his spiritual home was France and so, when he returned to Europe in 1804, he settled in Paris and soon became closely associated with the Arcueil group. In 1804 he published a joint paper with Biot on terrestrial magnetism. Gay-Lussac and Biot made useful magnetic observations in their balloon ascent in that year. Humboldt's main collaborator was to be Arago, and from 1809 the two made joint observations at the Observatory. Arago was to become one of the leading exponents of international cooperation on magnetic measurements, but it was principally as a member of the Bureau des Longitudes rather than of the Academy, to which he had been elected in 1809. Nevertheless, Arago's position in the Academy gave him greater influence over instructions given to the various French naval expeditions in the period after 1815. In the 1820s the French were in the forefront of geomagnetic research but by the early 1830s British and German observers became seriously interested, with the establishment, for example, of the *Magnetische Verein* by Gauss and Weber in 1834 and British government support for Sabine's 'Magnetic Crusade' in 1838.

There was a good opportunity for international collaboration on geomagnetism in 1834 when Arago attended the fourth meeting of the British Association. The president reported to the meeting that Arago and the Academy of Sciences would be happy to collaborate with their British colleagues and other magnetic experts.[22] Arago seemed particularly keen to set up a chain of observatories with suitably trained staff and instruments to make observations according to a common plan. Yet there was a delay of several years before further steps were taken in international collaboration, and when the British made further moves in 1838 it was to join the German *Magnetische Verein* and to ignore Arago and the French. Whether this exclusion of the French can be explained in terms of difference of 'national styles' or attributed simply to Arago's growing political concerns and his deep involvement in the Academy and particularly in the launching of the *Comptes rendus*, is open to question. As secretary of the Academy, Arago would have been ideally placed to represent

[21] The following two paragraphs are based mainly on the work of John Cawood, 'Terrestrial magnetism and the development of international collaboration in the early nineteenth century', *Annals of Science*, **34** (1977), 551–87.

[22] *Report of the 4th meeting of the British Association for the Advancement of Science, Edinburgh, 1834*, 1835, p. xxx.

France in international collaboration. Yet perhaps Arago himself was too much of a nationalist at heart really to believe in a fruitful partnership with other nations. Whatever the reason, France temporarily ceased to be among the leaders of international scientific collaboration.

4. *International scientific collaboration in the late nineteenth century*

A further occasion for international collaboration, or perhaps international competition, presented itself with the prospect of a very rare transit of Venus in 1874. The last time the disc of the planet Venus had passed across the face of the sun (as seen by an observer on earth) was in the 1760s, which had been the occasion of both French and British expeditions to the Orient, where it was expected that observation would be particularly favourable.[23] From an exact measurement of the time of transit the distance of the sun could be calculated and hence the scale of the solar system. Yet eighteenth-century results had been widely discordant and much better results were expected a century later.

The Academy started to consider the problem five years in advance. In fact a letter from the Minister of Public Instruction on the predicted transit of Venus was read out at the Academy meeting of 1 February 1869.[24] The Minister encouraged the Academy to make preparations, saying that a recent French expedition to Malaya to observe a total eclipse in 1868 had brought great credit to France. Where would the Academy like to establish observation posts? What instruments would they like? Should they collaborate with astronomers from other countries? The letter ends with a certain ambiguity:

> The astronomical expedition could also be used to benefit other sciences. The Emperor wishes to give this expedition the character of a long scientific campaign (*une longue campagne scientifique*) covering all questions, whose study can be carried out across the ocean and in the other hemisphere.

The Emperor had in mind a 'grand enterprise', which seemed to be using a scientific justification for an expedition that might well have political benefits for France. In the circumstances, it seemed as if the French government intended to be exceptionally generous in fitting out the expedition. Unfortunately for the Academy the imperial government had little more than a year to run. War intervened, and when the Academy returned to the question in 1871, the new Minister replied that since no allowance had been made in the current budget of the Academy no money would be available before 1873![25]

The lack of preparedness seems to have been one of the factors which caused the president of the commission on the transit of Venus, the astronomer Faye, to hand in his resignation; a second astronomer, Le Verrier, refused to accept the responsibility of the presidency. Finally J. B. Dumas agreed to serve, for although

[23] Harry Woolf, *The transits of Venus. A study of eighteenth-century science*, Princeton, N.J., 1959.

[24] *C.R.*, **68** (1869), 205–7.

[25] *C.R.*, **73** (1871), 1268.

his special expertise was in chemistry, he was now secretary of the Academy and felt strongly that the honour of the Academy and of France itself was at stake.[26] The question of national honour was particularly sensitive after the recent defeat in the war. The most urgent question was that of finance. The Minister of the Navy, vice-admiral Pothuau, seems to have been much more sympathetic to the needs of the expedition than Jules Simon, the Minister of Public Instruction, and Dumas went out of his way to thank the former publicly.[27]

The principal observation points of the French expedition were at Nagasaki, where the two Academicians Janssen and Tisserand took up positions, Peking in the northern hemisphere, and the island of Saint Paul in the Indian Ocean and Campbell Island (south of New Zealand) in the southern hemisphere, together with two other locations. This total of six stations for France compared with five for Germany, seven for the United States, ten for Britain and thirty for Russia. Yet these numbers do not necessarily reflect relative national investment in science since the Russian stations for example were all on her own territory, whereas the United States at considerable expense had deliberately chosen remote stations in the Southern seas.[28] The method used by the French expedition for calculating the transit was that of the eighteenth-century French *savant* J. N. Delisle, a contemporary of Edmond Hallé. The French also placed considerable reliance on photography, using the French daguerrotype, a fact remarked on by a British commentator.[29] The French observers kept in touch with the Academy in Paris by letter and telegram[30] and on their return the Academy undertook to publish an account of their work in three large volumes.[31]

The respective national expeditions were organised largely independently and in a spirit of competition, but there was international collaboration in the sharing of the data obtained.[32] In the winter of 1874–5 the columns of the British journal *Nature* regularly reported results from various teams of observers, and were happy to give full credit to scientists from other countries. *The Times* of London, often known for its strongly British view of the world, received publication of the first French results with great enthusiasm, obviously unaware of some of the previous French internal problems:

> From beginning to end it shows how a nation should set itself to work – how all the intellect of a nation can and must be utilised, when a great problem involving many kinds of special knowledge has to be attacked. It is often said that in France science is crushed by the dead weight of officialism, and that in England it is free. However true this may be of teaching, there is ample

[26] Marcel Chaigneau, *J. B. Dumas, chimiste et homme politique*, 1984, p. 269.

[27] *C.R.*, 78 (1874), 139.

[28] *Nature*, 11, 103 (10 December 1874).

[29] *Ibid.*, 10, 86 (4 June 1874).

[30] E.g. *C.R.*, 79, 1395–1400 (14 December 1874).

[31] *Recueil de mémoires, rapports et documents relatifs à l'observation du passage de Venus sur le soleil*, 3 vols., 1877–85.

[32] *Nature*, 11 (1874–5), e.g. 112, 121–3, 171, 234–5, 365–6.

> evidence in this volume, that in one branch of research at least, the very opposite of this statement is much nearer the truth.[33]

The tone of the *Times* report was perhaps symptomatic of an atmosphere of greater international sympathy and understanding brought about by the transit expedition.

Next we may mention an international congress held to obtain agreement and co-operation in the construction of a photographic map of the sky, which met at the Paris Observatory in April 1887.[34] The Academy was the host, capitalising on the important contributions made by the French to the use of photography in astronomy. Nevertheless this was obviously a task which cried out for international collaboration. Astronomers of different countries had to agree on the use of similar photographic equipment to obtain uniformity. They also had to agree on the extent and definition of their study. They accepted that their first task, using relatively short exposures, should be to try to include all stars down to the eleventh magnitude. This would probably include $1\frac{1}{2}$ million stars. Altogether fifty-six astronomers attended the congress, representing sixteen countries. The veteran Russian astronomer O. W. Struve (1819–1905) was elected president of the congress, but there was also discussion about setting up a permanent office, since the task would obviously take many years.[35] Although Paris was a convenient place to meet, even Parisians admitted that conditions for photographing the sky in that city atmosphere were far from ideal. Delegates were sent away to obtain funds from their respective governments for the necessary photography.

Another important international event in the 1880s was the International Meridian Conference held in Washington in October 1884.[36] Although the Academy of Sciences was not represented as such, it gave advice about the French position.[37] Already at the first International Geographical Congress in 1871 a resolution had been passed that there should be a generally accepted zero of longitude and that this should be the Greenwich meridian. One of the French delegates seemed quite happy with the idea as far as sea charts were concerned, since the majority of these were now British. At a further meeting in Rome in 1875 it was said that if Britain were to accept the metric system it might be courteous of France to accept the Greenwich meridian, although the French were known to be keen to place the Observatory in Paris at the centre

[33] *The Times*, 20 January 1875, p. 4.

[34] Mouchez, 'Présentation des "Procès verbaux des séances du Congrès astronomique international pour l'exécution de la carte photographique du ciel"', *C.R.*, **105** (1857), 89–92.

[35] See *Bulletin du comité international permanent pour l'exécution photographique de la Carte du Ciel*, 1er fasc., 1888.

[36] The following account depends heavily on Derek Howse, *Greenwich time and the discovery of longitude*, Oxford, 1980, Chap. 5, pp. 131ff.

[37] A. S., *Comité secret, 1882–1902*, p. 51. 'M. Faye, au nom des sections d'Astronomie et de Géographie et Navigation, donne lecture d'un rapport concernant l'établissement d'un méridien universel' (25 February 1884).

of the world, and this differed from Greenwich by approximately two degrees. On sea maps most nations accepted the Greenwich meridian but on land maps each nation went its own way.

Further discussion in Europe as well as a plan of the Americans and Canadians to adopt a series of time zones stretching across their vast territories, prompted the U.S. Congress in 1882 to call an international conference to decide on a common prime meridian for longitude and time throughout the world. At the international Congress of 1884, twenty-five countries were represented, admiral Rodger (U.S.) being elected president with three vice presidents representing respectively Britain, France (Janssen) and Brazil. The majority of the delegates favoured the Greenwich meridian, but France was totally opposed. As a compromise it therefore proposed a system which would be absolutely neutral,[38] thus favouring neither of the traditional rival naval and commercial powers. It was pointed out that over 70% of commercial shipping at that time used the Greenwich meridian and the conference finally voted to accept this by twenty-two votes to one with two abstentions (France and Brazil).

Another aspect of the international Meridian Conference was agreement on a common system of time. The idea of a time zone system based on the Greenwich meridian was gradually adopted in different countries. France was understandably resistant but, as the system became more generally accepted in other countries, a French deputy in 1896 introduced a Bill in the Chamber of Deputies to accept 'Greenwich Mean Time' as the legal time in France.[39] In February 1898 the Bill was passed with the amendment that the time should be expressed as 'Paris Mean Time' minus nine minutes and twenty-one seconds. This is exactly the same but avoids the use of the word 'Greenwich'! The Academy was naturally consulted and discussed the matter in secret session on 11 July 1898.[40] It agreed on the resolution that *in principle* the international recognition of a common meridian was desirable. On the other hand:

> the responsibility of pronouncing on the adoption of this system should be left to the government, which alone is competent to appreciate the implications of such a measure for the commercial, economic and political relations of the country.

The Academy thus refused to take any responsibility. So did the government. It seems that some ministries, notably Commerce and Posts and Telegraphs, supported the Bill whereas the Ministry of Public Instruction and the Naval Ministry opposed it. No action was therefore taken for twelve years. Finally in March 1911 the law was passed recognising Greenwich Mean Time in the disguise described above. Now that France had rejoined the international

[38] The French claim was that the metric system was so neutral that if historical evidence of the origins were wiped out, later generations would not be able to associate it with any one nation.

[39] Howse, *op cit.*, p. 153.

[40] A.S., *Comité secret, 1882–1902*, p. 365 (11 July 1898).

community with regard to time it could take an initiative. With the advent of radio, the French thought it would be a good idea to call an international conference in Paris in 1912 to unify radio time signals. This was the background of the Bureau International de l'Heure, to be established at the Paris Observatory. National honour was satisfied!

Finally, we should mention the International Association of Academies founded in 1899, largely from initiatives of the Royal Society of London, the Academy of Sciences of Vienna and several German academies.[41] The Royal Society wanted international collaboration for the compilation of an international catalogue of scientific literature, while the Germans and Austrians had for several years been organising meetings of their leading academies. When the secretary Darboux reported to his colleagues, he drew special attention to the German dimension.[42] Representatives of the French Academy of Sciences[43] were invited at short notice to Wiesbaden in October 1899 to discuss the establishment of an international organisation. Darboux was careful not to commit the French Academy in so far as he expected adherence to have major financial consequences. Although the Academy now had a large budget, mainly through prize legacies, most of this money was earmarked for special purposes. Nevertheless, the Academy was happy to join in 1900 and the first international meeting was held in 1901 in Paris, where foreign delegates were given a magnificent reception both by the French government and the municipality.

In a way academies were combining to reassert their former influence, which in many ways had been diminished by the foundation of scientific societies for individual disciplines. Their increased authority on the international plane might compensate for challenges to their authority at the national level. The International Association of Academies hoped that with the authority it acquired from international representation it would also be in a superior position to negotiate with governments. Half a century earlier one might have said that the French Academy of Sciences was doing well and did not need help from sister academies in other countries. By 1900 however, even the French Academy was showing signs of age, and support on the international front was welcome.

5. *The Academy's reward system applied outside France*

A few Academy prizes were restricted to French nationals but the majority were open to foreigners. If the Volta prize of 3000 francs, awarded to Davy in 1808, is particularly well known, it is largely because Davy was British and France was

[41] Brigitte Schröder, 'Caractéristiques des relations scientifiques internationales, 1870–1914', *Journal of World History*, 10 (1966), 161–77. Brigitte Schroeder-Gudehus, 'Tendances de centralisation dans l'organisation de la co-operation scientifique internationale', in E. G. Forbes (ed.), *Human implications of scientific advance*, Edinburgh, 1978, pp. 150–62 (152–5).

[42] A. S., *Comité secret, 1882–1902*, pp. 396–9 (5 February 1900).

[43] The Académie des Inscriptions et Belles Lettres and the Académie des Sciences morales et politiques were also invited but did not attend.

at war with Britain at the time. Even the *grand prix* for mathematical sciences of the Academy, originally the preserve of French scientists, came to be awarded to non-French nationals such as the Dane Oersted (1822), and the Swiss Sturm (1834). In 1830, the prize was shared between two distinguished foreign mathematicians, the Norwegian Abel and the German Jacobi. In the case of the Cuvier prize, founded in 1851 for work on comparative anatomy and zoology, the majority of prize-winners were non-French, the first prize-winners being the Swiss Agassiz (1851), the German Johannes Muller (1854) and the Englishman Richard Owen (1857).

In 1895 the Academy awarded the very large Lecomte prize of 50000 francs to the British scientists Rayleigh and Ramsey for their discovery of argon. The recognition of the merits of Wilhelm Röntgen for his major discovery of X-rays created greater problems, suggesting that some anti-German feelings were still present. When on 28 November 1898 the secretary reported to his colleagues in secret session that the prize commission was proposing to award the Lecomte prize to Röntgen there were many objections and when a vote was taken only twelve supported the decision with twenty-eight votes against.[44] However, the Academy had been happy to recognise Röntgen's discovery in 1897 with a smaller prize of 10000 francs.

Academy prizes could also be used to arbitrate between claimants for priority on the international stage. Two Americans, the chemist Charles T. Jackson and dentist William T. G. Morton, claimed to have discovered the use of ether as a general anaesthetic. In 1846 they both sought confirmation of their respective claims by competing for the Montyon medicine prize of the Academy. Each sent supporting documentary evidence but the medical section also insisted on trying out the treatment. After reports of more than a thousand patients' cases had been examined, which seemed to confirm that ether could be used without adverse effects, the section made a judgement. It decided that the two Americans had made complementary contributions.[45] Jackson had priority in discovering the physiological effects of ether vapour, but Morton deserved credit for applying the discovery on a large scale in Boston hospitals. The Academy therefore divided the prize in equal parts between the two contestants.

But the award of prizes was not the only means the Academy had of recognising the merits of foreign scientists. Indeed, despite the large sums of money which became available by the end of the nineteenth century, one could argue that a prize was a transitory honour whereas membership was a permanent honour for as long as the scientist in question lived. Of course, we are not speaking of full membership, an accolade reserved for French nationals, but rather election as a corresponding member or as an *associé étranger*. Let us consider each of these positions in turn.

Corresponding members could be either French (provincial) or foreign. Since we have already discussed French correspondents in Chapter 2, it is appropriate

[44] A.S., *Comité secret, 1882–1902*, p. 374. [45] *C.R.*, 30 (1850), 239–44.

Table 11. *Number of foreign corresponding members in different Academies over period 1795–1895*

Country	Academy				
	Sciences	Inscriptions	Fine Arts	Political Sci.	Total
Great Britain, Ireland and Colonies	81	19	12	29	141
German States*	63	34	26	20	143
Italian States	30	18	45	12	105
Sweden, Norway and Denmark	27	6	8	6	47
Switzerland	25	6	6	6	43
Belgium	5	5	24	6	40
Austria, Hungary, Bavaria	12	8	8	8	36
Russia	14	4	10	7	35
Spain and Portugal	9	6	7	6	28
Holland	8	5	4	4	21
U.S.A.	11	1	1	7	20
Turkey	1	3	—	2	6
Greece	—	4	2	—	6
South American States	1	—	—	1	2

Adapted from Franquet de Franqueville, *Le premier siècle de l'Institut*, 1895, p. 49.
N.B. The Académie Française is excluded since it did not have foreign correspondents.
* Except Bavaria, here classified with Austria.

here to consider only those from outside France. Table 11 shows the total number of foreign corresponding members from different countries in the Academy of Sciences in the nineteenth century and, for comparison, the number in the other Academies. It will be noticed at once that the Academy of Sciences was the most international in its outlook. Thus when in 1862 the mineralogy section belatedly recommended Charles Lyell (1797–1875) as its first choice for a vacancy for a corresponding member, the other eleven names on its list represented geologists from a wide range of locations from Boston, Mass. in the west to St Petersburg in the east. We will consider the representation of different countries among the correspondents of the Academy of Sciences.

The high proportion of British and German correspondents is not unexpected. The United States, although providing only a modest number of correspondents, nevertheless had very many more for science than for any other area, which probably reflects the early importance of field work and the environmental sciences on the American scene. Overall Russia had a larger number in all fields, which is probably not so much an objective index of merit as a reflection of cultural contacts with western Europe. Since there is a danger that these figures may be interpreted as having an absolute value in relation to the scientific eminence of the respective countries, it may be worth referring to a discussion held in 1851 on the election of a correspondent in the astronomy section. It was pointed out that the task of the Academy was not necessarily to appoint the most eminent foreign astronomer available but to consider also the countries represented already and to fill any important gaps.[46] Of course this was a consideration which had greatest weight in an observational science rather than in a laboratory science.

Turning now to consider the *associés étrangers*: since there were only eight such positions at any one time, this was the greatest honour that the Academy could bestow on a foreign scientist. Newton had held such a position in the old Royal Academy and it was to provide an embarrassment to the nineteenth-century Academy that it had never elected Darwin to this class of membership. It had however elected him in 1878 as a corresponding member, which might be regarded as the junior grade of distinction for foreign men of science. Normally, if they were distinguished enough and they lived long enough, foreign men of science who were corresponding members would eventually become *associé étranger*. Joule never went beyond the rank of correspondent, while Kelvin exceptionally was elected as *associé étranger* in 1877 without having first been elected as a corresponding member! The number of candidates at each election was considerable, drawn as they were from several disciplines and many countries. Poor Sir John Herschel was an unsuccessful candidate throughout the 1840s and early 1850s.[47] Finally, in 1855, after fifteen years of lack of success, he finally obtained a majority in the tenth election! It is clear that although his name was known to the Academy, possibly because of his famous father who had been one of a number of British men of science elected *associé étranger* in 1802 after the peace of Amiens, he did not have extensive contacts with French men of science. Gauss, elected correspondent in 1804, became *associé étranger* in 1820; Liebig had to wait nearly twenty years (1842, 1861), while the British astronomer Airy had to wait nearly forty years (1835, 1872). There was clearly a large element of chance involved, with different enthusiasms for personalities, subjects and countries surfacing at different times. The great majority of *associés étrangers* came from Britain and the German states and nearly all were European. The representation of the American continents came through Harvard professor Louis Agassiz (Swiss-born, 1872), Dom Pedro II, emperor of Brazil (1877), J. Lawrence Smith of Louisville, Kentucky, a correspondent of Dumas (1879), and the Canadian-born Simon Newcomb, professor of mathematics and astronomy at Johns Hopkins (1895).

The Academy was concerned not only to encourage relations with scientists in other countries but also to cultivate foreign patronage. In the case of Pedro II, emperor of Brazil (1825–91), there seemed to be many reasons for electing him as a corresponding member in 1875. In the words of the secretaries:

[46] *C.R.*, **32** (1851), 600.

[47] It was unfortunate that the results of successive elections, with details of the ballots, was made public in successive issues of the *Comptes rendus* from Vol. 10 to Vol. 41.

> The extensive geographical studies accomplished in the new world by Your Majesty, the services which he has rendered to all those who are concerned with the history of the globe, the enlightened protection which he bestows on scientists, who visit his vast territories, the constant interest which he deigns to bestow on our own work, has long inspired the members of the Academy with the same sentiments as their predecessors felt with respect to the Tzar Peter the Great[48]

It was part of the same policy in 1891 to elect as a corresponding member Albert 1, prince of Monaco and an enthusiast and patron of marine biology. Monaco was only a tiny state but with extremely close links with France.

6. *Communications with foreign men of science*

Although the vast majority of communications to the Academy were from France, there were also scientific figures in other countries who wanted to communicate with it, since they considered it as more than a purely national organisation. Already in 1808 a certain Dr. Friedlaender had volunteered to translate any letter in German which might be addressed to the First Class.[49] This prompted the latter to make two official appointments of translator, so that the official French body of science should not be limited to material in French. Yet this cosmopolitan provision does not seem to have been very permanent.

Sometimes when foreign scientists wrote to the Academy, their motive would be to claim priority but more usually it would be to register their ideas on the international stage. Such for example would be Joule's paper, written in French, on his final determination of the mechanical equivalent of heat.[50] In 1834 the British comparative anatomist Richard Owen sent a memoir to the Academy and asked for it to be read out. As it was in English, it was translated into French and read at a later meeting.[51] Owen had in fact visited Paris in 1831 and had met Cuvier, whom he acknowledged as having a major influence on his work. Another British contact was the astro-physicist Lockyer, who in 1868 sent a memoir on the spectrum of the sun to the Royal Society and at the same time sent a similar account to his friend Warren de la Rue to be read at the Academy.[52]

Relations between the Royal Society and the Academy, however, were not always so open and it is worth considering the earlier case of Faraday. Foreign members of the Academy were sometimes reminded that it would be interested to receive accounts of their research. Soon after Faraday had been elected *associé étranger* in 1844, Dumas encouraged him to send a paper to the

[48] A.S., *Copie de lettres, 1875–1902*, p. 5. Letter of 1 March 1875. When Pedro was forced to abdicate in 1889, he came to Europe and died in Paris.

[49] *P.V.I.*, **4**, 135. See also *ibid.*, 228, 251–2.

[50] *C.R.*, **25** (1847), 309–11 (23 August 1847); *D.S.B.*, 7, 182.

[51] *P.V.I.*, **10**, 252, 601.

[52] A. J. Meadows, *Science and controversy. A biography of Sir Norman Lockyer*, 1972, p. 53.

Academy. He received a detailed reply, which deserves to be reproduced here because it reveals the problem of a conflict of loyalties. Faraday evidently felt that his first loyalty was to the Royal Society:

> I am a little taken by surprise and startled by your expression of a wish so honourable to me as that I would send some account of my late experiments to the Academy. Our Royal Society, which is rather antique in some of its customs and whose policy I do not on the whole approve of, have a great jealousy of its Fellows sending communications anywhere but to their meetings, or if sent to the meeting sending them anywhere else before they are published in the Transactions. I received a few hints about the notes I sent you respecting condensed gases.[53] Otherwise I should have sent you before this some brief notice of my recent experiments. When I received your last letter therefore I resolved to ascertain from one of the Secretaries the feeling of the Council (having long since ceased to take part personally in the management of the affairs) and the evening before last I saw Dr Roget. He told me it was against all rule and quoted to me a case in which Fox Talbot, having sent a paper to the Royal Society, afterwards sent a communication to the Academy, which, appearing in the *Comptes Rendus*, caused that his paper was not printed in the Transactions. This may be all right as it is the old custom but it hardly consists [sic] with slow publication and it prevents me from doing that which would be a great delight to me and, which is more, if accepted by the Academy would be a great honour.[54]

The Royal Society obviously did not understand that the *Comptes rendus* might represent only a preliminary announcement or claim rather than full publication. Despite his private criticism of the Society, Faraday was prepared to accept its discipline and, when he wrote to Dumas three months later, it was to say that his work had now been published by the Royal Society and he was hoping to use its good offices to send copies of the paper to his friends in France.[55]

After Rayleigh and Ramsay had discovered argon in 1894, they continued their research on the new inert gases and by the following spring they had discovered a mixture of argon and helium in the uranium mineral, clevite. On Saturday 23 March 1895 they sent a telegram to Berthelot as secretary of the Academy to communicate the discovery at the next meeting. Ramsay's telegram read:

> Gaz obtenu par moi – Clévite – mélange argon helium – Crookes identifie spectre – Faites communication Académie lundi.[56]

[53] Faraday to Dumas, *c.* January 1845, in L. P. Williams (ed.), *Selected correspondence of Michael Faraday*, Cambridge, 1971, vol. 1, pp. 432–4. It should be said that Faraday had previously written to Hachette in 1831 about his discovery of electromagnetic induction. Hachette reported it to the Academy, which did not yet have the *Comptes rendus*. This gave rise to unfortunate rival claims of priority. See *ibid.*, pp. 212–13, 223.

[54] Faraday to Dumas, 9 January 1846, A.S., Dumas file; also reproduced in L. P. Williams, (ed.), *op. cit.*, vol. 1, pp. 480–1.

[55] Faraday to Dumas, 15 March 1846, A.S., Dumas file (not in Williams).

[56] *C.R.*, **120** (1895), 660.

At the meeting of Monday 25 March Berthelot accordingly read out the telegram, giving also an assessment of the importance of the communication (to be explained below). At 6 p.m. that evening Berthelot received a letter from Ramsay giving further details, particularly relating to the spectra of the gases which were, of course, vital for their identification. Berthelot was able to incorporate the text of the letter as an appendix in the *Comptes rendus* for that week.

The gas given off from clevite had already been examined by the American chemist Hillebrand, but he had imagined it was nitrogen.[57] Helium had previously been detected in the sun from an analysis of the spectrum but the (impure) sample obtained by Ramsay was the first to be handled by man and the excitement generated was understandable. The British public received news of the discovery in the issue of the journal *Nature* dated 28 March 1895.[58] As evidence of the international interest of the discovery, we may cite a letter of 8 April sent from Uppsala to the Academy confirming the presence of helium in the gas.[59] Ramsay was to send a further communication to the Academy in 1910 on the density of the gas given off from radium, which he called *niton* but which is now known as *radon*.[60] This work was not published in Britain until the following year.

We have given a few examples of communications from British scientists, who were rather more numerous than German correspondents for example, but it should not be thought that the Academy's influence was confined to Europe. We have described in the previous section the part played by the Academy in 1850 in deciding between rival American claimants for the discovery of ether anaesthesia. We have also mentioned the election in the second half of the nineteenth century of several corresponding members and *associés étrangers* from both north and south America.

Finally, it should not be thought that the Academy was content with being a receiving agency. Of course, the Academy always had fairly close relations with the Royal Society of London, to which it sent its *Mémoires* and was happy to receive in return the *Philosophical Transactions*. But once the *Comptes rendus* was established, a continual question which occupied the Academy was how widely it should be distributed. We shall see that the Academy was able to make itself known to scientific societies of varying degrees of eminence over several continents.

Some indication of the international influence of the Academy may be gained from the register of distribution of the *Comptes rendus* in foreign countries.[61] Copies of the first volume (1835) of the new journal went, for example, to the Literary and Philosophical Society of Manchester, the Academy of Turin and the

[57] J. R. Partington, *History of chemistry*, vol. 4, 1964, p. 917.
[58] *Nature*, **51** (1894–5), 512.
[59] *C.R.*, **120** (1895), 834 (16 April 1895).
[60] *C.R.*, **151** (1910), 126–8, Partington, *op. cit.*, p. 941.
[61] A.S., *Distribution des Comptes rendus*, vol. 1, 1836–68.

University of Athens. These were probably in response to personal contacts and specific requests. For several years there does not seem to have been a plan for the distribution of the journal to other institutions, and the distribution proceeded in an ad hoc manner. Thus volume 12 (1841) went among many other places to the Munich Academy of Sciences, the University of Coimbra (Portugal), the imperial library of St. Petersburg, the Canadian Institute at Montreal, the Smithsonian Institution in Washington, and the American Academy of Arts and Sciences at Boston. Soon the Central Society of Agriculture of the state of Wisconsin (U.S.A.) was to be added, as well as European institutions at Milan, Naples and Venice, Vienna, Rotterdam and Uppsala. By the time volume 65 was published (1867) the *Comptes rendus* were also going to Spain, Hungary, Finland, India and Chile. It should be stressed that these were gifts to institutions and the diffusion of the *Comptes rendus* over several continents also took place on the basis of individual subscriptions. But whether bought by private persons or taken by a foreign library, the *Comptes rendus* helped to present a perspective of Paris-centred science to the world. Of course within the British Empire London, and perhaps Edinburgh, would provide an alternative focus. Yet even in the English-speaking world, the proceedings of the French Academy were considered sufficiently important for the weekly journal *Nature* regularly to devote a full column of small print to summarising what had transpired at the meeting in Paris in the previous week.

In the late nineteenth century American science was becoming increasingly independent of European influence. Yet earlier, when the American National Academy of Sciences was founded in 1863/4, it had been happy to follow the French Academy in forming sections.[62] The National Academy even accepted the basic number of six in a section although, in a demonstration of flexibility more typical of the new world than the old, this number was increased to nine in a subject like astronomy, considered of major importance, and reduced to three or less in the case of some branches of natural history.

7. *Expeditions*[63]

In the eighteenth century one of the great expeditions sponsored by the French government had been that of Louis Antoine Bougainville (1766–9), which had challenged the British in the Pacific Ocean. The subsequent voyage of La Pérouse (1785–8) had ended in shipwreck. The revolutionary and Napoleonic wars

[62] A. Hunter Dupree, 'The National Academy of Sciences and the American definition of science', in A. Oleson and J. Voss (eds.), *The organisation of knowledge in America, 1860–1920*, Baltimore, 1976, pp. 342–63 (352–3).

[63] An invaluable source of information for this section was Maxine Taylor, 'Prologue to imperialism: Scientific expeditions during the July monarchy', Ph.D. thesis, University of Oklahoma, 1980. See also Numa Broc, 'Les grandes missions scientifiques françaises au 19e siècle (Morée, Algérie, Mexique) et leurs travaux géographiques', *Revue d'Histoire des Sciences*, **34** (1981), 319–58; and Jean-Paul Faivre, 'Savants et navigateurs: Un aspect de la co-opération internationale entre 1750 et 1840', *Journal of World History*, 10 (1966), 98–124.

considerably reduced French scientific expeditions abroad with the outstanding exception of the expedition to Egypt under general Bonaparte. Bonaparte took with him a Commission on Science and the Arts (1798–1801), which was intended to represent the best of French civilisation in a supposedly barbaric country but was also intended to gather scientific and archaeological information about this largely unknown country. We can see in this expedition several features of nineteenth-century French government overseas policy, which was concerned with the extension of commercial and political influence and also its cultural influence, sometimes described as a *mission civilatrice*. The Egyptian expedition contained several members of the Institute, notably Monge and Berthollet, who might be considered to have been on temporary leave of absence and they, like other men of science, were principally concerned to use such expeditions to widen their experience of the natural world and collect local information. There was also the exceptional voyage (1800–3) of the naval officer captain Baudin, who asked the First Class to approve his plan of circumnavigation of the globe and reached Australia.[64]

After 1815 the French government was able to organise a succession of naval expeditions and circumnavigations. For many who had sought glory on the battlefield under Napoleon, the period of the Restoration was one of anti-climax and boredom but there was still adventure to be found at sea. The Pacific Ocean in particular was a vast area in which explorers could make discoveries and bring 'glory' to France and themselves. Yet most of these expeditions also had a scientific input. Although it was not unusual for scientists to be included in the expedition, the navy would regularly approach the Academy of Sciences for instructions on the scientific data which might usefully be collected by naval officers.

There is some ambiguity about the role of science in many of these expeditions. Particularly in the later expeditions the main motive seems to have been political, with science being added largely to gain respectability for these exploits. When one of the expeditions landed in New South Wales, the British already established there found it difficult to believe that their French visitors were only interested in science. Yet they certainly collected scientific data. There is a record, for example, of the geomagnetic data collected in successive expeditions: that of Freycinet in the *Uranie* (1817–20), Duperrey in the *Coquille* (1822–4), Hyacinthe Bougainville in the *Thétis* (1824–6) and D'Urville in the *Astrolaube* (1826–9). Several expeditions produced a considerable volume of scientific information. Freycinet, for example, on his return, published a seven-volume description of his voyage and he was able to present many useful specimens to the Muséum d'Histoire Naturelle. D'Urville's voyage yielded many maps, portraits of different racial types and many specimens, including some 900 mineral samples from twenty-two countries.

[64] *P.V.I.*, **2**, 119; Faivre, *op. cit.*, 112.

The Academy played a rather remote and subsidiary role in most of these expeditions. Although it might seem to be in control by giving 'instructions', these were really no more than suggestions. Naval commanders preferred to have their own junior officers making observations and collecting specimens rather than civilian scientists, who were not subject to naval discipline.[65] Of course the Academy and science generally was able to benefit from the data provided but after several requests for specific instructions on what should be done on a particular voyage, the Academy in 1838 considered that it would be simpler to draw up 'general instructions, in which all the needs of zoology, botany, geology, "geophysics" (*physique du Globe*) and geography would be pointed out and explained'.[66] Blainville, known for his outspokenness, wanted to tell the Naval Ministry which parts of the world represented the greatest gaps in scientific knowledge in the hope that future expeditions would be sent there. But he was rather naive if he supposed that the Ministry had science at the top of its list of priorities. Increasingly what the French government was looking for were *points d'appuis*, or safe ports for French ships which might later become bases for colonial expansion in distant parts of the world.

One of the main scientific expeditions with which the Academy was concerned was that to Algeria. French interest in Algeria arose out of the activities of Barbary pirates, who harassed Mediterranean shipping and used that country as a base. Two minor incidents in 1827 and 1829, interpreted as insults to France, were the justification for a landing in Algeria and the capture of Algiers. This was to be the beginning of the eventual occupation of the whole country. Yet little was known about the territory. Before total conquest was attempted, therefore, a scientific expedition was planned to report on the resources of the new territory. Algeria was to become France's greatest colonial prize.

In 1837 general Bernard, Minister of War, wrote to the Academy asking it to draw up a set of instructions for the expedition which he proposed to send to Algeria. This time the Academy commission not only discussed the most relevant scientific considerations but it also proposed the names of some twenty men, including military and naval personnel, who would have the necessary expertise to make the relevant observations in zoology, botany, geology, hydrography and geophysics.[67] The Chamber of Deputies approved the allocation of 75 000 francs for preparations for the expedition. In charge of the scientific expedition was colonel Bory de Saint Vincent, who had experience of a previous Greek expedition and who had been recognised by the Academy in 1834 by election as an *Académicien libre*.

On 18 May 1840 a first report by Bory was read to the Academy on the work

[65] See M. Taylor, *op. cit.*, p. 19.

[66] A.S., *Comité secret, 1837–44*, p. 15 (11 June 1838).

[67] A.S., *Comité secret, 1837–44*, pp. 23–4 (19 November 1838).

of the commission during its first three months.[68] The zoologist Deshayes had collected over sixty species of molluscs and over ninety gasteropods. Unfortunately, drawing of these creatures was imperfect owing to delays in the delivery of microscopes and books promised by the Minister of War. New species of fish had been observed by Guichenot. Various other animals and plants were noted, paying particular attention to their relation to the different climate of north Africa. There were also sections of the report on agriculture, which held out promise for future cultivation, and on geology. There was a further brief report in May 1841.[69] In July 1843 Bory felt it necessary to explain to the Academy how it was that his expedition, which had returned to France more than a year ago, had so far published nothing. Meanwhile he presented a memoir on the flora.[70] Eventually the full scientific results of the expedition were published at government expense in twenty-five volumes of text and five volumes of plates as *Exploration scientifique de l'Algérie* (Paris, 1844–67). One writer on the subject has commented that:

> Of all the [French] expeditions, which carried out scientific work in the nineteenth century, this commission appears to have had the most impressive record.[71]

There was certainly very much more done than one might expect from a narrow utilitarian approach which might only have been concerned with the short-term exploitation of the natural resources of the country. On the other hand some members of the scientific expedition carried out preliminary reconnaisance towards Morocco and Tunisia. Africa presented ample scope for the colonial ambitions of many European nations.

We should also mention briefly the scientific expedition to Mexico. There seem to have been a number of reasons why Napoleon III decided to intervene in Mexico. He had long been interested in the new world and Mexico seemed a county rich in raw materials and ripe for development. He also wanted to support European settlers and had ambitions of a foreign empire. After the defeat of the first French expeditionary force Napoleon sent out 30000 men as evidence of his serious intent. By 1863 the French troops managed to occupy Mexico city and they were able to install the French candidate, Maximilian, on the throne. Yet by 1865 the American government (previously preoccupied with the Civil War) was able to convey to the French its hostility to interference in its hemisphere. Maximilian lost his patron and the Mexican expedition was recognised as a failure. Its most important effect on subsequent French history was to weaken France's military strength for the Prussian challenge of 1870.

But we must go back a few years in order to explain how science entered the story. In 1864 Duruy, as Minister of Public Instruction, called a meeting of

[68] *C.R.*, 10 (1840), 781–6 (18 May); 849–53 (1 June).

[69] *C.R.*, 12 (1841), 901–2.

[70] *C.R.*, 17 (1843), 19–26.

[71] M. Taylor, *op. cit.*, p. 110.

savants, including many members of the Academy of Sciences such as Marshall Vaillant, Maury, Milne-Edwards and Quatrefages.[72] Recalling the Egyptian expedition, in which Napoleon III's uncle had played a prominent part, Duruy said that he was proposing to organise a scientific commission in Paris which would plan a programme for (junior) scientists and army officers in Mexico. It would be principally concerned with geography, geology and mineralogy but it would also study the animal and vegetable kingdom and 'atmospheric phenomena'. As with the Algerian commission there was an immediate budget to cover costs, in this case 200 000 francs. The first field workers were sent out to Mexico in October 1864 to avoid the excessive heat and disease of the summer, which had earlier decimated the French army. Several ambitious young men were chosen and we may particularly mention the mineralogist Guillemin and the geologists Dollfus and Montserrat, who were later to publish their findings. Their studies of volcanoes were particularly valuable. The Academy played only a peripheral role in the whole story, which may be regarded as one of the more notorious of Napoleon III's blunders. He had sought to emulate the grand plans of Napoleon I but he lacked the military genius.

We might also mention other overseas developments in which the Academy was involved. When in 1859 the astronomer Faye heard that the government was proposing to send an expedition to China, he suggested to his colleagues in the Academy that they should take the initiative and ask the government to include a scientific commission in the expedition.[73] Science should not be forgotten! Some time before work was begun on the Suez canal in 1859 its chief promoter, the engineer Ferdinand Lesseps, had sought the approval of the Academy,[74] which had appointed a commission to examine the proposal. Reporting back in March 1857 the commission, in a major report, gave its wholehearted approval.[75] The Academy commission provided a second report in 1858, taking into account recent preparatory work by Lesseps.[76] This report was tempted to congratulate Lesseps (and itself) on this humanitarian project which would save lives by making unnecessary the long sea voyage to the east via the Cape of Good Hope and would bring nations closer together. Nevertheless the commission realised that its authority was scientific and gave the project its full support mainly on the basis of its knowledge of geography, geology and surveying. It could, understandably, take some pride in the opening of the canal in 1869.

But the rule of Napoleon III was coming to an end. There is a certain irony in the fact that it should have been the next government, that of the Third

[72] Broc., *op. cit.*, p. 332.

[73] *C.R.*, **49** (1859), 829.

[74] *C.R.*, **42** (1856), 1163. The success of Lesseps with the Suez canal may be contrasted with his later failure in French attempts to construct the Panama canal through much more difficult terrain.

[75] *C.R.*, **44** (1857), 417–49.

[76] *C.R.*, **151** (1910), 1180.

Republic, which did most to build up France's colonial empire.[77] French expeditionary forces, traders, explorers and missionaries extended France's rule in Africa. In Indochina the Second Empire had established a foothold. This was to be developed in the 1870s and particularly the 1880s. Even some deputies on the far left, like the Academician Paul Bert, supported French colonial expansion in Indochina. Given his scientific eminence, some slight ministerial experience and his firm political support for its foreign policy, the government appointed Bert in 1886 as the first civil governor of Annam and Tonkin. Yet as an Academician Bert had a commitment to live in Paris. He therefore asked the Academy for indefinite leave of absence in the national interest.[78] He said that he hoped to show eastern men of letters the superiority of western science. He would contribute to French 'moral influence' and thus serve at the same time both science and France. He even referred to his appointment as a 'superior duty', a phrase which rankled with fellow Academicians. In fact in Indochina he was to be cut down by tropical disease within that same year and in an obituary the Academy seemed to reproach him with excessive ambition, since he had not been content to remain as a working member of the Academy.[79] In the end he had sacrificed science for politics. After consultation with the Academy, the French government went on to establish a 'service of scientific exploration' in Indochina. But in 1908 it decided to close this down and the Academy sent a letter of protest to the Minister of Public Instruction, asking him to forward it to the Minister of the Colonies. The Academy claimed that Britain, as the major rival colonial power, had widespread scientific institutions in its colonies and France should not be in an inferior position.[80]

Some mention must also be made of Antarctic exploration. Probably the most famous episode in this history was the race to the South Pole, reached by the Norwegian Amundsen in December 1911 and a month later by Scott of Britain, who perished on the return journey. These were not the only nations involved in Antarctica and we must say something of the French contribution. The French were the first to make magnetic observations in the far south. In 1838 D'Urville spent two months at the edge of the pack ice south of the South Shetlands and the South Orkneys. Two years later, having fulfilled the main purpose of an expedition to the Pacific, D'Urville attempted to reach the southern magnetic pole. He discovered land south of the 66th parallel in 149°E, which he named Terre Adélie, by which title it is still known. Most subsequent explorers were British until in 1904 and 1908 Jean-Baptiste Charcot, son of the Academician J. M. Charcot, led two expeditions and was able to construct a valuable chart of 2000 km of coastline. At the public meeting of the Academy in 1910 Charcot's courage and skill received high praise.[81]

[77] Alfred Cobban, *A history of modern France*, vol. 3, 1965, p. 28.

[78] *C.R.*, **102** (1886), 287–8 (8 February 1886).

[79] *C.R.*, **103** (1886), 1292–3.

[80] A.S., *Comité secret, 1903–1912*, pp. 217–18.

[81] *C.R.*, **151** (1910), 1180.

In conclusion, we should perhaps correct an impression that the main benefit to science of these various expeditions was simply to accumulate data from remote parts of the world. There was an important 'natural history' element and certainly in the history of geomagnetism for example, it was vital to have data from different parts of the world. But among the masses of data collected there was crucial evidence which could be used to test scientific theories, whether of geology or physical anthropology and ethnography.[82] Some ideas, which might be acceptable for those whose horizons did not extend beyond France or Europe, would have to be modified from a wider knowledge of conditions in distant continents.

In their compilation of instructions for future voyages, Academicians might seem as no more than humble servants of the state, yet they were in many instances using naval personnel as their agents. Poincaré in a speech at the public meeting of 1906 referred to a geodesic expedition in an equitorial climate organised by the Academy and paid tribute to 'those officers and soldiers, who have died for science [sic]'.[83] Science was obviously one beneficiary. Yet if the Academy was to use agencies under more direct government control to gather data, there is no doubt that the government used the Academy and, more generally, the concept of science as an alibi for colonial exploration. Here we see Academicians filling out the role not simply of *savants* but of patriotic Frenchmen. As Darboux, secretary of the Academy, claimed at the public meeting of 1909:

> Our Company has always combined the religion of the Motherland (*le culte de la Patrie*) and scientific research.[84]

[82] For a similar argument about eighteenth-century French expeditions, see Hahn, *The anatomy of a scientific institution. The Paris Academy of Sciences, 1666–1803*, Berkeley, Cal., 1971, p. 90.

[83] *C.R.*, **143** (1906), 998.

[84] *M.A.I.*, **51** (1910), xxxviii.

— *Chapter 12* —

THE CONTROL OF THE ACADEMY AND OF SCIENCE

The man of science and the man of letters, who wishes to enjoy perfect and unlimited freedom, should renounce academies and [paid] positions.

(P. J. Pelletan, surgeon and member of the First Class in open letter to members of the Institute, n.d. but *c.* 1804.)

The regulations of the French Institute would appear...revolting and injurious to the feelings of an Englishman.

(G. Moll, *On the alleged decline of science in England*, 1831, p. 24.)

The Academy has decided, after a vote on the subject, that from now on it will only accept measurements according to the metric system in its publications.

(*C.R.*, **102** (1886), 187.)

1. *The power of modern science*

By the nineteenth century science had reached a period of adolescence. Following this metaphor, the sixteenth and seventeenth centuries would correspond to a period of childhood when, after a long infancy, the universe began to be thought of in non-geocentric terms and investigations began with new scientific instruments like the telescope and the microscope. The seventeenth century also witnessed the beginnings of serious scientific organisation. Enough natural wonders had been revealed for some governments to begin to think that this new activity should come under state control, or at least state supervision. There is no better example of this than the Royal Academy of Sciences, founded in Paris in 1666 under the patronage of Louis XIV.

Whatever one's view of the control of science, one may feel that the French were far-sighted to see in immature science an activity capable of enormous growth. Indeed the development of science in the eighteenth century was often cited as the clearest possible evidence that civilisations do make progress. In the science of electricity little more was known in 1700 than had been known to the ancient Greeks, but soon the conduction of electricity was discovered and the ability to generate and store electricity enabled enormous sparks to be drawn. The parallel between these sparks and lightning was pointed out by Benjamin Franklin, thus firmly transferring to the natural world a phenomenon which many had previously regarded as supernatural. Electricity in the eighteenth century was purely electrostatics but in 1800 Volta introduced the electric cell and battery, thus making possible the study of current electricity. We have had occasion to mention the prize offered by the Institute for new developments in electricity.[1] Electricity caught the imagination of Napoleon Bonaparte, as it did later his nephew Napoleon III. In 1864 the Academy was to award a prize of 50000 francs to Rühmkorff for his development of the induction coil.[2]

Another eighteenth-century development was the discovery of gases, some of the most important being carbon dioxide (or 'fixed air'), nitrogen ('azote') and oxygen ('vital air'). These gases had power over life and death, in so far as the first two could kill a human being while the latter was shown to be necessary for life. Joseph Priestley could boast that as a result of the new knowledge tyrants might have cause to 'tremble even at an air pump or an electrical machine'.[3] But Priestley's experimental science drew from the conservative political writer Edmund Burke the accusation that people like him 'consider man in their experiments no more than they do mice in an air pump, or in a recipient of mephitic gas'.[4] Partly because of the radical politics of Joseph Priestley, Burke saw experimental science as something associated with revolution. At the very least it was a danger to the stability of society.

Burke's fear of reductionism, the fear that the new science would make no distinction between human beings and animals, finds a strong echo in the writings of Mary Shelley. In her most famous novel, begun on a continental tour in 1816, she portrays Victor Frankenstein as a person for whom the human corpse is not something sacred but legitimate raw material for dissection and experimentation. A dominant theme in Mary Shelley's *Frankenstein* is 'how dangerous is the acquirement of [scientific] knowledge'.[5] Recognising no limits to the power of the new science, and being particularly interested in electricity, Frankenstein would attempt to revive the dead to create a new kind of human being, a monster.

Images of science are clearly of some importance. Although there were Romantic writers like Blake in England and Chateaubriand in France who saw the dangers, even the evils, of Newtonian science, there were many others who, following the ideas of the eighteenth-century Enlightenment, saw in science a great potential power for good. Science was valued for its apparent objectivity, bringing together people who would have disagreed violently on matters of politics or religion. Condorcet advocated the teaching of science in schools to

[1] On the voltaic prize see Maurice Crosland, *The Society of Arcueil*, 1967, pp. 20–5.

[2] E. Maindron, *Fondations de prix à l'Académie des Sciences*, 1881, pp. 131–2. The hesitation of the Academy in not rewarding Rühmkorff earlier is largely explained by the large size of the prize, and hence the extra responsibility involved.

[3] *Experiments and observations on air*, Birmingham, 1790, vol. i. preface.

[4] 'Letter to a noble lord', *Works*, Bohn's Standard Library, 1903, vol. 5, p. 142.

[5] Mary W. Shelley, *Frankenstein*, Chap. 4, Everyman edn, 1921, p. 46.

provide rational training for children at the same time as giving them knowledge of the natural world. Condorcet was enormously optimistic about the power of science to create a better world and his optimism was echoed in the Third Republic. Science would organise everything and explain everything. It was therefore an important ideological weapon and it was often used in politics by the left in attacks on the right.

But there is another completely different way in which science represented power and that is by its applications. One cannot assume that the 'industrial revolution' which began in Britain in the late eighteenth century was dependent on science in any simple way but, if only a part of it depended on science, this would be enough to make the subject of practical concern. If we think of new sources of power, the idea that Watt's steam engine depended on the application of the science of his one-time colleague Joseph Black is intriguing but not satisfactorily proven. We do know, however, that the efficient working of steam engines *in the nineteenth century* depended increasingly on the applications of science.[6] The first science to be applied to practical use was probably chemistry, and by around 1800 Lavoisier's chemistry had something to contribute to the developing heavy chemical industry (centred on the manufacture of soda and sulphuric acid), as well as in the new revolutionary method of bleaching with chlorine, which replaced the traditional slow method of exposure to air and sunlight.[7] Later in the century French chemists were to contribute to the manufacture of artificial dyestuffs, although it was the Germans who were to become the world leaders in this area. We shall be looking later in the chapter at some important examples of French applied science which were discussed in the Academy.

2. *Government control of the Academy*

The phrase 'government control of the Academy' is capable of many different interpretations. If we look at this in the abstract it could have meant, in the extreme, the issuing of directives day by day or, rather more reasonably, it could at least have involved the setting of long-term goals. One could imagine an academy of sciences being established which would be asked to justify itself purely in utilitarian terms, looking only for the *applications* of science. Turning to the economic sphere, one could imagine a scientific organisation which was required to make money, or at least to cover its running expenses by modest earnings. This after all is how the university system was organised in France, where the fees for examinations and degrees were supposed to pay most of the cost of teaching and administration. In the case of science, presumably commercial and industrial interests would pay for advice. But we do not need to stretch our imaginations any further, since none of these conditions were imposed by the French government on the Academy of Sciences in the

[6] D. S. L. Cardwell, *From Watt to Clausius. The rise of thermodynamics in the early industrial age*, 1971, esp. chaps. 7 and 8.

[7] John Graham Smith, *The origins and early development of the heavy chemical industry in France*, Oxford, 1979.

nineteenth century. Yet the government would almost certainly have claimed to control the Academy and might, for example, have cited Article VI of the original constitution of the Institute, that each class should present an annual report on its work to the legislative body.

Let us review the ways in which this control operated in practice. Firstly the Academy was given occasional tasks to perform. We have seen (Chapter 9, Section 11) how this potential burden in the early years soon declined. We have also seen how the revolutionary, Napoleonic and Restoration governments exercised a certain amount of political control on the Academy, either by nomination of members or by refusing to accept the results of elections, although this did not happen after the 1820s. The most important permanent feature of control was the general accountability of the Academy to the responsible minister, usually the Minister of Public Instruction. But there was some ambiguity about the degree of accountability of the Academy to the Minister. When in 1855–6 the authoritarian Minister of Public Instruction Fortoul asked the Academy to resume making the annual reports to the government as had been envisaged in its early years, the Minister received a sharp rebuff:

> if we put to one side the [specific] reports and instructions which are asked for by the government, in the composition of which the Academy has always applied itself with the most scrupulous and active attention, *the Academy does not have work which is specially entrusted to it*, it does not have other work than that which its members undertake spontaneously, work of research and discovery, and that consequently it does not owe [to the government] and cannot owe anything more than an account which is purely intellectual and scientific.[8]

The 'accountability' in the end might have meant little more than the Minister having to approve the annual accounts of the Academy. Certainly *in principle* every time the Academy wanted to spend money, even the private legacy from the Montyon fund, it had to obtain the approval of the Minister. In practice, the Minister was not too worried how the Academy spent a private legacy and allowed it considerable latitude, no doubt happy to save public money. Yet financial control was one of the main sanctions in the hands of the Minister to exercise authority over the Academy. As the Montyon fund grew, and the Academy became more adventurous in spending the accumulated interest from the fund, we have seen how it was able to award large prizes and grants and to launch the publication of the *Comptes rendus* – all largely independently of the Minister. The Montyon case was one in which the Academy clearly took the initiative, something which earlier ministers had not liked. For example, in 1813 the Minister Montalivet had rebuked the First Class for seeming to act as a patron of science, which he regarded as his own

[8] A.S., *Comité secret, 1845–56*, p. 461 (13 October 1856) (my italics).

prerogative.[9] Thus the Montyon fund may be seen as one of the most important factors in altering the balance of power between the government and the Academy. We shall be taking a look shortly at the section of *Académiciens libres* as a further example of the decline in government control over the Academy.

It is perhaps ironic that it was precisely at a time when science was becoming more powerful that the Academy came less under government control. The period of strictest control of the Royal Academy had been under Louis XIV. In the first few years of the National Institute, immediately after the Terror, men of science were grateful to be alive and to have some recognition from the state. There was no question of their objecting to the government regulations of 1795. The succeeding regime of Napoleon, although claiming to hold science in great esteem, nevertheless imposed a high degree of control on the activities of the Institute. At the Restoration there was a feeling that science had been over-valued during the Napoleonic period. The poet Lamartine could write that there was a league of mathematical studies, which suppressed free thought and poetry:

> Mathematics were the chains of human thought. Now they are broken and I breathe again.[10]

Not only was the Académie Française, representing literature and language, given its former precedence over the other academies but, more important, the old nobility was expected to resume its prominent place in government and the higher levels of administration at the expense of the middle class, which had taken a major part in such activities since the Revolution. Some scientists tried unsuccessfully to be courtiers but generally speaking the Academy of Sciences was very much on the defensive during the period of the Restoration, and the government of Charles X was particularly repressive to freedoms of thought and expression. After the revolution of 1830 the interests of the middle class were once more well represented in government, which was never again to lean so hard on the Academy, not even under Napoleon III.

Under Napoleon III it was the ministry of Fortoul which exercised the greatest control on higher education, and the Institute was not entirely immune. We shall have to describe a case of gross government interference in 1855 since, although it was directed at the Academy of Moral and Political Sciences (hereafter AMPS), it was to send shock waves through the whole Institute and ultimately to affect the Academy of Sciences. Under Napoleon I that part of the Institute had been suppressed as politically subversive. What would be its fate

[9] A.I., *Commission administratif de l'Institut, 1795–1831*, dossier 1812–13, letter of 14 September 1813, 'Les loix...ne lui attribuent pas le soin de distribuer des faveurs autres que les prix fondés par ses règlements'.

[10] 'C'était une ligue universelle des études mathématiques contre la pensée et la poésie. Le chiffre seul était permis, honoré, protégé, payé...Les mathématiques étaient les chaines de la pensée humaine. Je respire; elles sont brisées.' A. de Lamartine, 'Des destinées de la poésie'. *Oeuvres complètes*, 1862, vol. 1, p. 27.

under the Second Empire? In the first few years the new Minister of Public Instruction Fortoul, despite his rigorous disciplining of the whole educational system,[11] left the Institute alone. In 1854 the AMPS even seemed to be courting favour in actually electing the Minister to its ranks. He should have felt more favourably disposed to that Academy than to the Académie Française which, in a series of elections in that year, had elected three literary figures hostile to the government of the day. Yet it was the AMPS to which government interference was directed. By ministerial decree of 14 April 1855 a new section was to be added to the Academy, a section given the miscellaneous title of 'politics, administration and finance' and consisting of ten members, all nominated by Fortoul.[12] This sudden increase in numbers from thirty to forty seemed a transparent attempt to weaken the intellectual independence of the members and perhaps even to change the political balance. The men nominated were always regarded by their colleagues as Academicians of a lower rank and the nominees themselves applied for vacancies in the other sections at the earliest opportunity. Fortoul died soon after, and the section of politics, administration and finance was finally abolished in 1866 under the Ministry of Duruy, who distributed its members among the other sections.

Returning now to the Academy of Sciences, it was under Fortoul's immediate successor Rouland (1856–63) that an opportunity arose of rectifying an anomaly which went back some sixty years. We may recall that when the Second Class of the Institute was abolished in 1803, three members of the Geography section having some connection with science had been transferred to the First Class, thus effectively creating a new half section with the title geography and navigation. In 1863 the Minister Rouland asked the Academy if it would approve of the strength of the section being brought up to six.[13] Several meetings were held and, when a vote was taken, thirty-four out of fifty-three voted in favour of the increase. There was, however, a sting in the tail. The Minister also consulted the Academy on the possibility of adding 'military sciences' to the title of the section. In other words a change, which implied an increase in the size of the Academy and consequently a slight increase in the budget, was not being suggested out of pure love of science.

The government of Napoleon III looked back with nostalgia to the period of Napoleon I, when the French armies had conquered a large part of the continent of Europe. Dreams of empire had encouraged Louis Napoleon to send a military expedition to Mexico in 1861, and in 1862 a French force in the far east

[11] Paul Gerbod, 'L'université sous le joug, octobre 1852–juillet 1856', *La condition universitaire en France au XIXe siècle*, 1956, pp. 309–53.

[12] Emile Mireaux, *Le coup d'état académique du 14 avril 1855*, Académie des Sciences Morales et Politiques, Séance Publique Annuelle, 2 décembre 1963.

[13] A.S., *Comité secret, 1857–69*, e.g. pp. 261–2 (22 June 1863) and 293 (30 May 1864). Discussions continued over a long period. See also the journal edited by the abbé Moigno, *Les Mondes* 1 (1863), 339, 385–6, 435, 476, 504, 604.

annexed Cochin-China and established a French protectorate over Cambodia. Now the geography section of the Academy was the obvious place to reflect these new imperial aspirations, not only because it was the section professionally concerned with distant parts of the world, but because it included under the 'navigation' rubric a succession of senior naval officers. Why not introduce into this section some senior army officers?

The imposition was less crude than it might have been in many other countries, because the sort of military men who would have been considered suitable might well have been graduates of the Ecole Polytechnique, the school which had provided a large number of physical scientists in the Academy. Many Academicians, however, viewed the military dimension with some alarm, all the more so as there were already three generals (Poncelet, Piobert, Morin) in the Academy as well as a few other army officers and an admiral. When on 29 June 1863 it seemed as if a small majority of the Academy might be prepared to accept the Minister's suggestion, Dumas intervened to say that the proposal was so far-reaching that it merited a six-month postponement to allow for full consideration of the implications. When the discussion was resumed the following year under the more liberal Ministry of Duruy the proposal was defeated with a large majority. An alternative proposal that the words 'physique du Globe' (environmental sciences) should be added did not gain quite enough support to succeed.

The question of political control we have been discussing is one partly relating to membership of the Academy and partly to the question of the duties of Academicians. Although the government had the right to consult the Academy regularly on technical matters, it was in fact coming less and less to rely on the Academy for advice. But this was not because it needed less advice; on the contrary it needed more, but it set up its own agencies to deal with specific problems. If the government gave a problem to the Academy, it was the Academy which chose suitable experts and which set its own timetable for the study of the problem. The government, which in theory had all the control, in practice had little contribution to make as to how the problem was tackled. For the solution of any specific problem it was much more satisfactory for the government to appoint an individual Academician as a consultant. It then had much more control over the situation. Nor was the Academy totally excluded, since it had played its part by electing the Academician and thus designating him as a great expert. Alternatively for the investigation of a long-term problem a commission could be set up, which might include not only scientists (often Academicians drawn from relevant areas) but also some members with practical experience and a few administrators.

Already in the eighteenth century we have seen that the question had been asked of how science could help French industry and the Bureau de Commerce had approached individual members of the Royal Academy of Sciences to ask for expert advice.[14] One area where this was needed was in the dyeing of cloth and another was mining.[15] Like the administration of Ponts et Chaussées, the administration of mining was complemented by a school where training was given to future inspectors of mining. The Ecole des Mines in Paris was also to provide employment for many leading mineralogists and geologists, of whom some were Academicians.

In the nineteenth century the Conseil des Mines continued to administer the mining of coal and minerals throughout France. This was a full-time occupation and could not have been delegated to the mineralogical section of the Academy. Similarly there was a Conseil d'Agriculture and, during the Second Republic, an Institut national agronomique was established at Versailles specifically to undertake agricultural research. Just as the state had been involved in the building of roads and canals in the seventeenth and eighteenth centuries, so the Corps des Ponts et Chaussées was involved from the 1830s in the planning of railways, or at least of a basic national network of main lines.[16] Here is a reminder that what one might think from a knowledge of British nineteenth-century history to be essentially a matter of private enterprise was seen in France as a matter for planning and control by central government.

We find occasional consultation of the Academy by different ministries. Thus in October 1859 the Academy advised the Ministry of War on the subject of safety matches and the Ministry of Commerce on alcohometers.[17] But there were few such consultations. The French Ministry of Commerce had a permanent commission on lighthouses, which included suitable experts from the Academy of Sciences. An Academician sometimes had a national reputation for his expertise in a particular area and it is no surprise that, when in 1878 the Ministry of War established a Commission scientifique des substances explosives it should call on Berthelot, who had distinguished himself in this field in the Franco–Prussian war.[18] We should remember that the Academy was responsible to only one minister (public instruction) and if other ministries wanted specific tasks performed, it made great administrative sense to appoint their own commission rather than approach the Ministry of Public Instruction. Occasionally the Academy would set up its own commission to deal with an emergency, like the phylloxera affecting vines which qualified for funding from the Ministry of Agriculture. However, such Academy commissions to deal with practical problems became the exception rather than the rule.

[14] H. T. Parker, 'French administration and scientists during the old regime', in Richard Herr (ed.) *Ideas in History* (Durham, N.C., 1965), pp. 85–109.

[15] A. Birembaut, 'L'enseignement de la minéralogie et des techniques minières', in R. Taton (ed.), *Enseignement et diffusion des sciences en France au XVIIIe siècle*, 1964, pp. 365–418 (384, 403).

[16] J. R. Fleckles, 'The state and the beginnings of the railroad *grandes lignes* network in France, 1830–42', Ph.D. thesis, Blackburn College, 1958.

[17] A.S., *Copie des lettres, 1841–60*, p. 287.

[18] Maurice Crosland, 'Science and the Franco-Prussian war', *Social Studies of Science*, 6 (1976), 185–214 (p. 207).

The devolution of scientific consultancy from the Academy to special commissions can, therefore, be seen as a measure of administrative efficiency rather than as a slight on the Academy. For narrow utilitarian purposes the Academy was simply too wide in its scope and, arguably, too independent. By the late nineteenth century it could have been regarded as something of an insult to the greybeards of the Academy to ask them to put to one side their studies on the frontiers of knowledge to tackle some comparatively trivial practical problem. It was the very success of science which opened the way for other agencies, thus in some ways lessening the importance of the Academy. This may have been a pity but it was preferable to an alternative scenario in which the Academy would have held a monopoly in scientific consultancy. Quite apart from anything else, such pressure towards applied science would have had disastrous consequences for its possible contributions to pure science.

Returning to the question of government financial control of the Academy, one may wonder why the members of the Academy of Sciences (and more generally members of the Institute) did not rebel against the paltry allowance awarded in 1796. Surely after the first few years they would have had greater self confidence? Yet there is always some embarrassment about asking for extra payment for oneself, particularly if the request has to be made public. There was also quite a complex machinery involved in any increase in the budget of a minister. It would have to be debated in the legislature and although there would certainly have been politicians who would have supported it, there would have been others who would have used the occasion to pillory some of the activities of the Academy. Then there was the delicate question of whether all Academicians (or members of different former Classes of the Institute) were equally worthy of an increase. Also to ask for more money left members of the Academy of Sciences open to additional formal obligations to the government. The best strategy for Academicians was to approach the problem of a larger income by an indirect path. They sought out the better paid posts in higher education. Moreover, since any one post usually involved no more than an hour or two of lectures a week and no responsibility for the work of their students, they sought several such posts, the practice known as *cumul*. Thus *cumul*, so often understandably condemned as reducing opportunities for junior scientists, can in a way be seen as a response by Academicians to the low honoraria which they inherited from revolutionary times.

But to explain the acceptance of low honoraria in purely economic terms or as largely due to problems of administration is to miss an important dimension, the moral one. Academicians felt that by not asking for more they increased their authority. This may be difficult to understand in an age where it is sometimes claimed that the very size of a large research grant for a project is a measure of the value of that project. But nineteenth-century attitudes showed greater humility. Few men of science felt that society owed them a living. Academicians could take a positive pride in how *little* money they took from the state. A good illustration of this occurs in one of the *éloges* given by Dumas in 1881 as secretary of the Academy. He mentioned a long line of Academicians going back to the time of Lavoisier and including Gay-Lussac, Arago, Dulong and Regnault, who had undertaken work for the state. This work had brought great benefit at minimum cost. He continued:

> These great *savants* were all great patriots, generous with their science and their time but miserly with government money. They have established among us *a tradition of disinterestedness, of self-denial and of respect for public funds, of which the Academy is proud* and which it will never allow to be broken.[19]

Another way of speaking of the moral authority of the poorly-paid Academicians is in terms of their independence. By accepting a pittance from the government, Academicians effectively reduced government control of their activities. According to the old proverb 'he who pays the piper calls the tune', but if the payment is no more than a few pennies the piper need hardly consider himself obliged to produce more than a few notes.

3. *The changing role of the section of Académiciens libres*

The creation and development of the section of *Académiciens libres* (literally 'free Academicians') provides an interesting case study of changing relations between government and the Academy. When Louis XVIII came back to France in 1815 after the final defeat of Napoleon, he was determined to revive as many practices of the *ancien régime* as was feasible. Yet he could hardly destroy the many successful institutions built up by the revolutionary and Bonapartist governments. But in accepting this, he could nevertheless change their names and graft a royalist element on to them. Accordingly the First Class of the Institute became the Royal Academy of Sciences of the Institute and a new section of *Académiciens libres* was created, corresponding to the honorary membership of the old Royal Academy, which had been drawn from the nobility. The effect of this new section was all the greater since it was to consist of *ten* new members rather than the traditional six.

Understanding what was required of it, the Academy pointed out that the duc de Noailles (1739–1824) and the duc de Lauraguais (1733–1824) were the only honorary members surviving from the *ancien régime* and invited them to take their seats as the first two *Académiciens libres*. But it was only prepared to tolerate this additional section on the understanding that its former constitution would not be further tampered with. The minutes record with unusual frankness:

> The Academy decrees unanimously that its existing constitution seems to it to be dictated by reason and experience and it is desirable that no [further] changes are made.[20]

[19] J. B. Dumas (éloge of Regnault), *Discours et éloges académiques*, 1885, vol. 2, p. 184 (my italics).

[20] *P.V.I.*, 6, 44, 27 March 1816.

But when this sentiment was transmitted to the Minister, the language used was more diplomatic:

> the Academy resolved unanimously to express to you the satisfaction which it has always felt with the current regulations and which have been dictated by long experience. They are, generally speaking, those of the former [Royal] Academy of Sciences.[21]

Meanwhile the Academy went ahead with elections for further *Académiciens libres*, two at a time,[22] requesting however that this section should not have the right to vote in elections, since they had only amateur status.[23] By the end of 1816 this section had its full complement of ten members, comprising three dukes, four barons, a marquis and a count, together with one gentleman of 'good family', Gillet de Laumont, inspector general at the Conseil des Mines. In the years 1817–30, as members died, they were almost invariably replaced by members of the nobility. Even in the first years after the 1830 Revolution a succession of barons were elected, thus seemingly perpetuating the precedents of the early Restoration.

There was a three year gap between the election of Bory de Saint Vincent in 1834 and the next vacancy for an *Académicien libre*. These years were particularly eventful in the history of the Academy, since they were the time of the introduction of the *Comptes rendus*. The new secretary, Arago, was the main force behind this move to 'democratise' the publications of the Academy and his republican principles would seem to have been influential also in a radical reappraisal of the use of the section of *Académiciens libres*. It probably helped considerably Arago's case for reform that the early titled *Académiciens libres* had hardly distinguished themselves by their attendance at meetings; at least two, the duc de Lauraguais and the duc de Noailles never having condescended to visit the Academy. Such indifference contrasted with the respect with which so many professional scientists viewed membership of the Academy, with its severely restricted membership. On the death of baron Desgenettes in early 1837, the Academy drew up a list of candidates in order of preference, placing first the geologist Bonnard, second the medical chemist Orfila; with a titled candidate, the duc de Rivoli, only in fourth place.[24] Bonnard was elected by an overwhelming majority, with no votes at all going to the duke.

But it would be a mistake to see this and subsequent elections in which

[21] A. S. Carton 35, dossier: 'Section des académiciens libres'.

[22] This method of creating a section in stages and by election was more acceptable to the Academy than the alternative of royal nomination of all members. The method adopted followed the precedent established for the foundation of the National Institute in 1795.

[23] 'les académiciens libres devant être pris parmi les amateurs', *ibid.* This denial of voting rights was to cause trouble later in the century, when professional scientists were included in the section.

[24] *C.R.*, 4 (1837), 627, 637.

working scientists were elected as *Académiciens libres* in purely social and political terms. It was not simply that the old aristocracy seemed irrelevant in France after the 1830 Revolution, nor even that the bourgeoisie was taking over positions originally considered the preserve of the nobility. The Académie des Beaux Arts, more explicit in its regulations than its fellow Academy, had spelled out that people could be elected as *Académiciens libres* either on the basis of their rank or on the basis of their talent.[25] In many cases there was a happy combination of both. But one aspect in which the scientists differed from the artists was that their studies were growing in a way that could hardly have been foreseen at the time of the foundation of the Institute. There was a restricted number of places and no allowance for new disciplines. It is therefore significant that the first person elected under the new interpretation was a geologist, a subject not formally recognised in 1795. Also we find among successful candidates up to 1850 two pharmacists, Pelletier and Bussy. In the eighteenth century pharmacists had found a natural place in association with the chemists but, with the enormous expansion of chemistry in the early nineteenth century, chemists were increasingly reluctant to give one of their valuable places to a pharmacist, who was not welcome either in the medical section, which had a marked preference for practising physicians and surgeons.

There is a further purpose which was served by the category of *Académiciens libres*, that of the scientist who was not a specialist. Such was the mathematician and astronomer, L. B. Francoeur (1773–1849), who also had strong interests in education and applied science. Francoeur had sought admission to the scientific sections in vain and by 1840 was convinced, not without good reason, that his only chance of election was as an *Académicien libre*. In Francoeur's dossier in the archives of the Académie des Sciences we may read the poor man's frank description of his plight in a letter to a colleague. In 1840 he was competing with the pharmacist Pelletier, who had also been unsuccessful in his efforts to gain election as a specialist. Francoeur writes:

> When I present myself to the mechanics section, someone says that I should address myself to the astronomy section; this section sends me back to the other, or even to the mathematics section. Thus too much diversity in my work has deprived me of any well defined speciality. Is it not in such a position that one can stake a claim to the place of *Académicien libre*? Was it not created for this very case?

Historically, of course, the answer to this question is no, but by 1840 the ideals of the Bourbon regime had already been forgotten.

But we must not lose sight of the political dimension. Although the 1840s witnessed the use of the new section by the Academy to introduce into its midst

[25] Académie des Beaux Arts, Ordonnance du 9 juillet 1816, Art.VI: 'La classe des académiciens libres est composée de 10 membres. Ils sont choisi parmi les hommes distingués, soit par leur rang and leur gout, soit par leur connaissances théoriques ou pratiques dans les beaux-arts'.

men of talent rather than men of rank, the Second Empire was to see a further development. It was surely more than a coincidence that the *Académiciens libres* elected in the first four years of the rule of Napoleon III were respectively financiers and business men, Bienaymé and François Delessert (1852), and army and naval officers Vaillant (1853) and Du Petit Thouars (1855), who subsequently rose to the ranks of marshal and admiral respectively.[26] The ambiguity of the relevant criteria for *Académiciens libres* left the Academy free to ingratiate itself with the government of the Second Empire. When Vaillant was made a government minister in 1854, the Academy must have privately congratulated itself on its foresight. Later, under the Third Republic, it elected the engineer-politician and former *polytechnicien* Charles de Freycinet (1828–1923) at a time (8 May 1882) when he was not only a minister but prime minister.[27]

The development of the section of *Académiciens libres* thus provides us with a good example of the tension between socio-political values on the one hand and scientific values on the other. Originally the institution of the section constituted a blatant attempt to reimpose some of the social values of the *ancien régime* and thus, by implication, to dilute some of the new bourgeois professionalism of the early nineteenth-century Academy. The Academy meekly accepted this as the price of freedom to carry on their work much as before the Restoration. But this passive acceptance was not to last for more than twenty years. It was the political skills of Arago which transformed what he undoubtedly saw as an instrument of servitude into a rare opportunity to increase the *scientific* membership of the Academy. In the period 1837–50 there were eight elections in this section and not one person of high social rank was elected.

But although we have shown that this provided an admirable opportunity to bring into the Academy men whose achievements were interdisciplinary, perhaps Arago had gone too far. By 1852 not only was Arago in failing health (he died the following year) but the Second Empire was proclaimed, harking back in many ways to the rule of Napoleon I, who had had his favourites and had intermittently given his patronage to the official body of science. The Academy for the rest of the century considered that it was in its interest to elect as members of this section national heroes like Lesseps, the builder of the Suez

[26] Some 12 years before his election to the Academy Du Petit-Thouars had become something of a national hero in supporting the rights of French missionaries in Tahiti over British missionaries, finally using military force to drive out the latter. This produced some diplomatic embarrassment to Guizot's government, which finally agreed to pay compensation. Nevertheless, the contribution of Du Petit Thouars in building up a zone of French influence in the South Pacific was well appreciated by many Frenchmen.

[27] De Freycinet had already been *President du Conseil* in the government, which lasted from 28 December 1879 to 23 September 1880. In this period of short-lived governments Freycinet's second Ministry lasted from 30 January to 7 August 1882.

canal[28] who was elected in 1873, or men who combined a modest scientific accomplishment with access to political power like Adolphe Carnot (elected 1895), who came from a famous scientific family and whose brother had been president of the Republic.[29] What had begun therefore as a means of increasing royalist influence in the Academy became a valuable adjunct to reward men of some distinction, whose careers transcended those of the ordinary specialist. Finally the section was used by the Academy to increase its public position and its influence in government circles. By occasionally electing men whose main career was outside professional science, the Academy helped bridge the gap between it and other parts of French society.

4. *The control of science by the Academy*

If the Academy of the *ancien régime* was seen in the revolutionary period as tyrannical or despotic, it was usually because it was associated with passing judgement on poor unfortunates who submitted work to it, notably artisans and inventors who, it has been argued,[30] were often humiliated. It stood out against charlatans in science. In the nineteenth century the Academy of Sciences continued to pass judgement on matters scientific submitted to it and, although this was an important means of controlling science, it is by no means evident that it should be placed first in our consideration. We have argued that the Academy had other perhaps less obvious means of control.

First the very nature of the Academy established a distinction between 'official science' and other science. Although this might have depended at first partly on the enormous difference of status in France between institutions recognised by the government and those not so recognised, the establishment of the *Comptes rendus* in 1835 had the effect of strengthening the distinction on a totally different plane. The new journal by its very *centrality and comprehensiveness*, and certainly not because of its later government subsidy, became recognised as the principal French journal for all branches of science. We have argued that, in a sense, if something new in science was not reported in the *Comptes rendus*, it was as if it had never happened.

Secondly, we have placed emphasis on the subjects officially recognised in the Academy. The original constitution established what were the official subjects to be studied before the first meeting of the First Class of the Institute was held. Thus in a sense it was the subjects which chose the Academicians rather than the Academicians who chose the subjects. This said, however, it must also be admitted that once the members had been elected they took a particular view of

[28] Lesseps was less happy in his later involvement in the unsuccessful French attempts to build the Panama canal, a project beset by immense technical and health problems.

[29] Sadi Carnot (b. 1837) had been assassinated in 1894. He is not to be confused with the Sadi Carnot of the Carnot cycle.

[30] R. Hahn, *The anatomy of a scientific institution. The Paris Academy of Sciences, 1666–1803*, Berkeley, Cal., 1971, e.g. pp. 23–4, 189ff.

their own respective subjects; also they were later called on to take a view of other subjects presented to them from outside, for example, physiology and geology. The section of medicine was forced by such powerful figures as Laplace to take physiology on board in the person of Magendie[31] and, once the precedent had been established, this enabled his student Claude Bernard to take his place in due course. The subject of geology was introduced more discreetly, little by little, into the section officially called mineralogy.

The original constitution of the Institute of 1795 included a clause inviting the Institute to draw up the new regulations on the conduct of meetings. Thus the final regulations governing the First Class were a combination of what had been proposed by government and by the *savants* of the Institute. The balance of power moved slowly in favour of the *savants* as the question of *interpretation* of the rules became more important. The scientists would obviously tend to adopt an interpretation that was most in their favour or which gave them the greatest freedom of action.

There was another factor which needs to be taken into account, that is the creation of precedents. In its dealings with the Minister of Public Instruction, the Academy was from time to time allowed greater freedom than it might have expected and this latitude tended to be built up to provide a permanent argument for discretion. But precedents were also relevant to the internal conduct of affairs in the Academy. When a dispute arose about matters of procedure, the secretaries would look back in the records to see if such a problem had arisen before and if so, how it had been dealt with. But as any researcher who has gone through the archives of an institution knows, it takes a long time to examine reports going back many decades. Minutes of the weekly meetings were kept but few of the records were indexed. By the late nineteenth century, therefore, there was a case for collecting together notable decisions in a book with the title: *Règlements intérieurs de l'Académie des Sciences.*[32] A very important section of the book is on elections, a section which begins by printing anything relevant from successive constitutions. This is supplemented by excerpts from the decisions of meetings held in secret session, for example on the postponement of elections. Thus if ever it was proposed that an election should be held over because of the lack of suitable candidates, the secretary had to hand the case law which he could cite. Admirable in many ways, it nevertheless tied the hands of the Academy, adding a body of case law to the original statute law.

We have provided ample evidence in Chapter 7 that the report system gave

[31] John E. Lesch, *Science and medicine in France. The emergence of experimental physiology, 1790–1855*, Cambridge, Mass., 1984, pp. 117–18.

[32] Printed by Gauthier-Villars but with no date. There were several successive editions of this, beginning in the late nineteenth century. Early twentieth-century editions are distinguished by extra pages inserted after p. 10 and labelled 10 *bis* ... 10 *septies*, thus removing the necessity for changes in pagination in the later and unaltered part of the book. One would not expect a book on such a delicate subject ever to be published.

the Academy considerable powers. The Academy's evaluations were always considered authoritative. Although the report system steadily declined after the 1840s, it was to some extent superseded by an expanding prize system in the second half of the century. Prizes which were originally largely honorific came increasingly to provide considerable sums of money as a tangible form of approval. It could be argued that the Academy exercised even more control by the award of grants for research in certain specified areas. Whereas a prize rewarded a scientist after research was completed, a grant by its nature was related less to the evaluation of a person than of a project. Nevertheless early recipients of grants often tended to be hangers-on, exemplifying the great power of the Academy exercised through patronage.

Another way in which science was controlled by the Academy was by the institution of secret sessions. Strictly speaking, it was not so much science as scientists which were discussed, since the main business was usually elections. In the early days of the First Class it was assumed that the business was to consist entirely of discussion of the sciences and their applications. After a few years it became clear to everyone that there would also need to be more personal discussions in connection with candidatures for elections. At first these discussions seem to have taken place as an integral part of the ordinary meetings but by 1801 we find explicit references to an occasional secret session.[33] Finally in 1802 this was regularised:

> A member proposes that each time an election is to be discussed the Class should form in secret session (*comité secret*) to discuss the merits of the candidates.[34]

This proposal was adopted unanimously and, considering that elections were a regular part of the business of the Academy, most meetings came to be divided into the main (semi-public) part and the final part, which was private. Fortunately for the historian, minutes were kept of the *comité secret*, even if annoyingly the early registers have been lost. The minutes reveal that although elections were normally the main business and their privacy permitted a full discussion of the relative academic merits of the respective candidates, other matters, such as those of protocol and relations with the Ministry of the Interior or Ministry of Public Instruction, were also discussed. Without an institutionalised secret session many important decisions would have had to be made by the principal officers of the Academy or simply by the two permanent secretaries, rather than by the members of the Academy collectively. Exceptionally there was also discussion in secret session of scientific matters considered dangerous to the public, such as poisons.[35]

[33] *P.V.I.*, 2, 304, 21 *pluviose* an 9 (10 February 1801).

[34] *Ibid.*, 455, 26 *nivose* an 10 (16 January 1802).

[35] *P.V.I.*, 4, 275–7, 13 November 1809.

Although it was successive governments which imposed the respective constitutions of 1795, 1803 and 1816, many of the changes most important for the progress of science in the nineteenth century can be seen to be the result of initiatives coming from the Academy. Prominent examples of Academy initiatives that we have discussed were the exploitation of the Montyon legacy for general prizes and grants, the founding of the *Comptes rendus* and the revised conception of the membership of the section of *Académiciens libres*. The Academy was sometimes able to defy government attempts to impose greater control. Thus when the authoritarian Minister Fortoul asked the Academy in 1855 and again in 1856 to revive its tradition of annual reports, it boldly replied that such reports were not in keeping with the modern period,[36] since the Academy now had its *Comptes rendus*. Many of the changes introduced by the Academy might hardly have been noticed by successive ministers since, although they made important contributions to the advancement of science, they had little political significance. Thus, whereas it would have needed a revolution (or more) to change the number of members in each section, changes which did not have immediate political or financial consequences could sometimes be quietly introduced. It was not however in the interests of the Academy to call too much public attention to its innovations. The consequence of this situation for the historian is that he finds that the Academy was not quite as conservative as is sometimes supposed.

A further way in which the Academy helped to regulate and control science was through language. In considering the part played by the Academy in the foundation and establishment of the metric system, not the least important stage was its decision, albeit a belated one, that it would in future only accept papers for publication in the *Comptes rendus* if they gave measurements in metric units. This was announced on 21 June 1886[37] and was an important step in facilitating the comparison of experimental results, which is so difficult if expressed in different units. The published announcement said that the Academy particularly expected French scientists to conform but it obviously hoped that foreign men of science would also use the metric system. Not for publication was a second decision of the Academy, meeting in secret session, that if the units used were not metric, the secretaries would have the task of translating measurements into the metric form for publication.[38]

Although the naming of new elements was usually accepted as the privilege of the discoverer, in the case of the substance discovered by Balard, it was the Academy commission which proposed that it should be called *brome* (bromine) by analogy with the other recently discovered elements, *chlore* and *iode*, since it

[36] 'un pareil exposé ne pouvant en aucune façon répondre aux besoins de notre époque', A.S. *Comité secret, 1845–56*, p. 461 (13 October 1856).

[37] *C.R.*, **102** (1886), 187.

[38] A.S., *Comité secret, 1882–1902*, meeting of 21 June 1886.

clearly had very similar chemical properties.[39] The Academy commission could take a wider view of the subject and it had the authority.

One of the great achievements of the second half of the nineteenth century was the germ theory of disease introduced by Pasteur. Although Lister was applying his theory to surgery by 1867, using the drastic antiseptic method involving large quantities of carbolic acid, it is not easy to decide precisely when the theory gained acceptance. Pasteur himself considered that the greatest milestone along this road came in 1878 when the surgeon Sédillot formally introduced the term *microbe* at a meeting of the Academy of Sciences[40] not, it should be noted, the Academy of Medicine, although both he and Pasteur were members of both Academies. Sédillot said that previously there had been a multiplicity of names given to these 'microscopic organisms' but now that their existence had been clearly demonstrated to the Academy, there should be a single and simple term. He proposed *microbe* from the Greek *mikros* = small, *bios* = life, and his friend the lexicographer and linguist Littré had agreed on this new term. Although it was an individual member of the Academy who had proposed the term, rather than an official commission, the fact that it was introduced formally at the Academy and publicly welcomed by Pasteur[41] had the effect of providing an official sanction.

5. *The control of new sciences*

One of the many problems which arise in a study of the Academy is its relation to the various specialist scientific societies which were founded in the nineteenth century. With established sciences like physics and chemistry, the foundation of new societies did not constitute a threat to the control of these subjects by the appropriate sections of the Academy. That is to say that the Academy always contained the elite, and specialist societies were not intended to challenge the elite but rather to spread the study of a particular science and its applications more widely.[42] Given the very restricted membership of the Academy there was certainly a need to organise the subject on a wider basis. Often senior positions in these societies would be offered to Academicians in order to increase the status of the society and often when the Academicians accepted, it was in a spirit of condescension. They could not be expected to attend regularly meetings of the specialist societies like, say, 'ordinary' physicists or chemists. It is true that the Société chimique could often respond quite quickly to new developments in the science, particularly in organic chemistry, and it soon built up its own network

[39] *P.V.I.*, **8**, 414.

[40] 'De l'influence des découvertes de M. Pasteur sur les progrès de la chirurgie'. *C.R.*, **86** (1878), 634–40.

[41] *Ibid.*, 1038. For the parallel terminology of 'germs' see Pasteur, 'La theorie des germes', *ibid.*, 1037–43.

[42] Robert Fox, 'The *savant* confronts his peers: scientific societies in France, 1815–1914', in R. Fox and G. Weisz (eds.), *The organisation of science and technology in France, 1808–1914*, Cambridge, 1980, pp. 241–82.

among junior and middle-ranking chemists. But even in this particularly active society, the Academy remained as the cream on top of the milk.

The real problem came with *new* sciences and this area represents one of the greatest potential weaknesses of the Academy, since we could not expect the structure imposed by the constitution of 1795 (slightly revised in 1803) to be equally valid a hundred years later, when various new sciences had emerged. Perhaps it is too much to hope that the Academy should incorporate every new development in science, or even the main ones? Surely many were better catered for by the foundation of specialist societies? We may remember how worried Joseph Banks was in England, as president of the Royal Society at the beginning of the nineteenth century, to see the emergence of new specialist societies like the Geological Society (1807) and the Astronomical Society (1820). He is reported to have said that he feared 'that these new fangled Associations will finally dismantle the Royal Society and not leave the old lady a rag to cover her'.[43] But Banks was probably thinking less of the growth of science than its subdivision on a more static model. He felt that if the existing sciences were parcelled out and studied in specialist societies, there would be no need for a society representing the whole of science.

The Academy could not object to the foundation of specialist societies and new journals. Normally new sciences did not threaten the old-established sciences but rather extended the range of science, leaving the Academy with a central core. It was, of course, very much more than a core. It included a very wide spread of traditional sciences. To take an extreme hypothetical situation, there might have been the danger that the Academy would come to represent 'old' science, leaving 'new' science to other bodies. But in practice few of the pioneers of these new sciences wanted to by-pass the Academy entirely. It was always to their advantage to have their research mentioned in the *Comptes rendus* and, going beyond mere registration, if there was the occasional prize or other sign of recognition so much the better. Thus the involvement of new sciences depended not only on the goodwill of the Academy but the initiative of pioneers at the frontiers of knowledge. In a few cases of obvious incompetence or charlatanism we have seen the Academy reject ideas from outside. But its more usual stance was one of toleration.

In a few instances the Academy might go further and try to assimilate or incorporate a new science but in such cases it had to overcome two major barriers. First, traditional or 'official' science was sub-divided into named sections and a new discipline would have to find a section willing to incorporate it. Secondly, this incorporation would have to be *at the expense* of the traditional branch of science, since there could be no increase in numbers. Thus physical

[43] Sir John Barrow, *Sketches of the Royal Society and Royal Society Club*, p. 10, quoted by H. C. Cameron, *Sir Joseph Banks, K.B., P.R.S., The autocrat of the philosophers*, 1952, p. 177. Sir Henry Lyons, *The Royal Society, 1660–1940*, Cambridge, 1944, pp. 216–17.

anthropology, for example, might seem in some respects to have a claim to representation within the section of anatomy and zoology but it could only enter by excluding established anatomists and zoologists. In fact we shall see that one of the main contenders, Paul Broca, who was trained in medicine, stood as a candidate in that section but with severe competition from others well qualified in established branches of medicine and surgery.

It was not always obvious that the Academy should concern itself with work submitted in borderline areas. Thus in August 1846 Boucher de Perthes (1788–1868), an obscure customs official from Abbeville in Normandy, submitted a work, printed but not yet published, entitled *De l'industrie primitive*, in which he argued from the evidence of his own excavations that many thousands of years ago primitive man had fashioned stone axes and other tools.[44] The evidence of the antiquity of the artefacts depended on appreciation of his great skill and originality in employing a rigorous stratigraphic method, and at first little notice was taken of his work. Boucher possibly only communicated with the Academy in the first place because his father had been a corresponding member in the botany section. Indeed soon after he had submitted his first memoir to the Academy and it had gone through the motions of appointing a commission, Boucher wrote again to the official body of science suggesting that, as his work was as much archaeology as geology, any commission appointed should contain suitable representatives from the Academy of Inscriptions.[45]

Boucher's conclusion that he had found evidence of the existence of man before the Flood and contemporary with extinct species of mammoths was not easily accepted. Indeed it raised violent opposition. Resistance to the acceptance of these ideas has been attributed to the powerful posthumous influence of Cuvier among French scientists.[46] But this may have helped Boucher gain the patronage of Geoffroy Saint-Hilaire[47] and he was also supported by Elie de Beaumont.[48] It may well have been the interest shown in Boucher's discoveries by a group of British scientists, including Charles Lyell, which prompted the Academy finally to take some action. It sent Gaudry to Normandy to examine the evidence and accepted his favourable report.[49] However the Academy never gave Boucher proper recognition for his pioneering work. He never became a correspondent and the Academy would not even bend its prize system to honour him. Even after his many fundamental discoveries and publications he remained, as he had begun, a provincial amateur. The Academy reserved its highest honours for professional men who fitted into its structure.

This brings us to the general subject of anthropology, which can be

[44] *C.R.*, **23** (1846), 355 (17 August 1846).

[45] *Ibid.*, 527 (7 September 1846).

[46] Jacques Roger, 'Boucher de Crevecoeur de Perthes', *D.S.B.*, Supplement, **15**, 50–2 (p. 51).

[47] *C.R.*, **46** (1858), 903.

[48] *C.R.*, **49** (1859), 581–2; **57** (1863), 334ff.

[49] *C.R.*, **49** (1859), 465–7. Mention of Boucher was omitted in the printed version, which gave rise to a protest from Elie de Beaumont, *ibid*, 581–2, 636.

approached from several directions, including the zoological, the ethnological and the linguistic. Our main concern here must be to investigate the gradual acceptance of anthropology as a science. By 1841 Serres (1786–1868), a member of the medicine section of the Academy, was protesting that anthropology belonged as much to his section as it did to the section of zoology.[50] The term 'anthropologie' had been coined in 1838 by Serres to describe the subject he taught at the Muséum d'Histoire Naturelle as professor of anatomy and the natural history of man. On the retirement of Serres in 1855 this chair was given to the zoologist Quatrefages de Bréau (1810–92). Quatrefages already had a doctorate in physical science, a second doctorate in medicine and a third in zoology, but he was willing to begin a further career in the new subject of anthropology.[51] For him the fundamental problem was the unity of the human species and, defending the conception of a human realm distinct from the animal one, he consistently opposed Darwinian theory.

Anthropological arguments were increasingly used in discussions about slavery. In 1850 we find Serres as rapporteur on a commission on the negro races, in which he claimed that the rehabilitation of negroes was one of the accomplishments of that century. The achievement of negroes was limited when they were held in subjugation but, he suggested, when they were not held down, they could reach the same level as caucasians.[52] But it was one of the crises of nineteenth-century French history which drew the most extreme conclusions from some French anthropologists. Before the Franco–Prussian war the conservative Quatrefages refused to apply anthropological principles to any subject of a political nature. After it he considered it a patriotic duty to examine the race which had humiliated his own country. In a booklet published in February 1871[53] he set out to prove that there was no such thing as a German race and he ended by claiming that the conquerors were really a hybrid of Slavs and Finns, a conclusion which was not only questionable but deeply offensive to many Germans, who considered themselves of pure Aryan stock. In some ways, therefore, anthropology was one of the most politically sensitive branches of science, if it was a science at all. One claim would be based on its links with zoology. Another would be the quantification and measurement applied in an extreme way by some physical anthropologists.

Quatrefages had entered the Academy in 1852 as a zoologist. His anthropological work came later. The career of Quatrefages, therefore, differs significantly from that of his rival Paul Broca (1824–80), surgeon, anatomist, biologist and anthropologist.[54] Broca, perhaps more than anyone else, helped put physical anthropology on the map in France, yet he was never elected to the

[50] *A.S., Comité secret, 1837–44*, p. 72 (2 August 1841).

[51] Camille Limoges, 'Quatrefages', *D.S.B.*, **11**, 233–5.

[52] *C.R.*, **30** (1850), 679–90.

[53] 'La race prussienne', *Revue des deux mondes*, **91** (1871), 647–69; also published separately.

[54] Francis Schiller, *Paul Broca, founder of French anthropology, explorer of the brain*, Berkeley, Cal., 1979. Edwin Clarke, 'Broca', *D.S.B.*, **2**, 477–8.

Academy. Broca was an indefatigable worker and was the main founder of the *Sociéte d'Anthropologie de Paris* in 1859 as well as the founder of the journal *Revue d'anthropologie* (1872). With these Broca had more of a power base than he would ever have had in the Academy and it is by such means that new disciplines are established. On the other hand the funding of such institutions is usually precarious and, being peripheral to the formal structure of French higher education, they did not offer very certain prospects for a scientific career.[55] Hence only a small proportion of the members of the Société d'Anthropologie were professional academics; the majority were, strictly speaking, 'amateurs' who were attracted by the idea of classification and such programmes as making the study of man 'scientific' by the careful measurement of the skull. The Society, unlike the Academy, was principally concerned with attracting members and was not worried if they included the occasional charlatan. It is significant that the Society was so dependent on its secretary, Broca, that it fell apart after his death. Yet general interest in anthropology increased as a result of the publication of Charles Darwin's *Origin of Species* (1859) and particularly *The Descent of Man* (1871).

A useful way to approach the question of the Academy's recognition of a new subject is to see how it dealt with the leading exponent of that subject. Although Broca submitted most of his research papers to medical or anthropological journals (particularly his own), he did send a few communications to the Academy. In 1855 he received a commendation from the Academy and two years later he was awarded the Montyon prize for medicine and surgery of 2500 francs. His most famous discovery came in 1861, when he announced that the seat of speech is located in the left hand side of the frontal lobe of the brain, which became known as the convolution of Broca. By 1867 he felt sufficiently far advanced in his career to stand as a candidate for election to the Academy. In fact he stood three times and had the satisfaction of seeing his name steadily advancing in the order of recommendation by the section of medicine. Had he not died in 1880 at the age of fifty-six, he would probably have reached the top of the list in due course. That would not of course have guaranteed his election, and indeed many would have felt that his proper place was in the Academy of Medicine, of which he eventually became vice-president. Although the charge of materialism[56] was levelled against him, this would only have deflected a minority in the Academy of Sciences. In the 1880s his avowed republicanism would have been a distinct advantage in an Academy election. A few months before his death he was nominated by the government as a senator. It might be true to say, therefore, that he received ample recognition for his political ideas

[55] Elizabeth A. Williams, 'Anthropological institutions in nineteenth-century France', *Isis*, **76** (1985), 331–48.

[56] Although Broca, like Quatrefages, was a Protestant by birth, his name was struck off the electoral list of the reformed church in Paris in 1864 because of his alleged atheism. P. Huard, 'Paul Broca (1824–80)', *Revue d'Histoire des Sciences*, **14** (1961), 47–86.

and his contributions to medicine but that he failed within his lifetime to secure full recognition within the scientific community.

To take a final example of the problem of new branches of science, it might be too much to hope that the Academy should play a part in the development of a subject like psychology; but even a superficial study reveals it as the patron of two major figures in late nineteenth-century experimental psychology, Binet and his rival Janet. Alfred Binet (1857–1911) did much to put experimental psychology on the map in France. He even carried out experimental tests on the personalities of his two daughters. In 1888 Binet received an honourable mention from the Academy for his entry on animal magnetism for the Montyon (medicine) prize. In 1892 he was the joint winner of the Academy's Lallemand prize for his book on the alteration of personality. In 1895 he founded the first French journal devoted to psychology and in his later work he laid the foundations of the system of I.Q. tests.

Binet's contemporary Pierre Janet (1859–1947), did important work in establishing a connection between academic psychology and the clinical treatment of mental illness. Like Binet he was much influenced by Charcot, professor at the Salpetrière, whose protégé he became. Janet won an honourable mention for his work on the mental state of hysterical patients, submitted for the 1894 Lallemand prize competition of the Academy. In 1896 he won the Thore prize and in 1909 the Cuvier prize, both awarded for outstanding work in the area of anatomy and zoology. This encouraged him to stand as a candidate for that section, but without success. As always, the problem was how to fit pioneers in new disciplines into the existing framework where there was no shortage of brilliant, if sometimes less original, candidates.

6. *The control of the natural world*

If science is basically concerned with understanding the natural world rather than changing it and if the term 'academic', derived from 'Academy', tends to mean removed from practical affairs, then there is the danger of an institution called the Academy of Sciences being thought of as doubly remote from problems of everyday concern to the human race. But although the emphasis in this book, as in the Academy, has been on pure science, it would be a mistake to neglect applied science. Nor did the Academy neglect it entirely, despite the existence of several agencies more relevant to technology, such as the Conservatoire des Arts et Métiers and the Société d'Encouragement pour l'Industrie Nationale, which had not existed before the Revolution.

We will concentrate on two areas of applied science and say something briefly about the hydraulic turbine before discussing the subject of manned flight, which will be taken as a case study of the eternal quest to overcome natural forces, in this case the force of gravity. France continued in the nineteenth century to make considerable use of its natural resources of water power for industrial purposes while most of the pioneering work on the steam engine was done in Britain which had vastly superior supplies of coal. This helps us to understand how it was in France that the turbine was first developed. The Academy made a substantial contribution to the history of hydraulics and particularly of the turbine, developed by two 'outsiders' from the provinces. In 1822 Claude Burdin (1778–1873), a professor at the Ecole des Mines at Saint Etienne, sent a memoir to the Academy which, in retrospect, we can see as historic since it introduced the term 'turbine' into engineering. The Academy did not immediately appreciate the potential of Burdin's ideas. Indeed he was never able to construct a working model of a turbine, a feat achieved later by his very practical student Benoit Fourneyron (1802–67). Burdin had to write to the Academy twice to persuade it that his memoir deserved a formal report and finally in 1824 a commission consisting of Prony, Dupin and Girard presented their report. They concluded that he deserved the encouragement of the Academy.[57] Later Arago proposed as a tangible form of encouragement that his name should be added to the list of candidates in an election for a corresponding member in the mechanics section,[58] saying that already hydraulic turbines were making a contribution to industry. Later Fourneyron was to send a succession of memoirs on turbines to the Academy and it was to the Academy in 1841 that he submitted his claim to be the real inventor of the turbine.[59]

In the encouragement of such pioneers the Société d'Encouragement pour l'Industrie Nationale was also important and was obviously better suited to developing the purely practical aspects of the turbine. Yet the fact that the Academy also played an important part confirms its relevance even for provincial mathematicians and engineers who might have no direct connection with Paris.

The history of the two related subjects of ballooning and aeronautics relates particularly well to the period covered by this book. France, the country which gave birth to the balloon, was also, up to the early twentieth century, the focus of many pioneering trials with heavier-than-air machines which can be counted among the first aeroplanes. The first hot air balloon in 1783 (Etienne and Joseph Montgolfier) and the first hydrogen balloon in the same year. (J. A. C. Charles) marked the beginning of a balloon craze, which involved the Royal Academy of Sciences but soon came to be frowned upon by the authorities as a potential threat to public order.[60] The Revolution tended to redirect people's attention back to the terrestrial plane, although the revolutionary war did bring about the

[57] *P.V.I.*, 8, 70–3. A standard history of the subject (Hunter Rouse and Simon Ince, *History of hydraulics*, New York, 1963, p. 146) states that 'Burdin's mémoire was not accepted' by the Academy, but this is misleading. Had he submitted it after the foundation of the *Comptes rendus* in 1835, it is probable that it would have been published, if only in an abbreviated form.

[58] *P.V.I.*, 10, 233 (25 March 1833). Unfortunately Burdin was not elected then but was successful in 1842.

[59] *C.R.*, 12 (1841), 795.

[60] J. M. Hunn, *The balloon craze in France, 1783–99. A study in popular science*, Ph.D. thesis, Vanderbilt University, 1982. See especially Chapters 5, 6 and 7.

exceptional use of balloons for military observation. In 1804 Gay-Lussac made a spectacular use of a balloon for purely scientific observations, thus opening a new nineteenth-century chapter in the history of flight. Reaching an altitude of 7016 m, he not only took measurements of pressure and temperature but also took readings of the earth's magnetic field and samples of air for chemical analysis. This altitude was not exceeded until the ascent of Barral and Bixio in 1850 and then only by a few metres.[61] In 1868 we find the popular astronomer Flammarion making systematic meteorological observations with balloons[62] but the use of unmanned balloons to make meteorological observations at very great heights (*c.* 20000 m) really belongs to the 1890s.[63]

The trouble with balloons is that while their vertical movement could be controlled fairly easily their horizontal movement could not. They were at the mercy of prevailing winds and many of the early volumes of the *Comptes rendus* have suggestions for controlling the direction of motion of balloons. By the 1850s there are discussions about the use of propellers to make a dirigible balloon. It is interesting that the inventor, Mertens, who caused the question to be raised in the Academy in 1855, provoked others to intervene and claim similar plans,[64] a good illustration of the value of publicity in encouraging the development of new ideas. Ten years later gunpowder and compressed air were among the propellants proposed to control the direction of balloons.[65]

The Franco–Prussian war of 1870–1, and in particular the siege of Paris, gave a great stimulus to ballooning, as balloons were used to restore communications between the beleaguered inhabitants of Paris and the rest of France.[66] Many of the balloons were released at night, passing silently and invisibly over the heads of the encircling Prussian army. A member of the Academy, Dupuy de Lôme, was commissioned by the 'government of national defence' to construct a dirigible balloon with a grant of 40000 francs. In fact it took much longer than the few months of the war to produce. Only in 1872 was such a balloon (elongated, 36 m long with a diameter of 15 m) ready for flight.[67] Power was supplied by the muscles of eight strong men turning a huge propeller 9 metres in diameter. The results were rather modest. The speed of the wind might greatly exceed the 10 km.p.h. of the balloon and so, although not completely at the mercy of the wind, it might not deviate more than a few degrees from the wind's direction.

[61] 'Journal du voyage aéronautique fait le 27 juillet 1850 par MM. Barral et Bixio', *C.R.*, **31** (1850), 126–31.

[62] *C.R.*, **66** (1868), 1051, 1113, 1207.

[63] E.g. C. Renard, 'Sur l'emploi des ballons non montés à l'execution d'observations météorologiques à très grand hauteur', *C.R.*, **115** (1892), 1049–53.

[64] *C.R.*, **40** (1855), 193.

[65] *C.R.*, **60** (1865), 1217; **61** (1865), 92, 268.

[66] See Maurice Crosland, 'Science and the Franco-Prussian war', *Social Studies of Science*, **6** (1976), 185–214 (pp. 203–4). Victor Debuchy, *Les ballons du siège de Paris*, 1973.

[67] *C.R.*, **74** (1872), 337–54. The Academy considered the memoir of such great interest that the normal restriction on the number of pages permitted was lifted.

A practical dirigible would have to depend on the development of an engine. In 1883 the brothers Albert and Gaston Tissandier, who kept in close touch with the Academy, flew a small airship powered by an electric motor, but the power was too feeble for proper propulsion.[68] The following year two army officers, Renard and Krebs, produced a nearly practical airship with an electric motor. Paying tribute to the data supplied by Dupuy de Lôme, they were able to report to the Academy a flight in which they claimed as much control as with a ship in water.[69] Using an electric motor run by batteries and weighing several times the weight of the pilots, they made a flight of 7.5 km, taking 23 minutes, before returning to the starting point, the first time a round flight had been achieved. It was such flights as these which made possible later developments, such as those of the famous giant Zeppelins, the first of which flew in Germany in 1900.

One might question whether without the benefit of hindsight one could claim that the Academy of Sciences was the appropriate body to monitor the development of ballooning and aeronautics. For some sixty years successive generations of the famous Godard family were associated with ballooning as a public spectacle not very different from the circus. A significant step in the advancement of aeronautics was made by the construction of models and it should not surprise us that Alphonse Pénaud, who introduced the use of twisted elastic as a source of power, was by trade a toymaker. The founder of both the Société d'Encouragement pour l'Aviation (1863) and the periodical *L'Aéronaute* was not a scientist but the photographer, caricaturist and writer, Nadar (1820–1910). Much of the early history of aviation depended on a practical approach, requiring ingenuity and physical courage rather than mere ideas. It was certainly the case that many of the leading aviators who flew in France, such as the Brazilian Alberto Santos-Dumont and the English-born Henry Farman, never reported to the Academy. Much more important for them was the Aero-Club de France (founded 1898) and big aviation meetings such as that held at Béthemy, near Reims in 1909.

Yet the Academy because of its authority and its multi-disciplinary nature had much to offer aeronautics. Mathematics could be usefully applied to solve technical problems. Astronomers, physicists and chemists were all interested in the upper atmosphere. Renard's memoir of 1884 was presented to the Academy by Nervé Mangon, a member of the agriculture section with a special interest in meteorology. One of the pioneers of the dirigible balloon, Dupuy de Lôme, was a member of the section of geography and navigation, and another enthusiast was the physiologist Marey, a member of the medical section, who made detailed studies of the flight of birds. Probably more Academicians were interested in the data that aviation could provide, rather than in providing data to make aviation possible. Nevertheless many members of the Academy shared the popular fascination with early flight, which was becoming increasingly scientific after

[68] *C.R.*, 97 (1883), 831–3.

[69] *C.R.*, 99 (1884), 316–19. See also *C.R.*, 101 (1885), 1111–8.

the 1860s, and between them they had much relevant expertise. In 1909, after Bleriot's cross-Channel flight from Calais to Dover, the Academy decided to honour with a gold medal not only him but several other leading figures in international aviation, including the Wright brothers from the U.S.A. In the recent spectacular advances in aviation the Academy was in danger of being left behind. Aviation was now a practical reality and the Academy felt that it should be seen to give it its full support.

7. Changes in the Academy over the period 1795–1914

The Academy in the early 1900s was obviously very different from what it had been in the early 1800s. Some critics might want to say that in the later period it reflected a general malaise in French scientific education and research, which in many ways had become undistinguished and old fashioned.[70] They would contrast this with the early 1800s when, in many fields, French science led the world. But before discussing the question of a possible decline, it may be useful to review developments in the Academy to support the view that it was not the static institution that some may assume.

We have seen that the period 1795–1802 was an experimental period which came to an end with the definitive constitution of 1803. Although this provided the basis of the regulations of the Academy for the rest of the century, the introduction of *Académiciens libres* in 1816 served as a reminder to the senior scientists, if any was needed, that the constitution of the Academy depended ultimately on the government. But probably the most interesting development during the period of the Restoration depended on the legacy of baron de Montyon, who died in 1820. The share of the Academy in this legacy at first created a number of administrative headaches but by the late 1820s the Academy was beginning to appreciate the possibilities of this fund as an additional source of income. These possibilities were worked out in succeeding decades. We have suggested the most important single use of the fund was to finance the *Comptes rendus*, an entirely new type of scientific publication, which depended on the initiative of the newly-elected secretary, Arago. The accusation by Raspail, quoted at the beginning of the Introduction to this book, that the Academy always remained the same, was patently unjust. Things had changed in forty years. Raspail's greatest mistake, writing in 1838, was to make no allowance for the recently introduced *Comptes rendus*, which transformed the working of the Academy, giving its proceedings a new national and even international importance. By the mid century a senior Academician like Biot, elected as a full member in 1803, would scarcely have recognised the Academy as the same institution which he had joined as a young man.

The second half of the century was hardly to see innovations of equal importance, although it was marked by a spectacular growth in the prize system

[70] Mary Jo Nye, 'Nonconformity and creativity: A study of Paul Sabatier, chemical theory and the French scientific community', *Isis*, **68** (1977), 375–91.

which may have helped to make the Academy feel that it was increasing its control over science. The reality, however, was that the central place of the Academy in French science was beginning to be undermined by the creation of new agencies and the establishment of new priorities. Under Napoleon III, or rather under the ministry of Duruy, research was given tangible encouragement not through the Academy but by the creation in 1868 of the Ecole Pratique des Hautes Etudes, although this was really a federation of existing laboratories and research facilities rather than a separate institution with its own resources.[71]

The military defeat in the Franco–Prussian war did not leave the Academy unmoved and in March 1871 there was an unprecedented and potentially important debate in which it was agreed that the Academy should play a greater part in public life.[72] There was a widespread feeling that the defeat was symptomatic of the weakness of French science. Henri Sainte-Claire Deville argued that the Academy should take a greater responsibility for improving the standard of scientific education[73] in France and his ideas received considerable support, most notably from Dumas, recently elected secretary. Quatrefages argued that such far-reaching questions should be discussed only in secret session and there it was agreed that nothing prevented the Academy from widening its discussions to include education and the diffusion of scientific ideas in society.[74] The Academy agreed to adjourn the discussion for two weeks for further consideration, but by then interest seems to have evaporated. There is no further discussion recorded in the minutes of the secret sessions and the published *Comptes rendus* suggest that scientists were reverting to their respective research interests. The Academy did not seem able to reconstitute itself as a reform group, and it was to take a further war (1914–18) before it could approach seriously the whole question of science applied to industry.

Yet the Academy discussions of 1871 were evidence of a growing belief that the institution should widen its concerns. Members spoke of improving the university system, something which would happen largely independently of the Academy. In the proceedings of the Academy the greatest change one finds after 1871 is probably in its public meetings, where intense expressions of patriotism tended to be exhibited. Although Sainte-Claire Deville and others believed that scientists rather than politicians should have more control over education, in the end the Academy came to provide moral support for the government, particularly in the 1880s and 1890s when science (or scientism) came nearest to representing the ideology of the Third Republic. This was symbolised by the election of Berthelot as secretary in 1889.

[71] R. Fox and G. Weisz (eds.), *op. cit.*, p. 3. Jean Rohr, *Victor Duruy, ministre de Napoleon III*, 1967, pp. 116–21.

[72] H. Sainte-Claire Deville, 'De l'intervention de l'Académie dans les questions générales de l'organisation scientifique en France', *C.R.*, 72 (1871), 237–9; See also *ibid.*, 261–9.

[73] Others, notably General Morin, focused on *technical* education and industry; *ibid.*, 262–4.

[74] A.S. *Comité secret, 1870–1881*, p. 48 (20 March 1871).

Under the Third Republic the Faculties of Science assumed a new importance, while the Academy was left much as it was before. It became very much the exception rather than the rule for the government to consult the Academy, so it became more important for the Academy to use its public meetings to assert its utility. It would have wished to claim the work of Pasteur largely to its own credit but medical research, the aspect of Pasteur's work which had most influence on French public opinion, was siphoned off into the new Pasteur Institute, set up in 1888 as a result of a massive public subscription. The Academy certainly saw its prize funds grow considerably in the late nineteenth century, but its influence as a funding agency for science was undermined to some extent by the establishment of a new government agency in 1901, the *Caisse des recherches scientifiques*, which distributed grants for scientific research. Although the Academy was strongly represented on this new committee, it did not have control.

So it would be understandable if it were concluded that there was a moderate decline in the Academy. If it had been set up simply to provide a model for science then it certainly declined, but of course this role was never explicit and belongs more to the late seventeenth and eighteenth centuries than to the nineteenth century. The rise of Romanticism came to undermine the whole idea of established rules for intellectual production, accepted under the earlier classical model. In addition, growing political freedom undermined the idea of science being necessarily directed from above. The changes in the prize system reflect a growing democratisation of science, with the abandonment of the old prize competitions on a subject chosen by the Academy and the substitution of a grant system with money given for a subject chosen by the grant applicant.

Thus although some Academicians in the late nineteenth century might have looked back with nostalgia to the early century when the honour of a *grand prix* seemed sufficient to produce original research in physics at the highest level, it would be more reasonable to speak of a transformation of the prize system rather than a decline. Similarly, around 1800 it was unthinkable that the First Class would produce anything very different from the annual *Mémoires*, similar to the series which did credit to the former Royal Academy of Sciences. The *Mémoires* continued to be published at intervals up to the early twentieth century but the growing irregularity of their appearance did not matter much since they had become largely irrelevant. What the scientific community wanted to read was the weekly *Comptes rendus* with its full account of papers presented at the latest meeting of the Academy.

Yet in several other ways the Academy did decline. The accusation that the nineteenth-century Academy was little more than a hall of fame[75] comes nearest to the truth in the *late* century, when the age of election had risen to the extent that, for some candidates, election to the Academy became a final reward. It did not decline as badly as the Muséum d'Histoire Naturelle with its ageing

[75] R. Hahn, *op. cit.* (1971), p. 318.

collections and greater inbreeding,[76] but it was a bad sign in the late nineteenth century when it called increasingly on its traditions and less on new initiatives. Its method of dealing with new subjects was evolutionary rather than revolutionary since it was impossible to make major changes in the organisation of the Academy without ministerial approval. Thus one of the effects of government control was to reinforce the innate tendency of the Academy to conservatism.

It has to be admitted that sometimes the Academy was very slow to take up new ideas. In 1891, some thirty years after the publication of Darwin's most famous book, Berthelot as secretary still felt it necessary to apologise to some colleagues in the Academy for raising the subject of evolution. He claimed that 'the time for hesitation [about the open discussion of this issue] is past'. His principal justification for speaking on the subject in the course of an *éloge* was that:

> it would enfeeble the very authority of the Academy and of French science by excessive timidity if we avoided speaking of it[77]

a sentiment with which everyone would agree, the only surprise being that this had to be said in the 1890s rather than, say, in the 1860s or 1870s. It might be added that in addition to the usual objections to the theory of evolution in the nineteenth century, in France an influential group of positivists was opposed to it.[78]

In the early years of the twentieth century there were a number of signs that the members of the Academy needed to make modest changes in their organisation, particularly with regard to membership, although the Academy felt that one thing it should avoid was a general discussion of its constitution in the Chamber of Deputies.[79] Thus in 1909 it was remarked that the number of foreign associates had not been increased since the original foundation of the Royal Academy in 1666. Members now regretted that when Darwin had died, amidst eulogies from the international scientific community, the Academy did not have the satisfaction of having recognised him with this supreme distinction. It therefore intended to ask the Minister of Public Instruction to increase the number of foreign associates from eight to twelve, pointing out that no extra cost was involved!

A more important way in which science was opened up in France around the turn of the century was by allowing the provinces to make a greater

[76] C. Limoges, 'The development of the Muséum d'Histoire Naturelle of Paris, *c.* 1800–1914', in R. Fox and G. Weisz (eds.), *op. cit.*, pp. 155–240.

[77] 'Notice historique sur Henri Milne Edwards', *M.A.I.*, 47 (1904), i–xxxvi (iii).

[78] Harry W. Paul, *From knowledge to power. The rise of the science empire in France, 1860–1939*, Cambridge, 1985, pp. 67–71.

[79] A.S., *Comité secret*, 23 January 1914, 'M. A. Lacroix fait remarquer que ces deux projets, impliquant une demande de crédit, devraient être soumis au Parlement, ce que l'Académie est formellement décidée à éviter.'

contribution in a system notorious for its over-centralisation. Already in 1885 the creation of provincial universities had brought some relief to a situation in which all important decisions in French higher education had been made in Paris.

In 1912 the Academy was presented with a petition, signed by twenty-five corresponding members, who reminded it that the Institute had been founded in 1795 as a national body.[80] They claimed that correspondents in the provinces had been cut off from activities in Paris. Now however improvements in communications meant that one could live in the provinces, far from Paris, yet travel by train to the capital in a few hours. They pointed out that a government move towards decentralisation had given autonomy to the provincial faculties, making them into independent universities. They asked the Academy to discuss the introduction of a new category of full members who would not be required to reside in Paris. Among the signatories were the chemist Sabatier (1854–1941) from Toulouse, the entomologist Fabre (1823–1915) from the Midi and the bacteriologist Calmette (1863–1933) from Lille. After considerable discussion in the Academy the creation of a new section of six non-resident members was agreed in March 1913. The members of the new section were chosen from among the former correspondents and included Pierre Duhem, professor of theoretical physics at Bordeaux, although his name came near the bottom of those recommended by the Academy. The idea of having a new section for 'the applications of science to industry' received a boost through the practical problems posed by the world war, but it was not until January 1918 that the section came into existence.

Finally, one change we always have to keep in mind is the continual internal development of the different branches of science during the nineteenth century. We might begin by considering physics. Already in the 1820s the new science of electromagnetism was born from the marriage of the two separate sciences of electricity and magnetism, but the most important unifying concept in physics – that of energy – was introduced in the 1840s and helped to transform what had been a rather miscellaneous group of subjects, including heat, light and mechanics, into an integrated model science. But the greatest changes in physics came around 1900 and these will be discussed in the concluding section. Thus 'physics' means something rather different in each decade. Chemistry too developed in ways which could not have been foreseen at the time of the foundation of the Institute. In the biological sciences, Darwin's theory of evolution by natural selection was eventually to have the widest repercussions, but we may recall that the Academy was slow to take the theory on board so that it was not until the twentieth century that its full implications were worked out. Indeed it was not until the early twentieth century that the genetic basis for evolution began to be understood. As for medicine, Pasteur's germ theory

[80] A.S., *Comité secret, 1903–12*, pp. 464–5.

obviously had major implications for surgery and for the understanding of disease. Yet the clinicians in the medical section were often very conservative.

All the branches of science underwent major changes of content, of fashion and sometimes of theory and this led, among other consequences, to an embarrassing time-lag between the understanding of a subject by the oldest cohort of Academicians, whose training might go back two generations, and the younger men who were constantly knocking at the door of the Academy. In those cases, where a new theory supplanted an old one, it was necessary to make a break with the past and such a break is sometimes painful. Yet institutions like the Academy are important in providing continuity in science and, broadly speaking, despite enormous progress in the course of the nineteenth century, most of the sciences were still recognisable by their original labels.

8. *The ethos of the Academician*

Before we draw any final conclusions, we should mention two aspects of control of the Academy which we have not previously emphasised, the first being control of Academicians by other Academicians. Perhaps to speak of mutual 'control' would be to suggest something too rigid and one needs a gentler term like 'support' or 'encouragement', but there is definitely a social factor which needs to be considered. In joining the Academy, members accepted a certain ethos of responsibility and commitment. Of course a few would regard membership of the Academy simply as a final reward for good work but they would not be permitted to rest on their laurels for long. At the very least there were weekly meetings to attend and a certain amount of committee work. In the later nineteenth century the application of individual expertise to the judgement of prizes was an increasing burden.

Senior members of the Academy would tend to lead by example, although there would always be some who lacked the energy of their younger days or were perhaps excused by genuine illness. Most Academicians would demonstrate a fierce loyalty to the Academy and its traditions, and the main pressure would come through a collective ethos, an *esprit de corps*, which demanded certain minimum standards of its members. They had a strong sense of corporate identity. A newly elected Academician would normally strive to prove himself worthy of his election. Most felt that they owed a duty to the institution which had honoured them. However grand their other qualifications, the title '*Membre de l'Institut*' was always the highest of intellectual honours and the one by which they would be known in society. But more important to each Academician than the respect of society was the respect of his peers and to earn that respect he had to conform to the highest professional standards. He had, for example, to keep a certain distance from popularisers of science. To *explain* science *de haut en bas* on rare occasions to a wider public was worthy enough, but to *popularise* science was to debase it; an activity which would tend, through him, to throw discredit on the Academy itself.

One of the criticisms levelled in the mid-century at the professors in the Muséum d'Histoire Naturelle was that each pursued his own fancy without any control ('sans contrôle').[81] Referring to the collections, the same critic continued:

> Each professor has his little palace and governs as an autocrat in his speciality. He alone has the key to the collection.[82]

One of the features of the Academy was that it avoided this extreme of isolation of authority within a speciality. With the possible exceptions of Laplace and Cuvier, there were no autocrats in the Academy. Alternatively if these two were autocrats, it was not primarily *as Academicians*.[83] Members of the Academy were normally members of a specialist section which provided a professional peer group and commanded their primary loyalty. They would also feel a responsibility to the Academy as a whole and they would not wish there ever to be grounds for criticism of their conduct. They would not, for example, be expected to make any public statements about matters of religion or politics unless it was something which supported long-term government policy.

Academicians were expected to share a common allegiance not only to the cause of science (as seen through Academy eyes) but also to the government. From the 1880s onwards, this meant loyalty to the cause of the Republic. Thus, to take an extreme case, Duhem in the 1890s and early 1900s would have been unacceptable as a member of the Academy many times over, despite his great ability: first because of his outspoken disagreement with secretary Berthelot over scientific matters (thermochemistry); secondly because, as an ultra Catholic, his allegiance to science might be felt to be less than total; and thirdly because, as a known supporter of the fading Royalist cause, his complete loyalty to the Republic was questionable. Under the Third Republic, and aided by such men as Louis Liard, *Directeur de l'Enseignement supérieur* from 1884, a secular and rationalist philosophy was consolidated in schools, universities and, less directly, in the Academy. An independent thinker like Duhem might be tolerated as a professor at Bordeaux, at the periphery of a very centralised system, but at the centre there were strong social forces to produce conformity.

The social pressure on Academicians to conform would be reinforced rather than diminished by their additional allegiance to other prestigious state institutions of higher education in Paris and also by the social round. The majority of Academicians tended to live in the *quartier Latin*,[84] where they or

[81] Isidore Salle de Gosse, *Histoire naturelle, drolatique et philosophique des professeurs du Jardin des Plantes*, 1847, p. 7.

[82] *Ibid.*, p. 10.

[83] Cuvier exercised autocratic control at the Muséum and Laplace made considerable use of his authority at the Bureau des Longitudes.

[84] Charles Richet calculated that in the early twentieth century 70% of Academicians lived either in the 4th or 5th arrondissements. Charles Richet, *The natural history of a savant* (1923), trans. 1927, reprint, New York, 1975, p. 26n.

their families might meet incidentally from time to time as in a small town. They would often socialise exclusively with other intellectuals and we have seen that intermarriage between the families of Academicians was not uncommon.

9. *The possibility of control of knowledge*

A final aspect of 'control' which should be considered is the possibility that the Academy might have been in a position not only to select subjects for research but even to influence the outcome, that is it might in some way have 'controlled' *the knowledge produced.* It has been argued by sociologists that knowledge is always the product of a particular social context.[85] An extension of this view is that even knowledge of the natural world, previously regarded as comparatively objective, is really pliable to the extent that it can be manipulated in different directions in response to various social, political and religious pressures. Of course, it may be *influenced* but we cannot accept that it is *determined* by such considerations.

It may be worth commenting briefly on two cases relating to the Academy where the relativity or even subjectivity of scientific results has been claimed. Two further cases will be added for comparison. The first two cases concern Fresnel and Pasteur respectively, while the latter two cases involve the more marginal figures of LeBon and Blondlot and the subject of invisible radiation. Taken together they span the greater part of our period and a range of different subjects. Only the barest outline of each case can be given. For further information, readers should consult the references given.

The triumph of Fresnel's wave theory of light in the Restoration period has been presented as a case where social or cultural factors were predominant.[86] The story should really start with the fact that in 1815 and 1816 two talented young physicists had presented memoirs on the diffraction of light to the Academy, using alternative theories. Pouillet (in collaboration with Biot) had drawn on the corpuscular theory of light while Fresnel had made use of a wave theory. The Academy therefore decided in March 1817 that the subject of diffraction would be a good one to choose for the grand prix of 1819 and the names of both Pouillet and Fresnel were mentioned in the text of the prize announcement.[87] Fresnel won the prize with a brilliant memoir in which he reported a series of experiments to confirm the mathematics of the wave theory.

Although commentators have been right to stress the fact that Fresnel, a young engineer with no formal training in optics, was helped enormously by the patronage of Arago, it should be remembered that politically the innovators

[85] For a good introduction to the sociology of knowledge see Michael Mulkay, *Science and the sociology of knowledge*, 1979.

[86] Michael Mulkay, 'Cultural interpretation in science', *ibid.*, Chapter 3, pp. 63–95 (87–90). Mulkay's account is based on E. Frankel, 'Corpuscular optics and the wave theory of light: the science and politics of a revolution in physics', *Social Studies of Science*, 6 (1976) 141–84. See also Ivor Grattan-Guinness, *Convolutions in French mathematics* (3 vols., Basel and Berlin, 1990), vol. 2, pp. 854ff.

[87] *P.V.I.*, 6, 164–5.

were more than balanced by the traditionalists. Indeed the awarding committee contained a majority of scientists supporting the corpuscular theory, led by Laplace himself together with Biot and Poisson. It is to their credit and that of the Academy that they allowed themselves to be persuaded of the superiority of the alternative theory. The emergence of the wave theory, therefore depends less on social factors than on cognitive ones, namely the rigour of Fresnel's memoir. In the early years of the century the Laplace school had done much to advance physics by encouraging certain lines of research.[88] It had come fairly close to the control of physical science, but in the end neither it nor the Academy as a whole could actually control scientific knowledge.

The second case is the famous Pasteur-Pouchet debate, previously mentioned at the end of Section 7 of Chapter 3. It has been argued (and widely believed) that the reasons why the Academy resolved the dispute in favour of Pasteur were social, religious and political rather than experimental and were related to the atheistic implications of the concept of spontaneous generation.[89] Further examination of the debate has, however, thrown doubt on this conclusion.[90] One of the main criticisms made relates to the biased composition of the panel of judges, who are characterised as very traditionalist Catholics.[91] This is to misunderstand the position of the secular rationalist Claude Bernard as well as other prominent Academicians. If, for example, the powerful Dumas opposed spontaneous generation it was not primarily for religious reasons but because, before becoming a chemist, he had made a special study of fertilisation. Pasteur's critics have also failed to place the dispute in the proper context, that of a history of Academy prize questions on the subject going back some twenty years which had already seriously undermined the theory of spontaneous generation. It may be difficult to accept that judges are ever completely impartial but if we conclude that the debate was settled *mainly* (but not necessarily exclusively) on the basis of internal scientific evidence, we may not be too far from the truth.[92] On the other hand to consider the full ramifications of the debate would require a whole

[88] Robert Fox, 'The rise and fall of Laplacian physics', *Historical Studies in the Physical Science*, **4** (1974), 89–136.

[89] J. Farley and G. L. Geison, 'Science, politics and spontaneous generation in nineteenth-century France: The Pasteur–Pouchet debate', *Bulletin of the History of Medicine*, **48** (1974), 161–98.

[90] N. Roll-Hansen, 'Experimental method and spontaneous generation: the controversy between Pasteur and Pouchet, 1859–64', *Journal of the History of Medicine*, **34** (1979), 273–92. Also Antonio Galvez, 'The role of the French Academy of Science in the clarification of the issue of spontaneous generation in the mid-nineteenth century', *Annals of Science*, **45** (1988), 345–65. Much of Pasteur's work on spontaneous generation is to be found in vol. 2 of his collected works: *Oeuvres*, 7 vols., 1922–39. See also index in vol. 7.

[91] Farley and Geison, *op. cit.*, pp. 181–2. Previously on p. 166 a quotation is given describing Catholics as reactionaries at this time. The final claim in the paper (p. 198) about the *objectivity* of Pouchet is undermined by Galvez, *op. cit.*, pp. 361, 364.

[92] Of course, the most that Pasteur could show is that spontaneous generation did not take place *in his* (carefully controlled) *experiments*. Logically one cannot absolutely disprove spontaneous generation until one has examined all possible cases.

book and we may be certain that the last word on this matter has not yet been said.

We turn now to two not unrelated cases around 1900 where the claim of the social construction of knowledge might seem to be more plausible. The respective cases of LeBon and Blondel have been the subject of recent comprehensive studies[93] and some of the main points can therefore be readily summarised. When news of Röntgen's X-rays were discussed in the Academy in January 1896 it was the Catholic mathematician Henri Poincaré who suggested that it would be worth investigating phosphorescent substances to see whether they also emitted invisible rays, a suggestion taken up by Henri Becquerel. It was reports like this which encouraged the amateur scientist Gustave LeBon to carry out experiments in his small private laboratory. He soon made the claim to have discovered a further type of radiation, coming from an ordinary oil lamp, which could pass through a metal box and affect a photographic plate and which he called 'black light'.[94] His paper delivered in 1896 was followed by several others on the same subject. Obviously, radiation was very much 'in the air' about this time. After the investigation of cathode rays in the 1880s and the recent excitement about X-rays it was easy to accept that there were other types of radiation waiting to be discovered. Once LeBon had told experimenters what to look for, confirmation of the existence of 'black light' was obtained from several parts of France although others were more sceptical.

It is interesting that, apart from physicists, the greatest interest was shown in such claims of invisible rays by other scientists either of Christian persuasion or who were favourable to a belief in spiritualism and more generally by those opposed to the apparent materialism of nineteenth-century science. Many saw evidence of the new radiation as supporting religion. Indeed historians have not only commented on the claims made by LeBon but also on the language he used. LeBon interpreted black light 'in terminology that drew on the prevalent intellectual and philosophical trends of his time: an anti-rationalism and particularly an anti-materialism that emphasised intuition, spontaneity, evolution and action at the expense of the traditional emphasis in [physical] science on mechanism, determinism and materialism'.[95] One thinks of the rise in the 1890s of Henri Bergson with his intuitionism and mysticism. Henri Poincaré and Pierre Duhem helped to introduce a philosophy of conventionalism, which argued that scientific claims are largely artificial constructs. This emphasised the individual mind at the expense of a supposed universal truth discoverable by science.

[93] Mary Jo Nye, 'Gustave LeBon's Black light: A study in physics and philosophy in France at the turn of the century', *Historical Studies in the Physical Sciences*, **4** (1974), 163–95. Mary Jo Nye, 'N-rays: An episode in the history and psychology of science', *Historical Studies in the Physical Sciences*, **11** (1980), 125–56.

[94] G. LeBon, 'La lumière noire', *C.R.*, **122** (1896), 188–90.

[95] Nye, *op. cit.* (1974), p. 164.

Now, as previously mentioned in Chapter 10, LeBon was not only a personal friend of Academicians such as Henri Poincaré and Emile Picard but he had many ideas in common with this group. Poincaré actually encouraged LeBon to present his ideas on black light to the Academy and when they were criticised, he invited LeBon to reply. Eventually Becquerel was able to explain that LeBon's photographic plate had been affected by nothing more extraordinary than infra-red light.[96] But the fact that LeBon's ideas had some general currency for several years and were received sympathetically by the Academy can only be explained by philosophical trends at the end of the nineteenth century. There were people who *wanted* to believe in the dematerialisation of matter. It would be quite unhistorical to conclude that the Academy was stupid. At the frontiers of knowledge there is always room for doubt and in the history of radiation this is a period of some confusion.

One cannot conclude from the claim of 'black light' that the Academy was validating pseudo-science. On the contrary, by acting as a central agency for the presentation of claims of new knowledge, it enabled such claims to be examined by professional scientists and in this case to refute them by providing an alternative simpler explanation of the effect on the photographic plate. The Academy's reputation was never at stake, only that of LeBon, although, as he was not a professional scientist, he was not jeopardising his career to the same extent.

There was, however, a second case which did involve a professional scientist (albeit a provincial one and therefore rather more independent) and which has been held to have undermined the reputation of the Academy. That is the claim by a corresponding member of the Academy, René Blondlot in 1903, to have discovered 'N-rays' (so-called after his home town and university of Nancy).[97] The claim depended on the observation of a spark during the emission of X-rays by a cathode ray tube. When the voltage was reduced to a level when he thought no X-rays were being produced, he obtained changes in the brightness of the spark and he concluded that he had discovered a type of radiation which, like X-rays, could pass through paper, wood or aluminium foil, but which he claimed was distinct from X-rays. In the hope of recognition Blondlot naturally presented his work to the Academy (23 March 1903). Confirmation of N-rays was not slow in coming, although it was a pity that their detection always depended on the (subjective) estimation of difference in brightness of an electric spark. Among sceptics was the young physicist Jean Perrin, who was particularly critical of the method of detection. However when Albert Turpain, a physicist from Bordeaux, wrote to Mascart, a senior member of the physics section of the Academy, asking him to report to the Academy his inability to detect the rays, Mascart replied that after consulting several colleagues he did not think it

[96] H. Becquerel, 'Explication de quelques expériences de M. G. LeBon', *C.R.*, **124** (1897), 984–8 (p. 985).

[97] R. Blondlot, 'Sur une nouvelle espèce de lumière', *C.R.*, **136** (1903), 735–8.

appropriate to publicise a negative result. Perhaps Turpain's eyes were not sufficiently sensitive?[98]

Mascart was to be a member of the prize commission which met in June/July 1904 to decide on the award of the triennial Le Comte prize of 50000 francs. Henri Poincaré was another member who supported Blondlot but it was actually Becquerel who proposed that Blondlot, whose work had received considerable prominence, should receive the prize.[99] It was a pity that the committee had to make a decision then, since only fifteen months had elapsed since the first claim of the existence of N-rays. We also have to take into account the heavy administrative burden of prize awards which affected the Academy at this time.[100] When the recommendation was made Poincaré agreed but asked that the terms of the award should be left open in case further important evidence about N-rays emerged before the actual presentation of the award at the annual public meeting in December. With this reservation the Academy voted unanimously in favour of Blondlot.

In late September the American physicist R. W. Wood went to Nancy and showed the N-rays were a purely subjective creation of Blondlot and his group. An account of this refutation was published in the *Revue Scientifique* of 22 October in time for Poincaré to intervene and change the terms of the award. The award stated that the prize was being given to a senior provincial physicist for his life's work ('l'ensemble de ses travaux') rather than for one part of it. There is no doubt that Poincaré (whose family were prominent citizens of Nancy) was generally sympathetic to the idea of invisible radiation and another Academician, the physiologist d'Arsonval, was positively enthusiastic about N-rays, which he associated with the activity of the brain. Yet the Academy never actually endorsed the discovery.

The Academy's handling of N-rays, however, shows that it was influenced by a number of different considerations which lay outside the field of science. Theoretically the objective judges of merit, a few prominent members of the Academy, were clearly biased in favour of evidence which seemed to undermine the old materialism. All Academicians wanted French scientists to have the honour of discovering some important new phenomenon, being only too aware that in physics most of the recent major advances had been made by German scientists. The Nancy physicists for their part were keen to put their provincial university on the map and were reckless in pursuing fame. Finally the French system emphasised the authority of senior figures, and younger scientists were expected to see what their elders told them to see. Thus philosophical, national and social pressures reinforced each other. Temporarily they clouded the issue, but in the end the demand for independent confirmation of the claim by rigorous

[98] *Revue Scientifique*, 5 (1906), 492.

[99] A.S., *Comité secret, 1903–1912*, p. 66 (6 June), p. 70 (4 July).

[100] More than 50 different prize awards were made at the public meeting of December 1904. Each of these prizes required the appointment of a committee and careful deliberation.

scientific method (rather than by subjective visual interpretation of sparks) won the day. The Academy was in danger of validating a bogus claim, but in the end it drew back. If prejudice contributes to the interpretation of evidence, as surely it does, we can probably claim that the instincts of the majority of Academicians would have been *against* the acceptance of a multiplicity of invisible radiations without clear proof.

If the Academy had acted exclusively as the patron of one of several conflicting theories or if it had systematically suppressed evidence contradicting the claims of one of its members or protégés (we only know of one example, that of Turpain cited above), then it could have been said to have been trying to control scientific knowledge in an absolute sense. But in both cases of radiation discussed above there was open debate. It is interesting that both the claimants were in a sense 'outsiders', seeking fame. But fame could only come through publicity and publicity is a double-edged sword. The Academy, far from wishing to suppress any serious claim, was happy to publish it *on the authority of the author*[101] and then allowed others to dispute the claim. In a competitive world there would often be scientists who had an interest in discrediting a controversial claim, but even the most neutral observer could reasonably ask for independent confirmation. Experimental methods, reliability of apparatus, independent witnesses, and the track record of the experimenter would all be critically evaluated in open debate. In the end, therefore, it was not so much the Academy which controlled scientific knowledge as the international scientific community, using the procedures which had slowly been built up over the centuries. Perhaps the ultimate check on the possibility of authoritarian control of scientific knowledge within one particular country lies in the international nature of science. If experimental results are not capable of authentication in a foreign laboratory, working within a different political or social system, then they cannot be considered to be a part of science.

10. *Conclusion*

For most of this book we have taken the title 'Academy' largely for granted, using it as a convenient label for the whole period even though the institution was officially known by the more cumbersome title of 'First Class of the National Institute' for the first twenty years of our study. It was the Bourbon Restoration which introduced the title 'Academy' in a deliberate attempt to suggest continuity with the *ancien régime*. This may have been unobjectionable in the case of the Académie Française, but did the Academy of Sciences of the nineteenth century fully deserve to be called an academy? It would be a big mistake to regard the nineteenth-century Academy of Sciences as essentially an eighteenth-century academy which by accident or design was allowed to trespass into the modern world. On the contrary we have argued that it took several major initiatives and even helped to shape the modern world of science.

[101] Sometimes shared with an Academician as sponsor.

Keeping the title 'Academy', it turned its back on many of the traditions and characteristics of eighteenth-century academies. With one temporary exception, there was none of the old limitation of middle-class participation by noble and feudal privilege[102] and, when the Restoration did try to put the clock back in this way, the Academy managed to turn the section of *Académiciens libres* to its own advantage. The idea of a social hierarchy, which was reflected in the eighteenth-century Academy of Sciences, had no place in its successor. If there was any distinction between members accepted in the nineteenth-century Academy, it was one of seniority. The cultivation of eloquence, one of the hallmarks of the academies, was rejected by the scientists with the single exception of the *éloges* pronounced at annual public meetings. The great majority of Academicians rejected the tradition of leisurely publication of long and carefully crafted memoirs for an alternative system of rapid publication of brief accounts of recent research, a shift which marked a major advance in the organisation of modern science. Yet given its formal meetings, fiercely contested elections and major prizes, we would not wish to reject the title of Academy as inappropriate; rather we would claim that it was an Academy with several major differences. Even the prize system, a feature of many academies, was turned into something like a grant system, thus opening another gateway to the world of modern science.

The enemies of the Royal Academy of Sciences at the time of the Revolution had claimed that it represented an *aristocracy*, an intellectual elite. The members of the nineteenth-century Academy would have claimed rather that it represented a *meritocracy*, if only that term had been current at the time. While largely agreeing with this judgement, we have suggested that in many ways it developed into a *gerontocracy*, since its membership depended on filling dead men's shoes, with no other way of rejuvenating the institution.

Yet for all its faults the Academy of Sciences probably represented science better for most of the nineteenth century than any of the other Academies represented their respective fields; better than the Académie Française, for example, represented literature. Even so, largely because of the technicalities of science, it was the Académie Française rather than the Academy of Sciences in which the French public took the greatest interest. Unlike the Academy concerned with painting, which had tended to dictate styles as well as standards, the Academy of Sciences did not make the mistake of excluding works not conforming to the fashion of the age. In science, therefore, there was never a breakaway movement corresponding to the first exhibition of the Impressionist painters of 1874, which dispensed with a jury; all subscribers having the right to display their work.

The Academy of Sciences was the most active of all the French Academies. As Poincaré said, it might be the Academy where (given the broad range of specialised subjects covered) members paid the least attention to the papers

[102] James E. McClellan, *Science reorganised. Scientific societies in the eighteenth century*, New York, 1985, p. 13.

presented during meetings, but it was the Academy that worked the hardest.[103] It was also the one with the greatest recognition abroad. If there were some signs of decline in the final decades of this study there were also some high points, such as some of the early work on radioactivity. It is true however that although the Academy was always open to receive new work and indeed encourage it, it was not well adapted to embracing new disciplines wholeheartedly. Given the philosophy of an Academy based on a fixed and severely restricted membership, new disciplines could only be represented at the expense of traditional fields. In an age when scholars were trying to make a science out of everything, there was increasingly a good case for founding specialist institutions. Thus the Academy, which had originally around 1800 tried to include the whole of science, was understandably 100 years later not able to keep all recent developments to itself.

We have spoken earlier in general terms of the possible handicap to the Academy in not having its own laboratories, but we need to relate this potential criticism more specifically to the different sciences and their development in the nineteenth century. At the time of the foundation of the Institute the only science to make regular use of laboratories was chemistry. Physics at that time had little more than *cabinets de physique*, where instruments were stored and displayed almost in the same way as in natural history, which in the eighteenth century had become increasingly popular, with its cabinets used to exhibit specimens. Electricity was one subject which provided a promising field for experimentation. In the course of the nineteenth century with the development of current electricity and the concern for measurement and standardisation, physics laboratories came increasingly to be called for,[104] sometimes converted from crude workshops, sometimes purpose built, often following the model of earlier chemistry laboratories. Among the biological sciences physiology led the way as an *experimental* science drawing on both physics and chemistry, and Claude Bernard's classic book *Introduction to the study of experimental medicine* (1865) helped make a case for real science as depending essentially on experiment rather than on mere observation. There was a heated debate on this subject in the Academy in 1868.[105] Under the Third Republic the French state began to take more seriously the claim that laboratories were an *essential* part of science.

Critics might view the absence of Academy laboratories as a handicap, but it is possible later in the nineteenth century to pursue this argument much further. As laboratories grew in importance in the second half of the century, one may argue that laboratories themselves began to rival scientific societies, and even the Academy of Sciences itself, as the real focus of scientific research. In so far

[103] Quoted by Darboux in : Gaston Buissier *et al.*, *L'Institut de France* (2 vols., 1907), vol. 1, p. 44.

[104] R. Sviedrys, 'The rise of physics laboratories in Britain', *Historical Studies in the Physical Sciences*, 7 (1976), 405–36; G. Gooday, 'Precision measurement and the genesis of physics teaching laboratories in Victorian Britain', *B.J.H.S.*, **23** (1990), 25–51.

[105] *C.R.*, **66** (1868), 1278–84.

as these laboratories were not only the place where experimental investigations were carried out but also where students were trained and where frequent discussions would arise about the planning, execution and significance of the work done, it was increasingly becoming the case by the late nineteenth century that it was the major laboratories rather than any scientific society or academy which was the real focus of science or at least *of individual research problems*. The Academy of Sciences, of course, still had a place for the presentation of research findings which could then be compared with other findings, but its importance had diminished. But we have never claimed that the importance of the Academy lay simply in the formal presentation of papers on Monday afternoons. Informal discussion and multi-disciplinary contacts were no less important; the publication of the *Comptes rendus* alone was enough to justify the Academy into the early twentieth century.

In the nineteenth century it had been the custom to turn to the Academy to resolve any dispute in science. But by the end of our period there are already indications that the Academy was not considered to have sufficient authority. As part of the growth of democracy the scientific community as a whole, rather than a small elite, wanted to be involved. A good example is a poll conducted in 1904 by the journal *Revue scientifique* on the reality of the phenomenon of N-rays.[106] It had proved unusually difficult for physical scientists to confirm independently the existence of these rays, which the Nancy physicist Blondlot claimed to have discovered. Professors of physics in provincial faculties as well as in the Parisian institutions were asked for their opinion, together with a few more junior physicists and one or two chemists and physiologists. Nearly sixty scientists were approached and their replies were published in weekly issues of the *Revue scientifique* in the period 5 November–10 December 1904. Some scientists were embarrassed to be asked to place their opinion on public record; others applauded the enquiry but asked for more conclusive and objective experiments. Several members of the Academy were included in the survey and most were remarkably non-committal. Henri Becquerel, who had strongly supported Blondlot in the Academy, said that his views were well known and refused to add any further comment, obviously displeased with the idea of a journal conducting an open scientific enquiry. The Academician Moissan went further and asked indignantly:

> Do you think that scientific questions can be resolved by plebiscites?[107]

In other words, in a world where governments were elected by popular vote (or at least adult male suffrage), it was all too easy to assume that scientific matters could be resolved in a similar way. Moissan obviously believed that senior scientists, by virtue of their greater experience, should have the responsibility. This was the assumption on which the judgemental role of the Academy was

[106] *Revue Scientifique*, (5), **2** (1904), 590–1, 620–4, 656–60, 682–6, 718–22, 752–4.

[107] *Ibid.*, 657.

based. The episode provides what is probably an extreme example of the democratisation of science at the end of our period, but it is an indication that the scientific community looked less to the Academy as the ultimate source of authority in the early 1900s than it did a century earlier. In more recent times it has become more generally accepted that decisions on the validity of scientific work cannot be simply handed down by groups of senior scientists. Ultimately a consensus is necessary among the scientific community.

It was the new physics more than anything else which disturbed the traditional authority of the Academy. Young men like Jean Perrin (1870–1942) were in a position to do good original work independently of their seniors, thus undermining the gerontocracy of the Academy. It was not until 1923 (three years before his Nobel prize) that Perrin was considered senior enough to be elected to the Academy. This generation gap was particularly unfortunate in the years around 1900 when physical science was changing so rapidly. Perrin's doctoral thesis of 1897 had been based on research on cathode rays and X-rays. Radioactivity was even more of a puzzle to the Academicians, who had been brought up in the world of classical physics and had a more comfortable view of the solidity and permanence of matter. Now they could no longer feel that they had the same control over the physical world.

At the beginning of the book it was suggested that in order to provide a suitable focus, it might make sense to speak of a specifically 'French science'. The story we have told of the control of science and its leading institution in Paris has been closely related to French history. It could not be told without, for example, some knowledge of the French Revolution or of Napoleon Bonaparte's authoritarian rule. France was unique in the nineteenth century in combining a high level of scientific achievement with an authoritarian and centralised administrative tradition. We cannot imagine anything quite like the Academy existing in nineteenth-century Britain or the German states. In Britain a laissez-faire philosophy undermined any idea of a large government-controlled scientific institution. In any case, the British tradition was one of societies founded by private initiative and working from the ground upwards, rather than of academies founded by the state. Each of the German states had their own institutions of higher education which were in open competition with each other.[108] The French model of a centralised state was clearly inapplicable to the German situation at least until the 1870s. Yet the story we have told represents in broad outline a possible scenario for a later period as governments of other countries came to appreciate the importance of science.

The nineteenth century Academy had evolved from the Royal Academy of 1666, hierarchically arranged and taking its orders from the King, yet it was quite different. In the eighteenth century the King had made a final choice of

[108] The only pan-German scientific organisation in the early nineteenth century, the Gesellschaft Deutscher Naturforscher und Artze, founded in 1822, was a peripatetic body, similar in many ways to the British Association and quite unlike an Academy.

Academicians from a short-list drawn up by the Academy. But as science became more technical, there was more justification for the choice to become the prerogative of professional scientists. Thus the balance of power between government and Academy was altered by the internal development of science.[109] At the same time, in the movement towards parliamentary democracy, the power of the government or the relevant Minister was reduced. But this meant that the Academy had to justify itself not only to the Minister but also to the public.

During the nineteenth century there was, therefore, a definite shift in the balance of power between the government (or the responsible Minister) and the Academy. Academicians gradually came to appreciate their institution as one which, after the trauma of the 1790s, could survive changes of government and even revolutions. Ministers might come and go but the Academy lived on, gaining greater autonomy, particularly on the financial front. This came about first through the Montyon legacy and then, in the second half of the century, through a succession of prize funds, ever greater in size; the Loutreuil legacy of 1910 (first awarded in 1915) amounted to 3 500 000 francs. Another gain relating to finance came through having a friend at court. When Berthelot was Minister of Public Instruction in 1886–7 one of the minor reforms he introduced was to abolish the traditional requirement for ministerial authorisation for miscellaneous Academy expenditure.[110]

This relaxation of ministerial control, although naturally welcomed by the Academy, permitted a relaxation in standards of accounting. In 1908 the Cour des Comptes insisted that it should be able to oversee the budget of the Institute in the same way as it did that of the universities. When it came to audit the accounts, it complained of serious discrepancies.[111] It was not that the Academy had been dishonest but rather that it was lax in the control of its large private income. The government budget for the Academy had remained fixed for many years at 180 000 francs (of which one third was earmarked for the *Comptes rendus*) but this may be compared with a total annual income from private prize funds of over 400 000 francs which could be used for prizes, grants, publications and a variety of other purposes. No wonder that the internal Academy commission of 1912 should say that it regarded the financial situation as excellent in every respect.[112] Yet, although the Academy was glad of its greater financial independence, it should have realised that a static government budget in a changing world was an indication that an increasing proportion of what might be termed the state 'science budget' was being spent on higher education and research outside the orbit of the Academy. Under the early Third Republic

[109] Charles Gillispie, *Science and polity in France at the end of the old regime*, Princeton, N.J., 1980, p. 89.

[110] 'Actuellement depuis le passage de M. Berthelot au Ministère de l'Instruction Publique, l'autorisation du Ministère n'est plus nécessaire.' A.S., *Commission administratif, 1908–18*, p. 193.

[111] *Ibid.*, pp. 189–190.

[112] A.S., *Comité secret, 1912–18*, p. 92 (21 July, 1913).

the government spending on education rose from 51 million francs in 1876 to 180 million francs in 1894. Much of this increase was directed towards science.[113]

What this means is that the unique status of the Academy was undermined by several initiatives in both the public and private sectors. Not only the government but wealthy industrialists would set up alternative agencies to deal with particular problems. The Third Republic invested heavily in the Faculties of Science and by the 1890s there were important technical institutes not only in Paris but also at Grenoble, Nancy and Toulouse, the latter, for example, having a large and vigorous Institut Electrotechnique.[114] The Academy seemed to be losing its monopoly over science and technology. But in technology it had never held a monopoly and in science it was never a complete monopoly. What it had from the beginning was high prestige as the only comprehensive official body of science and much of this prestige continued into the early twentieth century. The leading light in any provincial or specialist scientific society would be conscious that he was as nothing compared to the most junior member of the Academy.

One of the effects of the existence of the Academy was to distance the science of an elite from that of more junior or less well-educated enthusiasts. A recent study of popular science in the nineteenth century has contrasted the respectable amateurism of much science in Britain, where popular journals encouraged participation of the public, with the situation in France, where science was 'the domain of a gifted, specially trained, professional elite'.[115] For most of the year the Academy certainly encouraged this image although, on the occasion of the annual public meeting it was often felt appropriate to relax it, or even occasionally to contradict it, by claiming that *in principle* membership of that elite body was open to all on the basis of merit and hard work.

A few words by way of a final conclusion. In considering nineteenth-century France it is obvious that the Eiffel tower, built to commemorate the Revolution of 1789, is much more than a platform giving a spectacular view over Paris. It had and still has, a great symbolic value, about which much could be written. Similarly in the French revolution the guillotine was not only a method of execution; it was also a symbol of the Terror. In this book we have described many of the functions of the Academy, including the reception and evaluation of knowledge. But the Academy too was also a symbol, a symbol of order and regulation and of the existence of standards. Some of the most independently minded *savants* and naturalists might have seen it as a symbol of repression and of the censorship of unregulated self expression. Such self expression may have a legitimate place in the arts, and the history of the Impressionist painters of nineteenth-century France comes to mind. But given the growing profession-

[113] Harry W. Paul, *op. cit.*, p. 40.
[114] *Ibid.*, pp. 143–8.
[115] Susan Sheets-Pyenson, *Low scientific culture in London and Paris, 1820–75*, Ph.D. thesis, University of Pennsylvania, 1976, p. 217.

alisation of science in the nineteenth century, of which France provides probably the earliest example, the role of the independent amateur became increasingly tenuous. Whatever the political complexion of the would-be scientist, whether of the left or the right, whether we are concerned with a Raspail, an Arago, a Dumas or a Duhem, all would have seen the Academy as a symbol of control. Whether that control was desirable and, if so, how the control should be exercised, are issues which would keep many people arguing for some considerable time.

In modern times it is almost impossible to argue that science should be autonomous, controlled only by scientists. Quite apart from the huge financial investment involved, the increasing power over the natural world gained by modern science and technology has necessitated the intervention of governments. More generally there is also the problem of public accountability of scientists. The question of the control of modern science is a problem which is even more with us today than formerly, not only with expenditure on a scale unimaginable in the nineteenth century but with the additional responsibilities introduced through the development of nuclear, space and genetic technologies.

Name index

Subject index